Head-Related Transfer Function and Virtual Auditory Display

SECOND EDITION

BOSUN XIE

J.ROSS PUBLISHING

English language copyright © 2013 by J. Ross Publishing

ISBN-13: 978-1-60427-070-9

All Rights Reserved; Authorized English language edition from the 1st edition Chinese language version published by National Defence Industry Press.

New materials have been added to this 2nd edition of the book. The 2nd edition was translated from a newly prepared Chinese manuscript into English by Dr. Xiao-li Zhong and Bo-sun Xie.

Printed and bound in the U.S.A. Printed on acid-free paper.

10 9 8 7 6 5 4 3 2 1

Library of Congress Cataloging-in-Publication Data

Xie, Bosun, 1960-
    Head-related transfer function and virtual auditory display/by Bosun Xie.—Second edition.
        pages cm
    Includes bibliographical references and index.
    ISBN 978-1-60427-070-9 (hardcover : alk. paper)
    1. Surround-sound systems. 2. Auditory perception. 3. Virtual reality.
I. Title.
    TK7881.83.X54 2013
    621.389′3—dc23
                        2013010896

Phone: (954) 727-9333
Fax: (561) 892-0700
Web: www.jrosspub.com

Contents

Abstract

This book systematically states the basic principles and applications of head-related transfer function (HRTF) and virtual auditory display (VAD), and reviews the latest developments, especially those from the author's research group. The book covers binaural hearing; the basic principles, experimental measurements, computation, physical characteristics analyses, filter design, and customization of HRTFs; and the principles and applications of VADs, including headphone and loudspeaker-based binaural reproduction, virtual reproduction of stereophonic and multi-channel surround sound, binaural room simulation, rendering systems for dynamic and real-time virtual auditory environments, psychoacoustic evaluation and validation of VADs, and a variety of applications of VADs. More than 600 references are listed at the end of the book, representing the main body of literature in this field.

The book aims to provide necessary knowledge and the latest results to researchers, graduate students, and engineers who work in the field of HRTF and VAD. Readers will become familiar with the forefront of the field and able to undertake corresponding scientific research or technical development projects.

Foreword

Head-related transfer functions (HRTFs) describe the acoustic effects of the head and torso on the sound signals impinging upon listeners' ears. They represent the "antenna gain" of the human auditory system. Consequently, reliable HRTF data are a decisive element in many technologies dealing with audio technique, hearing aids, room acoustics and virtual reality—just to name a few.

This book provides a thorough and comprehensive analysis of issues relevant to HRTFs. Many of them have not been covered in such a detailed, but nonetheless, clear way in any other book. For example, the discussion of the measurement and/or numerical calculation of HRTFs, their interpolation, decomposition and, last but not least, their perceptual evaluation is excellent. Further, the main application areas of HRTFs are explained and discussed in the general framework of human spatial hearing at large. This qualifies the book not only as a reference book, but also as a textbook.

Head-Related Transfer Function and Virtual Auditory Display is a valuable source of information for everyone dealing with HRTF. Professor Xie has devoted a considerable amount of time and effort to create a work that appeals to readers' demands. The international acoustical community is pleased to gain access to its rich content by means of this English translation. We wish the book in this international format finds a broad readership and achieves the significant scientific and technological impact that it deserves.

Jens Blauert
Bochum, Germany
Spring 2013

Preface

Hearing is a crucial means by which humans acquire information from exterior environments. Humans can subjectively perceive some spatial attributes of sound (i.e., spatial auditory perception), in addition to the loudness, pitch, and timbre of sound, among other attributes. Under certain conditions, hearing can localize sound sources using the spatial information in sound, as well as to create the spatial auditory experience of acoustical environments.

Humans receive sound information through their two ears. Thus, binaural sound pressures contain the main information in sound. From the physical point of view, sound waves that are radiated by sources arrive at a listener after undergoing direct and reflected/scattered propagations off boundary surfaces and scattering/diffraction/reflection off human anatomical structures such as the head, torso, and pinnae. The process of scattering/diffraction/reflection by human anatomical structures converts the information in sound fields into binaural pressure signals. In the case of a free field and a fixed listener, the transmission from a point sound source to each of the two ears can be regarded as a linear time-invariant (LTI) process, and characterized by head-related transfer function (HRTF). HRTFs contain the main spatial information of sound sources, and are thus vital for research on binaural hearing. As an important application of HRTF research, virtual auditory display (VAD) is a technique developed over the last two decades. In VADs, virtual auditory events are created by reproducing HRTF-based synthesized binaural signals through headphones or loudspeakers. VADs have been widely employed in a variety of fields, such as scientific research, engineering, communication, multimedia, consumer electronic products, and entertainment. Therefore, research on HRTF and VAD has aroused widespread concern, and raised popular research topics in the areas of acoustics, signal processing, and hearing, among others.

Until a few years ago, few research groups in China engaged in research concerning HRTF and VAD. The author and the group have undertaken a series of both academic and applied studies on HRTF and VAD, dating from the middle of the 1990s. Recently, issues surrounding HRTF and VAD have gradually received attention in China, and some research groups have also started, or are planning, to undertake relevant studies.

Internationally, there are some books on spatial hearing, especially the famous monograph, *Spatial Hearing*, written by Professor Blauert. However, a book dedicated to HRTF and VAD is rare, with the notable exception of 3-*D Sound for Virtual Reality and Multimedia*, written by Dr. Begault in 1994.

Our present book systematically states the basic principles and applications of HRTF and VAD, and reviews the latest developments, especially those from the author's research group. The original Chinese edition was published by the National Defense Industry Press (Beijing) in 2008. The present English second edition includes the Chinese edition and amendments that address the more recent developments from 2008 to 2012. The book consists of 14 chapters, covering the main issues in the research field of HRTF and VAD. Chapters 1-7 mainly present the basic principles, experimental measurements, computation, physical characteristics analyses, filter design, and customization of HRTFs; Chapters 8-14 are devoted to the principles and applications of VADs, including headphone and loudspeaker-based binaural reproduction, virtual reproduction of stereophonic and multi-channel surround sound, binaural room simulation, rendering systems for dynamic and real-time virtual auditory environments, psychoacoustic evaluation and validation of VADs, and a variety of applications of VADs. More than 600 references are listed at the end, representing the main body of literature in this field.

This book intends to provide necessary knowledge and the latest results to researchers, graduate students, and engineers who work in the field of HRTF and VAD. Because this field is interdisciplinary, reading this book assumes some prior knowledge of acoustics and signal processing. Relevant references for these concepts are also listed in the References section.

The publication of the present book has been supported by the National Nature Science Fund of China (Grant No. 11174087). The relevant studies on HRTF and VAD of the author and our group have been supported by a series of grants from the National Nature Science Fund of China (Grant Nos. 10374031, 10774049, 50938003, 11004064, 11104082), the ministry of education of China for outstanding young teachers, Guangzhou science and technology plan projects (No. 2011DH014), and State Key Lab of Subtropical Building Science at South China University of Technology. South China University of Technology, in which the author works, has also provided enormous support.

The author appreciates Professor Jens Blauert of Ruhr-University Bochum, who first suggested and encouraged the publication of the present English second edition and wrote the foreword for this book. The author also thanks Professor Ning Xiang of Rensselaer Polytechnic Institute, who is the Series Editor for the present book, for his cooperation, helpful suggestions and work on the publication process.

Thanks to Dr. Xiao-li Zhong for the successful cooperation in translating the Chinese edition into English and to Dr. Guang-zheng Yu for preparing all figures.

During the more than fifteen years of the author's research experience, Professor Shan-qun Guan of Beijing University of Posts and Telecommunications has generously provided much guidance and many suggestions, especially on the present book. The author has also received help and support, since the middle of the 1990s, from Professor Zuo-min Wang, the author's doctoral advisor at Tongji University.

The author is especially indebted to Dr. Xiao-li Zhong, Guang-zheng Yu, Dan Rao, Cheng-yun Zhang, Professor Zhi-wen Xie, and Mr. Zhi-qiang Liang, as well as graduate students over the years, who provided support and cooperation. Many colleagues in China also provided the author with various kinds of support and help during the author's research work, particularly Professors Shuo-xian Wu and Yue-zhe Zhao at the school of Architecture, South China University of Technology; Professors Jian-chun Cheng, Bo-ling Xu, Xiao-jun Qiu, and Yong Shen at Nanjing University; Professors Xiao-dong Li, Jun Yang, Ming-kun Cheng, Hao Shen in the institute of acoustics at China Academy of Sciences; Professor Dong-xing Mao and Dr. Wu-zhou Yu at Tongji University; Professor Chang-cai Long at Huazhong University of Science and Technology; Professor Zi-hou Meng at Communication University of China; Professors Bao-yuan Fan and Jin-gang Yang; senior engineers, Jin-cai Wu and Hou-qiong Zhong, of The Third Research Institute of China Electronics Technology Group Company; and senior engineer, Jin-yuan Yu, of Guoguang Electric Company. J. Ross Publishing also contributed enormously toward the publication of the present book.

The author would like to give a heartful thanks to all of these people.

The author's parents, Professors Xing-fu Xie and Shu-juan Liang, were also acoustical researchers, who cultivated the author's enthusiasm for acoustics. The author's mother gave great support on the preparation of the text and this book is in memory of the author's parents who have passed away.

Bo-sun Xie
Acoustic Laboratory, Physics Department, School of Science, South China University of Technology
The State Key Lab of Subtropical Building Science, South China University of Technology
Oct. 2012

Biography

Bo-sun Xie was born in Guangzhou, China, in 1960. He received a Bachelor degree in physics and a Master of Science degree in acoustics from South China University of Technology in 1982 and 1987, respectively. In 1998, he received a Doctor of Science degree in acoustics from Tongji University.

Since 1982, he has been working at South China University of Technology and is currently the director and a professor at the Acoustic Laboratory in the School of Science. He is also a member of The State Key Lab of Subtropical Building Science. His research interests include binaural hearing, spatial sound, acoustic signal processing, and room acoustics. He has published over 150 scientific papers and owns five patents in these fields. His personal interest lies in classical music, particularly classical opera.

He is a member of the Audio Engineering Society (AES), a vice chairman of the China Audio Engineering Society, and a committee member of the Acoustical Society of China. He can be reached at phbsxie@scut.edu.cn.

1

Spatial Hearing and Virtual Auditory Display

This chapter introduces the essential concepts, definitions, and principles of spatial hearing and virtual auditory displays, including human auditory systems and auditory filters; the principles and cues of spatial hearing and localization; the definition of head-related transfer functions; an outline of room acoustics and relevant spatial perceptions; artificial head models; and the principle of the binaural recording and playback system. It provides the reader sufficient background information to facilitate understanding succeeding chapters.

This chapter presents the essential concepts, definitions, and principles of spatial hearing and virtual auditory displays. It is intended to provide the reader with sufficient background information to facilitate understanding of succeeding chapters. To avoid ambiguity that may arise from the discussion, the coordinate systems used in this book are defined in Section 1.1. Section 1.2 provides a concise review of the anatomical structure and function of the human auditory system and introduces the concepts of auditory filters and critical bands. In general, spatial hearing consists of single sound source localization, summing localization, and other spatial auditory perceptions of multiple sound sources. The basic concept of spatial hearing is outlined in Section 1.3, and the localization cues of a single sound source are comprehensively discussed in Section 1.4. Section 1.5 introduces the head-related transfer function (HRTF), given that most localization cues are encoded in HRTFs. The measurement point of HRTFs is also discussed in detail. In Section 1.6, we detail the investigation of the summing localization of multiple sound sources, and the derivation of the corresponding localization formula for virtual stereophonic sound sources at low frequencies. Precedence and cocktail effects are also discussed. Given that daily auditory events often occur in a reflectively enclosed space, such as a room, some basic concepts in room acoustics and relevant spatial perceptions are briefly discussed in Section 1.7. Section 1.8 describes certain types of artificial head models and the principle of the binaural recording and playback system. The concept of virtual auditory displays is then introduced. Finally, a comparison between a virtual auditory display and multi-channel sound is presented.

1.1. Spatial Coordinate Systems

In spatial hearing research, the position of a sound source is specified in terms of its direction and distance with respect to the listener's head. Usually, the head center at the level of entrances of two ear canals is selected as the origin of a coordinate system. Additionally, the directional vector pointing from the origin to the sound source describes the exact position of a sound source in space.

Three specific planes in space are defined by the directional vectors: the *horizontal plane*, determined by the two vectors pointing to the front and right (or left) directions, the *median plane*, determined by the two vectors pointing to the front and top directions, and the *lateral plane* (or the frontal plane), determined by the two vectors pointing to the top and right (or left) directions. The three planes are perpendicular to one another and intersect at the origin.

In the literature, various coordinate systems are adopted in describing the position of a sound source according to specific issues concerned. Among these systems, spherical and interaural-polar coordinate systems are commonly used.

Coordinate system A. A clockwise spherical coordinate system is shown in Figure 1.1. The sound source position is described by (r, θ, ϕ), where the source distance with respect to origin is denoted by r with $0 \leq r < +\infty$. The angle between the directional vector of the sound source and the horizontal plane is denoted by elevation ϕ with $-90° \leq \phi \leq +90°$, in which $-90°$ and $90°$ represent the bottom and top positions, while $0°$ represents the horizontal direction. The angle between the horizontal projection of directional vector and the front (y−axis) is denoted by azimuth θ with $0° \leq \theta < 360°$ labeled clockwise, where $0°$, $90°$, $180°$, and $270°$ represent the front, right, back, and left directions in the horizontal plane, respectively. Meanwhile, the complementary angle of ϕ, namely, ϕ' in Figure 1.1, can also be used to denote the elevation with:

$$\phi' = 90° - \phi \qquad 0 \leq \phi' \leq 180°. \tag{1.1}$$

In some studies, θ varies in the range $-180° < \theta \leq 180°$, where $0°$, $90°$, $180°$, and $-90°$ represent the front, right, back, and left directions in the horizontal plane, respectively. Because θ is a periodic variable with $(\theta + 360°)$ equal to θ, the two kinds of θ variation range are equivalent.

Coordinate system B. A counterclockwise spherical coordinate system is shown in Figure 1.2. The sound source position is described by $(r, \theta_1, \phi_1, \text{ or } \phi_1')$, where distance r and elevation ϕ_1 or ϕ_1' are specified in the same way as in coordinate system A. The most distinct difference between the two coordinate systems is that azimuth θ_1 in coordinate system B is labeled counterclockwise, contrary to that in coordinate system A. Hence, in the horizontal plane, $\theta_1 = 0°$, $90°$, $180°$, and $270°$ represent the front, left, back, and right directions, respectively. The transformation between coordinate systems A and B is $\theta = 360° - \theta_1$, and $\phi = \phi_1$. Similar to coordinate system A, θ_1 varies in the range $-180° < \theta_1 \leq 180°$ in some studies. In addition, coordinate system B with $-180° < \theta_1 \leq 180°$ and $0° \leq \phi_1' \leq 180°$ is used in some studies on multi-channel surround sound.

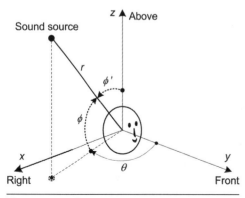

Figure 1.1 Coordinate system A

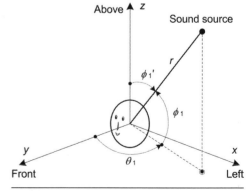

Figure 1.2 Coordinate system B

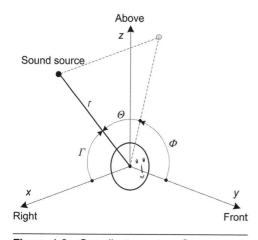

Figure 1.3 Coordinate system C

Coordinate system C. An interaural polar coordinate system is shown in Figure 1.3. The sound source position is described by (r, Θ, Φ), where the definition and range for r are the same as those in coordinate system A; the interaural polar azimuth Θ is defined as the angle between the directional vector of the sound source and the median plane with $-90° \leq \Theta \leq +90°$; and the interaural polar elevation Φ is defined as the angle between the projection of the directional vector of the sound source to the median plane and the y-axis (i.e., the frontal-axis) with $-90° < \Phi \leq 270°$ (in some studies, $0° \leq \Phi < 360°$). In coordinate system C, $(\Theta, \Phi) = (0°, 0°)$, $(0°, 90°)$, $(0°, 180°)$, $(0°, 270°)$, $(90°, 0°)$, and $(-90°, 0°)$ represent the front, top, back, bottom, right, and left directions, respectively. The transformation between coordinate systems A and C is:

$$\sin\Theta = \sin\theta\cos\phi \qquad \cos\Theta\sin\Phi = \sin\phi \qquad \cos\Theta\cos\Phi = \cos\theta\cos\phi. \tag{1.2a}$$

Particularly in the median plane,

$$\Phi = \begin{cases} \phi & \theta = 0° \\ 180° - \phi & \theta = 180° \end{cases}. \tag{1.2b}$$

Meanwhile, the complementary angle of Θ, i.e., Γ in Figure 1.3, is sometimes used instead of Θ with:

$$\Gamma = 90° - \Theta \qquad 0° \leq \Gamma \leq 180°. \tag{1.3}$$

All the coordinate systems listed above are adopted in the existing literature. However, ambiguity may arise if more than one coordinate system is used in a book without explanation. To avoid ambiguity, coordinate system A is adopted as the default in this book, and the results quoted from the literature have been converted into coordinate system A with specific explanations, unless otherwise stated. For convenience, coordinate system C is also used occasionally in this book. Because the variables in coordinate systems A and C are labeled by lowercase and uppercase letters, respectively, confusion in identifying the two coordinate systems is eliminated.

1.2. The Auditory System and Auditory Filter

1.2.1. The Auditory System and its Function

The human peripheral auditory system consists of three main parts: the external, middle, and inner ears (Figure 1.4). The external ear is composed of the pinna and ear canal. The shape and dimension of the pinna vary depending on each individual, with average length being 65 mm (52–79 mm) (Maa and Shen, 2004). The ear canal is a slightly curved tube, with an average diameter of 7 mm and an average length of 27 mm, and acts as a transmission path for sound. The pinna considerably modifies incoming sounds, particularly at high frequencies. In addition, its coupling with the ear canal leads to a series of high-frequency resonance modes. All these effects caused by the pinna play an important role in the localization of high-frequency sounds (see Section 1.4.4).

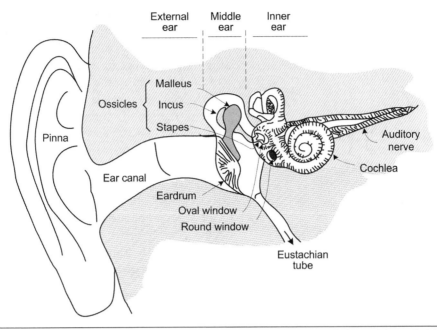

Figure 1.4   Cross section of the peripheral auditory system of humans (adapted from Geisler, 1998)

The middle ear consists of the eardrum, tympanic cavity, ossicles, associated muscles and ligaments, as well as the Eustachian tube. The eardrum lies at the end of the ear canal with an average surface area of approximately 66 mm^2 and a thickness of approximately 0.1 mm. The ossicles within the tympanic cavity consist of three small bones: the malleus, incus, and stapes. The malleus is attached to the eardrum, the stapes is terminated at the oval window of the inner ear, and the incus lies in between.

Essentially, the middle ear acts as an impedance transformer. If sound directly impinges onto the oval window (an interface with low-impedance air outside and high-impedance fluid inside), most of the energy is simply reflected back because of the distinct difference in acoustical impedance between air and fluid. Fortunately, before reaching the oval window, the sound is transduced into the mechanical vibration of the eardrum, ultimately resulting in the motion of the middle ear ossicles. Because the differences in area between the eardrum (about 66 mm^2) and the oval window (about 2 mm^2) as well as the lever effect of the ossicles facilitate the matching of equivalent acoustical impedances between air and fluid, the incoming sound can be effectively coupled into the inner ear.

The inner ear mainly consists of the cochlea, a bony-wall tube with a coiled (shelled) structure of approximately 2.75 turns. The length of the uncoiled cochlea is approximately 35 mm, with the end close to the oval window called the base, and the other end called the apex (Maa and Shen, 2004). Two membranes, the basilar membrane and Reissner's membrane, divide the cochlea into three chambers: the scala vestibuli, the scala media, and the scala tympani, as shown in Figure 1.5. The scala vestibuli and scala tympani, filled with perilymph, are connected by a small hole at the apex, that is, the helicotrema. Two openings exist between the middle ear and the cochlea: the oval window attached to the stapes, and the round window located at the bony wall of the scala tympani.

As described above, the sound pressure at the eardrum is transduced into a piston-like motion of the stapes, which couples sound energy into the scala vestibuli. The inward movement of the

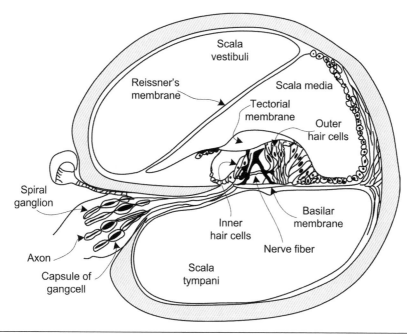

Figure 1.5 Cross section of the cochlea (adapted from Ando, 1998)

stapes increases the fluid pressure in the scala vestibuli, whereas the outward movement decreases fluid pressure. Such pressure variation results in the pressure differences across the basilar membrane, and ultimately leads to a traveling wave along the basilar membrane, from base to apex. At the same time, through the fluid in the scala vestibuli and scala tympani, the movement of the oval window causes an anti-directional movement in the membrane covering the round window, leading to a balance in fluid pressure.

In effect, the cochlea acts as a frequency analyzer. Because the mechanical properties of the basilar membrane vary continuously and considerably from base to apex, the response of the basilar membrane to the incoming sound is frequency dependent. As to tones, the displacement envelope (i.e., the maximum magnitude of displacement at each point along the basilar membrane) has a single peak at a location that is dependent on tone frequency. The basilar membrane around the base is relatively narrow and stiff, whereas that around the apex is wide and soft. As a result, the maximum displacement envelope for high-frequency tones appears near the base, then quickly fades at the remainder of the basilar membrane. On the contrary, the displacement envelope for low-frequency tones peaks at the more apical location, and then extends along the basilar membrane. Therefore, different tone frequencies correspond to different locations in the basilar membrane, where the displacement envelope of the basilar membrane oscillation is maximized. Figure 1.6 illustrates the envelope of the basilar membrane displacement for five low-frequency tones and the travelling wave of 200 Hz tone.

As described, the basilar membrane serves as a converter from frequency to location on the basilar membrane. Experimental results show that above 500 Hz, the linear distance from the apex to the location of the maximum displacement envelope is approximately proportional to logarithmic frequency. This directly results in a relatively fine frequency resolution at low frequencies, and rough frequency resolution at high frequencies. To illustrate the frequency resolution on the basilar membrane, the case in which two tones are simultaneously presented is considered. The two tones can be

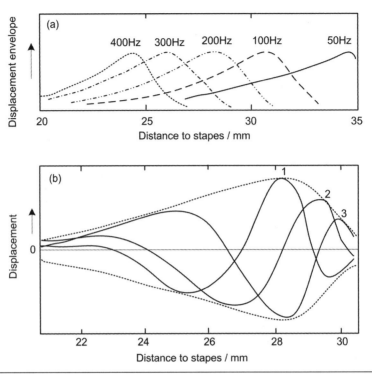

Figure 1.6 Envelopes of the basilar membrane displacement for five low-frequency tones and illustration of a traveling wave for 200 Hz tone: (a) envelope of the basilar membrane displacement; (b) the traveling wave for 200 Hz tone, with waveforms at successive times identified with the numbers "1" through "3" (adapted from Bekesy, 1960)

distinguished from each other if the frequency difference is distinct enough that the corresponding maximum displacement envelopes are located at two distinct positions on the basilar membrane. With decreasing frequency difference, the responses of the basilar membrane to the two tones may gradually overlap, resulting in a complicated response pattern. Ultimately, the two tones are indistinguishable if the frequencies are close to each other. Almost all auditory phenomena concerning frequency, such as the perception of pitch and masking, are related to the frequency selectivity of the inner ear. However, because the response width (i.e., resonance width) on the basilar membrane is so wide, it cannot be thoroughly responsible for the sharp frequency resolution of the auditory system. The active process of the auditory system should also be considered.

Another function of the inner ear is to convert mechanical oscillation into neural pulses or activity. Between the basilar membrane and the tectorial membrane are hair cells. When the basilar membrane moves up and down, a shearing motion is created between the basilar membrane and the tectorial membrane. Consequently, the stereocilia at the surface of the hair cells bend, causing kalium ions to flow into the hair cells, and thereby altering the potential difference between the inside and outside of the hair cells. This, in turn, induces a release of neurotransmitters and the initiation of action potentials in the neurons of the auditory nerve. In this way, the hair cells convert the basilar membrane oscillation into neural pulses, and the nerve system then conveys the neural pulses to the high-level nerve system for further processing. The conversion mechanism of the inner ear is complex and beyond the scope of this book. More details can be found in textbooks on physiological acoustics (e.g., Gelfand, 2010).

1.2.2. The Critical Band and Auditory Filter

Early in 1940, Fletcher investigated the masking of a tone by a band-pass noise (Fletcher, 1940) and reported that noise in a bandwidth centered at the tone frequency is effective in masking the tone, whereas the other noise component outside the bandwidth has no effect on masking. The bandwidth derived in this manner is called the *critical bandwidth* at the center frequency.

Fletcher attributed this phenomenon to the frequency analysis function of the basilar membrane. Given that each location on the basilar membrane maximally responds to a specific center or characteristic frequency, and the response decreases dramatically if the sound frequency deviates from the characteristic frequency, each location on the basilar membrane acts as a band-pass filter with a specific characteristic frequency. Correspondingly, the entire basilar membrane (strictly, the entire auditory system) can be regarded as a bank of overlapping band-pass or *auditory filters* with a series of consecutive characteristic frequencies.

The frequency resolution of the auditory system is related to the shape and width of the auditory filters. Fletcher simplified each auditory filter as a rectangular filter. If the bandwidth of the masking noise is within the effective bandwidth of the auditory filter, the noise effectively masks the tone at the characteristic frequency. If the bandwidth of the masking noise is wider than the effective bandwidth, the components of noise outside the effective bandwidth of the auditory filter have little effect on masking. The critical bandwidth provides an approximation of the bandwidth of the auditory filters. Analysis on the results of various psychoacoustic experiments indicated that the width of critical bandwidth (Δf_{CB}) in Hz is related to the center frequency f in kHz (Zwicker and Fastl, 1999) as:

$$\Delta f_{CB} = 25 + 75(1 + 1.4 f^2)^{0.69} . \tag{1.4}$$

Then, a new frequency metric related to auditory filters—that is, the *critical band rate* (in Bark)—can be introduced. One Bark is equal to the width of a critical frequency band and corresponds to a distance of about 1.3 mm along the basilar membrane. Critical band rate v in Bark is related to f in kHz as:

$$v = 13 \arctan(0.76 f) + 3.5 \arctan\left(\frac{f}{7.5}\right)^2 . \tag{1.5}$$

Some recent studies have indicated that an actual auditory filter is not exactly symmetric. An equivalent rectangular band-pass filter, however, is a good approximation for convenience, especially at moderate sound levels. The equivalent rectangular band-pass filter has a transmission in its passband equal to the maximum transmission of the specified auditory filter, and transmits the same power of white noise as the specified auditory filter. The bandwidth of the equivalent rectangular filter is known as *equivalent rectangular bandwidth* (*ERB*). Psychoacoustic experiments have indicated that the ERB in Hz is related to center frequency in kHz (Moore, 2003) as:

$$ERB = 24.7(4.37 f + 1) . \tag{1.6}$$

According to the concept of equivalent rectangular auditory filter, the *equivalent rectangular bandwidth number* (*ERBN*), a new frequency metric, can be introduced, which is related to frequency in kHz as:

$$ERBN = 21.4 \log_{10}(4.37 f + 1) . \tag{1.7}$$

According to the aforementioned formulae, the bandwidth in the critical bandwidth model is near 100 Hz within the frequency range of 100 to 500 Hz. Above this frequency, the bandwidth is approximately 20 percent of the center frequency; whereas, in the equivalent rectangular bandwidth model, a bandwidth above 500 Hz is similar to that in the critical bandwidth model and is frequency dependent below 500 Hz, rather than nearly a constant in the critical bandwidth model.

The bandwidths in both critical bandwidth and equivalent rectangular bandwidth models increase with rising frequency; hence, the frequency resolution of the auditory system decreases with increasing frequency. In this sense, both models can represent the nonuniform frequency resolution of the auditory system. Although both models are widely adopted, recent experiments have proven that the equivalent rectangular bandwidth model is more precise than the critical bandwidth model, particularly at bandwidths below 500 Hz. A variety of auditory phenomena, such as the perception of loudness and masking, are closely related to the properties of auditory filters.

1.3. Spatial Hearing

Aside from the perceptions of loudness, pitch, and timbre, human hearing also includes spatial perception, a subjective perception of the spatial attributes of sound. Using the sound coming from a sound source, the auditory system is capable of evaluating the spatial position of the sound source in terms of direction and distance, which is useful in assisting visual attention for seeking objects and warning humans (or animals) to evade potential dangers.

A variety of experimental results have demonstrated that the human ability to locate a sound source depends on many factors, such as sound source direction and source acoustical properties, and varies across individuals. On average, human acuity of localization is highest in front of the horizontal plane, that is, the *minimal audible angle* (*MAA*, or *localization blur*) $\Delta\theta$ reaches a minimum of 1° to 3° in front of the horizontal plane. In other regions, human acuity of localization decreases, with $\Delta\theta$ about threefold in lateral directions and twofold in rear directions. In the median plane, human acuity of localization is relatively low, with MAA varying from $\Delta\phi = 4°$ for white noise to 17° for speech.

In addition, the human ability to evaluate the distance of a single sound source depends on many factors, such as source properties and acoustic environment, and varies across individuals. In general, distance estimation is relatively easy, given prior knowledge of the sound source.

When two or more sound sources simultaneously radiate sound waves, the combined pressures in the two ears contain the spatial information of the multiple sound sources; the hearing can use this information to form spatial auditory events. The different acoustical attributes of the multiple sources lead to different spatial auditory events or perceptions. For incoherent sounds coming from multiple sound sources, the hearing may perceive each sound as a separate auditory event and then be able to locate each sound source. For coherent sounds coming from multiple sound sources, under certain situations, the hearing may perceive the sounds as a fused auditory event arising from the position of one real sound source, while ignoring the other real sources, as in the case of the precedence effect (discussed in Section 1.6.3). However, if the relative level and arrival time of each sound satisfy certain conditions, the hearing perceives a summing *virtual sound source* (shortened as *virtual source*, and also called *virtual sound image* or *sound image* in some references) at a spatial position where no real source exists. This is the basis of stereophonic and multi-channel surround sound reproduction, which will be discussed in Sections 1.6.1, 1.6.2, and 1.8.4. In some other situations, the hearing is unable to locate the source position or even likely to perceive an unnatural auditory event, such as the lateralization in a headphone presentation.

Here, two terminologies should be clarified: *localization* and *lateralization*. Localization refers to determining the position of an auditory event in three-dimensional space. By contrast, lateralization pertains to determining the lateral displacement of an auditory event in one-dimensional space, that is, along the straight line connecting the entrances of the two ear canals (Blauert, 1997; Plenge, 1974). Localization and lateralization judgments are both commonly used in psychoacoustic experiments. Generally, the former is usually employed in a natural or simulated natural auditory environment, whereas the latter is often used under an unnatural situation, such as in a headphone presentation where binaural signals are controlled or changed independently.

In an enclosed space with reflective surfaces, a series of reflections exist along with the direct sound, through which the hearing can acoustically perceive the spatial dimension and form a spatial impression of both environment and sound source. Thus, the spatial information encoded in reflections is vital to the design of room acoustics.

In summary, spatial hearing includes the localization of a single sound source, summing localization and other auditory events of multiple sound sources, as well as subjective spatial perceptions of environmental reflections. Research on spatial hearing is related to numerous disciplines, such as physics, physiology, psychology, and signal process. During the past century, a wide variety of studies on spatial hearing have been undertaken, which are reviewed at length in the famous monograph by Blauert (1997).

1.4. Localization Cues for a Single Sound Source

Auditory localization refers to ascertaining the apparent or perceived spatial position of a sound source in terms of its direction and distance in relation to the listener. Psychoacoustic studies have demonstrated that the directional localization cues for a single sound source include *interaural time difference* (ITD), *interaural level difference* (ILD), as well as *dynamic* and *spectral* cues (Blauert, 1997). Distance perception, another important aspect of auditory localization, is also based on the comprehensive effect of multiple cues. These localization cues are presented as follows.

1.4.1. Interaural Time Difference

ITD refers to the arrival time difference between the sound waves at the left and right ears, and plays an important role in directional localization. In the median plane, ITD is approximately zero because the path lengths from the sound source to both ears are identical. When the sound source deviates from the median plane, however, the path lengths to each ear are different; thus, ITD becomes non-zero. For example, in the horizontal plane [Figure 1.7(a)] the curved surface of the head is ignored and the two ears are approximated by two points in free space separated by $2a$, where a represents the head radius. Ideally, a plane wave is generated by a sound source at an infinite distance, but it approximately holds for a point source at a distance far larger than a in practice. The ITD for an incident plane wave from azimuth θ can be evaluated by:

$$\mathrm{ITD}(\theta) = \frac{2a}{c}\sin\theta,\tag{1.8}$$

where c is the speed of sound. Here, positive ITD indicates that the right ear is relatively closer to the sound source with the received sound leading in time, and vice versa.

If the curved surface of the head is considered, as shown in Figure 1.7(b), the head is approximated by a sphere with radius a, and the ears by two points on opposite sides. Accounting for the curved path of sound around the spherical head, the improved formula of ITD (Boer, 1940; Woodworth and Schlosberg, 1954) is:

$$\mathrm{ITD}(\theta) = \frac{a}{c}(\sin\theta + \theta) \quad with \quad 0 \leq \theta \leq \frac{\pi}{2},\tag{1.9}$$

which is usually called Woodworth formula in the literature. In accordance with the spatial symmetry of the spherical head, Equation (1.9) can be extended to the incident sounds in other quadrants.

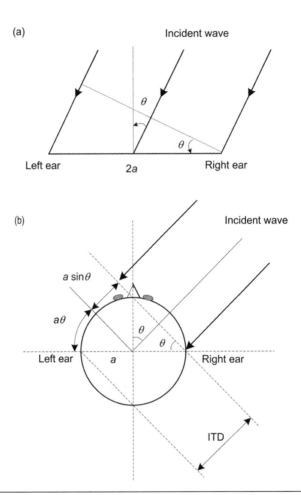

Figure 1.7 Sketch of calculating ITD in the horizontal plane: (a) neglecting the effect of the head; (b) considering the curved surface of the head

For frontal incident sound, $\sin\theta \approx \theta$, Equation (1.9) is approximately equivalent to Equation (1.8), whereas the results of the two equations are different for incident sound from other directions. Equations (1.8) and (1.9) show that ITD varies with azimuth θ and thereby acts as a directional localization cue. Moreover, because ITD is closely related to the anatomical dimensions of the head, which vary across individuals, ITD is individual dependent.

Calculated ITDs using Equation (1.9), are shown in Figure 1.8, where a equals 0.0875 m, as is generally used in the literature. Because of the symmetry of the spherical head, only the results for the right horizontal plane are given. In the figure, ITD is zero at an azimuth of 0° (i.e., the directly frontal direction), and gradually increases with sound source approaching the lateral region. At an azimuth of 90°, ITD reaches its maximum of 662 μ s. Subsequently, ITD decreases with sound source approaching the rear, and returns to zero at an azimuth of 180°.

Further psychoacoustic experiments (Blauert, 1997) revealed that *interaural phase delay difference* is an important cue for localization below approximately 1.5 kHz. Accordingly, ITD derived from the interaural phase delay is defined as:

$$\mathrm{ITD}_p(\theta, f) = \frac{\Delta\psi}{2\pi f} = -\frac{\psi_L - \psi_R}{2\pi f}, \tag{1.10}$$

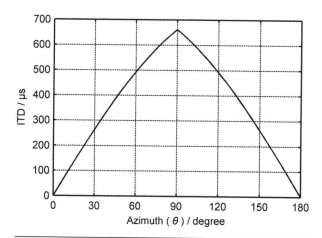

Figure 1.8 Calculated ITD from Equation (1.9) as a function of azimuth θ

where subscript p represents the phase delay, ψ_L and ψ_R represent the phases of sound pressures at the left and right ears, respectively, and $\Delta\psi = (\psi_R - \psi_L)$ is the interaural phase difference. The time factor of a sinusoidal wave is assumed equal to exp $(j2\pi ft)$; hence, $\psi > 0$ means leading in phase and $\psi < 0$ lagging in phase.

When the head dimension (path difference between two ears) equals half of a wavelength, roughly corresponding to a frequency of 0.7 kHz, the pressures at the two ears for lateral incidence are out of phase and, thus, interaural phase difference begins to provide ambiguous localization cues. Head or source movement may resolve this ambiguity. Above 1.5 kHz, when the head dimension is larger than the wavelength, however, the absolute value of interaural phase difference $\Delta\psi$ may exceed 2π, which gives rise to a completely ambiguous ITD_p.

Fortunately, for sound with complex wavefront (rather than sinusoidal sound), psychoacoustic experiments (Henning, 1974; Blauert, 1997) have proven that *interaural envelope delay difference*, termed ITD_e instead of ITD_p, is another localization cue above 1.5 kHz. Here, the sinusoidal modulation stimulus is presented as an example. Let us assume that there is a high-frequency sinusoidal carrier with frequency f_c of 3.9 kHz, whose amplitude is modulated by a low-frequency sinusoidal signal with a frequency f_m of 0.3 kHz (m is the modulation factor). Then, the binaural sound pressure signals associated with an ITD_e as τ_e are:

$$p_L(t) = \{1 + m\cos[2\pi f_m(t - \tau_e)]\}\cos(2\pi f_c t),$$
$$p_R(t) = [1 + m\cos(2\pi f_m t)]\cos(2\pi f_c t). \tag{1.11}$$

In this case, the localization (strictly speaking, the lateralization in headphone presentation) of sound is determined by τ_e, not by the ITD derived from the fine structure of the binaural signals. Some other studies, however, pointed out that ITD_e at mid and high frequencies is a relatively weak localization cue when compared with ITD_p at low frequencies (Durlach and Colburn, 1978).

In summary, the auditory system separately uses ITD_p and ITD_e as localization cues in different frequency ranges. Because Equations (1.8) and (1.9) are based on geometrical acoustics, the results are neither the ITD_p nor the ITD_e. However, accurate calculations of ITD_p and ITD_e are complex, whereas the calculation of the frequency-independent ITD using Equations (1.8) and (1.9) is relatively simple. Therefore, these two equations are often used for approximately evaluating ITD. We will see, in Section 4.1.2, that the ITD_p calculated from the scattering solution of a rigid spherical head model to the incident plane wave is about 1.5 times the result obtained from Equation (1.8) below 0.4 kHz, and nearly identical to that obtained from Equation (1.9) above 3 kHz.

1.4.2. Interaural Level Difference

ILD is another important cue to directional localization. When a sound source deviates from the median plane, the sound pressure at the farther ear (contralateral to the sound source) is attenuated (especially at high frequencies) because of the shadowing effect of the head, whereas the sound pressure at the nearer ear (ipsilateral to the sound source) is boosted to some extent. This acoustic phenomenon leads to direction- and frequency-dependent ILD as:

$$ILD(r,\theta,\phi,f) = 20 \log_{10} \left| \frac{P_R(r,\theta,\phi,f)}{P_L(r,\theta,\phi,f)} \right| \ (dB), \tag{1.12}$$

where $P_L(r, \theta, \phi, f)$ and $P_R(r, \theta, \phi, f)$ are the frequency-domain sound pressures at the left and right ears, respectively, generated by a sound source at (r, θ, ϕ).

ILD can be approximately evaluated from a simplified model, in which the head and the two ears are approximated by a rigid sphere and two opposite points on the spherical surface, respectively. In the far field with distance r far larger than head radius a, the pressures at the two ears, P_L and P_R, can be calculated as the scattering solutions of a rigid spherical head to the incident plane wave (see Section 4.1.1). The resultant ILD is irrelevant to distance r. Figure 1.9 shows the far-field ILD in the horizontal plane as a function of the azimuth of the incident plane wave at different frequencies ($ka = 0.5, 1.0, 2.0, 4.0$, and 8.0). Note that ka ($k = 2\pi f /c$ is the wave number), instead of f, is adopted as the metric for frequency in the figure.

At low frequencies with small ka values, ILD is small and varies smoothly with azimuth θ. For example, the maximum ILD is only 0.5 dB for a ka of 0.5, which indicates that the head shadowing effect and the resultant ILD are negligible at low frequencies in the far field. Moreover, for a ka larger than 1, ILD tends to increase with increasing frequency, exhibiting a complex relationship with azimuth and frequency. For example, the maximum ILDs are 2.9, 6.7, 12.0, and 17.4 dB for $ka = 1.0, 2.0, 4.0$, and 8.0, respectively. If $a = 0.0875$ m is selected, the frequencies corresponding to $ka = 0.5, 1.0, 2.0, 4.0$, and 8.0 are about 0.3, 0.6, 1.2, 2.5, and 5.0 kHz, respectively. Moreover, psychoacoustic experiments have demonstrated that ILD does not act as an effective directional localization cue until it varies with source direction, at above 1.5 kHz. In addition, because ILD is a multivariable function, it is both direction- and frequency-dependent for a given head radius a. From another point of view, ILD depends on the anatomical dimension for a sound source at given direction and frequency—that is, ILD is also an individual localization cue. Comparing Figures 1.8 and 1.9, the ILD for a narrow-band stimulus may be an ambiguous localization cue because the ILD does not vary monotonously with azimuth, even within the range of 0° to 90°.

ILD varies dramatically with azimuth at high frequencies, such as a ka of 4.0 and 8.0 (in Figure 1.9). Additionally, the maximum ILD for a sinusoidal sound stimulus (with a single frequency component) does not appear at an azimuth of 90° where the contralateral ear is exactly opposite the sound source. This is due to the enhancement in sound pressure at the contralateral ear by the in-phase interference of multipath diffracted sounds around the spherical head. For a complex sound wave with multiple frequency components, such as octave noise, ILD varies relatively smoothly with azimuth.

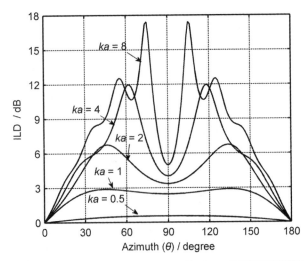

Figure 1.9 Calculated ILD as a function of azimuth at different ka, using the spherical head model

However, an actual human head is not a perfect sphere and includes pinnae and other fine structures. Therefore, the relationship between ILD, sound source direction, and frequency is far more complicated than that for a spherical head, as will be revealed in Chapters 3 and 4. Nevertheless, the results from the spherical head model are adequate for qualitatively interpreting some localization phenomena.

1.4.3. Cone of Confusion and Head Movement

The aforementioned ITD and ILD are regarded as two dominant localization cues at low and high frequencies, respectively, which were first stated in the classic "duplex theory" proposed by Lord Rayleigh in 1907. However, a set of ITD and ILD is inadequate for determining the unique position of a sound source. In fact, an infinite number of spatial positions exist; these possess identical differences in path lengths to the two ears (i.e., identical ITD). Assuming that the curved surface of the spherical head is disregarded, and that the two ears are approximated by two separated points in free space, the points form a cone around the interaural axis in three-dimensional space, which is called the "*cone of confusion*" in the literature (Figure 1.10). The cone of confusion can be conveniently expressed by a constant Θ or Γ in the interaural polar coordinate system specified in Figure 1.3. On the cone of confusion, ITD alone is inadequate for determining an exclusive sound source position. Similarly, for a spherical head model and at a far-field distance comparatively longer than the head radius, an infinite point set exists in space, within which the ILDs are identical for all points. For an actual human head, even when its nonspherical shape and curved surface are considered, corresponding ITD and ILD are still insufficient for determining the unique position of a sound source because they do not vary monotonously with the source position. In this case, the cone of confusion persists, but no longer as a strict cone.

An extreme case of the cone of confusion is the median plane, in which the sound pressures received by the two ears are nearly identical; thus, both ITD and ILD are zero. Another case is where two sound sources are located at the front-back mirror positions, as at azimuths of 45° and 135° in the horizontal plane. The resultant ITD and ILD for the two sound source positions are identical, as far as a symmetrical spherical head is concerned. Thus, ITD and ILD can determine only the cone of confusion in which the sound source is located, but not the unique spatial position of the sound source. Therefore, Rayleigh's duplex theory has only limited effectiveness, particularly in front-back and median plane localization.

To address this problem, early in 1940, Wallach hypothesized that the ITD and ILD changes introduced by head movements may be another localization cue (i.e., a dynamic cue) (Wallach, 1940). For example, when the head in Figure 1.11 is fixed, the ITDs and ILDs for sources at the front (0°) and rear (180°) in the horizontal plane are both zero, due to the symmetry of the head. Hence, the two source positions are indistinguishable by

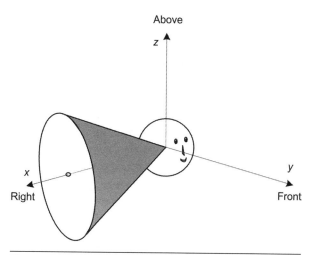

Figure 1.10 Cone of confusion in sound source localization

means of ITD and ILD cues. Head rotation, however, can introduce a change in ITD. If the head is rotated clockwise (to the right) around the vertical axis, the left ear comes closer to the front sound source, and the right ear comes closer to the rear sound source. That is, for the same head rotation, the ITD for the front sound source changes from zero to negative, whereas that for the rear sound source changes from zero to positive. If the head is rotated counterclockwise (to the left) around the vertical axis, a completely opposite situation occurs. Head rotation causes, not only a change in ITD but also changes in ILD and the sound pressure spectra at the ears. Taken together, this dynamic information aids localization. Previous experiments have preliminarily proven that the head rotation around a vertical axis is vital for resolving front-back ambiguity in horizontal localization. This conclusion has been further verified by some recent experiments (Wightman and Kistler, 1999) and has already been applied to virtual auditory displays (see Chapter 12).

In addition, Wallach hypothesized that head movement provides information for vertical localization. Although follow-up studies attempted to verify this hypothesis through experiments, because it was difficult to completely exclude contributions from other vertical localization cues experimentally (such as spectral cues, see Section 1.4.4), and because

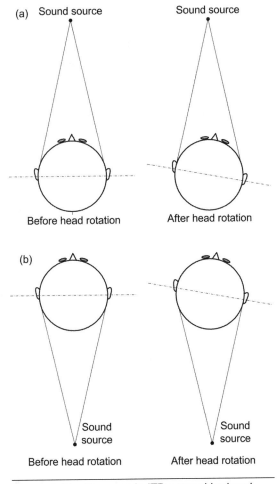

Figure 1.11 Changes in ITD caused by head rotation: (a) sound source in the front; (b) sound source in the rear

of lack of experimental support, Wallach's hypothesis received little attention for many years. Since the 1990's, however, the problem of vertical localization has attracted renewed attention with the need to develop a virtual auditory display. In 1997, Perrett and Noble first verified Wallach's hypothesis through experiments (Perrett and Noble, 1997). Our own work (Rao and Xie, 2005) further proved that the ITD change introduced by the head movement in two degrees of freedom (turning around the vertical and front-back axes, respectively, i.e., the so-called rotating and pivoting, or yaw and roll) provides information for localization in the median plane at low frequencies, and allowed a quantitative verification of Wallach's hypothesis to be given.

1.4.4. Spectral Cue

Many studies have suggested that the spectral feature caused by the reflection and diffraction in the pinna, as well as around the head and torso, provides helpful information for vertical localization and front-back disambiguity. In contrast to binaural cues (ITD and ILD), the spectral cue is a kind of monaural cue (Wightman and Kistler, 1997).

Batteau proposed a simplified model to explain the pinna effect (Batteau, 1967). Figure 1.12 shows that both direct and reflected sounds arrive at the entrance to the ear canal. The relative delay between the direct and reflected sounds is direction-dependent because the incident sounds from the different spatial directions are likely to be reflected by the different parts of the pinna. Therefore, the peaks and notches in the sound pressure spectra caused by the interference between the direct and reflected sound are also direction-dependent, thereby providing information for directional localization. In Batteau's model, the pinna effect is described as a combination of two reflections with different magnitudes, A_1 and A_2, and different time delays, τ_1 and τ_2. Hence, the transfer function of the pinna, including one direct and two reflected sounds, is expressed as:

$$H(f) = 1 + A_1 \exp(-j2\pi f \tau_1) + A_2 \exp(-j2\pi f \tau_2). \tag{1.13}$$

Batteau's model achieved only limited success because of its considerable simplification. Because the pinna dimension is about 65 mm, it functions effectively only if the frequency is above 2 to 3 kHz, where the sound wavelength is comparable to the pinna dimension. Moreover, the pinna effect is prominent at a frequency above 5 to 6 kHz. Because of its complex and irregular surface, the pinna cannot be regarded as a reflective plane from the perspective of geometrical acoustics within the entire audio frequency range, which is the inherent drawback of Batteau's model. Further studies pointed out that the pinna reflects and diffracts the incident sound in a complex manner (Lopez-Poveda and Meddis, 1996). The interference among direct and multi-path reflected/

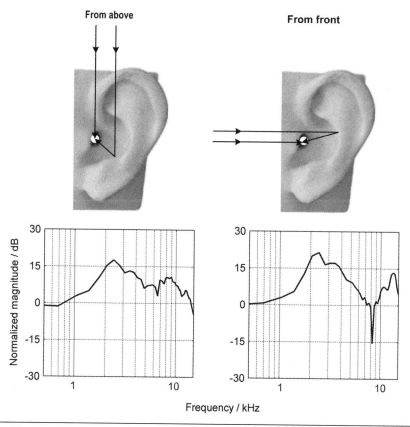

Figure 1.12 Pinna interacting with incident sounds from two typical directions

diffracted sounds acts as a filter and then modifies the incident sound spectrum as direction-dependent notches and peaks. In addition, the interference among the direct and multi-path reflected/diffracted sounds is highly sensitive to the shape and dimension of the pinna, which differ across individuals. Therefore, the spectral information provided by the pinna is an extremely individualized localization cue.

Shaw investigated the pinna effect in view of wave acoustics, and proposed a resonance model of the pinna, which states that the resonances within the pinna cavities and the ear canal form a series of resonance modes at mid and high frequencies of 3, 5, 9, 11, and 13 kHz (Shaw and Teranishi, 1968; Shaw, 1974). This model successfully interprets the peaks in the pressure spectra, among which the peak at 3 kHz for the first salient resonance is derived from the quarter wavelength resonance of the ear canal (note that the existence of the pinna extends the effective length of the ear canal), and hearing is also most sensitive around this frequency. Moreover, Shaw found that magnitudes of high-order resonance modes vary with the direction of incident sound, except for the first one. Hence, the resonance model also supports the idea that the spectral cue provided by the pinna is a directional localization cue.

Numerous psychoacoustic experiments have been devoted to exploring the localization cue encoded in spectral features. Thus far, however, no general quantitative relationship between the spectral features and the sound source position has been found because of the complexity and individuality of the shape and dimension of the pinna and the head. Blauert used narrow-band noise to investigate directional localization in the median plane (Blauert, 1997). Experimental results showed that the perceived position of a sound source is determined by the directional frequency band in the ear canal pressures, regardless of the real sound source position. That is, the perceived position of a sound source is always located at specific directions, where the frequency of the spectral peak in ear canal pressure caused by wide-band sound coincides with the center frequency of narrow-band noise. Hence, peaks in ear canal pressure caused by the head and the pinna are important in localization, as verified by other studies (Middlebrooks et al., 1989).

However, some researchers argued that the spectral notch, especially the lowest frequency notch caused by the pinna (called the *pinna notch*), is more important for localization in the median plane, even for vertical localization outside the median plane (Hebrank and Wright, 1974; Butler and Belendiuk, 1977; Bloom, 1977; Kulkarni, 1997; Han, 1994). In front of the median plane, the center frequency of the pinna notch varies with elevation in the range of 5 (or 6) kHz to about 12 (or 13) kHz. This is due to the interaction of the incident sound arriving from different elevations to the different parts of the pinna, which leads to different diffraction and reflection delays relative to direct sound. Thus, the shifting frequency notch provides vertical localization information. Moore found that shifting in the central frequency of the exquisitely narrow notch could easily be perceived, although hearing usually is more sensitive to the spectral peak (Moore et al., 1989).

Other researchers contended that both peaks and notches (Watkins, 1978) or spectral profiles are important in localization (Middlebrooks, 1992a). Although many fine structures in the pressure spectra (or HRTF, discussed later) disappear after smoothing by the auditory filter, the direction-dependent spectral feature remains (Carlile and Pralong, 1994). Moreover, psychoacoustic experiments have shown that the high-frequency acoustical information at the contralateral ear contributes less than that at the ipsilateral ear, although the information at the contralateral ear is useful in localizing the sound sources outside the median plane and resolving front-back ambiguity (Humanski and Butler, 1988; Morimoto, 2001; Macpherson and Sabin, 2007). Because of the head-shadowing effect, high-frequency sound is dramatically attenuated when it reaches the contralateral ear. In addition, Algazi et al. (2001a) suggested that the change in the ipsilateral spectra below 3 kHz caused by the scattering and reflection of the torso (especially the shoulder) provides vertical localization information for a sound source outside the median plane. Analysis on an elliptical torso model was also given.

In brief, the spectral feature caused by the diffraction and reflection of an anatomical structure is an important and individualized localization cue. Although a clarified and quantitative relationship between the spectral feature and the direction of sound source is far less complete, everyone is undoubtedly capable of using his/her own spectral feature to localize sound source.

1.4.5. Discussion on Directional Localization Cues

In summary, the directional localization cues can be classified as follows:

1. For frequencies approximately below 1.5 kHz, the ITD derived from interaural phase delay difference (ITD_p) is the dominant localization cue.
2. Above the frequency 1.5 kHz, both ILD and ITD, derived from interaural envelope delay difference (ITD_e), contribute to localization. As frequency increases (approximately above 4 to 5 kHz), the ILD gradually becomes dominant.
3. Spectral cue is important to localization. In particular, above frequencies of 5 to 6 kHz, the spectral cue introduced by, for example, the pinna is vital to vertical localization and the disambiguation of front-back confusion.
4. In addition, the dynamic cue introduced by slight head movements is also helpful in resolving front-back ambiguity and for vertical localization.

The aforementioned localization cues are individual dependent, due to unique characteristics of anatomical structures and dimensions. The auditory system determines the position of a sound source based on a comparison between obtained cues and patterns stored from prior experiences. Even for the same individual, however, anatomical structures and dimensions vary with time, especially from childhood into adulthood, albeit slowly. Therefore, comparison with prior experiences may be a self-adaptive process, and the high-level neural system is automatically able to modify stored patterns by using auditory experiences.

Different kinds of localization cues work in different frequency ranges and contribute differently to localization. For sinusoidal or narrow-band stimuli, only localization cues existing in the frequency range of the stimuli are available; hence, the resultant localization accuracy is likely to be frequency-dependent. Mills investigated localization accuracy in the horizontal plane using sinusoidal stimuli (Mills, 1958). Results showed that the localization accuracy is frequency-dependent, and the highest accuracy is $\Delta\theta = 1°$ in front of the horizontal plane for frequencies below 1 kHz ($\theta = 0°$). Assuming an average head radius a of 0.0875 m, the corresponding variation in ITD evaluated from Equation (1.8) is about 10 μ s, and the variation in low-frequency interaural phase delay difference is about 15 μ s (see discussion in Section 4.1). This is consistent with the average value of just noticeable difference in ITD derived from psychoacoustic experiments (Blauert, 1997; Moore, 2003). Conversely, localization accuracy is poorest around the frequency of 1.5 to 1.8 kHz (the so-called range of difficult or ambiguous localization). This may be because the cue of ITD_p becomes invalid within this frequency range; unfortunately, ITD_e is a relatively weak localization cue and the cue of ILD only begins to work and does not vary significantly with direction.

In general, the more localization cues contained in the sound signals at the ears, the more accurate the localization of the sound source position because the high-level neural system can simultaneously use multiple cues. This factor is responsible for numerous phenomena, including:

(1) the accuracy of binaural localization is always much better than that of monaural localization; (2) the localization accuracy in a mobile head is usually better than that in an immobile one; and (3) the localization accuracy for wideband stimulus is usually better than that for narrow-band stimulus, especially when the stimulus contains components above 6 kHz, which can improve accuracy in vertical localization. Because the information provided by the multiple localization cues may be somewhat redundant, the auditory system is able to localize the sound source, despite the absence of some cues. For example, the auditory system can localize a sound source in the median plane using the spectral cue provided mainly by the pinna without the dynamic cues provided by head movement. Under some situations, when some cues conflict with others, the auditory system appears to determine the source position according to the more consistent cues. This indicates that the high-level neural system has the ability to correct errors in localization information. Psycho-acoustic experiments performed by Wightman et al. proved that, as long as the wideband stimuli include low frequencies, the ITD is dominant—regardless of conflicting ILD (Wightman and Kistler, 1992a). Nevertheless, if too many conflicts or losses exist in localization cues, accuracy and quality are likely to be degraded, as proven by a number of experiments.

Aside from acoustical cues, visual cues have also been shown to have dramatic effects on sound localization. The human auditory system tends to localize sound from a visible source position. For example, when watching television, the sound usually appears to come from the screen, although it actually comes from the loudspeakers. However, an unnatural perception may occur when the discrepancy between the visual and auditory location is too large, such as when a loudspeaker is positioned behind the television audience. This phenomenon further indicates that sound source localization is the consequence of a comprehensive processing of a variety of information received by the high-level neural system.

1.4.6. Auditory Distance Perception

Although the ability of the human auditory system to estimate sound source distance is generally poorer than the ability to locate sound source direction, a preliminary—but biased—auditory distance perception can still be formed. Experiments have demonstrated that the auditory system tends to significantly underestimate distances for sound sources at a physical source distance farther than about 1.6 m, and typically overestimates distances for sound sources at a physical source distance closer than about 1.6 m. This suggests that the perceived source distance is not always identical to the physical one. Zahorik examined experimental data from a variety of studies and found that the relationship between perceived distance r' and physical distance r could be well approximated by a compressive power function using the linear fit method (Zahorik, 2002a):

$$r' = \kappa r^{\alpha}, \tag{1.14}$$

where κ is a constant, whose average value is slightly greater than one (approximately 1.32); and a is a power-law exponent, whose value is influenced by various factors—such as experimental conditions and subjects—and thereby varies widely, with a rough average value of 0.4. In the logarithmic coordinate, the relationship between r' and r is expressed by straight lines with various slopes, among which a straight line through the origin with a slope of 1 means r' is identical to r, that is, the case of unbiased distance estimation. Figure 1.13 shows the relationship between r' and r, obtained using linear fit for a typical subject (Zahorik, 2002b).

Auditory distance perception, which has been thoroughly reviewed by Zahorik et al. (2005), is a complex and comprehensive process based on multiple cues. Subjective loudness has long

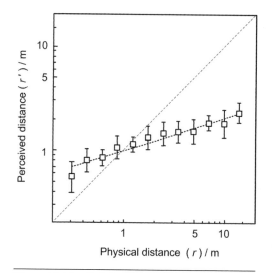

Figure 1.13 Relationship between r' and r obtained using linear fit for a typical subject, with $\alpha = 0.32$, $\kappa = 1.00$ (Zahorik, 2002b, by permission of Zahorik)

been considered an effective cue to distance perception. Generally, loudness is closely related to sound pressure or intensity at the listener's position (usually, strong sound pressure results in high loudness). In the free field, the sound pressure generated by a point sound source with constant power is inversely proportional to the distance between the sound source and the receiver (the $1/r$ law); that is, the sound pressure level decreases by 6 dB for each doubling of source distance. As a result, a close distance corresponds to high sound pressure and subsequent high loudness, which makes using loudness to estimate distance possible. However, the $1/r$ law is held only in the free field and deviates in reflective environments. Moreover, sound pressure and, therefore, loudness at the listener's position also depends on source properties, such as radiated power. Prior knowledge of the sound source or stimulus also influences distance estimation for loudness-based cues. In general, loudness is regarded as a relative distance cue, unless the listener is highly familiar with the pressure level of the sound source.

High-frequency attenuation caused by air absorption may be another cue to auditory distance perception. For a far sound source, air absorption acts as a low-pass filter, and thereby modifies the spectra of sound pressures at the received point. This effect is important only for an extremely far sound source and is negligible in an ordinary-sized room. Moreover, prior knowledge of the sound source may also influence the performance of distance estimation when using high-frequency attenuation-based cues. Generally speaking, high-frequency attenuation provides information for relative distance perception, at most.

Some studies have demonstrated that the effects of acoustic diffraction and shadowing by the head provide information on evaluating distance for nearby sound sources (Brungart et al., 1999b, 1999c). As shown in Section 1.4.2, ILD is nearly independent of the source distance in the far field, but varies considerably as source distance changes within the range of 1.0 m (i.e., in the near field), especially within the range of 0.5 m (as will be shown in Sections 3.6 and 4.1.4). Because ILD is defined as the ratio between the sound pressures at the two ears, it is irrelevant to source properties. A near-field ILD is, therefore, a cue to absolute distance estimation. In addition, the pressure spectrum at each ear changes with near field source distance, which potentially serves as another distance cue. Note that these cues are reliable only within a source distance of 1.0 m. Reflections in an enclosed space are also believed to be effective cues to distance estimation (Nielsen, 1993), which are extensively described in Section 1.7.2.

In summary, auditory distance perception is derived from the comprehensive analyses of multiple cues. Although distance estimation has recently received increasing attention with the advent of virtual auditory displays, thus far, knowledge regarding its detailed mechanism remains incomplete.

1.5. Head-Related Transfer Functions

As presented in Section 1.4, in the free field, the sound radiated from a point source reaches the two ears after interacting with anatomical structures, such as being diffracted and reflected by the head, torso, and pinnae. The resultant binaural sound pressures contain various types of localization information, such as ITD, ILD, and spectral cues. Finally, the auditory system comprehensively uses these cues to localize the sound source. In nature, the transmission process from a point sound source to each of the two ears on a fixed head can be regarded as a linear time-invariant (LTI) process (see Figure 1.14). *Head related-transfer functions, or HRTFs*, which describe the overall filtering effect imposed by anatomical structures, are introduced as the acoustic transfer functions of the LTI process. With regard to an arbitrary source position, a pair of HRTFs, H_L and H_R, for the left and right ears, respectively, is defined as:

$$H_L = H_L(r,\theta,\phi,f,a) = \frac{P_L(r,\theta,\phi,f,a)}{P_0(r,f)},$$

$$H_R = H_R(r,\theta,\phi,f,a) = \frac{P_R(r,\theta,\phi,f,a)}{P_0(r,f)},$$

$$(1.15)$$

where P_L and P_R represent the complex-valued sound pressures in the frequency domain at the left and right ears, respectively; and P_0 represents the complex-valued free field sound pressure in the frequency domain at the center of the head with the head absent, which takes the form (Morse and Ingrad, 1968):

$$P_0(r,f) = j\frac{k\rho_0 cQ_0}{4\pi r}\exp(-jkr),$$

$$(1.16)$$

where ρ_0 is the density of the medium (air), c is the speed of sound, Q_0 denotes the intensity of the point sound source, and $k = 2\pi f/c$ represents the wave number. The time factor of harmonic waves is denoted by $\exp(j2\pi ft)$ in this book; thus, the transmission of outward waves with distance is denoted by factor $\exp(-jkr)$.

As mentioned, for a certain spatial position of the sound source, a pair of HRTFs exists, one for each ear. Generally, an HRTF is a function of frequency and sound source position in terms of distance r, azimuth θ, and elevation ϕ. Furthermore, the HRTF is individual dependent due to the unique anatomical structure and dimension of each human, which is denoted by the variable a. Strictly speaking, a refers to a set of parameters specifying the dimensions of relevant anatomical structures. Although the HRTF is a multi-variable function, some variables that are irrelevant to the succeeding discussions can be disregarded for simplicity. For example, distance r is disregarded when discussing the far field HRTF because the far field HRTF is approximately distance independent, as will be proven in Section 4.1.4. Elevation ϕ is ignored when the discussion is restricted to the horizontal plane, whereas a is omitted when the individual feature of the HRTF is not addressed. Additionally, the subscript denoting the left or right ear is often omitted when the discussion is general to both ears.

The measurement point for P_L and P_R in Equation (1.15) varies across the literature, among which the eardrum is the most natural choice. Møller (1992) investigated this issue theoretically and

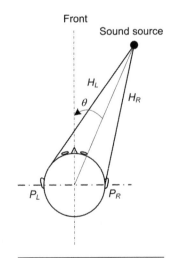

Figure 1.14 Diagram of sound transmission from a point sound source to the two ears

experimentally. Results indicated: when the ear canal is approximated by a tube with a diameter of 8 mm, slightly larger than the average 7 mm (Maa and Shen, 2004), the entire ear canal could be regarded as a one-dimensional (i.e., direction independent) transmission line for sound waves whose wavelengths are much larger than the diameter. A wavelength of 8 mm corresponds to a frequency of 42.5 kHz, and a quarter of the wavelength approximately corresponds to a frequency of 10 kHz. Therefore, the approximation of one-dimensional ear canal transmission is held below 10 kHz. Relevant experiments (Hammershøi and Møller, 1996) demonstrated that the entire ear canal could be regarded as a one-dimensional transmission line below 12 to 14 kHz.

Figure 1.15(a) is a sketch of the anatomy of the human external ear, and Figure 1.15(b) shows the analogue model for a one-dimensional sound transmission through the ear canal (Møller, 1992). In the model, the complete sound field outside the ear canal is described by two variables from Thevenin's theorem: open-circuit pressure P_1 and generator impedance Z_1. Z_1 is the radiation impedance, as seen from the ear canal into free air. P_1 does not physically exist in a natural listening situation. However, if the ear canal is blocked to make volume velocity zero (by analogy to electric current), P_1 can be measured just outside the blockage. The natural sound pressure P_2 at the open entrance to the ear canal is related to P_1 as:

$$\frac{P_2}{P_1} = \frac{Z_2}{Z_1 + Z_2}, \tag{1.17}$$

where Z_2 is the input impedance, looking into the ear canal. Moreover, sound pressure P_3 at the eardrum is related to P_1 and P_2 by:

$$\frac{P_3}{P_2} = \frac{Z_3}{Z_2}; \quad \frac{P_3}{P_1} = \frac{Z_3}{Z_1 + Z_2}, \tag{1.18}$$

where Z_3 is the eardrum impedance.

The entire sound transmission from a sound source to the eardrum is then divided into two consecutive parts: a direction-dependent part, describing the transmission from a sound source to the ear canal entrance; and a direction-independent part, describing the transmission along the ear canal to the eardrum. These suggest that sound pressures P_1, P_2, and P_3, or even those measured at

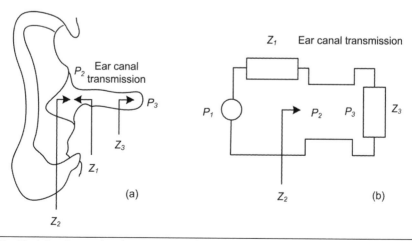

Figure 1.15 Models of sound transmission through the human external ear: (a) anatomical sketch; (b) analogue model (adapted from Møller, 1992)

other points in the ear canal, contain the same spatial information for the sound source. Therefore, they can all be used as the binaural sound pressures in the HRTF definition, except in a special notation. In the literature, the measurement point for binaural sound pressures is called the *reference point*. Moreover, the binaural sound pressures measured at different reference points can be converted into any of the three (P_1, P_2, or P_3), by cascading a direction-independent function. In practice, measuring P_1 is the most convenient, as will be shown later in this book.

In Equation (1.15), HRTFs are defined in the frequency domain. Their time-domain counterparts are known as *head-related impulse responses* (*HRIRs*), which are the impulse responses from a point sound source to two ears in the free field. HRTFs and HRIRs are related by Fourier transform:

$$h_L(r,\theta,\phi,t,a) = \int_{-\infty}^{+\infty} H_L(r,\theta,\phi,f,a)e^{j2\pi ft}\,df\,, \quad h_R(r,\theta,\phi,t,a) = \int_{-\infty}^{+\infty} H_R(r,\theta,\phi,f,a)e^{j2\pi ft}\,df\,;$$

$$H_L(r,\theta,\phi,f,a) = \int_{-\infty}^{+\infty} h_L(r,\theta,\phi,t,a)e^{-j2\pi ft}\,dt\,, \quad H_R(r,\theta,\phi,f,a) = \int_{-\infty}^{+\infty} h_R(r,\theta,\phi,t,a)e^{-j2\pi ft}\,dt. \tag{1.19}$$

As with HRTF, HRIR is also a multi-variable function that varies with sound source position, time, and anatomical structures and dimensions.

If the HRTFs or HRIRs are known, then the binaural sound pressures can be calculated by:

$$P_L(r,\theta,\phi,f,a) = H_L(r,\theta,\phi,f,a)P_0(r,f),$$
$$P_R(r,\theta,\phi,f,a) = H_R(r,\theta,\phi,f,a)P_0(r,f), \tag{1.20}$$

or equivalently expressed in the time domain as:

$$p_L(r,\theta,\phi,t,a) = \int_{-\infty}^{+\infty} h_L(r,\theta,\phi,\tau,a)p_0(r,t-\tau)d\tau = h_L(r,\theta,\phi,t,a) * p_0(r,t),$$

$$p_R(r,\theta,\phi,t,a) = \int_{-\infty}^{+\infty} h_R(r,\theta,\phi,\tau,a)p_0(r,t-\tau)d\tau = h_R(r,\theta,\phi,t,a) * p_0(r,t), \tag{1.21}$$

where p_L and p_R are sound pressures in the time domain at the left and right ears, respectively, and p_0 is the sound pressure in the time domain at the center of the head with the head absent. They are related to the corresponding sound pressures in the frequency domain by Fourier transform. The sign "*" refers to convolution.

Given that HRTFs or HRIRs determine the binaural sound pressures caused by a point sound source and contain the most important localization cues, they are a core issue in binaural spatial hearing and virtual auditory displays (Zhong and Xie, 2004), and are therefore given considerable attention in this book. We believe the statement of HRTFs, containing all the localization cues, may be biased in some studies. Strictly speaking, HRTFs do not include all the localization cues discussed in Section 1.4, or, at least, they lack the dynamic cue introduced by head movements. Thus, stating that HRTFs include the main, or most, localization cues seems more reasonable and accurate.

1.6. Summing Localization and Spatial Hearing with Multiple Sources

The localization of multiple sound sources, as for the localization of a single sound source presented in Section 1.4, is another important aspect of spatial hearing (Blauert, 1997). When two or more sound sources simultaneously radiate coherent sounds, under a specific situation, the auditory system may perceive a sound coming from a spatial position where no real sound source exists. Such kind of illusory or phantom sound source, also called virtual sound source or virtual sound image (shortened as sound image), results from summing localization of multiple sound sources. In summing localization, the sound pressure at each ear is a linear combination of the pressures generated by multiple sound sources. The auditory system then automatically compares the localization cues, such as ITD and ILD, encoded in the binaural sound pressures, with patterns stored from prior experiences of a single sound source. If the cues in the binaural sound pressures successfully match the pattern of a single sound source at a given spatial position, then a convincing virtual sound source at that position is perceived (Guan, 1995). However, this is not always the case. Under some situations, such as in the case of the precedence effect to be described in Section 1.6.3, the listener perceives a sound coming from one of the real sources, regardless of the presence of other sound sources. On the other hand, when two or more sound sources simultaneously radiate partially coherent sounds, the auditory system may perceive an extended, or even diffusely located, spatial auditory event. These phenomena are related to the summing spatial hearing of multiple sound sources, which is the subjective consequence of comprehensively processing the spatial information of multiple sources by the auditory system. The overall subjective spatial perception of the environmental reflections is also closely related to the spatial hearing of multiple sound sources, and will be presented later in this chapter.

1.6.1. Summing Localization of Two Sound Sources and the Stereophonic Law of Sine

In 1940, Boer conducted the earliest experiment investigating the summing localization of two sound sources, that is, two-channel stereophonic localization (Boer, 1940). The typical two-channel stereophonic loudspeaker configuration is illustrated in Figure 1.16. The listener symmetrically locates between the left and right loudspeakers, which are separated by an angle of $2\theta_0$ (usually, $2\theta_0 = 60°$). When both loudspeakers provide identical signals, the listener perceives a single virtual source at the midpoint of the two loudspeakers, that is, directly in front of the listener. When adjusting the level difference between the loudspeaker signals, the virtual source moves toward the direction of the loudspeaker with the larger signal level. An interchannel level difference larger than approximately 15 dB is sufficient to position the virtual sound source at either of the loudspeakers. Then, the position of the virtual sound source no longer changes, even with increasing level difference.

Strictly, this phenomenon can be theoretically interpreted by analyzing the combined sound pressures at the left and right ears if the HRTFs from each loudspeaker to the ears are known. In fact, in the 1950s, Clark successfully clarified the phenomenon in two-channel stereophonic localization using a simplified model, in which the shadow of the head is neglected

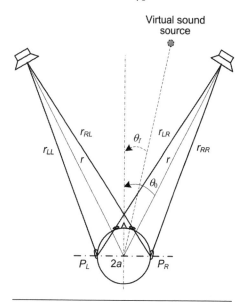

Figure 1.16 Two-channel stereophonic loudspeaker configuration, with θ_I denoting the azimuth of the perceived source with respect to the midline, and θ_0 denoting the azimuth of the loudspeaker with respect to the midline

and the two ears are approximated by two points in free space separated by $2a$, where a is the head radius (Clark et al., 1957).

According to Figure 1.16, the left and right loudspeakers are placed at azimuths of $(360° - \theta_0)$ and θ_0, respectively, and fed with the same sinusoidal signals of amplitudes proportional to L and R. In this case, when the distance between the listener and either of the loudspeakers is far longer than head radius a (i.e., in the far field), the pressures at the two ears below 0.7 kHz (denoted by P_L and P_R for the left and right ears, respectively) can be regarded as a linear combination of the two plane waves coming from the left and right loudspeakers, taking the forms:

$$P_L = L\exp[j(2\pi ft - kr_{LL})] + R\exp[j(2\pi ft - kr_{LR})],$$
$$P_R = L\exp[j(2\pi ft - kr_{RL})] + R\exp[j(2\pi ft - kr_{RR})], \tag{1.22}$$

where $k = 2\pi f/c$ is the wave number and:

$$r_{LL} = r_{RR} \approx r - a\sin\theta_0 \quad and \quad r_{LR} = r_{RL} \approx r + a\sin\theta_0$$

denote the distances from the loudspeaker to the ipsilateral (near) and contralateral (far) ears, respectively. When omitting common phase $\exp[j(2\pi ft - kr)]$, interaural phase difference $\Delta\psi$ can be calculated as

$$\Delta\psi = \psi_R - \psi_L = -2\arctan\left[\frac{L-R}{L+R}\tan(ka\sin\theta_0)\right]$$

Additionally, the corresponding interaural time difference, ITD_p, is:

$$ITD_p = \frac{\Delta\psi}{2\pi f} = -\frac{1}{\pi f}\arctan\left[\frac{L-R}{L+R}\tan(ka\sin\theta_0)\right]. \tag{1.23}$$

Because ITD_p is considered a dominant cue to localization at low frequencies, a comparison between the combined ITD_p in Equation (1.23) and the single-source ITD_p derived from prior auditory experiences [Equation (1.8)] enables the determination of azimuthal position θ_I of the summing virtual source as:

$$\sin\theta_I = -\frac{1}{ka}\arctan\left[\frac{L-R}{L+R}\tan(ka\sin\theta_0)\right]. \tag{1.24}$$

At low frequencies, where $ka \ll 1$, Equation (1.24) can be expanded as a Taylor series of ka. If only the first expansion term is retained, then this equation can be simplified as:

$$\sin\theta_I = -\frac{L-R}{L+R}\sin\theta_0 = \frac{R/L-1}{R/L+1}\sin\theta_0. \tag{1.25}$$

This is the famous *stereophonic law of sine*, demonstrating that the spatial position of the virtual sound source (i.e., summing sound source) is completely determined by the amplitude ratio between the two loudspeaker signals and the span angle between the two loudspeakers with respect to the listener, but is irrelevant to frequency and head radius.

Thus, Equation (1.25) suggests:

1. when L and R are identical, $\sin\theta_I$ is zero, indicating that the summing virtual source is positioned at the midpoint between the two loudspeakers;
2. when R is larger than L, $\sin\theta_I$ is positive, meaning the summing virtual source is positioned close to the right loudspeaker;
3. when R is far larger than L, $\sin\theta_I$ is approximately equal to $\sin\theta_0$, indicating that the summing virtual source is positioned at the right loudspeaker; and

4. because of the left-right symmetry in configuration, similar results obtain when L is larger than R.

Substituting $\theta_0 = 30°$ in Equation (1.25), Figure 1.17 illustrates the relationship between the position of the virtual source and the level difference between the loudspeaker signals (i.e., the interchannel level difference), denoted by $d = 20 \log_{10}(R/L)$ dB. In Figure 1.17, θ_I varies continuously from 0° to approximately 30° with increasing d starting at 0 dB. The stereophonic law of sine is quite effective below 700 Hz, as has been proven by numerous experiments.

Some remarks on the two-channel stereo and the stereophonic law of sine are the following:

1. The stereophonic law of sine is based on the principle of the summing localization of multiple sound sources. In stereophonic localization, the auditory system uses the ITD_p, encoded in the combined binaural pressures, to locate the position of the virtual source at low frequencies. This indicates that a transformation occurs, from the interchannel level difference at the two loudspeaker signals to the ITD_p at the two ears. The interchannel level difference regarding the loudspeaker signals should not be confused with the ILD at the two ears (introduced in Section 1.4.2).

2. Because the combined binaural pressures at the two ears contain only the localization cue of ITD_p, which is an effective cue below 1.5 kHz, the approach of creating virtual sources by adjusting interchannel level difference is invalid above 1.5 kHz for two-channel stereophonic reproduction.

3. According to the stereophonic law of sine, the absolute value of $\sin\theta_I$ is directly proportional to the difference between L and R, and inversely proportional to the sum of L and R. Therefore, the larger the difference between the amplitudes of the loudspeaker signals is, the larger the absolute value of $\sin\theta_I$, (i.e., the larger the angle deviation of the virtual source from the midpoint of the two loudspeakers) will be. If the signals L and R are in phase, the absolute value of $\sin\theta_I$ is less than $\sin\theta_0$; thus, the virtual source is positioned within the span of the two loudspeakers. Whereas, if they are out of phase, the absolute value of $\sin\theta_I$ is greater than $\sin\theta_0$, and, if $\sin\theta_0 < |\sin\theta_I| \leq 1$, then the virtual source may be positioned outside the span of the two loudspeakers (Xie, 1981). Of course, this is only valid at low frequencies.

4. The existence of the head leads to an increase in ITD_p, when compared with the simplified model in which the head effect is disregarded. Therefore, the ITD_p evaluated from Equations (1.8) and (1.23), which are based on the simplified model, is less than the actual value when a is taken as the actual head radius. As was mentioned in Section 1.4.1 and will be proven in Section 4.1, the exact

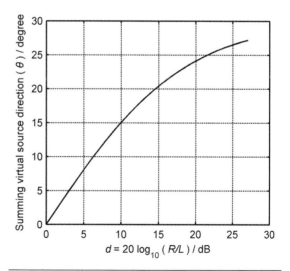

Figure 1.17 Relationship between the position of the virtual source and the level difference between the loudspeaker signals

value of the ITD_p for frequencies below 0.4 kHz is frequency independent and approximately 1.5 times the value resulting from either Equation (1.8) or (1.23) using the actual head radius. This indicates that ITD_p can be evaluated accurately from Equations (1.8) or (1.23), if a is 1.5 times the actual head radius. Fortunately, such correction of a is unnecessary in the derivation of the stereophonic law of sine [i.e., Equation (1.25)] because of its inherent independence from a. However, the stereophonic law of sine is valid only below 0.7 kHz because of the various approximations introduced into its derivation. If a is taken as 1.2 times the actual head radius, however, Equation (1.24) exhibits reasonable results below approximately 1.5 kHz (He et al., 1993).

5. Given that the counterclockwise coordinate system is commonly used in the literature on stereophonic and multi-channel surround sound, the clockwise coordinate system is employed in this book; the stereophonic law of sine that appears in other studies lacks the negation in Equation (1.25).

Despite considering only two sound sources, the two-channel stereophonic localization experiment provides important insights into the basic principle of summing localization with multiple sound sources. In practice, generating a virtual sound source by adjusting the interchannel level difference of the two loudspeaker signals is called *intensity stereo*, and the corresponding system is called the two-channel intensity stereophonic sound system, or simply stereophonic sound system. The two-channel signals can be either recorded with a combination of directional microphones or synthesized with pan-pot control (Xie, 1981; Rumsey, 2001).

1.6.2. Summing Localization Law of More Than Two Sound Sources

The stereophonic law of sine can be generalized in a case with more than two sources in the horizontal plane. The shadow of the head is ignored and the two ears are approximated by two points in free space separated by $2a$ at low frequencies. M loudspeakers are arranged in a circle around a listener with the same distance relative to the listener's head center. The azimuth of the ith loudspeaker is denoted by θ_i, and the amplitude of the ith loudspeaker signal is proportional to A_i. Then, similarly to the derivation of Equation (1.25) for the two-channel stereophonic sound, the position of the summing virtual source with multiple sound sources can be evaluated by comparing the resultant ITD_p with that of a single sound source. Finally, the position of the summing virtual source is related to the positions of multiple sound sources as (Xie, 2001a):

$$\sin\theta_I = \frac{1}{ka}\arctan\left[\frac{\displaystyle\sum_{i=0}^{M-1} A_i \sin(ka\sin\theta_i)}{\displaystyle\sum_{i=0}^{M-1} A_i \cos(ka\sin\theta_i)}\right], \tag{1.26}$$

where k is the wave number. When ka is considerably less than 1 (i.e., $ka \ll 1$), then Equation (1.26) can be expanded as a Taylor series of ka. If only the first expansion term is retained, the equation can be simplified as (Bernfeld, 1975):

$$\sin\theta_I = \frac{\displaystyle\sum_{i=0}^{M-1} A_i \sin\theta_i}{\displaystyle\sum_{i=0}^{M-1} A_i}. \tag{1.27}$$

Equation (1.27) can be regarded as an extension of the stereophonic law of sine under an arbitrary number of multiple sources (loudspeakers are used as sources in the derivation) in the horizontal plane. Furthermore, the formula for the summing localization can also be extended to the case of multiple sound sources in three-dimensional space (Xie, 1988; Rao and Xie, 2005). These summing localization formulae are only valid at low frequencies. In practice, creating virtual sources in the horizontal plane, or even in three-dimensional space, is highly possible by employing appropriate loudspeaker layouts and signal panning or mixing methods, which is the basis of multi-channel horizontal or spatial surround sound (Xie and Guan, 2002a).

1.6.3. Time Difference between Sound Sources and the Precedence Effect

The previous sections dealt with the summing localization of sounds with merely amplitude differences among them. Under some situations, however, sounds radiating from multiple sound sources are likely to exhibit arrival time differences in wavefronts as well, which may arise from differences in the transmission paths from each source to the listener, or the relative delay among electrical signals fed to each loudspeaker (used as sound source). In this case, the summing spatial auditory event depends upon arrival time differences and signal properties.

The same experimental configuration shown in Figure 1.16 can be used to investigate the spatial hearing of two sound sources with time differences, in which an electrical signal and its delayed version are fed into the two loudspeakers. Studies have shown that the perceived position of the summing virtual source moves toward the loudspeaker reproducing the leading signal (Leakey, 1959). With regard to signals with transient characteristics, such as music and speech, when the amount of time difference (i.e., interchannel time difference) reaches a value varying from some hundred microseconds to about one millisecond, the summing virtual source is perceived at the direction of the loudspeaker reproducing the leading signal (Blauert, 1997). We also examined the summing localization with interchannel time difference, both theoretically and experimentally and found that the perceived position of the virtual sound source is frequency dependent (Xie, 2002b) and that the interchannel time difference could not work effectively until above 0.7 kHz.

When the relative arrival time difference between two sounds exceeds a lower limit, a kind of spatial auditory event that is completely different from summing localization occurs. This is known as the *precedence effect,* or the *Hass effect.* The precedence effect refers to the phenomenon where, for two coherent sounds with transient characteristics (such as click, music, and speech), the auditory system always perceives a sound as coming from the position of the leading sound (regardless of the existence of lagged sounds) when the difference in arrival time falls within a certain boundary, which is defined by a lower-limit τ_L and an upper-limit τ_H. The two sounds appear to be perceived as a fused spatial auditory event, and localization is dominated by the wavefront first reaching the two ears; that is, the ability to perceive the second sound as a separate spatial auditory event is suppressed. Thus, the precedence effect is also called the *law of the first wavefront.* Conversely, if the difference in arrival time between the sounds exceeds τ_H, then a separate echo is perceived.

The values of τ_L and τ_H depend on some acoustical factors, such as the source position, and properties and relative amplitudes of the sound signals. Generally, τ_L is approximately 1 to 3 ms, whereas τ_H is several milliseconds for a single pulse and approximately 50 ms for speech. Some studies have shown that the value of τ_H is relevant to the time duration of the autocorrelation function of the sound signal. Ando (1985) suggested that the time duration, when the autocorrelation function of the sound signal decays to 10% of its maximum, be used to evaluate τ_H. The precedence effect is robust, even when the lagged sound intensity is moderately higher than the leading sound. Finally, the increasing lagged sound (higher than 10–15 dB and within an appropriate range of delay compared with the leading sound) is perceived separately.

In the precedence effect, the lagged sound contributes in a specific way to the overall auditory event (including timbre, loudness, and other spatial attributes), although it is not perceived as a separate auditory event from the leading sound. In this sense, the suppression of the lagged sound in the precedence effect is restricted to the spatial location information, and does not include all perceived information. Numerous studies on the mechanism of the precedence effect have been conducted; generally, the precedence effect is seen as originating from comprehensive processing of the binaural sound information by the high-level neural system. Some models of the precedence effect have been proposed (see Zurek, 1987). In addition, some hearing phenomena associated with the precedence effect are meaningful in room acoustics and spatial hearing, which are discussed in the following section.

1.6.4. Cocktail Party Effect

If a desired speech sound source and one or more interfering sound sources (e.g., competitive speech sources) simultaneously exist, the auditory system can take advantage of the spatial separation of the desired and interfering sound sources to more effectively detect the desired speech information. This phenomenon is known as the *cocktail party effect*. In daily life, the cocktail party effect facilitates the detection of the desired speech information even in a noisy environment.

The cocktail party effect is a kind of binaural auditory effect associated with the spatial hearing of multiple sources. If the spatial information of the sources is lost, such as in monaural presentation via headphone or monophonic reproduction by a loudspeaker, benefits from the cocktail party effect are subsequently lost. Since the pioneering work of Cherry (1953), a large number of investigations on the mechanism of the cocktail party effect have been carried out, but no final conclusion has thus far been achieved. The cocktail party effect is generally considered the consequence of comprehensive processing on binaural sound information by the high-level neural system (e.g., auditory stream segregation). Bronkhorst (2000) provided a thorough review of this issue.

1.7. Room Acoustics and Spatial Hearing

The preceding sections focused on spatial hearing in the free field where no reflection exists. In an enclosed space, however, reflections from surroundings, such as walls, ceilings, and floors, are common and have considerable influence on spatial hearing in a room. Analyses of the enclosed sound field and its influence on hearing are the main aspects of room acoustics, which is beyond the scope of this book. This section will include only an overview of some basic principles, with an emphasis on the issue related to spatial hearing in a room. For more details (such as formula derivations), refer to relevant books (Kuttruff, 2000; Beranek, 1996; Wu and Zhao, 2003).

1.7.1. Sound Fields in Enclosed Spaces

In an enclosure (usually a room) with reflective boundaries, not only direct sound but also reflected sounds reach a receiver. Figure 1.18(a) shows a typical example of sound transmission paths from

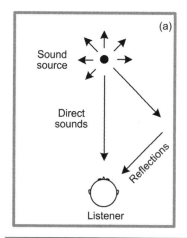
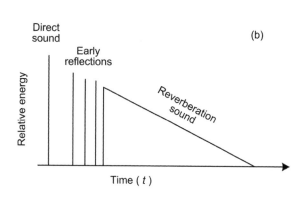

Figure 1.18 (a) Sound transmission in an enclosed space; (b) a typical example of room energy impulse response (energy vs. time curve)

a sound source to a receiver (i.e., a listener) in a room, and Figure 1.18(b) shows corresponding energy impulse response (the square of room impulse response) on a logarithmic scale. After the direct sound arriving earliest, early reflections from reflective surfaces (or boundaries) arrive discretely with different time delays, in which the first reflection with the shortest time delay comes from the reflective surface with minimum difference in the acoustical path relative to the direct sound. After these, the number of reflections rapidly increases, leading to a dramatic increase in the temporal density of reflections. However, the total energy of reflections decreases because of surface absorption and high-frequency air absorptions, leading to a reverberant sound field. Finally, the reflection level drops to a level below background noise.

During the past century, a number of methods for investigating the enclosed sound field have been developed, and can be classified into two categories, namely, statistical or geometrical acoustics-based and wave acoustics-based methods. The statistical acoustics-based method is relatively simple but less accurate when compared with wave acoustics, and is invalid at low frequencies and in small rooms. By contrast, the wave acoustics-based approach surveys the enclosed sound field by solving the wave equation imposed on a variety of boundary conditions. The corresponding calculation is complicated, and analytic solutions are usually impossible except with some regular room shapes. In the last three or four decades, various numerical solutions based on the wave acoustics method have been developed with advances in computer technology.

In statistical acoustics, the concept of a *diffuse field* is of major importance, and must satisfy the following requirements:

1. In a way, sound propagates as a sound ray with a constant speed c along straight lines, and the energy carried by the sound ray is completely uniform with respect to directional distribution.
2. Sound rays are mutually incoherent; thus, the superimposed phase varies randomly.
3. Average sound energy density is constant everywhere in the region under consideration.

Reverberation time, T_{60}, is an important parameter for describing the sound field and corresponding perceived attributes in a room. It is defined as the time duration for the sound pressure level to decay by 60 dB after the source is suddenly switched off. Therefore, it can be used to evaluate the decaying speed of reverberation sound. A variety of calculation techniques and resultant formulae with different applicability have been developed, among which a widely used formula is:

$$T_{60} = 0.161 \frac{V}{-S\ln(1-\alpha)}. \qquad (1.28)$$

Equation (1.28) is called *Eyring's formula*, with V denoting the room volume, S encompassing the total absorption area, and α representing the average absorption coefficient. Furthermore, it can be simplified as the famous *Sabin's formula* for $\alpha < 0.2$:

$$T_{60} = 0.161 \frac{V}{S\alpha}. \qquad (1.29)$$

Sound energy in a room includes two components: direct sound and reverberant (or reflected) sound. For a point source, the direct sound pressure is inversely proportional to the source distance relative to the received point ($1/r$ law); or, equally, the relationship between energy density and distance obeys an inverse-square law. Therefore, the closer to the sound source, the larger the direct-sound energy density is, and vice versa. The ratio of direct sound energy density to reverberant sound energy density is called *direct-to-reverberant energy ratio* (K):

$$K = \frac{D_s S}{16\pi r^2} \frac{\alpha}{1-\alpha}, \qquad (1.30)$$

where D_S is the directivity factor of the source, and r is the distance between the source and the receiver. Distance r_c, at which direct sound energy density is equal to that of reverberant sound (i.e., $K = 1$), is called the *reverberation radius, room radius*, or *critical distance*, and is given by:

$$r_c = \frac{1}{4} \sqrt{\frac{D_s S\alpha}{\pi(1-\alpha)}}. \qquad (1.31)$$

According to Equation (1.31), r_c depends on total absorption area S, average absorption coefficient α, and directivity factor D_S. When locating for a distance that is less than the reverberation radius r_c (i.e., $K > 1$), the listener receives more direct sounds than reverberant sounds. By contrast, when the distance relative to the sound source is larger than the reverberation radius r_c (i.e., $K < 1$), more reverberant sounds than direct sounds are received. Strictly speaking, this analysis is valid only in the diffuse reflected field. As to common rooms, wherein the actual reflected sound field may not be ideally diffused, this analysis can be used only as a rough, but reasonable, approximation.

When the sound source and the receiver are in fixed positions, the sound transmission between them is a linear time-invariant process, and can be described by *room impulse response (RIR)*. Because RIR contains completely temporal characteristics of direct and reflected sounds, many important room acoustical parameters, such as reverberation time, can be evaluated through RIR (Schroeder, 1965).

RIR is related to room properties, in which the receiver is simply treated as a point in space. In actual conditions with a human listener, however, the transmission process from source to receiver immediately becomes complicated for two reasons. First, humans use two ears to listen, that is, there are two closely spaced receivers rather than one. Second, and more importantly, the presence of human anatomical structures in the transmission path inevitably modifies the sounds reaching the two ears through diffraction and reflection. Thus, the binaural pressure signals in a room contain the spatial information of both sound source and environment. Correspondingly, the impulse responses are called *binaural room impulse responses (BRIRs)* and can be regarded as generalized HRIRs, from a free field without reflections to a room sound field with reflections. As with HRIRs, the measurement point of BRIRs can be chosen at any point along the ear canal, even at the blocked entrance to the ear canal. BRIRs depend on the absolute positions of the sound source and receiver rather than the relative positions in HRIRs. Similarly to Equation (1.21), the binaural pressures in a room can be calculated as:

$$p_L(t) = h_L(t) * p_0(t) \quad \text{and} \quad p_R(t) = h_R(t) * p_0(t), \tag{1.32}$$

where $p_0(t)$ denotes the free-field sound pressure at the head center when the head is absent, h_L and h_R denote the BRIRs for the left and right ears, respectively, and the sign "*" denotes convolution.

1.7.2. Spatial Hearing in Enclosed Spaces

Reflections in enclosed spaces create significant auditory effects, and therefore become a key issue in the acoustic design of rooms. In this section, a great deal of attention is given to how reflections influence the spatial hearing in rooms.

First, we address the problem of how reflections influence the directional localization of the direct sound. From the perspective of geometrical acoustics, reflections in a room can be regarded as multipath sounds originating from a series of image sound sources. Thus, the spatial hearing in a room can be viewed as an extreme case of that for multiple sound sources (discussed in Section 1.6). Numerous experiments have indicated that reflections have little influence on the perceived direction of the direct sound, as long as the time interval and the relative intensity difference among the direct sound and the reflections satisfy the prerequisites of the precedence effect. With increasing reflected energy, however, reflections begin to impair the localization accuracy for the direct sound. In particular, at the position outside the reverberation radius of the room, where the reverberant sound energy density is larger than that of the direct sound, the directional localization of the direct sound becomes difficult and even impossible. In this case, the existence of the reverberation sound decreases the degree of coherence in the binaural sound pressures, thereby leading to the incapability to localize.

These results are of considerable importance in practice. In most daily activities, such as talking in a room, the directional localization is not influenced by room reflections because of the precedence effect. In domestic sound reproduction, because the size of the listening room is relatively small, a reasonable loudspeaker arrangement and absorption treatment to the room surfaces are required to ensure that the time interval and relative intensity difference among the direct and reflected sounds satisfy the prerequisites of the precedence effect, so that the influence of reflections on the summing localization of the stereophonic sound source can be avoided. This also allows for free field calculation [as used in deriving Equation (1.25)] in the study of two-channel stereophonic localization. On the other hand, in the acoustical design of halls, special attention should be paid to ensure that the time interval and relative intensity difference among the direct sound and reflections satisfy the prerequisites of the precedence effect to avoid perceived echoes. In a distributing sound reinforcement system with multiple loudspeakers, the perceived sound direction can be controlled by applying appropriate delays to parts of the loudspeaker signals, according to the rule of the precedence effect.

Our second problem of interest concerns the ways in which reflections influence auditory distance perception. As mentioned in Section 1.4.6, reflections in a room are important cues to absolute source distance estimation. According to Equation (1.30), the direct-to-reverberant energy ratio is inversely proportional to the square of distance. Therefore, it can be used as a distance cue, although a real reflected sound field may somewhat deviate from the diffuse sound field, from which Equation (1.30) is derived. Studies from Bronkhorst et al. (1999) indicated that a simple model based on a modified direct-to-reverberant energy ratio could accurately predict the auditory distance perception in rooms. Moreover, frequency-dependent boundary absorption modifies the power spectra of reflections, and the proportion of reflected power increases with increasing sound source distance. Thus, the power spectra of binaural pressures vary with sound source distance. This also provides information for auditory distance perception.

Spatial impression is another spatial auditory effect introduced by reflections. When satisfying the prerequisites of the precedence effect, reflections in rooms are not perceived as separate echoes and have little influence on the perceived direction of the direct sound. However, they do contribute to the perceived size of the sound source, along with the perceived spatial and acoustical features of the room. Hence, reflections create comprehensive auditory impressions of the sound source and environment.

Spatial impression is an important subjective attribute of sound quality in concert halls. It consists of at least two components, namely, auditory source width (ASW) and envelopment. The ASW, or apparent source width, is the auditory perception of sound source broadening compared with the visual width of the actual source. A large number of studies have indicated that early lateral reflections, especially frequency components with band centers below 2.0 kHz, are vital to generating ASW (Beranek, 1996; Barron, 1971). According to Ando (1985, 1998), ASW is related to the spatial distribution of the discrete reflections from various directions, especially from the preferred azimuths of $55° \pm 20°$. The interaural cross-correlation coefficient (IACC) has been suggested as an objective index for evaluating ASW and some other spatial auditory perceptions (Damaske, 1969/1970; Ando, 1985). IACC is defined as the value that maximizes the cross-correlation function of the binaural sound pressures in the time domain:

$$\Phi_{LR}(\tau) = \frac{\displaystyle\int_{t_1}^{t_2} p_L(t+\tau)p_R(t)dt}{\left[\displaystyle\int_{t_1}^{t_2} p_L^2(t)dt \int_{t_1}^{t_2} p_R^2(t)dt\right]^{1/2}}. \tag{1.33}$$

$$IACC = \max\left[\Phi_{LR}(\tau)\right] \quad with \quad |\tau| \le 1\,ms. \tag{1.34}$$

In Equation (1.33), if the moment when the earliest direct sound arrives is chosen as the time origin (i.e., zero in time), the reflections can be treated separately by the choice of integration interval. When selecting $t_1 = 0$ and $t_2 = 80$ ms, the result of Equation (1.33) is called the early IACC, denoted by $IACC_E$. When selecting $t_1 = 80$ ms and $t_2 = 1–2$ s, the result of Equation (1.33) is called the late IACC, denoted by $IACC_L$. Finally, when selecting $t_1 = 0$ and $t_2 = 1–2$ s, the result of Equation (1.33) is denoted by $IACC_A$. Based on the definition of IACC, its value varies from 0 to 1.

Psychoacoustic experiments demonstrate that ASW is closely related to $IACC_E$, and a low-value $IACC_E$ leads to a favorable ASW. $IACC_E$ describes the similarity between the binaural sound pressures that are generated by the direct and early reflected sounds arriving within the initial 80 ms. Because the head shadowing effect is maximized in the lateral region, lateral reflections lead to marked differences between the sound pressures at the two ears; therefore, a low $IACC_E$ subjectively causes a wide ASW.

Envelopment (or listener envelopment) is another important component of auditory spatial impression. Envelopment describes the subjective sensation of being surrounded by reverberant sounds and is closely related to the extent of reverberant sounds diffusion (Beranek, 1996).

1.8. Binaural Recording and Virtual Auditory Display

1.8.1. Artificial Head Models

According to the discussion in previous sections, binaural sound pressures or signals contain most of the spatial information of sound events, such as localization information for the sound source and spatial information for environmental reflections. Therefore, the spatial information of sound events can be obtained by analyzing binaural sound signals. The most direct and natural way to obtain binaural signals is to record the pressures at the entrances of human ear canals by a pair of miniature microphones or at a certain point along human ear canals by a pair of probe microphones. In this manner, the characteristics introduced by individual anatomical shapes and dimensions are preserved in the resultant binaural signals. However, slight movements and noise caused by human subjects during the recording session are likely to degrade the accuracy of signal measurements. Moreover, if auditory research needs an average result across multiple subject groups, a large group of human subjects is required, which is time-consuming and difficult for experimentation.

An *artificial head* (or dummy head, head and torso simulator) is a kind of model made of specific materials and used to simulate anatomical structures, such as the head, torso, and pinnae of a real human in perspective of acoustics (Vorländer, 2004). Its design is based on the average dimensional properties from certain populations, or on the anatomical features of a "standard" or "typical" human subject. Moreover, the acoustic properties of the materials used in an artificial head are similar to those of a human. Artificial heads can be used in a variety of acoustic measurements and research, such as binaural recording, sound quality evaluation, headphone testing, and hearing aid testing, instead of a human subject. Especially in binaural hearing, the effects of diffractions and reflections caused by the head, pinnae, and torso can be simulated using the artificial head. Additionally, the resultant binaural sound pressures can be recorded by a pair of microphones placed at the ear canal entrance or at a certain point along the ear canal of the artificial head.

Among various artificial head products, KEMAR (Knowles Electronics Manikin for Acoustic Research), originally produced by Knowles Electronics and now acquired by G. R. A. S. Sound & Vibration in Denmark, has been used extensively in the field of binaural hearing. This may be due to two reasons: (1) KEMAR has a humanlike appearance, and (2) the HRTF database from KEMAR was probably the first available on the internet, and consequently used in a substantial number of relevant works. We mainly refer to the specification of KEMAR from Knowles Electronics because many existing works, including ours, are based on this product.

KEMAR was originally designed for hearing aid measurement and relevant acoustic research (Burkhard and Sachs, 1975). Its design was based on the average head and torso dimensions from anatomical measurements of Western male and female adults from the 1950s to the 1960s, and satisfies the requirement of IEC 60959:1990 and ANSI S3.36/ASA58-1985. Some parameters in the anatomical measurement and main dimensions for KEMAR are listed in Table 7.1 in Chapter 7.

Figure 1.19 shows the KEMAR with detailed head, torso, and pinnae. Basically, two sizes of pinnae are available (note that there are slight variations in the two sizes of pinnae for different applications). As suggested in the specifications of KEMAR, the pair of small pinnae (DB-060/061 for the right and left ears, respectively) is typical of American and European females, as well as Japanese males and females, and the pair of large pinnae (DB-065/066 for the right and left ears, respectively) is typical of American and European males. Figure 1.19(b) shows small pinna DB-061 on KEMAR. In addition, KEMAR is also equipped with a pair of Zwislocki occluded-ear simulators, satisfying the requirements of ANSI S3.25/ASA80-1989 for simulating sound transmission through the human ear canal to the eardrum. One end of the occluded-ear simulator is connected to the ear canal entrance by a canal extension, and a standard 12.7 mm (1/2 in) pressure field microphone is built-in at the other end (i.e., inside the head). Hence, a microphone diaphragm, analogous to the eardrum, is used to convert sound pressures into electrical signals.

In addition to KEMAR, some other commercial artificial head products are available (Figure 1.20), such as B & K 4100 and 4128 (from Brüel & Kjær), Head Acoustics HMS IV (from Head Acoustics GmbH), MK2B (from 01 dB-metravib), and Neumann KU-100 (from Georg Neumann GmbH). Researchers at Aalborg University in Denmark also designed an artificial head, called VALDEMAR (Christensen et al., 2000). The B & K 4100 is constructed for binaural recording, whereas the B & K

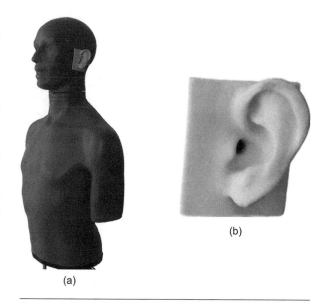

(a)

(b)

Figure 1.19 KEMAR and DB-061 pinna: (a) KEMAR; (b) DB-061 pinna

4128, including the head, torso, and detailed pinnae, is widely used for electroacoustics measurements, such as telephone and headphone testing. The HMS IV, consisting of head and shoulder with highly simplified pinnae, is usually applied in the evaluation of sound quality and noise. The MK2B is used for sound quality in the automobile industry and psychoacoustics. The KU-100 includes only the head, and is often employed in binaural recording. VALDEMAR is designed especially for research on binaural and spatial hearing.

Note that almost all available artificial heads were designed based on published data on the average anatomical dimensions from Western human subjects, or a replica of a "standard" or "typical" Western human subject. Given that anatomical dimensions are likely to vary across different populations and undergo alterations over time, the available artificial head models may not be

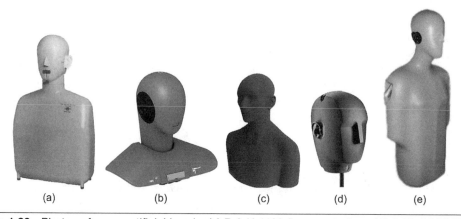

(a)　　　　(b)　　　　(c)　　　　(d)　　　　(e)

Figure 1.20 Photos of some artificial heads: (a) B & K 4128 (by permission of B & K); (b) HMS IV (by permission of Genuit); (c) MK2B; (d) Neumann KU-100 (by permission of Georg Neumann GmbH); (e) VALDEMAR (Christensen et al., 2000, by permission of AES).

appropriate for current Chinese (or other Eastern) subjects. This should be considered when undertaking relevant studies.

For future developments of artificial heads, some researchers have argued that one should design a great variety of artificial heads, so that each "anthropometric (anatomical) group" will have its own optimized and matched artificial head (Genuit and Fiebig, 2007). Another possible approach is to design and standardize a "universal" artificial head in order to obtain consistent results and provide the comparability among artificial head recordings and measurements. However, a "universal" artificial head may not match each individual and, in turn, may cause errors in recorded binaural signals.

1.8.2. Binaural Recording and Playback System

As previously presented, the binaural sound signals recorded in the ear canals of a human subject or an artificial head contain the main spatial information of sound events (Møller et al., 1996a). If the pressures at the eardrums generated by reproducing the recorded binaural signals via a pair of headphones are exactly equal to those in the original sound field where the sound event actually occurs, the overall spatial auditory event or experience, including sound source localization and those caused by environmental reflections, can be replicated to the listener. This is the basic principle of the *binaural recording and playback system* (or *technique*). In practice, an artificial head is often used in recording the binaural signals. That is, an artificial head is often used, instead of the potential listener, to "listen" in the original sound field. For this reason, the binaural recording and playback system is sometimes called the *dummy head stereophonic system*. Figure 1.21 shows the block diagram of the binaural recording and playback system. An article by Paul (2009) reviewed the full history of binaural recording and playback technique and the development of the artificial head.

As with HRTFs, binaural signals can be recorded at any point from the ear canal entrance to the eardrum, even at the entrance of the blocked ear canal. The binaural signals recorded at all these reference points capture the main spatial information of the sound event. Because of the one-dimensional transmission in the ear canal, however, the binaural signals recorded at different reference points also differ, although the difference is direction independent. Therefore, directly rendering the recorded binaural signals without accounting for the measurement position may lead to incorrect sound pressures at the eardrums. Moreover, the nonideal transfer characteristics of the recording and playback chain, originating from uneven frequency responses of the recording microphone and reproducing headphone, as well as unwanted coupling between the headphone and external ear, inevitably cause linear distortions in both magnitude and phase of the reproduced sound pressures at the eardrums. The overall nonideal transfer characteristics of the recording and playback chain can be represented by a pair of transfer functions, $H_{pL}(f)$ and $H_{pR}(f)$, one for each ear. Ideally, if the recorded binaural signals are equalized by the inverse of $H_{pL}(f)$ and $H_{pR}(f)$ [see Equation (1.35)] before being

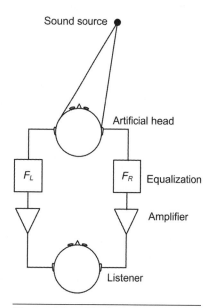

Figure 1.21 Block diagram of the binaural recording and playback system

presented to the headphone, the linear frequency distortion in the signal chain can be eliminated, or at least reduced as much as possible.

$$F_L(f) = \frac{1}{H_{pL}(f)} \quad \text{and} \quad F_R(f) = \frac{1}{H_{pR}(f)}. \tag{1.35}$$

Chapter 8 will provide more details on the signal processing algorithm of equalization.

Similarly to HRTFs, binaural signals are individual dependent. If binaural signals are recorded from the listener's own ears, they match the individual characteristics of the listener and result in an authentic subjective performance in reproduction. Otherwise, if the binaural signals are recorded from an artificial head (or other human subject), the subjective performance in reproduction depends on the extent of similarity in the anatomical shapes and dimensions between the recording artificial head (or human subject) and the listener. Generally, the more similar they are, the more authentic the subjective performance that is obtained. Hence, the subjective performance in non-individual binaural recording and playback varies across listeners. Because an artificial head is merely an approximate model of a real human head, localization performance using the binaural recording from an artificial head is expected to be generally inferior to that using individual binaural recording, as revealed by many subjective listening tests (Møller et al., 1996b, 1999; Minnaar et al., 2001).

The binaural recording and playback technique was introduced in the early stages of spatial sound research. However, compared with conventional stereophonic or multi-channel surround sound techniques, this technique received little attention in the field of domestic sound reproduction for at least two reasons. First, the artificial head technique was immature at that time. Second, recorded binaural signals, originally intended for headphone reproduction, were not compatible with loudspeaker reproduction without further signal processing. On the other hand, because it is capable of capturing and recreating the main spatial information of a sound event, it has long been a useful tool for evaluating the sound quality in halls and for some psychoacoustic experiments. In the recent two or three decades, the binaural recording and playback technique has been increasingly used in scientific research.

1.8.3. Virtual Auditory Display

Another approach to obtaining binaural signals is to synthesize them by signal processing. Some researchers practiced this technique in the early 1980s (Morimoto and Ando, 1980). Since 1989, when Wightman and Kistler (1989b) simulated free field virtual sound source in three-dimensional space by synthesizing binaural signals for headphone presentation, this technique has been tremendously developed and applied to extensive fields (Begault, 1994a; Bronkhorst, 1995). Wightman et al. (2005) provided a thorough description of this issue.

The detailed synthesis is as follows. As discussed in Section 1.5, the free field binaural sound pressures, created by a point source at position (r, θ, ϕ), are determined by a pair of HRTFs, $H_L(r, \theta, \phi, f)$ and $H_R(r, \theta, \phi, f)$, or HRIRs, $h_L(r, \theta, \phi, t)$ and $h_R(r, \theta, \phi, t)$. Provided that the HRTFs or HRIRs have been measured or calculated, the free field binaural signals can be synthesized by filtering mono stimulus $E_0(f)$ with a pair of HRTFs. The resultant signals can then be expressed in the frequency domain as:

$$E_L(r,\theta,\phi,f) = H_L(r,\theta,\phi,f)E_0(f) \quad \text{and} \quad E_R(r,\theta,\phi,f) = H_R(r,\theta,\phi,f)E_0(f). \tag{1.36}$$

Or, equivalently, they can be expressed in the time domain as:

$$e_L(r,\theta,\phi,t) = h_L(r,\theta,\phi,t) * e_0(t) \quad \text{and} \quad e_R(r,\theta,\phi,t) = h_R(r,\theta,\phi,t) * e_0(t). \tag{1.37}$$

When the resultant binaural signals are presented over a pair of headphones, the pressures inside a listener's two ears are directly proportional to those generated by a real source at spatial position (r, θ, ϕ), leading to an authentic virtual sound source. Compared with the binaural sound pressures generated by a real source shown in Equation (1.20), the synthesized binaural sound pressures (or signals) shown in Equation (1.36) lack a scaling factor and a frequency-independent linear delay. This is because the magnitude and phase variations of $P_0(r, f)$ with source distance (in detail, the magnitude attenuation with the inverse of the source distance and the linear delay introduced by free propagation) are omitted in the synthesis. For rendering a single virtual sound source, the magnitude variation of P_0 with source distance can be implemented by scaling the gain of input stimulus $E_0(f)$ with $1/r$, whereas omitting the linear delay in $P_0(r, f)$ has no influence on auditory effects. If multiple virtual sound sources at different spatial positions are rendered simultaneously, however, the gain should be scaled, and different linear delays, $\tau = r / c$, should be added for each input stimulus in turn. For simplicity, the steps of scaling input gain and adding linear delay are disregarded in the succeeding discussion, unless otherwise stated.

In reflective environments, corresponding binaural signals $e_L(t)$ and $e_R(t)$ can be synthesized by convoluting mono stimulus $e_0(t)$ with a pair of binaural room impulse responses h_L and h_R, according to Equation (1.32).

$$e_L(t) = h_L(t) * e_0(t) \quad \text{and} \quad e_R(t) = h_R(t) * e_0(t). \tag{1.38}$$

When the resultant binaural signals are presented over a pair of headphones, the pressures inside a listener's two ears are directly proportional to those generated by the real source in the reflective environment, leading to an authentic spatial auditory perception of the sound event caused by the direct and reflected sounds.

In the literature, this technique of replicating the spatial auditory perception of sound events by HRTF- or HRIR-based binaural synthesis is known as *virtual auditory display* (VAD). Occasionally, it is called virtual audio, virtual auditory space, virtual sound, 3-D sound or 3-D audio, binaural technique (technology), et cetera. Binaural technique is generally taken to include virtual auditory display and the binaural recording and playback technique.

Because spatial hearing in nature consists of the localization of a single sound source, as well as the overall spatial auditory perception of multiple sound sources and environmental reflections, a virtual auditory display that aims to authentically replicate real spatial hearing should also include recreating all the spatial perceptual aspects mentioned above. In this sense, virtual auditory display is sometimes called virtual auditory environment display or virtual acoustic environment display. However, the terms virtual auditory environment display and virtual acoustic environment display are always employed in cases where primary emphasis is placed on recreating or synthesizing the overall spatial information of acoustical environments.

As with the binaural recording and playback technique in Section 1.8.2, a virtual auditory display also emphasizes the exact reproduction of binaural sound signals based on the hypothesis that the same binaural pressures lead to the same auditory perception. However, the binaural signals in a virtual auditory display are synthesized by signal processing, whereas the binaural signals in the binaural recording and playback technique are recorded directly from the ears of an artificial head or human subject. This is the essential distinction between a virtual auditory display and the binaural recording and playback technique.

Similar to the binaural recording and playback technique [see Equation (1.35)], the nonideal transfer characteristics in the playback chain and in the HRTF measurement for a virtual auditory display should be equalized before headphone presentation. Incorporating equalization, the signal processing algorithm of Equation (1.36) becomes:

$$E_L(r,\theta,\phi,f) = F_L(f)H_L(r,\theta,\phi,f)E_0(f),$$
$$E_R(r,\theta,\phi,f) = F_R(f)H_R(r,\theta,\phi,f)E_0(f),$$

(1.39)

or, in the time domain, as:

$$e_L(r,\theta,\phi,t) = f_L(t) * h_L(r,\theta,\phi,t) * e_0(t),$$
$$e_R(r,\theta,\phi,t) = f_R(t) * h_R(r,\theta,\phi,t) * e_0(t).$$

(1.40)

where $f_L(t)$ and $f_R(t)$ are the inverse Fourier transforms of $F_L(f)$ and $F_R(f)$, respectively, and the sign "*" denotes convolution.

Figure 1.22 shows a block diagram of frequency-domain signal processing in synthesizing a free field virtual sound source. Theoretically, equalization (or compensation) to the nonideal transfer characteristics in the playback chain is necessary for accurately recreating the binaural pressures at the eardrums. However, because the transfer characteristics in the playback chain is highly dependent upon headphone type and listener, which makes the signal processing more complicated, equalization is often omitted in some applications where accuracy is noncritical. Given that the processing of equalization is independent of sound source position, it is disregarded in most of the succeeding discussions for simplicity—except in Chapter 8, where the method of equalization processing will be addressed extensively.

Because HRTFs are individual dependent, the subjective performance in a virtual auditory display varies across listeners, as it does in the binaural recording and playback technique. Results of psychoacoustic experiments indicated that subjective performance is optimal when using individualized HRTFs in synthesizing a virtual sound source (Wenzel et al., 1993a), whereas, subjective performance varies depending on the similarity between the HRTFs used and those of the listener when using nonindividualized HRTFs (Wightman and Kistler, 1993). This problem will be thoroughly addressed in Chapters 7 and 8.

1.8.4. Comparison with Multi-channel Surround Sound

Two commonly used but different approaches are employed in spatial sound reproduction. Accordingly, spatial sound reproduction systems are classified into two categories. The first aims to reconstruct the original sound field around the listener to a certain extent by controlling the loudspeaker signals, as in the two-channel stereophonic technique and other kinds of surround sound reproduction techniques. The other approach emphasizes the exact reproduction of the binaural sound pressures based on the hypothesis that the same binaural pressures lead to the

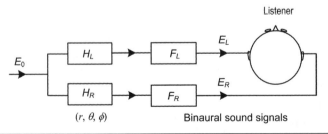

Figure 1.22 Block diagram of frequency-domain signal processing in synthesizing a free field virtual sound source

same auditory perception, as in binaural recording and playback techniques and in virtual auditory display techniques.

From a physical point of view, a sound field is entirely described by the spatial and temporal distributions of sound pressure $p(x, y, z, t)$. For the same listener, identical sound fields should lead to the same auditory perception. The starting point for a multi-channel sound is to reproduce a sound field that is identical, or close to, an ideal or desired one (at least within some spatial regions), and thus to provide a listener with the desired spatial information in the reproduced sound field (Xie and Guan, 2002a). According to the Huygens-Fresnel principle and the Kirchhoff-Helmholtz boundary integral equation, the pressure and its derivative on the surface of an arbitary closed space without sources determine sound field within the space. Thus, a desired sound field can be reproduced by the combination of sound waves from an array of monopole and dipole loudspeakers arranged on the boundary surface. This is the basic principle of the *wave field synthesis method*, which has been developed over the past two decades (Berkhout et al., 1993; Boone et al., 1995). Another, more straightforward, method arranges many loudspeakers in space and feeds each with an independent signal. This allows the sound from an arbitrary direction to be reproduced by the loudspeaker at a corresponding direction. This way, the direct and early-reflected sounds in a room can be reproduced, respectively, in their intended directions, and the diffuse reverberation sounds can be reproduced by a number of loudspeakers in all directions. Hence, the desired spatial information of the sound field is accurately reproduced. The auditory perception created by this method is authentic and natural. In addition, the listening area is wide, so that subjective performance in reproduction does not markedly vary with alterations in listener position. This method has been applied in research on room acoustics (Kleiner et al., 1993). However, it was rarely used for prevalent sound reproduction because of its complexity.

Taking advantage of some psychoacoustic principles, the multi-channel sound system is usually simplified into several channels, that is, sampling the spatial sound field in several discrete directions. Considering the limited capacity of transmission or recording media, the simplified multi-channel sound system attempts to preserve the spatial information important to subjective perception, while discarding information that is irrelevant or relatively unimportant. The extent of simplification varies according to the aim and application of reproduction. For example, multi-channel spatial surround sound preserves the directional information in three-dimensional space, and horizontal surround sound preserves, at most, the directional information in the horizontal plane. Many practical systems are unable to completely accomplish this aim, only retaining the directional information from the front and some from the rear. Conventional stereophonic sound only preserves the directional information from the front at low frequencies. Within the audio frequency range, the more spatial information is reserved, the better the reproduction performance that is achieved. However, more channels are required to preserve more spatial information, which leads to a more complicated system. The need to simplify the system often conflicts with the requirement for better reproduction performance. In the development of multi-channel sound, the number of required channels has long been an open issue.

The simplification of multi-channel sound often requires the use of the psychoacoustic rule of summing spatial hearing of multiple sound sources (discussed in Section 1.6). According to the rule of summing localization, reproducing an illusory or virtual sound source at an arbitrary direction may be accomplished by adjusting the relative amplitudes of multi-channel signals (i.e., interchannel level difference) for some appropriate loudspeaker configurations. Generating spatial auditory experience, similar to what is perceived in a diffuse reverberation sound field, may also be accomplished by feeding the uncorrelated (partially coherent) signals to a pair, or more, of loudspeakers (Damaske and Ando, 1972). Although the reproduced sound field of a multi-channel sound may be different from the original or desired sound field in terms of physics, it is a good approximation in hearing because the subjective auditory experience in the reproduced sound field is reasonably similar to that in the original sound field.

A variety of multi-channel sound systems have been developed. Early in the 1970s, quadraphonic sound systems were extensively developed (Xie, 1981; Rumsey, 2001). Quadraphonic sound systems consist of four loudspeakers, normally arranged in left-front, right-front, left-back, and right-back positions, to reproduce the spatial information of sound in the horizontal plane. Because of their inherent defects, such as unstable virtual sound source and narrow listening area, quadraphonic sound systems are not widely used. Gerzon (1985) proposed another kind of spatial surround sound system known as Ambisonics. Ambisonics offers flexibility in loudspeaker configuration, and is able to recreate three-dimensional virtual sound sources within certain frequency limits, using sound field signal mixing.

Currently, the 5.1 channel surround sound system is the most widely used. Originally developed for digital film sound at the end of the 1980s, it was subsequently applied to domestic sound reproduction. In 1994, 5.1 channel surround sound was recommended by the International Telecommunication Union (ITU) as the standard for multi-channel stereophonic sound systems, with and without accompanying pictures (ITU-R, BS775-1, 1994). The 5.1 channel surround sound system is now extensively employed in digital video discs (DVDs), domestic theaters, and digital televisions (TVs). Five independent channels, with full audio bandwidth in the 5.1 channel system are available, and the ITU-recommended configuration of loudspeakers is illustrated in Figure 1.23. The three front channels, including left (L), right (R), and center (C), are used to recreate stable virtual sound sources in the front region to match the picture. The two surround channels, including left surround (LS) and right surround (RS), are used to recreate ambience and other spatial effects. An optional low frequency effect channel (called .1 channel) operates in the frequency range of 20 to 120 Hz and drives a subwoofer to convey such specific effects as explosion sound.

For application in the cinema, however, LS and RS in 5.1 channel surround sound are reproduced through a series of loudspeakers arranged at the side and rear (Figure 1.24). The loudspeakers aim

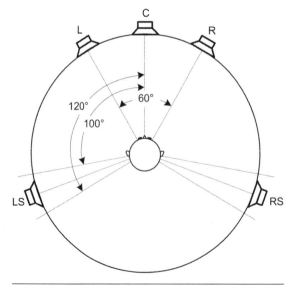

Figure 1.23 Configuration of loudspeakers for the 5.1 channel surround sound recommended by ITU

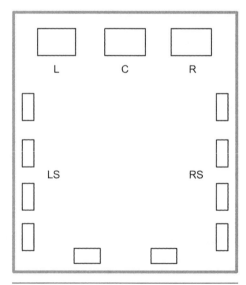

Figure 1.24 5.1 channel loudspeaker configuration in cinema

to improve subjective envelopment. In addition, LS and RS are sometimes decorrelated before being rendered to the loudspeakers. The 5.1 channel surround sound system was introduced as a compromise, between system complexity and desired perceived effect, and its reproduced sound field is only a rough approximation of the original or ideal sound field. More details on multi-channel sound can be found in the literature, such as Rumsey (2001).

Comparing the multi-channel surround sound with virtual auditory display (or more extensively, binaural technique), some distinctions are as follows:

1. The multi-channel sound is originally devised to reproduce, as accurately as possible within some areas, a desired (or intended) sound field. In practice, however, some kinds of simplification are incorporated, such as discarding relatively unimportant spatial information and using the psychoacoustic rule of summing spatial hearing. For this reason, the realism and genuineness of spatial auditory perception in reproduction are inevitably degraded. Even when simplification is considered, multi-channel sound still requires more than two channels, making its recording, transmission, and playback stages complex.

2. The aim of a virtual auditory display is to accurately replicate binaural pressures, rather than to reproduce a sound field around a listener. A virtual auditory display is, relatively, simpler because only two independent channels are required for recording, transmission, and playback. Because the reproduced binaural pressures preserve the main spatial information of sound within a wide frequency range, the resultant auditory experience is authentic and natural.

3. When a listener is located inside the reproduced sound field of multi-channel sound, the ears automatically receive the binaural sound pressures containing the spatial information. A well-designed multi-channel sound promises a relatively large listening area, and subjective performance depends little on the listener's individual anatomical structures and dimensions. Because the multi-channel sound signals are originally intended for loudspeaker reproduction, an additional signal conversion process is needed before they are presented over headphones (for details, see Section 10.1). Otherwise, the spatial information in sound reproduction is destroyed, resulting in an unnatural auditory experience.

4. A virtual auditory display, whose performance highly depends on the individual anatomical characteristics of a listener, recreates binaural sound pressures. Moreover, because the binaural signals are originally intended to be presented over headphones, an additional signal conversion process is needed before they are reproduced by loudspeakers (for details, see Section 9.1). However, even with appropriate signal conversion processing, the size of the listening area remains limited because a virtual auditory display inherently aims to generate the sound pressures at two points (two ears) in space accurately.

5. Traditionally, multi-channel sound was widely used for domestic sound reproduction, whereas virtual auditory display was mainly used for scientific purposes. During the past decades, the situation has dramatically changed. The succeeding chapters will show that virtual auditory display has been used extensively in the fields of domestic sound reproduction and multimedia computer technology.

6. So far, we have emphasized the differences between multi-channel sound and virtual auditory display. In Sections 6.4 and 6.5, however, both multi-channel sound and virtual auditory display will be relevant to the spatial sampling of a sound field. In this sense, the principles of the two reproduction methods are closely related and can be mutually referred to.

1.9. Summary

The clockwise spherical coordinate system is specified in this book, while the interaural polar coordinate system is used occasionally for convenience in discussion. The peripheral part of the human auditory system includes the external, middle, and inner ears. The external ear is composed of the pinna and the ear canal. The pinna, whose size and shape are highly individual dependent, plays an important role in high-frequency sound localization. The middle ear acts as an impedance-matching device, facilitating effective transmission of sound waves from the external into the inner ear. The inner ear behaves as a frequency analyzer and converts mechanical vibrations into neural pulses. In view of function, the frequency analysis of the auditory system can be modeled, using a combination of a series of overlapping auditory filters, whose bandwidth is approximated by a critical bandwidth or equivalent rectangular bandwidth. Correspondingly, Bark and equivalent rectangular bandwidth number, two new frequency units, can be introduced.

Spatial hearing is a subjective sensation of the spatial attributes of sound, which covers the localization of a single sound source, summing localization and auditory event of multiple sound sources, and subjective spatial perception of environmental reflections. Localization determines two aspects: direction and distance of sound sources.

The directional localization cues for a single sound source include interaural time difference (ITD), interaural level difference (ILD), dynamic and spectral cues, and others. Different cues take effect within different frequency ranges. Auditory distance perception is a comprehensive consequence resulting from multiple cues.

HRTFs are acoustic transfer functions, from the sound source to the two ears in the free field. They represent the overall filtering effect of the anatomical structures and are highly individual dependent. HRTFs, or the time version of HRIRs (head-related impulse responses), contain most of the localization cues, and can be measured at an arbitrary point along the ear canal, even at the entrance to the blocked ear canal.

The principle of summing localization for multiple sound sources is similar to that for a single sound source, which provides a basis for intensity stereo and multi-channel surround sound. The position of a summing sound source at low frequencies can be evaluated by localization formulae, such as the stereophonic law of sine. The precedence effect, which is important in room acoustics and hearing, is a special auditory phenomenon that concerns the spatial hearing of multiple sound sources. The cocktail party effect is another kind of binaural auditory phenomenon related to the spatial hearing of multiple sources.

In enclosed spaces with reflective boundaries, sounds reflected from boundaries exist in addition to those from the direct sound of the sound source. Reflections create comprehensive auditory impressions of the sound source and surrounding environment. Binaural room impulse responses contain the main spatial information of the direct and reflected sounds.

An artificial head is an acoustic model, which acoustically simulates the effects of anatomical structures (such as the head, torso, and pinnae) of a real human. A binaural recording and playback system reproduces the spatial information of sound events by presenting the binaural signals recorded on an artificial head or a human subject over a pair of headphones. By contrast, a virtual auditory display reproduces the spatial information of a sound event by presenting the binaural signals synthesized by HRTF-based filtering. In addition, the binaural pressures at the eardrums are reconstructed in a virtual auditory display (as well as in binaural recording and playback), rather than the sound field around a listener in a multi-channel surround sound reproduction.

2

HRTF Measurements

This chapter discusses issues concerning HRTF measurements. The principles for the measurement of time-domain impulse response, or frequency-domain transfer function of a linear-time-invariant system, are reviewed. The basic principles and technology of HRTF measurement, as well as measurement error analysis, are discussed. Some existing HRTF databases and their features are outlined. Some specific HRTF measurement approaches, particularly for near-field HRTF measurement, are provided.

As mentioned in Chapter 1, head-related transfer functions (HRTFs) represent the acoustic transmission functions from a sound source to the two ears, and play an important role in spatial hearing and virtual auditory displays. Nowadays, empirical measurement is the most important and accurate approach to obtaining an HRTF. In principle, all methods used to determine the transfer function or impulse response of a linear-time-invariant (LTI) system can be applied to an HRTF measurement. In the past two decades, the techniques used for HRTF measurement have gradually become more sophisticated, with the advent of computer technology and digital signal processing. Numerous research groups have measured HRTFs and established databases, mainly for research purposes. In Section 2.1, the underlying principle for the measurement of time-domain impulse response or frequency-domain transfer function of an LTI system, along with the characteristics of a variety of measurement signals, is reviewed. Details on signal processing can be found in textbooks (Oppenheim et al., 1999). Section 2.2 presents the basic principles, design methods, and technology details of HRTF measurement, as well as measurement error analysis. Section 2.3 outlines some existing HRTF databases and their features. Finally, some specific and near-field HRTF measurement approaches are provided in Section 2.4.

2.1. Transfer Function of an LTI System and its Measurement Principle

2.1.1. Continuous-Time LTI System

Under certain conditions, many electroacoustic or room acoustic systems, such as a loudspeaker or a hall, can be approximately regarded as a continuous-time LTI system, whose time-domain output $y(t)$ is related to input $x(t)$ as:

$$y(t) = \int_{-\infty}^{+\infty} h(\tau)x(t-\tau)d\tau = h(t) \star x(t), \tag{2.1}$$

where the sign "*" denotes convolution, and $h(t)$ denotes the impulse response of the system, that is, the system response to the Dirac delta impulse signal $\delta(t)$ under zero initial condition. For a causal and stable system, $h(t)$ should fulfill the following requirements:

$$h(t)=0 \quad t<0 \quad and \quad \int_{-\infty}^{+\infty}|h(t)|dt<\infty \tag{2.2}$$

Equally, Equation (2.1) can be expressed in the frequency domain as:

$$Y(f)=H(f)X(f), \tag{2.3}$$

where $X(f)$ and $Y(f)$ represent the frequency-domain input and output, respectively, and $H(f)$ represents the transfer function of the system. Time-domain $h(t)$ and frequency-domain $H(f)$ are related by Fourier transform thus:

$$H(f)=\int_{-\infty}^{+\infty}h(t)e^{-j2\pi ft}dt \quad and \quad h(t)=\int_{-\infty}^{+\infty}H(f)e^{j2\pi ft}df. \tag{2.4}$$

A similar relationship holds between $x(t)$ and $X(f)$ or $y(t)$ and $Y(f)$.

$H(f)$ is a complex-valued function, with $|H(f)|$ and $\psi(f)$ referring to the magnitude and phase characteristics of the system, respectively:

$$H(f)=|H(f)|\exp[j\psi(f)],$$

$$with \quad |H(f)|=\sqrt{[\operatorname{Re}H(f)]^2+[\operatorname{Im}H(f)]^2} \quad and \quad \psi(f)=\arctan\left[\frac{\operatorname{Im}H(f)}{\operatorname{Re}H(f)}\right], \tag{2.5}$$

where Re and Im denote the real and imaginary parts of the complex-valued function, respectively.

The physical property of a continuous-time LTI system is completely determined by $h(t)$ or $H(f)$. Generally, system identification methods are implemented to determine $h(t)$ or $H(f)$. The following are the three commonly used methods.

Method 1: Impulse method. According to the definition of $h(t)$, the most straightforward approach is to excite the system by a Dirac delta impulse $\delta(t)$, and the system output can then be recorded as $h(t)$. The limitations of this method are shown in later discussions.

Method 2: Fourier analysis method. According to Equation (2.3), $H(f)$ can be calculated as the ratio between the output and input in the frequency domain, provided that both the input and output are known.

$$H(f)=\frac{Y(f)}{X(f)}. \tag{2.6}$$

This method is generally accepted in traditional measurement. It is, however, suitable only for deterministic input signals with finite energy, not stationary random input signals, because the latter have infinite energy, and thus cannot be analyzed by Fourier transform.

Modern acoustic measurements commonly implement the correlation analysis method, which is suitable for both deterministic and stationary random input signals. The autocorrelation function of a real input $x(t)$, with infinite energy but finite power, is defined as:

$$R_{xx}(\tau)=\lim_{T\to\infty}\frac{1}{2T}\int_{-T}^{+T}x(t)x(t+\tau)dt. \tag{2.7}$$

Additionally, the cross-correlation function between input $x(t)$ and output $y(t)$ is defined as:

$$R_{xy}(\tau) = \lim_{T \to \infty} \frac{1}{2T} \int_{-T}^{+T} x(t)y(t+\tau)dt. \tag{2.8}$$

For an input with finite energy, the denominator $2T$ should be discarded.

For a continuous-time LTI system, $R_{xx}(\tau)$ is related to $R_{xy}(\tau)$ thus:

$$R_{xy}(\tau) = h(\tau) * R_{xx}(\tau) = \int_{-\infty}^{+\infty} h(t)R_{xx}(\tau - t)dt. \tag{2.9}$$

If selecting a specific input signal, whose autocorrelation function is characterized by:

$$R_{xx}(\tau) = \delta(\tau), \tag{2.10}$$

then Equation (2.9) yields:

$$R_{xy}(\tau) = h(\tau). \tag{2.11}$$

A third method for system identification is then derived.

Method 3: Correlation analysis method, that is, choosing an input signal whose autocorrelation function is exactly a Dirac delta function. The cross-correlation function between the input and output is, hence, equal to the system impulse response.

This method is also valid for an input signal whose autocorrelation function is not a Dirac delta function. Suppose that the auto- and cross-spectral densities of $R_{xx}(\tau)$ and $R_{xy}(\tau)$ are:

$$S_{xx}(f) = \int_{-\infty}^{+\infty} R_{xx}(\tau)e^{-2\pi ft}d\tau \quad and \quad S_{xy}(f) = \int_{-\infty}^{+\infty} R_{xy}(\tau)e^{-2\pi ft}d\tau. \tag{2.12}$$

According to Equation (2.9), the transfer function of the system can be calculated as:

$$H(f) = \frac{S_{xy}(f)}{S_{xx}(f)}. \tag{2.13}$$

The prominent property of the correlation analysis method is noise immunity. Assuming that there is a disturbance, $n(t)$, in the measurement, then output $y'(t)$ takes the form:

$$y'(t) = y(t) + n(t). \tag{2.14}$$

Provided that $n(t)$ is uncorrelated to input $x(t)$, the influence of $n(t)$ can be automatically removed in the calculation of Equation (2.8), ensuring the final result of Equation (2.11).

2.1.2. Discrete-Time LTI System

Traditionally, the transfer function of an acoustical system is measured by using analog methods. The aforementioned signal processing techniques for a continuous-time LTI system can therefore be implemented. In the recent two decades, however, the analog method has been replaced by the digital method, due to advancements in computer technology and digital signal processing. The digital method is akin to the discrete-time LTI system, whose basic principle is similar to that of the continuous-time LTI system.

A discrete-time signal is a time sequence sampled from the continuous-time analog signal. Discrete time t_n takes the form:

$$t_n = nT = \frac{n}{f_s}, \quad with \quad -\infty < n < +\infty \tag{2.15}$$

where T and f_s denote sampling period and sampling frequency, respectively. Usually, discrete time is expressed by the integer, n, for simplification. The original analog signal can be completely recovered from the discrete signal as long as the Shannon-Nyquist sampling theorem is fulfilled.

For discrete-time LTI systems, input signal $x(n)$ is related to output signal $y(n)$ thus:

$$y(n) = \sum_{q=-\infty}^{+\infty} h(q)x(n-q) = h(n) * x(n), \quad \text{with} \quad -\infty < n < +\infty \tag{2.16}$$

Here, "*" denotes discrete-time convolution, and $h(n)$ represents the impulse response of the system—that is, the zero initial-state response to the unit sampling sequence $\delta(n)$, with:

$$\delta(n) = \begin{cases} 1 & n = 0 \\ 0 & n \neq 0 \end{cases}. \tag{2.17}$$

With regard to a causal and stable system, $h(n)$ should fulfill the following requirements:

$$h(n) = 0 \quad \text{with} \quad n < 0 \quad \text{and} \quad \sum_{n=-\infty}^{+\infty} |h(n)| < \infty \tag{2.18}$$

In a discrete-time system, the Z-transform of $x(n)$, $y(n)$, and $h(n)$ is defined as:

$$X(z) = \sum_{n=-\infty}^{+\infty} x(n)z^{-n}, \quad Y(z) = \sum_{n=-\infty}^{+\infty} y(n)z^{-n}, \quad H(z) = \sum_{n=-\infty}^{+\infty} h(n)z^{-n}. \tag{2.19}$$

Here, $H(z)$ denotes the system function of the discrete-time LTI system. Then, we set $z = \exp(j\omega)$, where ω is the digital (angular) frequency that is related to conventional analog frequency, f, and angular frequency, $\omega' = 2\pi f$, thus:

$$\omega = \frac{2\pi f}{f_s} = \frac{\omega'}{f_s}. \tag{2.20}$$

Then, $X[\exp(j\omega)]$ and $Y[\exp(j\omega)]$ are the Fourier transforms of $x(n)$ and $y(n)$, respectively. Additionally, $H[\exp(j\omega)]$—the so-called system transfer function—is the Fourier transform of impulse response $h(n)$, which is the value of $H(z)$ on the unit circle of $|z| = 1$ in the complex Z-plane, and, hence, a periodic function of ω with period 2π. The physical property of the system is completely determined by $h(n)$, $H(z)$, or $H[\exp(j\omega)]$. For simplicity, $H[\exp(j\omega)]$ is shortened to $H(\omega)$.

Analogous to Equation (2.3), Equation (2.16) is expressed in the Z-domain as:

$$Y(z) = H(z)X(z). \tag{2.21}$$

Any practical measurement is conducted within a finite time duration, resulting in a finite length sequence. Under such a condition, the N-point discrete Fourier transform (DFT) of finite length input signal $x'(n)$, output signal $y'(n)$, and impulse response $h'(n)$ can be defined as:

$$\begin{aligned} X'(k) &= DFT[x'(n)] = \sum_{n=0}^{N-1} x'(n)\exp(-j\frac{2\pi}{N}kn), \\ Y'(k) &= DFT[y'(n)] = \sum_{n=0}^{N-1} y'(n)\exp(-j\frac{2\pi}{N}kn), \\ H'(k) &= DFT[h'(n)] = \sum_{n=0}^{N-1} h'(n)\exp(-j\frac{2\pi}{N}kn), \text{where } k = 0,1,2...(N-1). \end{aligned} \tag{2.22a}$$

Correspondingly, the N-point inverse discrete Fourier transform (IDFT) of $X'(k)$, $Y'(k)$, and $H'(k)$ are

$$x'(n) = IDFT[X'(k)] = \frac{1}{N}\sum_{k=0}^{N-1} X'(k)\exp(j\frac{2\pi}{N}kn),$$

$$y'(n) = IDFT[Y'(k)] = \frac{1}{N}\sum_{k=0}^{N-1} Y'(k)\exp(j\frac{2\pi}{N}kn), \quad (2.22b)$$

$$h'(n) = IDFT[H'(k)] = \frac{1}{N}\sum_{k=0}^{N-1} H'(k)\exp(j\frac{2\pi}{N}kn), where\ n = 0,1,2...(N-1).$$

Although $x'(n)$, $y'(n)$, $h'(n)$, $X'(k)$, $Y'(k)$, and $H'(k)$ are all defined as finite length sequences, Equations (2.22a) and (2.22b) inherently contain periodicity with a time period, N.

On the unit circle, $H(z)$ in Equation (2.19) is sampled with N uniform points, and if $z = \exp(j2\pi k/N)$—with $k = 0,1 \dots N-1$—then:

$$H'(k) = H(z)\Big|_{z=\exp(j\frac{2\pi}{N}k)} = \sum_{n=-\infty}^{+\infty} h(n)\exp(-j\frac{2\pi}{N}kn),\ k = 0,1....N-1. \quad (2.23)$$

Considering $H'(k)$ as an N-point discrete sequence in the frequency domain, it can be inversely Fourier transformed to:

$$h'(n) = \frac{1}{N}\sum_{k=0}^{N-1} H'(k)\exp(j\frac{2\pi}{N}kn) = \left[\sum_{m=-\infty}^{+\infty} h(n+mN)\right]_N. \quad (2.24)$$

The $h'(n)$ is the principal value of the system impulse response $h(n)$, with periodic extension and a succedent N-point truncation. Although $h'(n)$ deviates from $h(n)$, the latter can be recovered from the former, provided that the actual length of $h(n)$ is equal to, or less than, N. Otherwise, time-domain aliasing arises. This is the so-called frequency-domain sampling theorem.

The uniformly sampled $H(z)$ on a unit circle with N points is equal to uniform sampling system transfer function $H[exp(j\omega)]$ with N points in the digital angular frequency range of 0 to 2π, resulting in $H[exp(j\omega_k)]$ with $\omega_k = 2\pi k/N$ ($k = 0, 1\dots N-1$). According to Equation (2.20), this can also be interpreted as N-point uniform sampling of the continuous system transfer function in the analog frequency range from 0 Hz to f_s with $f_k = kf_s / N$ ($k = 0, 1\dots N-1$). Hence, the discrete presentation of the system transfer function can be written as $H[exp(j\omega_k)]$, $H(f_k)$, or $H(k)$— depending on the cases discussed in this book.

Another issue introduced by the N-point truncation of the time-domain sequence is frequency resolution. The time duration of an N-point time-domain sequence is $\Delta t = N/f_s$. According to the uncertainty principle of time and frequency, that is, when $\Delta t\,\Delta f \geq 1$, the frequency resolution is:

$$\Delta f \geq \frac{f_s}{N}. \quad (2.25)$$

For a certain sampling frequency, the shorter the sequence length of N is, the lower the frequency resolution will be. The frequency resolution problem, which is an inevitable consequence introduced by time-domain truncation, also exists in a continuous-time system.

As with a continuous-time system, various measuring and processing methods are available for system impulse response $h(n)$, or the transfer function $H[exp(j\omega)]$, of a discrete-time system.

Method 1: Impulse method. The system is probed by a unit sampling sequence $\delta(n)$, and then the output of the system $h(n)$ is recorded. Conversely, according to Equation (2.21), $H(z)$ can be evaluated by directly dividing the output signal by the input signal in the Z-domain. In Equation (2.19), the time duration of summation is infinite. For many practical problems that are relevant to acoustics, however, $h(n)$ can be approximated as a finite length sequence, similar to the actual input and output signals.

Method 2: Fourier analysis method. The N-point DFT is applied to input signal $x'(n)$ and output signal $y'(n)$, and then divided:

$$H'[\exp(j\omega_k)] = H'(k) = \frac{Y'(k)}{X'(k)}, \quad k = 0, 1, 2 \dots (N-1). \tag{2.26}$$

The $h'(n)$ in Equation (2.24) can be calculated by IDFT to $H'(k)$. To successfully reconstruct $h(n)$ from $H'(k)$, N should be equal to, or larger than, the effective length of $h(n)$, according to the frequency-domain sampling theorem.

The correlation method can also be applied to the discrete-time system. The normalized N-point circular autocorrelation function $R'_{xx}(n)$ and cross-correlation function $R'_{xy}(n)$ for two N-point finite real value sequences, $x'(n)$ and $y'(n)$, are defined respectively as:

$$R'_{xx}(n) = \frac{1}{N+1} \sum_{q=0}^{N-1} x'(q)x'(q+n),$$

$$R'_{xy}(n) = \frac{1}{N+1} \sum_{q=0}^{N-1} x'(q)y'(q+n), \quad n = 0, 1, 2 \dots (N-1), \tag{2.27}$$

where $x'(q + n)$ and $y'(q + n)$ are the principal values of original sequences $x'(n)$ and $y'(n)$, with periodic extension, and a succedent N-point truncation, after shifting q-point to the left.

Similar to Equation (2.9), considering $R'_{xx}(n)$ and $R'_{xy}(n)$ as two N-point finite sequences, they are related to each other thus:

$$R'_{xy}(n) = \sum_{q=0}^{N-1} h'(q)R'_{xx}(n-q), \quad n = 0, 1, 2 \dots (N-1), \tag{2.28}$$

where $h'(n)$ is defined in Equation (2.24), and $R'_{xx}(n - q)$ is the principal value of $R'_{xx}(n)$, with a periodic extension and a succedent N-point truncation after shifting q-point to the right. If we choose a specific input signal, $x'(n)$, whose N-point autocorrelation function (with extension) is a unit sampling sequence $\delta(n)$, then:

$$R'_{xx}(n) = \delta(n) \quad n = 0, 1 \dots (N-1). \tag{2.29}$$

An inherent periodicity exists in Equation (2.29). Then:

$$R_{xy}(n) = h'(n) \quad n = 0, 1 \dots (N-1). \tag{2.30}$$

Method 3: Correlation method. A specific input signal, $x'(n)$, whose N-point autocorrelation function (with extension) is a unit sampling sequence, $\delta(n)$, is selected. Then, the resultant output signal, $y'(n)$, is recorded. Finally, $h'(n)$ is calculated as the cross-correlation function of $x'(n)$ and $y'(n)$. To avoid time-domain aliasing, N should be equal to, or larger than, the effective length of $h(n)$. This should be kept in mind.

For input signals that do not fulfill Equation (2.29), the auto- and cross-spectral densities can be calculated by applying N-point DFT to $R'_{xx}(n)$ and $R'_{xy}(n)$ thus:

$$S'_{xx}(k) = \sum_{n=0}^{N-1} R'_{xx}(n)\exp(-j\frac{2\pi}{N}kn),$$

$$S'_{xy}(k) = \sum_{n=0}^{N-1} R'_{xy}(n)\exp(-j\frac{2\pi}{N}kn), \quad k = 0, 1, 2 \dots (N-1). \tag{2.31}$$

Then, dividing the latter by the former, $H'(k)$ is obtained as:

$$H'(k) = \frac{S'_{xy}(k)}{S'_{xx}(k)}, \quad k = 0,1,2...(N-1). \tag{2.32}$$

Finally, $h'(n)$ is obtained by IDFT to $H'(k)$. Similarly, the problems of time-domain aliasing and frequency resolution should always be considered.

2.1.3. Excitation Signals

Generally, the input signal used to probe the system impulse response should contain all frequency components of interest. In theory, all types of signals containing frequency components of interest can be employed as the input signal to the system under evaluation. However, the instability caused by the denominator in Equations (2.6) and (2.26) as it approaches zero is likely to arise at frequencies with less energy. In this sense, the input signal should have an even energy (or power) spectrum across all frequencies. Emphasizing some frequencies in advance improves the signal-to-noise ratio in measurements.

A measuring system inevitably suffers from interference noise. A direct way to achieve a desired signal-to-noise ratio is to increase the power of the input signal. However, because of the limited dynamic range of a typical electroacoustic system, such as a loudspeaker and power amplifier, an excessively strong signal level may cause the system to overload, leading to nonlinear distortion. Therefore, to increase the input power without violating the linear property of the measuring system, an ideal input signal should have a low crest factor (ratio of the peak value to the effective value of the signal).

In past decades, a wide variety of signals have been applied in electroacoustic and room acoustic measurements. A number of techniques for measuring impulse response or transfer function, associated with various signal-processing methods, have also been created; some of these techniques have been introduced in HRTF measurements. Especially in the past two decades, digital measuring techniques have developed quickly and have already replaced traditional analog measuring techniques. A summary and comparison of different measuring techniques is given by Stan et al. (2002). Some commonly used input signals in measuring system response or transfer function are as follows:

(1) Sinusoidal and sweep signals

A sinusoidal signal with a crest factor of 1.414 is a deterministic signal commonly used in analog measurements. Probing the system under evaluation with a sinusoidal signal $X(f)$ and recording output signal $Y(f)$, the value of the system transfer function at frequency f can be calculated according to Equation (2.6). Therefore, this method is a point-by-point measurement. In actual audio measurements, sweep sinusoidal signals are usually used. A sinusoidal signal is sometimes used to measure electroacoustic devices, such as loudspeakers. Although a sinusoidal signal was also employed in HRTF measurements in early years (Blauert, 1997), it has been replaced by other signals because of its operational complexity. But the sweep signal is still used. In particular, a multiple exponential sweep signal obviously accelerates HRTF measurement (Majdak et al., 2007).

(2) Impulse signal

The ideal impulse signal is a Dirac delta function, a deterministic signal with a flat magnitude spectrum and linear phase. The impulse signal used in actual measurement is only an approximation of the ideal Dirac delta function. Starting guns, sparks, and popping air balloons have been used as traditional impulse sources in room acoustic measurements. However, the physical properties of these sound sources are difficult to control. Moreover, the excessive transient sound pressure is likely to cause a nonlinear effect in the air.

The impulse signal can also be created on a computer by generating a rectangular signal of a certain width. Generally, the wider the rectangular signal is, the more energy it contains, but the narrower the spectral range. For an impulse width of 10 μs, the flat spectral range can reach 20 kHz ($f \leq 20$ kHz). Because of its simplicity for signal processing, the impulse signal has often been used in measurements of electroacoustic systems (such as loudspeaker systems), in the early stages of digital measuring techniques, as well as in HRTF measurements (Blauert, 1997; Mehrgardt and Mellert, 1977). However, because of the impulse signal's high crest factor, the energy of the measuring signal, which is limited by the dynamic range of the electroacoustic system, is relatively low, leading to a low signal-to-noise ratio. An averaging method can alleviate this problem to some extent, as M number of measurements can improve signal-to-noise ratio theoretically as:

$$\Delta\left(\frac{S}{N}\right) = 10\log_{10} M \quad (dB). \tag{2.33}$$

However, the signal-to-noise ratio of the impulse signal remains insufficient, when compared with other signals. Recently, the impulse signal has seldom been used, except in specific cases.

A time-stretched pulse signal is also used (Aoshima, 1981). In this case, a high signal-to-noise ratio is achieved by stretching the impulse signal in time, which increases signal energy without exceeding the system's dynamic range. Then, the actual impulse response of the system under evaluation can be obtained by compressing the recorded impulse response in time. This kind of signal has also been used in HRTF measurements (Takane et al., 2002).

(3) Random noise signal

Stationary random signals, including white and pink noises, can also be used to probe system response. The autocorrelation function of a white noise is a Dirac delta function [see Equation (2.7)]; which renders it suitable for the correlation method, with:

$$R_{xx}(\tau) = \sigma^2 \delta(\tau), \tag{2.34}$$

where σ^2 is a constant. According to Equation (2.12), the auto-spectral density of white noise is a constant, expressed as:

$$S_{xx}(f) = \sigma^2. \tag{2.35}$$

Thus, white noise has an even power spectrum and a low crest factor, due to its random phase. It also has the advantage of noise immunity. Given that the power of white noise is infinite, it cannot exist in reality. The white noise in practical use is always band limited, so that it has an even power spectrum and finite power in the frequency range of interest.

Pink noise is another kind of random noise, whose power spectrum density function decreases by −3 dB per octave. It can be generated from white noise using a −3 dB/octave filter. Moreover, filtering white or pink noise with a band-pass filter leads to band-pass noise signals, such as a 1/3 octave noise signal.

The principle for measuring system transfer function by random signals is given in Equations (2.9) and (2.13). Because white and pink noises have suitable power spectrum characteristics, they are widely used in electroacoustic and room acoustic measurements. However, because of complex signal processing, the pseudorandom signal has recently replaced the random signal in HRTF measurements.

(4) Pseudorandom noise signal

Pseudorandom noise is a deterministic discrete time sequence. Therefore, it can be easily generated and averaged to improve signal-to-noise ratio. Moreover, pseudorandom noise has a property similar to that of random noise and can therefore be designed as a signal with an ideal power spectral characteristic and a low crest factor. Given that pseudorandom noise combines the advantages of both deterministic and random signals, it is widely used in HRTF measurements (Xiang, 2007).

The random-phase flat spectrum signal is a discrete time sequence of finite length. It has even spectral characteristics in a certain time window, and random phase in a certain frequency range. If the length is 2^L, a fast Fourier transform can then be applied easily. A given section of white noise can be approximately regarded as a random-phase flat spectrum signal. In HRTF measurements, a pre-weighted random-phase flat spectrum signal is sometimes used to improve signal-to-noise ratio (Wightman and Kistler, 1989a).

The *maximal-length sequence* (MLS) is a popular pseudorandom noise, and was first introduced into room acoustic measurement by Schroeder (Schroeder, 1979). Details can be found in Rife and Vanderkooy (1989) and, more recently, in Xiang (2007).

An MLS is a binary sequence consisting of a series of integer, 0, or 1. In practice, it is usually shifted in level, appearing as a bipolar (symmetrical) impulse sequence including -1 and $+1$. The period of an L-stage MLS is:

$$N = 2^L - 1. \tag{2.36}$$

The MLS can be generated by a series of binary shift registers and modulo 2 summators with proper connection of feedback tap, as shown in Figure 2.1. The L-stage binary shift register includes an L sequent register, whose state is either 0 or 1. The initial state of the ith register, numbered from left to right, is $a_i(0)$, $i = 1, 2, \ldots L$. Under the control of a clock pulse, the state of register moves toward the right, step by step. Assuming that the state of register is $a_i(n)$, $i = 1, 2, \ldots L$ after n clock pulses, and the state of register after $(n + 1)$ clock pulses is $a_i(n + 1) = a_{i-1}(n)$, $i = 2, 3 \ldots L$, the state of the first register is determined by linear modulo 2 the sum of states of L registers after n clock pulses, as:

$$a_1(n+1) = c_1 a_1(n) \oplus c_2 a_2(n) \oplus \ldots \oplus c_{L-1} a_{L-1}(n) \oplus c_L a_L(n), \tag{2.37}$$

where $c_1, c_2 \ldots c_{L-1} = 0$ or 1, $c_L = 1$. If the aforementioned method is repeated and the state of the Lth register is used as output, then the output signal after n clock pulses is:

$$a_L(n) = a_1(n - L + 1). \tag{2.38}$$

Finally, a bipolar impulse sequence, consisting of -1 and $+1$, is obtained by shifting the electronic level of $a_L(n)$. The MLS can then be generated according to Equation (2.38) or by software.

The MLS has a flat power spectrum and the lowest crest factor because its phase is evenly distributed in the range $[-\pi, +\pi]$. According to Equations (2.27–2.30), the normalized N-point autocorrelation function of the bipolar MLS with a period of N is:

$$R'_{xx}(n) = \delta(n) - \frac{1}{N+1} \qquad n = 0,1\ldots(N-1) \tag{2.39}$$

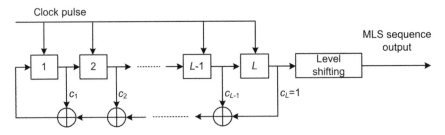

Figure 2.1 Generation of maximum length sequence signal

The N-point cross-correlation function of the system output and bipolar MLS input is:

$$R'_{xy}(n) = h'(n) - \frac{1}{N}\sum_{n=0}^{N-1} h'(n) + \frac{1}{N(N+1)}\sum_{n=0}^{N-1} h'(n), \quad (2.40)$$

where $h'(n)$ is defined in Equation (2.24). The second and third terms on the right side of Equation (2.40) represent direct current components of $h'(n)$ with the third term damping $1/(N+1)$ times. For a large N, the third term on the right is negligible. Then, the difference between $R'_{xy}(n)$ and $h'(n)$ is a direct current term, which can be disregarded for an alternating current coupling system, i.e. $R'_{xy}(n) \rightarrow h'(n)$. The MLS has excellent properties, which are similar to those of ideal white noise. Therefore, the MLS is popular in HRTF measurements. In addition, Xiang and Schroeder (2003) proved that a reciprocal MLS pair, consisting of a bipolar MLS and its reversed time order sequence, possesses a cross-correlation function with a low value. This favorable characteristic allows for simultaneously exciting more than one separate sound source using a reciprocal MLS pair and, thereby, accelerating HRTF measurement.

On the other hand, the length of an MLS is not exactly a power of 2; hence, it cannot be calculated by fast Fourier transform. Fast Hadamard transform can be applied instead (Rife and Vanderkooy, 1989). In recent years, some other new algorithms for MLS have been proposed (Xiang and Schroeder, 2003; Daigle and Xiang, 2006). Another disadvantage of MLS is its sensitivity to the nonlinear property of the system under evaluation, leading to artificial peaks in the results (Vanderkooy, 1994).

Golay codes are another kind of pseudorandom noise signal used in HRTF measurements (Zhou et al., 1992). The L-order Golay codes are a pair of complementary sequences, of length $N = 2^L$ and consisting of -1 and $+1$. Therefore, fast Fourier transform operations can be directly applied in the calculations of Golay codes.

Golay codes can be generated by the following recursive method. Let the complementary sequences in Golay codes be expressed by $a_L(k)$ and $b_L(k)$ with $k = 0, 1, \dots 2^L - 1$. For $L = 1$, $a_1 = (+1, +1)$ and $b_1 = (+1, -1)$ with a length of 2^1 for each sequence. For $L = 2$, a_2 can be achieved by appending b_1 to a_1, whereas b_2 can be achieved by appending $-b_1$ to a_1, in which case $a_2 = (+1, +1, +1, -1)$ and $b_2 = (+1, +1, -1, +1)$ with a length of 2^2 for each sequence. The rest can be done in the same manner, allowing a pair of Golay codes of length 2^L to be obtained after an $L - 1$ recursive operation, taking the form:

$$a_L(n) = \begin{cases} \pm 1 & 0 \le n \le N-1 \\ 0 & other \end{cases} \quad and \quad b_L(n) = \begin{cases} \pm 1 & 0 \le n \le N-1 \\ 0 & other \end{cases}. \quad (2.41)$$

The most important property of Golay codes is that the sum of the autocorrelations of both sequences is a two-valued function,

$$\sum_{q}[a_L(q)a_L(n+q)+b_L(q)b_L(n+q)] = \begin{cases} 2N & n=0 \\ 0 & other \end{cases}. \quad (2.42)$$

The equation refers to a linear autocorrelation, and it also holds for circular autocorrelation. Applying N-point DFT, Equation (2.42) becomes:

$$DFT[a_L(n)]DFT^*[a_L(n)] + DFT[b_L(n)]DFT^*[b_L(n)] = 2N \quad (2.43)$$

where the superscript "*" denotes a complex conjugate.

The procedure of using Golay codes to measure the transfer function $H'(k)$ of a discrete LTI system is summarized as follows: (1) the system is separately excited with $a_L(k)$ and $b_L(k)$; (2) the N-point DFT is applied to the system output, yielding $H'(k)$ DFT(a_L) and $H'(k)$ DFT(b_L); and (3) the $H'(k)$ DFT(a_L), multiplied by DFT*(a_L), and $H'(k)$ DFT(b_L), multiplied by DFT*(b_L), is summed as:

$$H'(k)DFT[a_L(n)]DFT^*[a_L(n)] + H'(k)DFT[b_L(n)]DFT^*[b_L(n)] = 2NH'(k). \quad (2.44)$$

The time-domain impulse response $h'(n)$ can be obtained by the N-point IDFT of resultant $H'(k)$ in Equation (2.44).

The disadvantage of using Golay codes is that they are time-consuming because the two complementary sequences require sequential use. This is critical in HRTF measurements for human subjects because subjects are inclined to make small, unconscious, movements during a long-term measurement, which inevitably destroys the complementary property of Equation (2.42) (Zahorik, 2000). In this sense, Golay codes are more sensitive to the subject's small movements than MLS.

2.2. Principle and Design of HRTF Measurements

2.2.1. Overview

As presented in Chapter 1, HRTFs are the transfer functions of the LTI process for sound waves, from a sound source to the two ears. Thus, all the measurement principles and methods for the various LTI systems mentioned in the previous section can be applied to HRTF measurement. Analog methods, which suffer from complex processing procedures and poor accuracy, were frequently employed in the early stages of research (Blauert, 1997). Moreover, considerable differences existed among the measurement methods used in different studies. Later, with the advancement of digital measurement techniques, measurements with impulse signals were commonly used (Blauert, 1997; Mehrgardt and Mellert, 1977). In recent decades, with the further development of digital signal processing and computer technology, a wide variety of digital measurement techniques have been implemented. Observing the most recent works, the measurement principles and methods are basically similar; although there are detailed differences among experimental arrangements, apparatuses, and so forth (Wightman and Kistler, 1989a; Gardner and Martin, 1995a; Møller et al., 1995b; Genuit and Xiang, 1995; Riederer, 1998a; Bovbjerg et al., 2000; Algazi et al., 2001b; Cheng, 2001a; Takane et al., 2002; IRCAM Lab, 2003; Grassi et al., 2003; Wersenyi and Illenyi, 2003; Kim and Kim et al., 2005; Begault et al., 2006; Xie et al., 2007a; Fukudome et al., 2007; Dobrucki et al., 2010; Begault et al., 2010).

Figure 2.2 shows a typical block diagram of an HRTF measurement. The measuring signal generated by a computer is rendered to a loudspeaker after passing through a digital to analog (D/A)

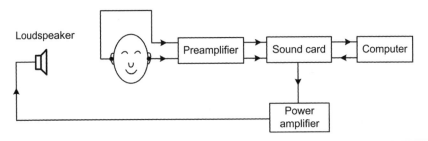

Figure 2.2 Block diagram of HRTF measurements

converter and a power amplifier. The resultant signals are recorded by a pair of microphones at the subject's two ears, and then delivered to the computer after amplification and analog to digital (A/D) conversion. Finally, HRTFs or head related impulse responses (HRIRs) can be obtained, after the necessary signal processing is implemented with the computer.

2.2.2. Subjects in HRTF Measurements

Measured subjects include both artificial heads and humans. Generally, the measurement error of an artificial head is relatively small and signal-to-noise ratio is high. However, because an artificial head is based on the mean, or typical, anatomical dimensions of a certain population, resultant HRTFs can only represent the common characteristics of that population, and not the individualized characteristics of each person. In particular, existing artificial heads are all constructed according to the anatomical dimensions of a certain Western population; thus, they may not be suitable for other populations, such as the Chinese, even in the averaging sense. Meanwhile, some types of artificial heads were constructed with a simplified head and pinna. Hence, some important localization information that is associated with details of the head and pinna may be lost in resultant HRTFs.

Measuring a human subject enables obtaining individualized HRTFs. In the measurement, however, human subjects are inclined to make small movements of the head and body, especially during a long-term measurement, which destroys the time-invariant characteristics of the measuring system and leads to measurement error. Additionally, human subjects may unconsciously generate noises, which are harmful to measurements.

HRTF is closely related to anatomical dimensions. Given that the anatomical dimensions of adults are relatively stable, subjects over 18 years old are usually chosen in HRTF measurements. For the convenience of conducting listening tests in follow-up studies, many require human subjects with normal hearing. Evidently, children are chosen as subjects in some child-related studies, but cooperative obedience is difficult to achieve. Moreover, animals, such as monkeys and cats, can also be selected as subjects in research related to physioacoustics (Musicant et al., 1990; Spezio et al., 2000; Young et al., 1996).

2.2.3. Measuring Point and Microphone Position

HRTFs are defined as the free-field transfer functions, from a sound source to the two ears. According to the discussion in Section 1.5, the measurement of binaural signals can be undertaken at arbitrary points along the line from the ear canal entrance to the eardrum because the ear canal can be approximated as a one-dimensional transmission line. However, because results measured at different points differ considerably from one another, and the length of the ear canal is individual dependent, many researchers select a point close to the ear canal entrance or the eardrum as their measuring point, for the purpose of standardization.

For convenience, the HRTFs of an artificial head are often measured at the equivalent position of the eardrums, by fixing pressure field microphones at the end of the occluded-ear simulator with their diaphragm analogues at the eardrums. The resulting measurements contain the information of directional localization, as well as of ear canal resonances. Gardner and Martin (1995a) adopted this technique in the KEMAR measurement. For human subjects, the measurement is commonly performed at the position of about 1 to 2 mm from the eardrum, to avoid uncertainties from standing-wave interference. To minimize the influence of the microphone on the sound field in the ear canal, a probe microphone (i.e., a microphone coupled to a probe tube with an outer diameter of 1 to 2 mm) should be used in HRTF measurements for humans (Wightman and Kistler, 1989a). However, incorrect operation of this technique may harm subjects, and the measuring position is difficult to fix. Moreover, a probe microphone usually suffers from poor frequency response and low sensitivity, which directly diminish measurement accuracy and signal-to-noise ratio. To address this problem, based on the transmission line model of the ear canal and an energy-based estimation, Hiipakka et al. (2012) proposed a method to estimate the HRTF magnitude at the eardrum, based on the pressure and velocity measurement at the ear canal entrance, using a miniature sensor.

Since the blocked ear canal technique was first introduced (Møller, 1992), it has been widely applied in human HRTF measurements because of its convenience and safety—a finding also verified by experiments (Algazi et al., 1999). A pair of miniature microphones, with relatively high sensitivity and a wide frequency response range, is used in the blocked ear canal technique. Usually, a miniature microphone, embedded into an elastic silicon rubber earmold or a modified swimming earplug, is inserted into the subject's ear canal, with its diaphragms flush with the entrance to the ear canal or slightly inward. Then, releasing the elastic material around the miniature microphone will automatically occlude the ear canal. Figure 2.3 shows a type of miniature microphone, DPA 4060, with a pressure sensitivity of 20 mV/Pa and a frequency response ranging from 20 Hz to 20 kHz (±2 dB). Figure 2.4 shows the blocked ear canal configuration with this type of microphone that was used by our research group.

2.2.4. Measuring Circumstances and Mechanical Devices

In accordance with the definition of HRTF, measurement should be undertaken in an anechoic chamber, which is the case in many relevant works. However, restrictions stemming from the structure of an anechoic chamber need careful design and installation of various measuring apparatuses in the chamber. Moreover, HRTF measurement is a time-consuming task, especially with a large number of subjects and measuring directions. In practice, occupying an anechoic chamber for long periods may present difficulties. Therefore, some measurements were carried out in a non-anechoic room (Algazi et al., 2001b; Begault et al., 2006). The unavoidable reflections from the boundaries of a non-anechoic room should be removed from measured results. Usually, the arrival time of the reflections to the ears can be controlled, to later than the time duration of the HRIR—on the order of several milliseconds. This

Figure 2.3 Photo of the miniature microphone (DPA 4060)

can be achieved by selecting an appropriate measuring localization and placing sound-absorbing materials on floors and ceilings between the subject and the sound source. Under this condition, the unwanted reflections can be removed by applying a time window to the raw measurements in signal processing. To this end, checking whether all the reflections appear behind the duration of HRIR is a necessary step conducted in the preparation of HRTF measurement. Checking is conducted by examining the impulse response between the sound source and the measuring location. In addition, the specific treatment of sound insulation and absorption should be undertaken to ensure a relatively low background noise.

During the measurement session, subjects can sit or stand. Sitting helps subjects remain stable, minimizing measurement error. However, posture-related reflections from the subject's knee may be introduced, along with reflections from the chair. Without appropriate treatment, these reflections are likely to influence results. In this sense, standing is preferred. However, long-term standing is uncomfortable and may lead to unconscious movements.

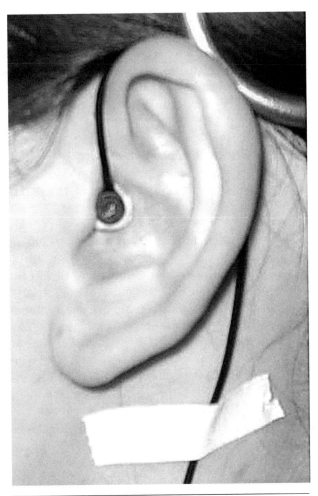

Figure 2.4 Blocked ear canal configuration with positioned microphone

In HRTF measurements, one or more loudspeakers are arranged on a virtually spherical surface, with the head center of the subject coincident with the sphere center. In the far field, HRTF is nearly distance independent when distance is relative to the loudspeaker (i.e., spherical radius) $r >$ 1.0 to 1.2 m. To collect HRTFs at various spatial directions, altering the relative position between the subject and the loudspeaker is required. This can be achieved through three approaches. The first approach alters the position of the loudspeaker using mechanical devices with the subject fixed. The second adjusts the elevation angle by altering the position of the loudspeaker while adjusting the azimuth angle by altering the position of the subject with a turnable chair (sitting) or a turnplate (standing). The third approach arranges multiple, identical, loudspeakers in elevation angles of interest, with each loudspeaker for each elevation plane. Then, only the azimuth of the subject alters when the measurement is performed. Generally, the third method is the most convenient, but a careful adjustment is required because of the number of loudspeakers involved. A representative example for the third method from our laboratory is shown in Figure 2.5. Lasers attached to the turnable chair and an overground coordinate system is also helpful in denoting the

Figure 2.5 Layout with many loudspeakers positioned at different elevation angles in a non-anechoic room

relative position between the subject and the loudspeaker. The backrest of the turnable chair and a head holder facilitate fixing the position of the subject. Moreover, a laser, visional, or an electromagnetic tracking system is usually used to monitor head position.

Usually, it is difficult to measure HRTFs at low elevations, such as $\phi < -40°$, which is closely related to the sitting position of the subject. Hence, many studies excluded HRTF measurements at low elevations.

To minimize the influence from reflections, few objects, or none, should be allowed in the measuring room. Moreover, reflective objects, such as the loudspeaker and turnable chair, should be wrapped with sound-absorbing materials. The multiple reflections, or scattering, among loudspeakers should also be considered when more than one loudspeaker is used (Yu et al., 2012a).

2.2.5. Loudspeaker and Amplifier

Ideally, a measuring sound source should fulfill two requirements: a reasonable frequency response in the entire audio frequency range, and approximating a point source. In practice, a single loudspeaker unit often suffers from a limited frequency range. Moreover, a loudspeaker with a good low-frequency response usually has a large volume, whereas the sound radiation center of a commonly used two- or three-way loudspeaker system may not be spatially coincident. From this viewpoint, they cannot be approximated as a point source.

In actual measurements, a single small loudspeaker, or a coaxial loudspeaker system, is typically used to ensure the coincidence of the sound radiation center. The upper frequency limit of the two types of sources can reach, or approximate, 20 kHz, whereas their low frequency limit ranges from 100 to some hundreds Hz and is no less than dozens of Hz (on the same order of the low frequency limit of an ordinary anechoic chamber). Limited by the frequency characteristics of these measuring apparatuses, the measured HRTF is less accurate at low frequencies.

The loudspeaker system used in HRTF measurement should have as flat a frequency response as possible and have less nonlinear distortion. Moreover, attention should be paid to consistency in loudspeaker properties when using multiple loudspeakers. To reduce the scattering of sounds from the loudspeakers, some researchers employ a tailor-made, approximately spherical loudspeaker chest coated with sound-absorbing materials. No specific requirement is imposed for power and microphone amplifiers, and common amplifiers used in acoustical measurements are suitable.

2.2.6. Signal Generation and Processing

Signal generation, collection, and processing are performed by the D/A and A/D converters of a sound card associated with a computer, or a specific digital signal processor (DSP). Some commercial electroacoustic measuring systems are currently available. In recent decades, the signal sampling frequency has reached 44.1 kHz or higher, and a resolution of 16 bit or higher.

In principle, the various measuring signals and methods presented in Section 2.1 can be used in HRTF measurements, in which the measuring method using pseudorandom noise (such as MLS) is widely applied because of its superior properties. According to Equation (2.24) and the succeeding discussion, the time length N/f_s of the periodic signal used in HRTF measurement should be longer than the duration of HRIR (several milliseconds) to avoid time aliasing. As far as non-anechoic measurements are concerned, N/f_s should be longer than the length of the room impulse response or reverberation time. Moreover, increasing N allows for fewer estimation errors, as shown in Equation (2.40). In actual measurements, a high signal-to-noise ratio can be achieved by averaging repeated measurements of the periodic measuring signal. Increasing the signal length and averaging repeated measurements improves measuring accuracy at the cost of increasing measuring time. In this sense, it is applicable in the measurement of artificial heads. However, for human subjects, excessive measuring time is very likely to induce movements and thus destroy the time-invariant characteristics of the measuring system. All these factors should be comprehensively considered in the choice of signal length and number of repetitions.

As to signal processing, the first step is to reconstruct a raw binaural impulse response $h'(n)$, with length N, from the output of a pair of recording microphones. The chosen duration of N is longer than that of an actual HRIR; thus, $h'(n)$ is redundant. In detail, a series of near-zero samples, corresponding to the transmission delay from the loudspeaker to the ear, exist at the beginning of $h'(n)$. A series of reflections, caused by boundaries and mechanical devices, are also observed at the end of $h'(n)$.

To remove the redundant samples from $h'(n)$, a time window is needed to truncate $h'(n)$ to $h(n)$, that is, the actual HRIR. However, the truncation may reduce frequency resolution and introduce distortions at low frequencies (Gardner, 1997). According to Equation (2.25), the frequency resolution of an N-point time-domain sequence is $\Delta f = f_s/N$. On the other hand, because of the nondirect-current-coupled property of the measuring system and the lower limit of the measuring loudspeaker's frequency response, the magnitude of the measured HRTF drops quickly below the lower limit of the measuring system. This may be worsened by a brief time window, which can trigger a decrease in the magnitude of the measured HRTF, even above the lower limit of the measuring system. All these factors should be comprehensively considered in selecting the appropriate length of the time window. In practice, the length of the time window varies from several to 20 ms.

2.2.7. HRTF Equalization

In practical applications, to minimize changes in the subjective timbre in virtual auditory displays, HRTF equalization is commonly implemented (Møller, 1992; Jot et al., 1995). Section 8.1.2 (Chapter 8) may be referred to for further details.

Measurement-equalized HRTFs are defined by Equation (1.15), which is rewritten here, neglecting the subscripts denoting the left and right ears and source distance r, as well as anatomical parameter a:

$$H(\theta,\phi,f) = \frac{P(\theta,\phi,f)}{P_0(f)}. \tag{2.45}$$

Free-field-equalized HRTFs are equalized with respect to one of the measured HRTFs at a specific direction (θ_0, ϕ_0), which is usually selected as the direct front $(\theta_0 = 0°, \phi_0 = 0°)$:

$$H_{free}(\theta,\phi,f) = \frac{H(\theta,\phi,f)}{H(\theta_0,\phi_0,f)}. \tag{2.46}$$

Diffuse-field-equalized HRTFs are equalized with respect to the diffuse-field average. The diffuse-field average is obtained by the root-mean-square value of HRTF magnitudes across all directions:

$$H_{diff}(\theta,\phi,f) = \frac{H(\theta,\phi,f)}{\sqrt{\dfrac{1}{M}\sum_{i=0}^{M-1}|H(\theta_i,\phi_i,f)|^2}}, \tag{2.47}$$

where the summation in the denominator is across all measuring directions, M.

As discussed in Section 1.5, the measurement-equalized HRTF varies with the reference point in measurements, whereas free-field-equalized HRTFs are irrelevant to the reference point because both the numerator and denominator in Equation (2.46) contain the one-dimensional (direction-independent) transfer function in the ear canal. The one-dimensional transfer function can be cancelled out during division. For diffuse-field-equalized HRTFs, only the magnitudes are irrelevant to the reference point. Moreover, for free- and diffuse-field equalization, all common (direction-independent) poles and zeros in HRTFs can be eliminated.

All parts of the electroacoustic measuring system, including the sound source, microphone, amplifier, and such, are assumed ideal in these theoretical analyses. However, this does not hold true in actual practice. Assuming that the transfer function of the electroacoustic measuring system is $H_0(f)$, the binaural pressures recorded are $H_0(f)P(\theta, \phi, f)$. Using the same system, the pressure recorded at the head center with the head absent is $H_0(f)P_0(f)$. Then, the negative effect arising from the nonideal transfer characteristics of the electroacoustic measuring system can be automatically eliminated, using the measurement equalization expressed in Equation (2.45). In practice, the transfer function $H_0(f)$ of the measuring system is measured first. Then, the combined transfer function of the measuring system and HRTF—that is, $H(\theta, \phi, f)H_0(f)$—is evaluated by recording the binaural signals and calculating according to the methods in Section 2.1. Finally, the measurement-equalized HRTF can be obtained by dividing $H(\theta, \phi, f)H_0(f)$ by $H_0(f)$.

Similarly, the nonideal transfer characteristics of the electroacoustic measuring system can also be automatically eliminated in free-field equalizations, by division in their definitions. For diffuse-field equalization, the effect of the nonideal phase characteristics of the electroacoustic measuring system cannot be removed, in contrast to the nonideal magnitude characteristics. Evidently, for frequencies outside the frequency range of the electroacoustic measuring system, all equalization methods are invalid.

All three equalizations are related to the inverse operation of a transfer function. According to the theory of signal processing, only the inverse of a minimum-phase function is causal and stable (Oppenheim et al., 1999). Unfortunately, the actual denominator of Equation (2.46) is sometimes non-minimum phase, while the denominator of Equation (2.47) is always non-minimum phase. In these cases, the following methods can generate an approximately causal and stable equalized HRTF: (1) Implementing minimum-phase reconstruction to the denominator prior to calculating the inverse function, and (2) Adding an appropriate linear delay to the equalized HRTF. Only nonideal magnitude characteristics of the electroacoustic measuring system can be eliminated using these methods, whereas nonideal phase characteristics are retained.

Finally, assuming left-right symmetry, the denominators of Equations (2.45), (2.46), and (2.47) are identical for the left and right ears. If left-right asymmetry exists, such as in the characteristics of the left and right microphones, then equalization should be carried out separately for the left and right ears.

2.2.8. Example of HRTF Measurement

The early HRTF measurement systems were time-consuming. Recently, some well-designed measurement systems have been developed, which obviously accelerate the measurement (Begault et al., 2010; Pollow et al., 2012). As examples, this section describes the two generations of HRTF measurement systems in our laboratory (Xie et al., 2007a; Yu et al., 2012b).

The first generation of measurement was conducted in a listening room with a reverberation time of 0.15 s. The background noise was less than 30 dB(A). To eliminate reflections from the floor, a pile of four 0.1 m thick absorption boards made of glass wool, with increasing density from the top to the bottom, were placed on the floor between the loudspeaker and the subject. Meanwhile, some sound absorption materials were also attached to the ceiling. Both the response from the sound source to the measuring location (i.e., the receiving point) and the calculation of the transmission path difference between the direct and reflected sounds demonstrate that the earliest reflection appears 13 ms later than the direct sound when using the acoustic treatments mentioned. Therefore, using an appropriate time window can completely remove room reflections.

A KEF-Q1 coaxial loudspeaker was used as the sound source, with a distance of 1.5 m relative to the head center, that is, the far-field measurement. The subject was seated on a rotating chair, and constrained by the chair's backrest and a special iron hoop against the head. The surfaces of the backrest and iron hoop, as well as the subject's knee, were covered with sound absorption materials to weaken reflections. The height of the chair was adjustable. For all subjects, the height of the entrance to the ear canal was adjusted to 1.15 m. Elevation of the sound source could be changed by adjusting the suspender from which the loudspeaker hangs, and azimuth θ could be modified by rotating the chair. A circle coordinate of radius 2.0 m was fixed on the floor, and its center was consistent with the axes of the rotating chair. The circle coordinate and two laser pointers fixed on the chair indicated azimuth θ between the sound source and the subject. Measurements were carried out at nine elevation planes, with 493 spatial directions for each subject.

Figure 2.2 is a block diagram of the measurement system. An 8191-point MLS with a 44.1 kHz sampling frequency and a length of 0.186 s longer than the reverberation time of the room was used as stimulus. The MLS was repeated 8 times to improve the signal-to-noise ratio to about 9 dB. After passing the D/A converter in the sound card (Echo Layla 24) and the power amplifier (B&K 2716C), the signal was fed into the loudspeaker. The sound pressure at the position of the head center was about 90 dB when the head was absent.

The blocked ear canal technique was employed in the measurements (Figure 2.4). The output of the miniature microphones was passed through a microphone-conditioning amplifier (B&K 2690A 0S4) and the A/D converter in the sound card, and then fed into the computer. After deconvolution using software written in MATLAB, a raw HRIR with a length of 8191 points was obtained. The raw impulse response was truncated by a rectangular time window to remove room reflections, and then equalized to compensate for loudspeaker and microphone responses, resulting in a 512-point HRIR. Furthermore, the 512-point HRTF could also be obtained using DFT.

Limited by its mechanical apparatus, the first-generation HRTF measurement system is time-consuming. Several hours are required to measure HRTFs, at 493 source directions per subject. This problem hinders HRTF measurement with higher directional resolution, especially for human subjects. In 2010, we designed and constructed a second-generation HRTF measurement system to accelerate measurement. The system is illustrated in Figure 2.5. The circumstances and block diagram of measurement are similar to those of the first-generation system. The maximum number of potential sound sources (small loudspeakers) is 64, whereas the actual number used can be flexible, depending on the required elevation resolution. A computer-controlled horizontal turntable with a 0.1° azimuthal resolution was adopted, onto which a rod was installed to support the artificial head or a seat was provided for a human subject. The source distance, relative to the head center, was adjustable, with 1.2 m as the maximum distance. The position accuracy of the system was evaluated and adjusted by a head-tracking device (Polhemus FASTRAK). The measurement and control were implemented by software written in C++ language. This updated system was more efficient than the previous one and suitable for both far- and near-field HRTF measurements with various spatial resolutions. For example, less than 30 minutes is necessary to measure far-field HRTFs at 493 source directions with nine sources fixed at different elevations.

2.2.9. Evaluation of Quality and Errors in HRTF Measurements

In HRTF measurements, both quality and error have been evaluated by some authors (Wightman and Kistler, 1989a; Bovbjerg et al., 2000), and a detailed discussion was given by Riederer (1998b).

The first issue concerns signal-to-noise ratio. In measurements, certain background and inherent noises from the electroacoustic measuring system inevitably exist. To improve signal-to-noise ratio, these noises should be minimized. An effective way is to increase the sound pressure of the measuring stimulus, which may cause discomfort for the human subject and lead to a nonlinear distortion if the pressure is too high. Usually, the desired free-field sound pressure at the position of the head center, with the subject absent, is about 75 dB. Slightly higher pressure is allowed when using the blocked ear canal technique or for brief sound stimuli. Even so, a pressure of not more than 90 dB is always selected in actual measurements to avoid the nonlinear distortion of the electroacoustic system.

The quantitative evaluation of the actual signal-to-noise ratio is difficult to perform. Although the background noise (usually on the order of 15–30 dB) can be evaluated using measurements, the spectral characteristics of the background noise differ from those of the measuring signal. Moreover, the binaural pressure signals recorded at the two ears vary with sound source position and frequency, and are different from the free-field pressure signals recorded at the head center because of the head's scattering, diffracting, or shadowing effect. As shown in Section 3.1.2, the high-frequency sound pressure at the ear that is ipsilateral to the sound source rises to approximately 6 dB (compared with the free-field pressure); whereas, for the ear that is contralateral to the sound source, high-frequency pressure is considerably attenuated—up to approximately 20 dB at some frequencies—leading to a low signal-to-noise ratio. Correlation and averaging methods can suppress the influence of noise to some extent. All these factors determine the final signal-to-noise ratio. In actual HRTF measurement, the highest signal-to-noise ratio should not exceed 60 to 70 dB.

There are many factors leading to measurement error, such as linear and nonlinear distortions in the electroacoustic measuring system, accuracy of the D/A and A/D converters, and reflections from the environment. Moreover, the subject's hairstyle and glasses may affect measured results to some extent.

Excluding all these, the remaining error factors are subject movement and microphone position. Head translation and turning during the session destroys the time-invariant characteristics of the system under evaluation. Moreover, if the subject deviates from the predefined location, even with the subject's position fixed during the measurement session, it can lead to measurement error. For example, according to Equation (1.9), a head rotating about 1° results in a change of approximately 9 µs in the ITD for a sound source at ($\theta = 0°$, $\phi = 0°$), which is close to the average just noticeable difference of ITD (i.e., 10 µs). Hirahara et al. (2010) used a head tracker to detect head turning during human HRTF measurement. The results indicated that the angle of head turning around the left-right axis (pitch) and vertical axis (rotation) reached 10°, which caused as much as a 4 to 6 dB spectral distortion in measured HRTF magnitudes. On the other hand, in the far field, a slight left-right or front-back translation of the subject causes slight change in the sound source direction relative to the subject. For example, for a sound source at ($\theta = 0°$, $\phi = 0°$, $r = 1.5$ m), a left-right head translation of 0.02 m causes a less than 0.8° change in azimuth of sound source with respect to subject. However, this may create an alternation in the ITD and the onset of HRIR. A well-designed backrest and head support can prevent head deviation during measurements, but their accompanying sound reflections or diffractions should be avoided. This can be achieved by, for example, wrapping appropriate sound absorbing material on the surface of the backrest and head support.

Because binaural signals vary with the measuring point along the ear canal (particularly at high frequencies), results strongly depend on the microphone's position. The blocked ear canal technique, which has been proven to have high repeatability, is generally adopted in HRTF measurements for human subjects. Here, we focus on the change in the high-frequency characteristic pinna notch in the HRTF spectra at 5 to 12 kHz caused by a small deviation in the microphone position. The notch is formed from the out-of-phase interference between the direct and pinna-reflected sounds at the ear canal entrance (Section 1.4.4). When transmission path difference, x, fulfills $2\pi f x / c = \pi$, a notch positioning at $f = c/2x$ occurs, where c is the speed of sound, with a value of 340 m/s. If a change in path difference occurs, namely Δx, because of a deviation in microphone position, the error in the frequency position of the characteristic notch can be estimated as:

$$|\Delta f| \approx |\frac{\partial f}{\partial x} \Delta x| = |\frac{2f^2}{c} \Delta x|. \tag{2.48}$$

From Equation (2.48), $\Delta f \approx 588$ Hz at 10 kHz when $\Delta x = 0.001$ m. Therefore, a slight deviation in the microphone's position at the ear canal entrance causes a detectable change in the measured result of pinna notch frequency. Furthermore, a deviation in the microphone position may cause the measured notch depth to vary because the actual interaction between the pinna and incident sounds is far more complicated when multiple-path scattering is included. Therefore, the microphone position, which is essential to measurement quality, should be carefully adjusted in actual blocked ear canal measurements. Selecting a microphone position slightly inside the ear canal entrance may alleviate this problem.

Repeatability in HRTF measurements can be described in terms of the magnitude difference for two or more repeated measurements. Riederer (1998b) investigated the repeatability of HRTF measurement with a high frequency resolution of about 23 Hz (at a sampling frequency of 48 kHz, a length of 2048 points) when using the blocked ear canal technique. Results indicated that the variation is about 1 to 2 dB below 6 kHz and less than 3 to 5 dB below 10 kHz. Moreover, the deviation increased with frequency: measuring error occurs at high frequencies. This phenomenon is attributed to the fact that narrow high-frequency peaks and notches are highly sensitive to microphone placement when the sound wavelength is very short. Because the error in HRTF magnitudes is related to the frequency resolution of measurement, decreasing the frequency resolution (such as at a sampling frequency of 48 kHz, a length of 512 points) can smooth HRTF magnitudes, resulting in an apparent reduction in magnitude error.

Katz et al. (2007) initiated a round-robin to compare the HRTFs of a unique artificial head, Neumann KU 100 [with pinnae but without a torso, in Figure 1.20(d)], measured by different laboratories in the world. The authors aimed to compare both the physical differences among measurement systems and the differences in measured HRTF magnitudes from a single test specimen. Five research groups contributed HRTFs until 2007. Preliminary results demonstrated that, although measuring conditions—including measuring circumstance, apparatus, and method— vary among laboratories, a relatively strong coherence exists between the different measurements using internal microphones, and some important features of HRTFs (such as spectral peaks and notches, to be discussed in Chapter 3) are stable. However, for HRTFs measured using the blocked ear canal technique, datasets demonstrate considerable variability in center frequency and depth of characteristic notches at high frequencies. Given that no universally accepted or established criteria are in place for determining the quality of a given HRTF measured, no quantitative analysis of measuring quality and accuracy can be conducted.

2.3. Far-field HRTF Databases

Early in the 1940s, studies on HRTF measurements increased (Wiener and Ross, 1946; Wiener, 1947). However, until the middle of the 1970s, most measurements were confined to the horizontal or median plane with few subjects (Shaw, 1966; Searle et al., 1975; Burkhard and Sachs, 1975). Refer to a review by Shaw (1974) for more details. In 1985, Shaw tabulated HRTF data for several azimuths in the horizontal plane, as well as for some frequencies (Shaw and Vaillancourt, 1985). The monograph of Blauert listed relevant results and references before 1972 in detail (Blauert, 1997). In the early stages, measurements using analog technology suffered from low accuracy and inconvenient data storage. Thus, the results were merely applied in some specific studies, and were not used as general-purpose data. Moreover, limited by the measuring technology available at that time, most HRTF data included only magnitude characteristics; phase characteristics were neglected.

The situation changed dramatically, beginning in the middle of the 1970s—especially in the middle and late 1980s—with the advent of computer technology and the development of digital measurement and signal processing techniques. A number of research groups measured far-field HRTFs using the methods presented in Section 2.2, and constructed some HRTF databases with different spatial resolutions. Corresponding results are partially available in the public domain, and can be used for scientific purposes. Table 2.1 lists the main far-field HRTF databases constructed in recent decades. Some explanations and comments are as follows:

1. Database number 3 was measured on KEMAR (Figure 1.19) by the MIT Media Lab, and is available on the Internet at (http://sound.media.mit.edu/KEMAR.html). This database may be the most popular in recent research. The database contains far-field ($r = 1.4$ m) 512-point HRIRs of 710 spatial directions, whose elevation varies from −40° to 90°. In each elevation plane, measuring number M and azimuthal interval $\Delta\theta$ are given in Table 2.2. In the measurements, a pair of microphones were fixed at the end of the occluded-ear simulator DB-100, that is, at the eardrums. The left ear of KEMAR was mounted with a small pinna, DB-061, and the right ear with a large pinna, DB-065. Thus, data for the left and right ears are asymmetric, or different. In practice, the unmeasured data for both pinnae can be conveniently derived using the left-right symmetrical characteristics of KEMAR. For example, the result for the right ear with a small pinna at ($\theta = 330°$, $\phi = 0°$) is identical to that for the left ear with a small pinna at ($\theta = 30°$, $\phi = 0°$). Moreover, the diffuse-field-equalized data are also included in the database.

Table 2.1 Far-field HRTF databases

No.	Reference	Measuring conditions	Subject and features
1	Wightman and Kistler, 1989a.	Anechoic chamber, far-field with $r = 1.38$ m, pre-weighted wideband and noise-like signal, sitting, probe microphone, and 1–2 mm from the eardrum.	10 humans (4 males, 6 females); 144 directions ($\phi = -36° \sim +90°$, $\Delta\phi = 18°$, in the horizontal plane $\Delta\theta = 15°$); 50 kHz sampling frequency, 16 bit, and 1024-point.
2	Møller et al., 1995b.	Anechoic chamber, far-field with $r = 2.0$ m, MLS, standing, miniature microphone, and blocked ear canal.	40 humans (22 males, 18 females); 97 directions ($\phi = -67.5° \sim +90°$, $\Delta\phi = 22.5°$, $\Delta\theta = 22.5°$, except for $\Delta\theta = 45°$ at $\phi = \pm67.5°$); 48 kHz sampling frequency, and 256-point.
3	Gardner and Martin, 1995a (MIT database).	Anechoic chamber, far-field with $r = 1.4$ m, MLS, and at the end of occluded-ear simulator.	KEMAR artificial head, DB-061 small pinna for the left ear and DB-065 large pinna for the right ear; 710 directions ($\phi = -40° \sim +90°$, $\Delta\phi = 10°$, in the horizontal plane $\Delta\theta = 5°$); 44.1 kHz sampling frequency, 16 bit, and 512-point.
4	Genuit and Xiang, 1995.	Anechoic chamber, far-field with $r = 2.0$ m, and pseudo-random noise.	Head Acoustics HMS I and HMS II artificial heads; the full space, $\Delta\phi = 1° \sim 5°$, in the horizontal plane $\Delta\theta = 0.9°$; 48 kHz sampling frequency, 16 bit, and 4096-point.
5	Riederer, 1998a.	Anechoic chamber, far-field with $r = 1.89$ m, a random-phase flat-spectrum signal, sitting, miniature microphone, and blocked ear canal.	51 humans (39 males, 12 females); $\phi = -30°, -15°, 0°, 15°, 30°, 60°$ and $90°$, $\Delta\theta = 10°$; 48 kHz sampling frequency, 16 bit, and 512-point.
6	Blauert, et al., 1998 (AUDIS database).	Anechoic chamber, far-field with $r = 2.4$ m, sitting, random-phase pseudo noise, miniature microphone, and blocked ear canal (inward 4 mm).	About 20 human subjects, data for 12 subjects were published; 122 directions ($\phi = -10° \sim +90°$, $\Delta\phi = 10°$, $\Delta\theta = 15°$, with some lateral elevation being left out); 44.1 kHz sampling frequency, and roughly 3 ms in length.
7	Bovbjerg et al., 2000.	Anechoic chamber, far-field with $r = 2.04 \sim 2.13$ m, MLS, 1/2 inch microphone, and blocked ear canal.	VALDEMAR artificial head; 11975 directions ($\phi = -90° \sim +90°$, $\Delta\phi = 2°$, in the horizontal plane $\Delta\theta = 2°$); and 48 kHz sampling frequency.
8	Nishino, et al., 2001, 2007.	Non-anechoic room with reverberation time of 0.15 s, far-field with $r = 1.52$ m, time stretched pulse, miniature microphone, and not entirely blocked entrance to the ear canal.	96 humans published; 72 directions in the horizontal plane at intervals of 5° per subject, 1 human measured at full spatial directions; 48 kHz sampling frequency, and 512-point.
9	Algazi, et al., 2001b (CIPIC database).	Non-anechoic room, far-field with $r = 1.0$ m, Golay code, sitting, probe microphone, and blocked ear canal.	43 humans (27 males, 16 females); 1250 directions (interaural polar coordinates, 50 directions in the horizontal plane, non-uniform distribution); 44.1 kHz sampling frequency, 16 bit, and 200-point.
10	Takane, et al., 2002.	Anechoic chamber, far-field with $r = 1.2$ m, time stretched pulse, sitting on the floor, miniature microphone, and blocked ear canal.	3 males; 454 directions ($\phi = 0° \sim +90°$, $\Delta\phi = 10°$, in the horizontal plane $\Delta\theta = 5°$); 44.1 kHz sampling frequency, float-point, and 512-point.

No.	Reference	Measuring conditions	Subject and features
11	IRCAM Lab, 2003	Anechoic chamber, far-field with r = 1.95 m, logarithmic sweep signal, sitting, miniature microphone, and blocked ear canal.	51 humans; 187 directions (ϕ = −45° ~ +90°, $\Delta\phi$ = 15°, in the horizontal plane $\Delta\theta$ = 15°); 44.1 kHz sampling frequency, 24 bit, and 8192-point (raw) or 512-point (compensated).
12	Grassi E. et al., 2003.	Non-anechoic room with absorption treatment, loudspeaker response range 700 Hz ~ 20 kHz, linear sweep signal, miniature microphone, and blocked ear canal.	7 humans; 1132 directions (interaural polar coordinates); 83.3 kHz sampling frequency and 540-point.
13	Xie, et al., 2007a (Chinese human subject database).	Non-anechoic room with reverberation time of 0.15 s, far-field with r = 1.5 m, MLS, sitting, miniature microphone, and blocked ear canal.	52 humans (26 males and 26 females); 493 directions (Table 2.3); along with KEMAR artificial head with small pinna DB-060/061 and without pinna, 72 directions in the horizontal plane; 44.1 kHz sampling frequency, 16 bit, and 512-point.
14	Majdak et al., 2010; (ARI HRTF database, 2012), or see the webpage of ARI, Austrian Academy of Sciences: http://www.kfs.oeaw.ac.at/content/view/608/606/	Semi-anechoic chamber, far-field with r = 1.2 m, exponential sweep signal, sitting, miniature microphone, blocked ear canal, and behind the ear microphones placed in hearing-aid devices.	77 humans in total, with blocked ear canal for 68 subjects (40 males and 28 females) and microphones placed in hearing-aid devices for 9 other subjects (5 males and 4 females); 1550 directions (ϕ = −30° ~ +80°, $\Delta\phi$ = 5°, in the horizontal plane, $\Delta\theta$ = 2.5° within the front-azimuthal region from θ = 315° to 45° and $\Delta\theta$ = 5° for other region); 48 kHz sampling frequency, 50 ms duration (long) or 5.33 ms (short) duration.
15	Yu, et al., 2012b.	Non-anechoic room with reverberation time of 0.15 s, far-field with r = 1.0 m, MLS, and at the end of occluded-ear simulator	KEMAR artificial head; DB-060/061 small pinnae; 3889 directions (ϕ = −45° ~ +90°, $\Delta\phi$ = 5°, $\Delta\theta$ = 2.5°); 96 kHz sampling frequency, 24-bit, 1024-point.

Table 2.2 Number of measurements M in each elevation plane and azimuthal intervals $\Delta\theta$ of the MIT HRTF database

ϕ (°)	−40	−30	−20	−10	0	10	20
M	56	60	72	72	72	72	72
$\Delta\theta$ (°)	6.43	6.00	5.00	5.00	5.00	5.00	5.00
ϕ (°)	30	40	50	60	70	80	90
M	60	56	45	36	24	12	1
$\Delta\theta$ (°)	6.00	6.43	8.00	10.00	15	30	−

2. Database number 7 was obtained from VALDEMAR using the blocked ear canal technique. Measurements were conducted along an elevation of −90° to +90° at an interval of 2°. In each elevation plane, the azimuthal interval varied, with $\Delta\theta$ = 2° in the horizontal plane. Thus, the database contains far-field HRTFs at 11975 directions, making it the database with the highest spatial resolution to date. The highest spatial resolution is termed with respect to the full space. As far as a specific elevation plane is concerned, the resolution of database number 4 is higher than that of database number 7.

3. Database number 9 consists of the data of 43 human subjects, and the blocked ear canal technique was used. Unlike other databases, an interaural polar coordinate was used in this database (Figure 1.3). Measurements were conducted at 1250 spatial directions, in which there were 50 nonuniformly distributed directions in the horizontal plane. This database is characterized by a high spatial resolution for human subjects and a large number of subjects. Moreover, the anatomical parameters of human subjects were also measured. This database is also available in the public domain, at (http://interface.cipic .ucdavis.edu/CIL_html/CIL_HRTF_database.htm).

From the databases tabulated above, we can see:

1. Databases 3, 4, and 7 were all measured from artificial heads. However, artificial heads were designed based on mean or typical anatomical structures and dimensions of certain populations, and thus cannot represent the individual characteristics of human subjects.
2. Naturally, HRTF is a continuous function of direction. In practice, however, HRTF can usually be measured in finite directions, that is, spatial sampling of the continuous function of HRTF. Furthermore, considering the efficiency of early measurement systems and the subject's tolerance for long-term measurements, the simplification of measured directions has been implemented for human subject measurements. Therefore, the directional resolution of some early human subject databases is low. For example, azimuthal resolutions in the horizontal plane are 22.5°, 15°, and 10° for databases 2, 11, and 5, respectively. Of course, current well-designed measurement systems can alleviate this problem. A commonly used HRTF database of human subjects with high directional resolution is the CIPIC, database number 9. However, because the interaural-polar coordinate was used in the measurement, the resolution in the lateral directions remains insufficient. In particular, the HRTFs at the direct left and right directions were not measured.
3. In some databases, the number of subjects was so small that it seems insufficient for obtaining reasonable statistical results. For example, only 3 and 7 subjects were included in databases 10 and 12, respectively. Moreover, for all existing human HRTF databases except database number 2, the number of male and female subjects is apparently unequal. Under this situation, the overall statistical mean of gender-related variables is likely to be biased.
4. Most available data mainly measure Western subjects. For a long time, these data were used for relevant research in China. However, HRTFs are strongly related to anatomical structures and dimensions. Because some differences in anatomical structures and dimensions exist across populations, the data from primarily Western subjects appear to be unsuitable for Chinese subjects.

In 2003, with funding from the National Natural Science Foundation of China, our group launched a project to construct an HRTF database using Chinese subjects (number 13 in Table 2.1). This work was completed in November of 2005 (Xie et al., 2007a), and the measured data were analyzed (Chapter 3). The measurement conditions are described in Section 2.2.8 (using the first generation measurement system). This database contains 52 human subjects, 26 males and 26 females; the blocked ear canal technique was used in measurement. For each subject, far-field HRTFs were measured at 493 spatial directions, along an elevation of $-30°$ to $90°$ at a distance of 1.5 m. The data were stored in forms of time-domain 512-point HRIRs, with a sampling frequency of 44.1 kHz and 16-bit quantization. Table 2.3 shows a measurement number, M, and azimuthal interval, $\Delta\theta$, in each elevation plane.

Table 2.3 Number of measurements M in each elevation plane and azimuthal intervals $\Delta\theta$ from our group

ϕ (°)	−30	−15	0	15	30	45	60	75	90
M	72	72	72	72	72	72	36	24	1
$\Delta\theta$ (°)	5	5	5	5	5	5	10	15	—

Aside from human subjects, the HRTFs of KEMAR without pinnae and with a pair of small pinnae of DB-060/061 were also measured at 72 azimuthal directions in the horizontal plane. Unlike the MIT database, the blocked ear canal technique and a pair of small pinnae were employed in our measurements.

Using the second-generation measurement system (Section 2.2.8), our laboratory also established a fine directional resolution HRTF database for the KEMAR artificial head, with DB 60/61 small pinnae (number 15 in Table 2.1). The binaural pressures were recorded at the ends of a pair of Zwislocki occluded-ear simulators. The resulting database contains far-field binaural HRIRs, with a source distance $r = 1.0$ m and at 3889 spatial directions, including elevations varying from −45° to 90°, at an interval of 5° and a fine azimuthal resolution of 2.5° (except at 90° elevation). Each HRIR has a 1024-point length with 96 kHz sampling frequency and 24-bit quantization. This database is open for public access for scientific research purposes (by sending an e-mail request to the author of this book).

2.4. Some Specific Measurement Methods and Near-field HRTF Measurements

2.4.1 Some Specific HRTF Measurement Methods

Although the HRTF measurement methods presented in Section 2.2 are general, other specific measurement methods are employed under specific situations.

Traditionally (Section 2.2), microphones are positioned at the two ears. The far-field HRTFs at various spatial directions are then obtained by changing the relative direction between the sound source and subject, one by one. This measuring method suffers, as a lengthy process with low efficiency. To address these problems, Zotkin et al. (2006) proposed an efficient method that applies the acoustical principle of reciprocity, in which a microspeaker is inserted into the subject's ear, and a series of microphones are placed around the subject. HRTFs can be obtained from the output of the microphones because of the reciprocal property of the transfer function between the sound source (microspeaker) and the receiver (microphone). Moreover, this method simultaneously acquires HRTFs at all microphone positions. However, two main disadvantages limit the reciprocal method. One disadvantage is that the microspeaker's size may lead to poor low-frequency performance, and the reliable measuring frequency usually ranges from 1 to 16 kHz. The other disadvantage is a low signal-to-noise ratio, due to the relatively low sound pressure level radiating from the microspeaker. Thus, an appropriate signal processing method is needed to compensate for these phenomena. In practice, even if the microspeaker is able to generate a higher sound pressure level, it remains unsuitable for human HRTF measurements because excessively high sound levels at the ear canal cause subjects discomfort.

Incidentally, a reciprocal method, similar to that used in HRTF measurements, can be used to measure the directivity of human (or artificial mouth) speech (Halkossari et al., 2005), which is important in communication and room acoustics. The results are useful in simulating sound source and reflections inside a room, as will be discussed in Chapter 11. In measurements of speech directivity, a microphone is placed at the mouth, rather than at the entrance to the ear canal in HRTF measurements, and the sound source is positioned at different spatial directions. According to the acoustical principle of reciprocity, the output of microphones, varying in source direction, represents the directivity of human speech.

All of these methods can measure only HRTF data at one or a few spatially discrete directions in a run. To speed up measurements—and considering that HRTF is naturally a continuous function of spatial directions—some researchers proposed a continuous measurement method for obtaining an azimuthally continuous HRTF in a run using a servo swiveled chair (Fukudome et al., 2007). In essence, measuring with a turnable chair, with the subject rotating continuously, is an issue relevant to identifying a linear time-variant system. In that research, an MLS was used as the excitation signal, and a low, constant, angular speed of the turnable chair was chosen. Under such a condition, the relative change in azimuth between the subject and the sound source is negligible in a period of the MLS signal. The HRTF at an arbitrary azimuth can then be extracted from the measured signal by applying an appropriate signal-processing algorithm. Similarly, Enzner (2008, 2009) used white Gaussian noise and a multi-channel adaptive filtering algorithm to measure azimuthally continuous HRTF data at different elevations.

Ajdler et al. (2007) also proposed a similar technique for measuring the azimuthally continuous HRTF, using a moving microphone. This method is characterized by accounting for the Doppler effect in the recording, caused by high-speed movement relative to a fixed sound source. With this technique, the measurement duration can be dramatically reduced. For example, HRTFs sampled at 44.1 kHz can be measured at all angular positions along the horizontal plane in less than 1 s. However, to correctly reconstruct HRTF (and HRIR), the moving motor and the emitted sound must be perfectly synchronized. Moreover, these researchers provided only a theoretical setup and analysis regarding HRTF measurements, without experimental verification.

2.4.2 Near-field HRTF Measurement

The discussions so far have focused on far-field HRTF measurements, and the most existing databases are for far-field HRTFs. Increasing interest has been given to near-field HRTFs recently because of their source distance-dependent characteristics. However, two issues related to near-field HRTF measurement have to be resolved. These two problems hinder the development of near-field HRTF measurements.

First, a sophisticated sound source is urgently needed. In far-field HRTF measurements, the point sound source needed can be approximated by a common, small, loudspeaker system, where measurement errors caused by the size and directivity of the loudspeaker, as well as the multiple scattering between subject and loudspeaker, are negligible. By contrast, sound source size and multiple scattering between subject and sound source strongly influence measurements at a near distance. Under such conditions, a commonly used loudspeaker can no longer be regarded as a point sound source.

Second, the overall workload for near-field HRTF measurement is rather heavy because HRTFs should be measured, not only at various spatial directions but also at distances ranging less than 1.0 m. This problem is especially critical when using human subjects. Thus, constructing near-field HRTF databases with individual human subjects is very difficult.

Even so, some authors and groups have designed various sound sources and measured the near-field HRTFs with artificial heads or spherical head models. Duda et al. (1998) measured the HRTFs of a bowling ball (used to simulate a human head) at 7 different distances, from 0.135 to 2.18 m, with a 6.4 cm in diameter Bose Acoustimass™ loudspeaker. Some other researchers used specific sound sources to measure near-field HRTFs. For example, Brungart et al. (1999a) measured near-field HRTFs on a KEMAR artificial head using a periodic sine-wave sweep and an acoustic point source mainly composed of an electrodynamic horn driver and a Tygon tube with an internal diameter of 1.2 cm (Figure 2.6). Measurements were obtained at 12 equally spaced azimuths in the horizontal plane, at distances of 0.125, 0.25, 0.50, and 1.00 m.

Nishino et al. (2004) used a spark noise to measure the near-field HRTFs on an artificial head (B&K 4128) in the horizontal plane, at distances of 0.15, 0.20, 0.25, and 0.30 m. At each distance, the HRTFs at 24 equally spaced azimuths were measured. Results can be partially downloaded from http://db.ciair.coe.nagoya-u.ac.jp/. In another study (Hosoe et al., 2005, 2006), a small dodecahedral loudspeaker system, with a diameter of 38 mm, was designed, in which each piezoelectric unit had a diameter of 18 mm and a thickness of 1 mm. Measurements showed that the dodecahedral loudspeaker system was approximately omnidirectional, from 1 to 8 kHz. At higher frequencies, however, the directivity became obvious, and magnitude response considerably dropped below the 1 kHz frequency. Using the dodecahedral loudspeaker system, the near-field HRTFs of an artificial head (B&K 4128) were collected at 9 distances, ranging from 0.2 to 1.0 m at an interval of 0.1 m. At each distance, the HRTFs were measured in 29 elevations with 72 azimuths in each elevation. The total number of measurement positions was 18153, including a position directly above ($\theta = 0°$, $\phi = 90°$) (Hosoe et al., 2006). This is the most complete near-field HRTF database in the literature. In 2007, this research group further designed a dodecahedral loudspeaker system composed of micro electrodynamic loudspeaker units to improve low-frequency characteristics (Hayakawa et al., 2007). Moreover, Gong et al. (2007) and Qu et al. (2009) also constructed a near-field HRTF database of KEMAR using a specialized spark gap as the sound source. The database consisted of 8 distances, from 0.2 to 1.6 m, with 793 directions at each distance.

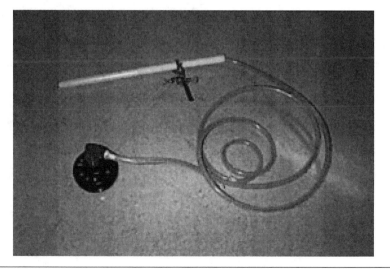

Figure 2.6 Acoustic point source, mainly composed of an electrodynamic horn driver and a 1.2 cm diameter tube (from Brungart and Rabinowitz, 1999a, by permission of J.Acoust.Soc.Am.)

The low-frequency characteristics of a near-field HRTF are particularly important. The actual measured results, however, are limited by the low-frequency performance of the sound source. Generally, if the low-frequency response of the sound source used in near-field HRTF measurement drops from approximately 1 kHz, a signal processing method to compensate is then required. However, if the low-frequency radiated energy of the sound source dramatically decreases more than 15 dB, then compensation is difficult. Although an improved low-frequency performance can be achieved using a relatively large loudspeaker, the influence of multiple scattering between subject and source becomes significant. This means that improving low-frequency performance conflicts with reducing or avoiding multiple scatterings between subject and source. Recently, an acceptable range of source size in near-field HRTF measurement has been evaluated by analyzing the multiple scattering between sound source and the head, based on a spherical head, and a pulsating spherical sound source model (similar to what will be discussed in Section 4.2; Yu et al., 2008). Results showed that the magnitude error in the measured HRTF, caused by multiple scattering, varied as a function of source direction, distance, and frequency. To ensure a magnitude error within 1.0 dB up to 20 kHz, the radius of the sound source should not exceed 0.05 m (or 0.03 m) for a source distance not less than 0.20 m (or 0.15 m). For our research, a spherical dodecahedron sound source consisting of 12 micro electrodynamic loudspeaker units was made with a radius of 0.035 m [Figure 2.7(a)] and lower frequency limit of approximately 350 Hz. The error in the measured near-field HRTF at low frequencies below 350 Hz can be corrected by theoretical calculation (Section 4.1.5). Both calculations and measurements indicated that the directivity and multi-scattering characteristics of the source satisfy the requirements for near-field HRTF measurements (Yu et al., 2009). Based on this sound source and a modification of the first generation measurement system in our laboratory, near-field HRTFs from KEMAR were measured at 10 distances (0.2, 0.25, 0.3, 0.4, 0.5, 0.6, 0.7, 0.8, 0.9, and 1.0 m). At each distance, measurements were conducted at 493 spatial directions, which is identical to those given in Table 2.3. Figure 2.7(b) shows the scene of measurements (Yu et al., 2010a, 2012d).

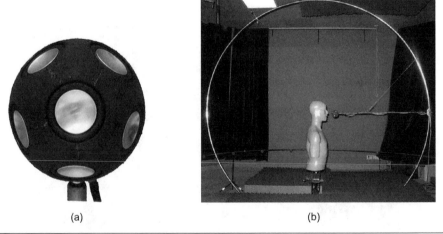

(a)　　　　　　　　　(b)

Figure 2.7　Photos of the sound source and the scene of near-field HRTF measurements: (a) spherical dodecahedron sound source; (b) scene of measurements

2.5. Summary

The general principles regarding the measurement of the transfer function or impulse response of a linear time-invariant system can be applied in HRTF measurement. A number of acoustic measuring methods, such as impulse, Fourier analysis, and correlation methods, can be employed in measuring HRTF. In HRTF measurements, a wide variety of signals can be used as exciting stimuli, such as sine, sweep, and impulse signals, as well as random and pseudorandom noises. In the past 20 years, traditional analog measuring techniques have been replaced by modern digital measurements and signal-processing techniques, in which the digital measuring technique employing the pseudorandom noise signal is widely adopted because of its unique advantages.

Although various HRTF measurement methods exist, the underlying principles and design methodologies are similar. Generally, measuring subjects include artificial heads and human beings, and the measuring position can be chosen at an arbitrary point along the line between the entrance to the ear canal and the eardrum. In human HRTF measurement, the most convenient and generally accepted method is the blocked ear canal configuration. In practice, special attention should be paid to the treatment of the measuring environment and the adjustment of the mechanical device to avoid or eliminate reflections and measurement errors. Moreover, some specific requirements on measuring sound source and length of the stimuli should be considered. In postprocessing, adding a time-domain window and frequency-domain compensation are helpful, to some extent, in removing unwanted reflections and compensating for the influence of the measuring system. During a measurement session, ensuring the accurate position of the subject and the measuring microphone is an important issue because a slight head movement, as well as minimal change in microphone position, leads to considerable measurement errors.

Many research groups have devoted their efforts to HRTF measurement and constructed some far-field HRTF databases for scientific research. These databases differ in measurement setup, subjects, and spatial resolution, among others. Our research group has also constructed an HRTF database with high spatial resolution, using Chinese subjects. Under some specific conditions, specific measuring methods are also presented.

Increasing interest has been given to near-field HRTFs because of their source distance-dependent characteristic. Two issues related to near-field HRTF measurement have to be resolved, namely, the problems of sound source and workload. Some researchers and groups have designed various sound sources and measured the near-field HRTFs from artificial heads or spherical head models.

3

Primary Features of HRTFs

This chapter presents the primary features and localization cues of HRTFs. Some common or basic characteristics of time-domain HRIRs and frequency-domain HRTFs in far fields, as well as the minimum phase characteristic of HRTFs, are outlined. The directional localization cues encoded in far-field HRTFs, including interaural time difference, interaural level difference, and spectral cue, are analyzed in detail. The spatial symmetry of HRTFs is examined. The distance dependence of near-field HRTFs and some HRTF-relevant issues for binaural hearing are briefly discussed.

Chapter 2 described the empirical measurements of head-related transfer functions (HRTFs), and provided an overview of some existing HRTF databases in Section 2.3. On the basis of the measured HRTF data, we can analyze a variety of HRTF features, thereby acquiring insight into the localization cues encoded in HRTFs. In this chapter, Section 3.1 examines the basic characteristics of time-domain head-related impulse responses (HRIRs) and frequency-domain HRTFs in far fields, as well as the minimum-phase approximation of HRTFs. Section 3.2 analyzes interaural time difference (ITD), a dominant interaural localization cue at low frequencies. Different definitions and methods of evaluating HRTF-derived ITD are presented. The corresponding results for KEMAR and Chinese subjects are shown as examples and then compared with the findings of previous studies. Section 3.3 presents the investigation of interaural level difference (ILD), another important interaural localization cue. The variations in ILD, with source position and frequency, are analyzed. The spectral cues in HRTFs are illustrated in Section 3.4, with an emphasis on pinna-related high-frequency notches and torso-related spectral features below 3 kHz. The relationship between HRTF symmetry and anatomical structures is discussed in Section 3.5, where the applicable frequency range of the front-back and left-right symmetrical model of HRTFs is also investigated. Section 3.6 analyzes the distance dependence of near-field HRTFs, as possible cues for auditory distance perception. HRTF-relevant issues for binaural hearing are briefly discussed in Section 3.7.

3.1. Time- and Frequency-domain Features of HRTFs

3.1.1. Time-domain Features of Head-related Impulse Responses

HRIRs are the time-domain counterpart of HRTFs. Figure 3.1 shows the far-field HRIRs of KEMAR, with small pinnae measured by the MIT Media Lab (see Section 2.3) at azimuths of 0°, 30°, 60°, 90°, 120°, 150°, and 180° in the horizontal plane. Because two different sizes of pinna were mounted on KEMAR (a DB-061 small pinna on the left ear and a DB-065 large pinna on

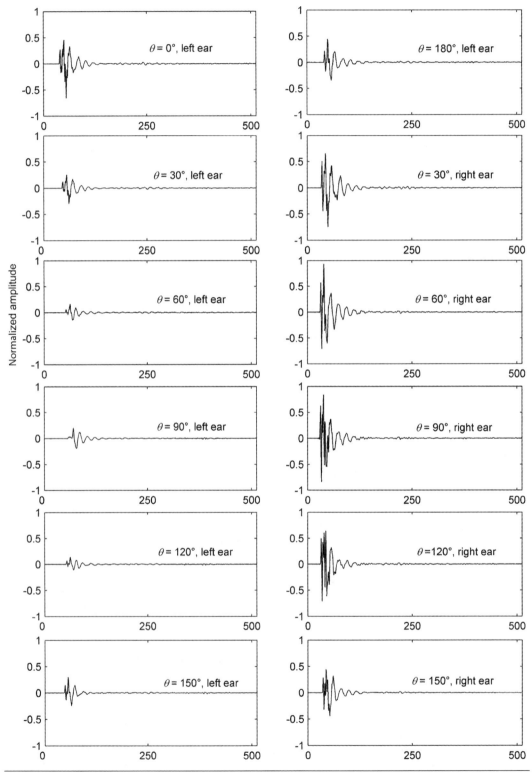

Figure 3.1 KEMAR far-field HRIRs for several different source azimuths in the horizontal plane

the right ear) during the measurement, a symmetric version of right ear HRIRs with small pinnae is obtained by applying a left-right mirror reflection to the corresponding left ear HRIRs. For example, the right ear HRIR at azimuth 30° corresponds to the measured left ear HRIR at azimuth 330°. Given that the left and right ear HRIRs obtained in this manner are identical at azimuths of 0° and 180°, only the left ear data for these two azimuths are provided in Figure 3.1.

In Figure 3.1, the HRIR amplitude at the preceding 30 to 58 samples is approximately zero. These zero-amplitude samples correspond to the propagation delay from sound source to ear, and is caused by the neglect of the term $\exp(-jkr)$ in P_0 in Equation (1.15). In practice, a time window is usually applied to measured HRIRs; thus, the initial delay has only relative significance. The main body of the HRIRs, which reflects the complicated interactions between incident sound waves and the head, torso, and pinna, persists for about 50 or 60 samples (i.e., about 1.1–1.4 ms at a sampling frequency of 44.1 kHz). Subsequently, the HRIR amplitudes return to nearly zero. When the sound source deviates from directly front and back directions, the initial delay difference in the left and right ear HRIRs reflects the propagation time difference from the sound source to the left and right ears; this difference is known as interaural time difference (ITD). At an azimuth of 90°, for instance, the left ear HRIR lags behind the right ear HRIR with a relative delay of 28 samples (approximately 635 µs at a sampling frequency of 44.1 kHz). Moreover, when the sound source is located contralateral to the concerned ear, for example, at an azimuth of 90° for the left ear, the HRIR amplitude is visibly attenuated because of the head's shadowing effect.

Figure 3.2 shows the left ear HRIRs for a typical Chinese subject, number 25 (see Section 2.3), varying with time and azimuth in the horizontal plane. Similar results are observed for the right ear HRIRs. An initial delay at a range of 55 to 85 samples is observed in the HRIRs, a result that differs from that shown in Figure 3.1. This discrepancy is attributed to different onset periods of the truncation window used in the two studies. The main body of the HRIRs persists for about 50 samples, and then the amplitude returns to nearly zero. For contralateral sources at azimuths of 60° to 120°, the impulses are visibly attenuated. These results are valid for our 52 Chinese subjects

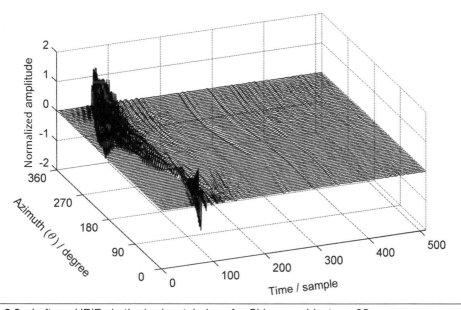

Figure 3.2 Left ear HRIRs in the horizontal plane for Chinese subject no. 25

and are consistent with those of other studies, at least from a qualitative perspective (Gardner and Martin, 1995a; Møller et al., 1995b; Riederer, 1998a).

3.1.2. Frequency-domain Features of HRTFs

Figure 3.3 presents the normalized (logarithmic) HRTF magnitudes of KEMAR with small pinnae at azimuths of 0°, 30°, 60°, 90°, 120°, 150°, and 180° in the horizontal plane. Similarly to the case of HRIRs, left ear data and their left-right mirror reflections are used in the figure.

At frequencies below 0.4–0.5 kHz, the normalized magnitudes of HRTFs $20\log_{10}|H|$ approach 0 dB, and are roughly independent of frequency because the scattering and shadowing effect of

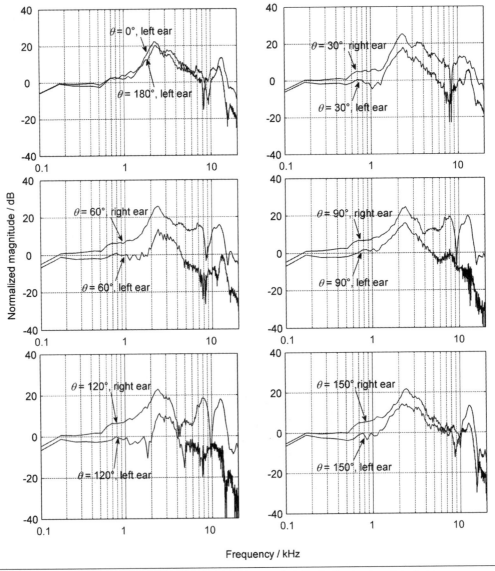

Figure 3.3   Magnitudes of KEMAR HRTFs at various azimuths in the horizontal plane

the head are negligible. The decrease in magnitude below 150 Hz is caused by the low-frequency limit of loudspeaker response used in HRTF measurement, rather than by the HRTF itself (Section 2.2.5). Because of the finite source distance relative to the head center ($r = 1.4$ m, Section 3.6), a 2–4 dB difference between the left and right ear HRTF magnitudes is observed at a lateral azimuth of 90°, even at low frequencies. This difference is larger than that observed in an infinitely distant source (plane wave incidence; Figure 1.9).

As frequency increases, the normalized magnitudes of the HRTFs vary with frequency and azimuth in a complex manner. This complexity is attributed to the overall filtering effects of the head, pinna, torso, and ear canal. The apparent peak in HRTF magnitude at 2–3 kHz results from the resonance of the occluded-ear simulator of KEMAR. Above 4 kHz, the contralateral HRTF magnitudes (such as the magnitude of the left ear at an azimuth of 90°) are visibly attenuated because of the low-pass filtering properties of the head shadow. The ipsilateral HRTF magnitudes (such as the magnitude of the right ear at an azimuth of 90°) increase to a certain extent, although some notches occur. This phenomenon is partially attributed to the approximate mirror-reflection effect of the head on ipsilateral incidence at high frequencies, thereby leading to increased pressure for ipsilateral sound sources. Actually, a 6 dB increase in pressure is obtained on the surface of a rigid plane with infinite dimensions, when compared with the case of a free field.

The difference between high-frequency HRTF magnitudes at azimuths of 0° and 180° is also observed in Figure 3.3. This difference is caused by the front-back asymmetry in the head shape and ear location, and the diffraction effect of the pinna. This front-back difference in HRTF magnitudes is regarded as a cue for resolving front-back confusion in localization, and will be further investigated in Sections 3.4 and 3.5. With increasing elevation, the variations in HRTF magnitudes, with azimuth, decrease and smoothen (the corresponding figure is omitted for brevity).

For comparison, Figure 3.4 plots the normalized HRTF magnitudes of the left ear of a typical Chinese human subject, number 25, as a function of frequency and azimuth in the horizontal

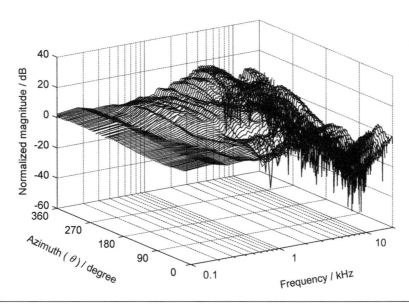

Figure 3.4 Normalized magnitudes of left ear HRTFs in the horizontal plane for Chinese subject no. 25

plane. The human and KEMAR HRTF magnitudes share some common features. Generally, these results are valid for both ears of all 52 Chinese subjects, at least qualitatively; the findings are also consistent with those of other studies (Gardner and Martin, 1995a; Møller et al., 1995b; Riederer, 1998a).

To demonstrate the individuality of HRTFs, we plot the normalized magnitudes of left ear HRTFs at ($\theta = 0°$, $\phi = 0°$) for 10 subjects in Figure 3.5. These magnitudes are randomly selected from the Chinese subject HRTF database. The intersubject differences in HRTF magnitudes are clearly visible above 6–7 kHz. Similar results are generated for other source locations, and are consistent with those of

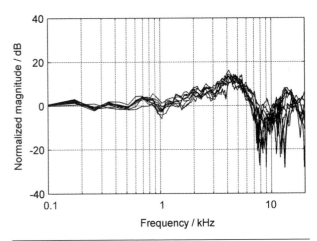

Figure 3.5 Left ear HRTF magnitudes for 10 subjects at azimuth 0° in the horizontal plane

other studies (Wightman and Kistler, 1989a; Møller et al., 1995b; Riederer, 1998a).

Finally, HRTFs are multivariate functions of azimuth, elevation, and frequency—even in the far field. A variety of graphical representations of HRTFs, which differ according to specific context, are presented in literature. For instance, far-field HRTFs vary only with azimuth and elevation at a given frequency, and can therefore be fully described by the spatial frequency response surface proposed by Cheng and Wakefield (1999, 2001b). Some other graphical representations of HRTFs will be introduced in Section 3.4.1.

3.1.3. Minimum-phase Characteristics of HRTFs

For a given source direction, HRTFs are complex-valued functions of frequency, including magnitude and phase components. According to the principles of signal processing (Oppenheim et al., 1999), HRTFs, as transfer functions of a linear-time-invariant system, can be represented by the product of its minimum-phase function $H_{min}(\theta, \phi, f)$, all-pass function $\exp[j\psi_{all}(\theta, \phi, f)]$, and linear-phase function $\exp[-j2\pi f\,T(\theta, \phi)]$:

$$H(\theta,\phi,f) = H_{min}(\theta,\phi,f)\exp\left[j\psi_{all}(\theta,\phi,f)\right]\exp\left[-j2\pi f\,T(\theta,\phi)\right]$$
$$= H_{min}(\theta,\phi,f)\exp[j\psi_{cess}(\theta,\phi,f)], \tag{3.1}$$

where $T(\theta, \phi)$ is the pure time delay caused by propagation from sound source to ear; it roughly corresponds to the initial delay of the HRIRs in Figure 3.1. The overall phase caused by all-pass and linear-phase components is called excess phase $\psi_{cess}(\theta, \phi, f)$, and minimum-phase function $H_{min}(\theta, \phi, f)$ can be expressed as:

$$H_{min}(\theta,\phi,f) = |H_{min}(\theta,\phi,f)|\exp[j\psi_{min}(\theta,\phi,f)]$$
$$= |H(\theta,\phi,f)|\exp[j\psi_{min}(\theta,\phi,f)]. \tag{3.2}$$

The phase of a minimum-phase function is related to the logarithmic magnitude by Hilbert transform:

$$\psi_{min}(\theta,\phi,f) = -\frac{1}{\pi}\int_{-\infty}^{+\infty}\frac{\ln|H(\theta,\phi,x)|}{f-x}dx. \tag{3.3}$$

Thus, minimum-phase function $H_{min}(\theta, \phi, f)$ can be accurately reconstructed by its magnitude. For a causal minimum-phase function, all the zeros and poles are located inside the unit circle in the Z-plane. According to Equations (3.1) and (3.2), excess phase component, $\exp[j\,\psi_{cess}(\theta, \phi, f)]$, can be evaluated from the ratio of $H(\theta, \phi, f)/H_{min}(\theta, \phi, f)$. The minimum-phase function of HRIRs in the time domain can be obtained by applying inverse Fourier transform to frequency-domain counterpart $H_{min}(\theta, \phi, f)$:

$$h_{min}(\theta,\phi,t) = \int_{-\infty}^{+\infty} H_{min}(\theta,\phi,f)e^{j2\pi ft}df . \tag{3.4}$$

Figure 3.6 shows $h_{min}(\theta, \phi, t)$ for the left and right ears of KEMAR at an azimuth of 30° in the horizontal plane. Compared with Figure 3.1, Figure 3.6 shows that the energy of $h_{min}(\theta, \phi, t)$ is concentrated around $t = 0$, with disappearing initial delay and changing response shape.

Equation (3.1) indicates that the overall phase of HRTFs generally comprises a minimum phase, an all-pass phase, and a linear phase, expressed thus:

$$\psi(\theta,\phi,f) = \arg[H(\theta,\phi,f)] = \psi_{min}(\theta,\phi,f) + \psi_{all}(\theta,\phi,f) - 2\pi fT(\theta,\phi). \tag{3.5}$$

If the contribution of the all-pass phase component is negligible (or the excess phase can be approximately fit by a linear function of frequency), Equation (3.1) can be approximated as:

$$H(\theta,\phi,f) \approx H_{min}(\theta,\phi,f)\exp\left[-j2\pi fT(\theta,\phi)\right]. \tag{3.6}$$

Equation (3.6) is known as the *minimum-phase approximation of HRTFs*, in which an HRTF is approximated by its minimum-phase function cascaded with a linear phase or a pure delay.

Some researchers have investigated the minimum-phase approximation of HRTFs. After analyzing the HRTFs of 20 human subjects, at 20 directions in the horizontal plane and 10 directions in the median plane, Mehrgardt and Mellert (1977) revealed that HRTFs are approximately of minimum phase up to 10 kHz. Kistler and Wightman (1992) applied the minimum-phase approximation to HRTF reconstruction.

Kulkarni *et al.* (1999) investigated the similarity between original and minimum-phase approximated HRTFs by cross-correlation analysis. Given a source direction (θ, ϕ) and concerned ear, the normalized cross-correlation function of the original HRIR $h(\theta, \phi, t)$ and its minimum-phase version $h_{min}(\theta, \phi, t)$, derived from Equations (3.3) and (3.4), is defined as:

$$\Phi(\tau) = \frac{\displaystyle\int_{-\infty}^{+\infty} h(\theta,\phi,t+\tau)h_{min}(\theta,\phi,t)dt}{\left\{\left[\displaystyle\int_{-\infty}^{+\infty} h^2(\theta,\phi,t)dt\right]\left[\displaystyle\int_{-\infty}^{+\infty} h_{min}^2(\theta,\phi,t)dt\right]\right\}^{1/2}}. \tag{3.7}$$

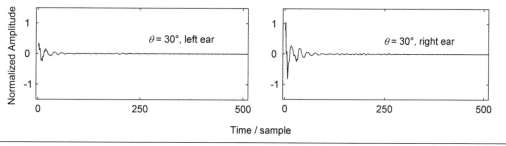

Figure 3.6 Minimum-phase components of KEMAR HRIRs at azimuth 30° in the horizontal plane

By the definition of Equation (3.7), $0 \leq |\Phi(\tau)| \leq 1$. The maximal value of $\Phi(\tau)$, $\Phi(\tau_{max})$ reflects the similarity between $h(\theta, \phi, t)$ and $h_{min}(\theta, \phi, t)$. The closer $\Phi(\tau_{max})$ is to unity, the more accurate the minimum-phase approximation of Equation (3.6). When $\Phi(\tau_{max}) = 1$, $h(\theta, \phi, t)$ is precisely represented by $h_{min}(\theta, \phi, t)$ with a linear delay τ_{max} added at the beginning; thus, Equation (3.6) is accurate. The τ_{max} derived from the cross-correlation calculation can also be used as the linear delay $T(\theta, \phi,)$ in Equation (3.6).

The statistical results for the HRTFs of two human subjects show that at 144 measured source directions for one subject, the $\Phi(\tau_{max})$ for 92% of the left ear HRTFs and 75% of the right ear HRTFs is greater than 0.9; at 505 measured source directions for the other subject, the percentage reaches 96% and 97% for the left and right ears, respectively. Hence, Kulkarni et al. concluded that the minimum-phase approximation of the HRTF in Equation (3.6) is adequate. Further analysis indicates that phase error often occurs at low source elevations of $\phi < 0°$ and at contralateral source directions. This phase error possibly results from the complex diffraction and scattering of anatomical structures (such as the head and torso) before sound reaches the shadowed ear; this phenomenon diminishes the minimum-phase characteristics of HRTFs. Kulkarni et al. also employed a 4-interval, 2-alternative forced choice discrimination paradigm (4I/2AFC; see Section 13.2.1), to verify the perceptual adequacy of the minimum-phase approximation of HRTFs. The authors implemented this verification by comparing the measured HRTFs and their minimum-phase versions. The results indicate that the minimum-phase approximation is reasonable for most sound source positions.

Nam et al. (2008) further analyzed the relationship between the phase error caused by minimum-phase approximation and $\Phi(\tau_{max})$. Equation (3.7) can be converted into the frequency domain by Fourier transform:

$$\Phi(\tau) = \frac{\int_{-\infty}^{+\infty} H(\theta,\phi,f) H_{min}^*(\theta,\phi,f)\exp(j2\pi f\tau)df}{\int_{-\infty}^{+\infty} |H(\theta,\phi,f)|^2 df}. \tag{3.8}$$

Substituting Equations (3.1) and (3.5) into Equation (3.8) yields:

$$\Phi(\tau) = \frac{\int_{-\infty}^{+\infty} |H(\theta,\phi,f)|^2 \exp\{j[\psi_{all}(\theta,\phi,f) - 2\pi f T(\theta,\phi) + 2\pi f\tau]\}df}{\int_{-\infty}^{+\infty} |H(\theta,\phi,f)|^2 df}. \tag{3.9}$$

If the all-pass phase term in Equation (3.9) equals zero, that is, $\psi_{all}(\theta, \phi, f) = 0$, $\Phi(\tau_{max})$ reaches its maximum 1 at $\tau = \tau_{max} = T(\theta, \phi)$, then $h(\theta, \phi, t)$ and $h_{min}(\theta, \phi, t)$ are completely coherent. If $\psi_{all}(\theta, \phi, f) \neq 0$, then $\Phi(\tau_{max}) < 1$ and the coherence between $h(\theta, \phi, t)$ and $h_{min}(\theta, \phi, t)$ decreases. The numerator of Equation (3.9) shows that, at a given frequency f, the contribution of $\psi_{all}(\theta, \phi, f)$ to the integral is weighted by the power spectra $|H(\theta, \phi, f)|^2$ of HRTFs. The larger the power spectra at this frequency, the more considerable the contribution of $\psi_{all}(\theta, \phi, f)$ to the integral will be. Because the power of HRTFs is low at the notch frequency or at the contralateral source direction at high frequencies, even a large deviation from minimum-phase approximation with $\psi_{all}(\theta, \phi, f) \neq 0$ may not cause an obvious reduction in $\Phi(\tau_{max})$. By contrast, the influence of $\psi_{all}(\theta, \phi, f)$ on $\Phi(\tau_{max})$ is clearly observable at the frequency in which HRTFs exhibit high power spectra. The results provided by Nam et al. suggest that, in contrast to phase error, low coherence often occurs at ipsilateral directions rather than at contralateral directions. This difference should be considered in practical analysis.

Some other studies (Møller et al., 1995b; Jot et al., 1995; Gardner, 1997), however, revealed that HRTFs often contain non-minimum-phase zeroes at high frequencies, especially above 10 kHz. These non-minimum-phase zeros are caused by the delayed and focused reflections from anatomical structures (such as pinnae); such reflections are more energetic than are direct responses. The validity of minimum-phase approximations of HRTFs varies with each individual.

The phase characteristics of an HRTF measurement system (such as a sound source) inevitably influences measured results (Yu et al., 2012c). If the measurement system is of nonlinear phase, phase equalization to original measured results is required to compensate for the phase response of the measurement system. Otherwise, errors may occur in the evaluation of the non-minimum-phase characteristics of HRTFs. As will be discussed in Chapter 5, HRTF-based binaural synthesis benefits from the minimum-phase approximation of HRTFs because approximation simplifies signal-processing procedures.

3.2. Interaural Time Difference Analysis

3.2.1. Methods for Evaluating ITD

ITD can be estimated from a pair of HRTFs. As stated in Section 1.4.1, interaural phase delay difference (ITD_p) is a dominant directional localization cue below 1.5 kHz, whereas interaural envelope delay difference (ITD_e) is useful in directional localization above 1.5 kHz. Accordingly, the ITD_p and ITD_e encoded in HRTFs should be used for localization analysis. However, the resultant ITD_p is complicated because of its frequency dependence. Similarly, because ITD_e depends on signal types, it is seldom directly analyzed. A variety of definitions and methods for evaluating ITD have been proposed in previous studies for different purposes; the differences in definitions and methods may cause confusion. Although the ITD that is evaluated via certain methods cannot be directly applied to analyzing sound source localization, it is strongly related to ITD_p and ITD_e, and therefore meaningful and applicable in practice. In the following, we outline the definitions and methods for evaluating ITD. In the following discussion, arguments θ and ϕ are disregarded from the denotation of an HRTF and its phase for simplicity.

Definition 1: ITD_p pertains to interaural phase delay difference. According to Equation (1.10), ITD_p is defined as:

$$ITD_p(\theta,\phi,f) = \frac{\Delta\psi}{2\pi f} = -\frac{\psi_L - \psi_R}{2\pi f}, \tag{3.10}$$

with:

$$\psi_L = \arg[H_L(f)] = \arctan\left[\frac{\operatorname{Im} H_L(f)}{\operatorname{Re} H_L(f)}\right],$$
$$\psi_R = \arg[H_R(f)] = \arctan\left[\frac{\operatorname{Im} H_R(f)}{\operatorname{Re} H_R(f)}\right]. \tag{3.11}$$

The equations indicate that ITD_p can be computed from the HRTF-derived frequency-dependent phase functions of both ears.

However, several problems may arise when calculating ITD_p by using Equations (3.10) and (3.11). First, for frequencies above 1.5 kHz, when head dimension is greater than sound wavelength, interaural phase difference may exceed 2π, leading to ambiguity in determining ITD. Second, the measured HRTFs include the propagation delay from sound source to ear, which corresponds to the linear-phase function in Equation (3.1); thus, the ambiguity in the HRTF phase is likely to occur, even at low frequencies. In this case, an unwrapping processing of the HRTF phase is often employed in practical calculations, in which the principal values of ψ_L and ψ_R within $[-\pi, +\pi]$ are first computed from Equation (3.11). Then, the discontinuous variations in phase are corrected by adding multiples of 2π at discontinuous points, to ensure the continuity of ψ_L and ψ_R with frequency variations. Nevertheless, this processing is valid only in cases wherein HRTFs do not have sharp resonant peaks.

As indicated by Equations (3.10) and (3.5), ITD_p accounts for the contributions of the minimum-phase, all-pass phase, and linear-phase components of HRTFs, and is directly related to sound source localization below 1.5 kHz. Generally, ITD_p varies with frequency and is, therefore, complicated. Kuhn (1977) investigated ITD_p at different frequency ranges.

Definition 2: ITD_{ave} pertains to interaural average (central) time difference. The average time (i.e., energy gravity) of HRIRs is defined as:

$$\bar{t}_L = \frac{\int_{-\infty}^{+\infty} t\, h_L^2(t)\, dt}{\int_{-\infty}^{+\infty} h_L^2(t)\, dt}, \quad \bar{t}_R = \frac{\int_{-\infty}^{+\infty} t\, h_R^2(t)\, dt}{\int_{-\infty}^{+\infty} h_R^2(t)\, dt}, \tag{3.12}$$

whose difference is ITD_{ave}:

$$ITD_{ave}(\theta,\phi) = \bar{t}_L - \bar{t}_R. \tag{3.13}$$

ITD_{ave} reflects the propagation time difference in energy between left and right ear HRIRs; it varies with sound source direction, but is independent of frequency. Equation (3.12) can be converted into the frequency domain by Fourier transform:

$$\bar{t}_L = \frac{\int_{-\infty}^{+\infty} \left[-\frac{1}{2\pi} \frac{d\psi_L}{df} \right] |H_L(f)|^2\, df}{\int_{-\infty}^{+\infty} |H_L(f)|^2\, df}, \quad \bar{t}_R = \frac{\int_{-\infty}^{+\infty} \left[-\frac{1}{2\pi} \frac{d\psi_R}{df} \right] |H_R(f)|^2\, df}{\int_{-\infty}^{+\infty} |H_R(f)|^2\, df}. \tag{3.14}$$

The definition and calculation method given by Equation (3.13) were used in early studies on stereophonic reproduction (Mertens, 1965), and occasionally in studies on HRTFs (Sontacchi et al., 2002).

Definition 3: ITD_g refers to interaural group delay difference, which is defined as the interaural difference in the slope of HRTF phase divided by (-2π):

$$ITD_g(\theta,\phi,f) = -\frac{1}{2\pi} \left(\frac{d\psi_L}{df} - \frac{d\psi_R}{df} \right). \tag{3.15}$$

ITD_g is generally a function of source direction and frequency. In some specific cases, however, it is approximately independent of frequency, with $ITD_g \approx ITD_p$.

ITD$_g$ is closely related to ITD$_{ave}$. Frequency-domain binaural pressures, P_L and P_R, generated by a point source, can be calculated by Equation (1.20). If H_L and H_R [in Equation (3.14)] are replaced by P_L and P_R, then the term in the square bracket in the numerator can be drawn out from the integral for narrowband signals, whose bandwidths are minute, compared with their central frequencies. Under this condition, ITD$_g$ ≈ ITD$_{ave}$ for narrowband signals. Accordingly, Equation (3.13) can be regarded as a weighted mean of interaural group delay difference across frequencies, for wideband or impulse signals. ITD$_g$ is also closely related to ITD$_e$, the interaural envelope delay difference. For narrowband signals with bandwidths that are significantly smaller than their central frequencies, ITD$_g$ ≈ ITD$_e$, which can be verified using the sinusoidal modulation signal in Equation (1.11). Given its unique characters, ITD$_g$ is often used in analyzing HRTFs and binaural signals (Cooper, 1987).

Definition 4: ITD$_{corre}$ is obtained from the cross-correlation calculation of left and right ear HRIRs. The normalized interaural cross-correlation function of an HRIR pair is calculated by:

$$\Phi_{LR}(\tau) = \frac{\int_{-\infty}^{+\infty} h_L(t+\tau) h_R(t)\,dt}{\left\{\left[\int_{-\infty}^{+\infty} h_L^2(t)\,dt\right]\left[\int_{-\infty}^{+\infty} h_R^2(t)\,dt\right]\right\}^{1/2}}. \tag{3.16}$$

By definition, $0 \le |\Phi_{LR}(\tau)| \le 1$. Time τ_{max} that maximizes $\Phi_{LR}(\tau)$ in the range $|\tau| \le 1$ ms is defined as ITD$_{corre}$; that is,

$$ITD_{corre}(\theta,\phi) = \tau_{max}. \tag{3.17}$$

If the HRIR is normalized as:

$$h_{A,L}(t) = \frac{h_L(t)}{\left[\int_{-\infty}^{+\infty} h_L^2(t)\,dt\right]^{1/2}} \qquad h_{A,R}(t) = \frac{h_R(t)}{\left[\int_{-\infty}^{+\infty} h_R^2(t)\,dt\right]^{1/2}},$$

then the difference between $h_{A,L}(t)$ and $h_{A,R}(t)$ can be evaluated by the integral:

$$I(\tau) = \int_{-\infty}^{+\infty} [h_{A,L}(t+\tau) - h_{A,R}(t)]^2\,dt$$

If the shape of $h_{A,L}(t)$ is similar to that of $h_{A,R}(t)$, but with a time delay $\tau = \tau_{max}$, integral $I(\tau)$ reaches its minimum when $\tau = \tau_{max}$. The maximization of $\Phi_{LR}(\tau)$ is mathematically equivalent to the minimization of $I(\tau_{max})$, with $\Phi_{LR}(\tau) = 1$ corresponding to $I(\tau_{max}) = 0$. Therefore, ITD$_{corre}$ is evaluated on the basis of the similarities between the left and right ear HRIRs.

Some notes on ITD$_{corre}$ are worth mentioning:

1. The result of Equation (3.17) can be regarded as a type of weighted-mean ITD across an entire frequency range, and is therefore independent of frequency. According to actual conditions, however, some preprocessing methods—including low-pass, high-pass, and band-pass filtering—are often applied to $h_L(t)$ and $h_R(t)$ prior to calculating cross-correlation. Thus, the resultant ITD is a type of weighted mean across the concerned frequency range.

2. In practice, only the discrete time samples of HRIRs, $h_L(n)$ and $h_R(n)$, are available. Hence, the integral on continuous time t in Equation (3.16) is replaced by the summation on discrete time n. To improve the calculation resolution of the resultant ITD, $h_L(n)$ and $h_R(n)$ are often upsampled before calculation. For example, the time resolution of HRIRs is only 23 μs at a sampling frequency of 44.1 kHz. The time resolution in calculation can be improved to 2.3 μs by 10-time HRIR upsampling. A similar process can be applied to Equations (3.7) and (3.12).

3. With Fourier transform, Equation (3.16) can be converted into the frequency domain as follows:

$$\Phi_{LR}(\tau) = \frac{\int\limits_{-\infty}^{+\infty} H_L(f)\, H_R^*(f) \exp(j 2\pi f\tau)\, df}{\left\{\left[\int\limits_{-\infty}^{+\infty} |H_L(f)|^2\, df\right]\left[\int\limits_{-\infty}^{+\infty} |H_R(f)|^2\, df\right]\right\}^{1/2}}, \tag{3.18}$$

where "*" denotes complex conjugation. Some researchers, such as Gardner (1997) and Riederer (1998a), used ITD_{corre}.

Definition 5: The onset-delay-based ITD, ITD_{lead}, is obtained from the difference between the onset delays of left and right ear HRIRs. Figure 3.1 shows some zero-valued samples at the initial HRIR, which correspond to the propagation delay (which may be truncated by a time window) from source to ear. Therefore, the time difference between the rising onset of left ear HRIR and right ear HRIR can be defined as ITD. However, some small, non-zero, artifacts may exist in the initial samples of a measured HRIR because of slight measurement errors and non-causality in signal processing. To reduce influence from the artifacts, the onset delays can be evaluated by leading-edge detection, that is, detecting instants, $t_{L,\eta}$ and $t_{R,\eta}$, at which the HRIRs first reach a certain percentage η (5%–20%) of maximum peak amplitudes. Then, ITD_{lead} is calculated by:

$$ITD_{lead}(\theta,\phi) = t_{L,\eta} - t_{R,\eta}. \tag{3.19}$$

Similar to the time resolution of the ITD_{corre} calculation, that of the ITD_{lead} calculation can be improved by preprocessing through HRIR upsampling.

A variety of threshold η for leading-edge detection were chosen in different studies. To reasonably select η, Zhong (2007b) evaluated the stability of the ITD_{lead} calculated from the HRTFs of a rigid sphere, a KEMAR artificial head, and human subjects with η = 5%, 10%, 15%, and 20%. The results indicate that reasonably selecting η depends on certain aspects, including the type of object tested and method for obtaining HRIRs; the stability of the resultant ITD_{lead} is determined primarily by the onset gradient of HRIRs. The diffraction caused by complicated and asymmetric anatomical structures decreases the onset gradient of contralateral HRIRs, thereby causing instability in leading-edge detection. This problem is prominent at lateral directions. Taking all aspects into account, Zhong suggested a value of η = 10% for practical calculation.

According to Equation (3.5), the HRTF phase comprises the minimum phase, all-pass phase, and linear phase. ITD_{lead} only approximately accounts for the interaural group delay difference (equal to phase delay difference) caused by the linear-phase component. Under the minimum-phase approximation of HRTFs, in which the all-pass phase component is negligible, ITD_{lead} is important in analyzing HRTFs and simplifying the signal processing of virtual source synthesis because the interaural time difference in this case is characterized only by the contributions of the linear-phase and minimum-phase components. The minimum-phase component is completely determined by HRTF magnitude according to Equation (3.3). Provided that HRTF magnitude and ITD_{lead} are known, HRTFs can therefore be recovered, and the corresponding signal processing of virtual source synthesis (Chapter 5) can be implemented. ITD_{lead} has been employed in a number of studies (Sandvad and Hammershøi, 1994; Møller et al., 1995b; Algazi et al., 2001b, 2001c). The calculation in Equation (1.9) only considers the contribution of linear phase to the ITD of a spherical head model.

Excluding the all-pass phase component from the overall ITD calculation may cause errors when the contribution of this component is non-negligible. Minnaar et al. (1999), however, found inconsistent results when they investigated the all-pass phase of the HRTFs of 40 subjects with 97 spatial directions per subject (Table 2.1, HRTF database no. 2). The authors found that below 1.5 kHz, the contribution of the all-pass phase component to interaural group delay difference is nearly independent of frequency. As shown in Equation (3.20), if the interaural group delay difference caused by the all-pass phase component is replaced by its value at 0 Hz, the error caused by approximation is less than 30 μ s and is inaudible (Plogsties et al., 2000a):

$$ITD_{all}(\theta,\phi,0) = \lim_{f \to 0}\left[-\frac{1}{2\pi}\left(\frac{d\psi_{all,L}}{df} - \frac{d\psi_{all,R}}{df}\right)\right]. \tag{3.20}$$

Here, $\psi_{all,L}$ and $\psi_{all,R}$ denote the all-pass phases of the left and right ear HRTFs, respectively.

If the contributions of linear and all-pass phases are taken into account, we have another definition of ITD:

Definition 6: ITD_{cess} is the interaural group delay difference of the excess phase component. This definition accounts for the entire excess phase component at 0 Hz [see Equation (3.1)]:

$$ITD_{cess}(\theta,\phi,0) = \lim_{f \to 0}\left[-\frac{1}{2\pi}\left(\frac{d\psi_{cess,L}}{df} - \frac{d\psi_{cess,R}}{df}\right)\right]$$
$$= ITD_{all}(\theta,\phi,0) + [T_L(\theta,\phi) - T_R(\theta,\phi)]. \tag{3.21}$$

The bracketed term in the second equality refers to the contribution of the linear-phase component and is approximately equivalent to the value in Equation (3.19). Minnaar et al. (2000) confirmed that virtual source localization can be improved by replacing the ITD_{lead} in Equation (3.19) with ITD_{cess}. They also outlined some other ITD calculation methods.

Definition 7: ITD_{fit} is obtained from linear curve fitting. First, the $\psi_{cess,L}(f)$ and $\psi_{cess,R}(f)$ of each pair of HRTFs are evaluated by Equation (3.1) as a function of frequency, and separately fitted by a linear curve over a certain frequency range. Then, the time delay for each ear can be obtained by multiplying the slope of the linear curve by $-1/(2\pi)$. Finally, the difference in the time delay between the two ears is regarded as ITD_{fit}.

ITD_{fit} is a type of group delay mean, whose value depends on the frequency range used in linear fitting. Under the minimum-phase approximation of HRTFs, ITD_{fit} is equivalent to ITD_{lead} in Definition 5. Below 1.5 kHz, ITD_{fit} is nearly equivalent to ITD_{cess} in Definition 6. In previous studies, a variety of frequency ranges was selected for linear curve fitting. A frequency range of 1–5 kHz was chosen by Jot et al. (1995), whereas 500 Hz–2 kHz was used by Huopaniemi and Smith (1999a). The ITDs for various HRTF data are comparable only when they are evaluated according to the same definitions and conditions. In practice, appropriate results can be obtained by adopting one or some of these methods, depending on the context.

3.2.2. Calculation Results for ITD

Figure 3.7 plots the variations in ITDs with horizontal azimuths of 0°–180°. The ITDs are calculated from MIT-KEMAR HRTFs with DB-61 small pinna (Table 2.1, database no. 3), and left-right symmetric HRTFs are assumed. The ITDs in Figure 3.7, including ITD_p at 0.35 and 2.0 kHz, ITD_{lead} with threshold $\eta = 10\%$, and ITD_{corre}, are evaluated by four different methods. Before ITD_{corre} is calculated, a pair of HRIRs is subjected to low-pass filtering below 2.0 kHz, to avoid the influence of resonance from the occluded-ear simulator (Figure 3.3).

The ITDs derived by different methods generally vary with azimuth θ in a similar manner. The ITDs are zero at azimuths of 0° and 180°, then gradually increase as the source deviates from the median line, maximizing at directions close to the lateral. However, some numerical differences exist among the ITDs at a given azimuth that are derived by different methods. Among four types of ITDs, ITD_{lead} is the smallest because leading-edge detection accounts only for the linear-phase component, regardless of the minimum-phase and all-pass phase components of HRTFs. The ITD_p at 0.35 kHz is the largest. With increasing frequency, the contributions of the minimum-phase and all-pass phase components decrease; ITD_p at 2.0 kHz is close to ITD_{corre} below 2.0 kHz, and is less than ITD_p at 0.35 kHz. The maximal ITDs that are associated with the corresponding azimuth are (778 μs, 95°), (716 μs, 90°), (710 μs, 90°), and (630 μs, 85°) for ITD_p at 0.35 kHz, ITD_p at 2.0 kHz, ITD_{corre} below 2.0 kHz, and ITD_{lead}, respectively. Except for ITD_{lead}, the other three ITDs smoothly vary with azimuth around 90°, suggesting that a small variation in ITD corresponds to a large change in azimuth. As will be discussed in Chapter 9, this correspondence is important for analyzing the instability of lateral virtual sound sources.

Figures 3.8(a) and 3.8(b) show the mean ITD_{corre} across 26 males and 26 females from the Chinese subject HRTF database, respectively (Xie et al., 2007a). The mean ITD_{corre} is plotted as a function of azimuth at various elevations. For simplicity, the subscript "*corre*" is excluded from the denotation of ITD. Because of the pinna effect, high-frequency HRTFs exhibit considerable intersubject differences. Moreover, ITD is a low-frequency localization cue. To obtain stable results in ITD calculation, an HRIR pair is subjected to low-pass filtering, with a cutoff frequency of 2.7 kHz prior to cross-correlation calculation.

In Figures 3.8(a) and 3.8(b), the mean ITDs of males and females vary with direction in a similar manner. At a given elevation, ϕ, the |ITD| for azimuth θ in the right hemisphere is almost identical to that for the mirror azimuth $(360° - \theta)$ in the left hemisphere. Particularly, ITD ≈ 0 at $\theta = 0°$ or 180° in the median plane. These results are due to the relatively long sound wavelength below a frequency of 2.7 kHz, so that the asymmetric details of anatomical structures impose minimal influence on ITDs. Furthermore, the effect

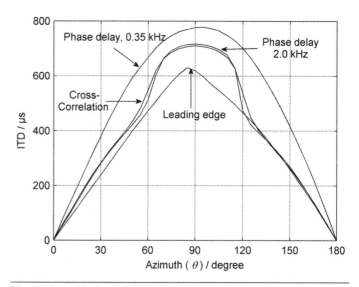

Figure 3.7 Horizontal ITDs of KEMAR evaluated by various methods

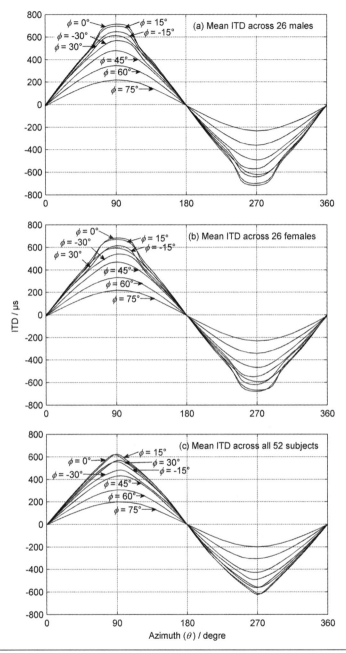

Figure 3.8 Mean ITDs across 52 Chinese subjects vary with azimuth and elevation: (a) Mean ITD$_{corre}$ across 26 males; (b) Mean ITD$_{corre}$ across 26 females; (c) Mean ITD$_{lead}$ across all 52 subjects

of the asymmetric details of anatomical structures on ITD, if any, can be effectively canceled by averaging over multiple subjects. As the source approaches lateral directions, the absolute value of ITD increases and reaches its maximum near $\theta = 90°$ and 270°. In comparison with the results for other elevations, the range of ITD variation is maximal at $\phi = 0°$. As ϕ deviates from the horizontal plane, the range of ITD variations decreases. These results qualitatively agree with those of existing studies, even though a variety of methods for evaluating ITDs were adopted (Møller et al., 1995b; Gardner, 1997; Riederer, 1998a; Algazi et al., 2001b). For comparison, Figure 3.8(c) shows the mean ITDs across all 52 Chinese subjects; these ITDs are calculated by leading-edge detection with $\eta = 10\%$ (Zhong and Xie, 2007a). Before the ITDs are calculated, the HRIRs are upsampled 20 times to improve the time resolution in the calculation to about 1 μs. The ITDs evaluated by the two methods in Figure 3.8 share common characteristics with those in Figure 3.7.

The maximal absolute value of ITD (denoted as $|ITD|_{max}$) and associated source direction (θ_{max}, ϕ_{max}) can be determined by analyzing the ITDs at 493 measured directions for each subject. Some results are worth noting. The cases when the ITDs that are evaluated by cross-correlation method are taken as examples.

For most of the subjects (47 out of 52), ϕ_{max} is 0° (in the horizontal plane) and θ_{max} is within the range 90° ± 10° (or 270° ± 10°). Moreover, θ_{max} is located at 90°–95° (or 265°–270°) for 39 of 47 subjects. A few subjects (5 of 52) exhibit a ϕ_{max} of 15° (that is, a little higher than the horizontal plane) and θ_{max} is located at 90°–95° (or 265°–270°). For these subjects, however, the ITDs near the lateral azimuth $\theta = 90°$ (or 270°) are almost identical at elevations $\phi = 0°$ and $\phi = 15°$, with a difference of less than 10 μs. Therefore, the maximal absolute value of ITD occurs near the lateral direction in the horizontal plane. Some abnormal observations are likely to be caused by anatomical structures or measurement errors. The mean θ_{max} across 52 subjects is 91.9°, with a standard deviation of 3.6°, or 267.9°, with a standard deviation of 3.9°. This finding slightly differs from previously published results, and will be further discussed in the conclusion of this section.

Table 3.1 lists the statistical results for the $|ITD|_{max}$ of males and females. The Analysis of Variance (ANOVA) results show that, at a significance level of 0.05 ($F = 43.6$, $p = 2.54 \times 10^{-8}$), the mean $|ITD|_{max}$ of males and females significantly differs, with the male mean $|ITD|_{max}$ greater than the female mean $|ITD|_{max}$. As confirmed in Chapter 7, this difference is caused by variances in anatomical dimensions. Figures 3.8(a) and 3.8(b) indicate that a small change in ITD corresponds to a large change in θ at the lateral direction. Hence, the difference in $|ITD|_{max}$ between males and females is significant enough to cause a perceivable directional distortion in lateral virtual sound sources. Statistical analyses and corresponding models should, therefore, be separately constructed for males and females—or, at least, an equal number of male and female subjects should be included. Otherwise, results may suffer from gender bias—a point underestimated in some studies.

As mentioned in Section 2.3, of all the public HRTF databases for human subjects, the Center for Image Processing and Integrated Computing (CIPIC) database (Table 2.1, no. 9) is a commonly used database with a high directional resolution and a sufficient number of subjects (27 males and 16 females). The results for our database are compared with those for the CIPIC database.

Algazi et al. (2001b, 2001c) calculated the statistical results for the maximal ITDs of the 43 subjects of the CIPIC database with a mean $|ITD|_{max} = 646$ μs and a standard deviation = 33 μs. Some notes for these statistical results are as follows:

1. The ITD was calculated by leading-edge detection with a threshold of 20%, prior to which the HRIRs were subjected to high-pass filtering at 1.5 kHz. Because ITD_{lead} includes only

Table 3.1 Statistical results for the $|ITD|_{max}$ of the Chinese subject HRTF database

Gender	Number of subjects	Maximum (μs)	Minimum (μs)	Mean (μs)	Standard deviation (μs)
Male	26	761	671	722	22
Female	26	734	659	686	17

the contribution of the linear-phase component, the resultant ITD is less than that evaluated by other methods.

2. An interaural-polar coordinate was employed for the CIPIC database, and the spatial resolution in the lateral direction is insufficient. In particular, the HRTFs at the left and right source directions were not measured. Because the maximal ITD often occurs at the left and right directions, the calculated maximal ITDs may be inaccurate.

3. The gender difference between males and females was not addressed because of the unequal number of male and female subjects.

For reliable comparison, the maximal ITDs of the CIPIC database are recalculated by cross-correlation method (identical to that used in the calculation for Chinese subjects). First, the ITDs at all directions are calculated from the measured HRTFs of each subject, and the maximal absolute value of ITD is then determined as the lower bound of the $|ITD|_{max}$ of the subject (the actual $|ITD|_{max}$ never falls below this value, if the ITDs at horizontal lateral azimuths are available in the calculation). Table 3.2 lists the statistical results for males and females. The ANOVA results confirm that, at a significance level of 0.05 ($F = 17.9$, $p = 1.26 \times 10^{-4}$), the mean lower bound of the $|ITD|_{max}$ of males and females significantly differs, confirming the importance of considering statistical results for males and females.

Compared with the findings presented in Table 3.1, the ANOVA results for males and females confirm that, at a significance level of 0.05, the mean $|ITD|_{max}$ of the Chinese subjects significantly differs from that derived for the CIPIC database ($F = 15.2$, $p = 2.88 \times 10^{-4}$ for males; $F = 20.4$, $p = 5.46 \times 10^{-5}$ for females), with the Chinese mean $|ITD|_{max}$ lower than the CIPIC mean $|ITD|_{max}$. This difference is caused by the statistical variance in anatomical dimensions across different populations. Some Asian subjects were included in the CIPIC database. When statistical results are taken completely from Western subjects, the difference is even larger. Therefore, the results presented here fully demonstrate the necessity of establishing an HRTF database for Chinese subjects; directly employing data collected from Western subjects in binaural research on Chinese subjects are problematic.

A different coordinate was adopted in the CIPIC database, and the measured data do not allow exact evaluation of the direction (θ_{max}, ϕ_{max}) associated with the maximum $|ITD|$. Thus, the results for the database are difficult to compare with ours. The calculation for another HRTF database (Table 2.1, no. 5) indicates that the maximal absolute value of ITD occurs at an azimuth of θ_{max} = 110°–120° or 250°–260° in the horizontal plane (Riederer, 1998a). Some authors attribute this result to the backward-offset ear locations from two sides of the head. Our results, however, show that θ_{max} is near 90° or 270°. The discrepancy is caused by the different methods used for ITD calculation. The full-bandwidth HRTFs that include the pinna effect were used by Riederer in the cross-correlation calculation of ITD, whereas the present work uses low-pass filtered HRIRs with a cutoff frequency of 2.7 kHz. Our choice was prompted by our desire to alleviate the uncertainty factor introduced by pinnae. Conversely, the backward-offset ear locations appear to cause the maximal ITD_{corre} to occur at a forward azimuth (Section 4.1.3).

Table 3.2 Statistical results for the $|ITD|_{max}$ of the CIPIC HRTF database

Gender	Number of subjects	Maximum (µs)	Minimum (µs)	Mean (µs)	Standard deviation (µs)
Male	27	≥814	≥706	≥748	27
Female	16	≥758	≥681	≥714	24

3.3. Interaural Level Difference Analysis

As stated in Section 1.4, ILD is an important localization cue above 1.5 kHz. According to Equations (1.12) and (1.20), ILD is defined as:

$$ILD(r,\theta,\phi,f) = 20\log_{10}\left|\frac{P_R(r,\theta,\phi,f)}{P_L(r,\theta,\phi,f)}\right| = 20\log_{10}\left|\frac{H_R(r,\theta,\phi,f)}{H_L(r,\theta,\phi,f)}\right| \ (dB). \tag{3.22}$$

Hence, ILD is a multivariate function of source position (including direction and distance) and frequency. Far-field ILD is approximately independent of source distance.

Figure 3.9 shows ILD varying with horizontal azimuth at several different frequencies. This ILD is calculated using the MIT-KEMAR HRTFs associated with the DB-061 small pinna (i.e., Table 2.1, database no. 3). The comparison of Figures 3.9 and 1.9 shows a similar variation trend for both head models. If the head radius a of KEMAR is 0.08 m, which is slightly lower than the standard head radius of 0.0875 m for a spherical head, then 0.35 kHz corresponds to $ka = 0.5$. Figure 3.9 illustrates that the ILD at low-frequency 0.35 kHz is small and changes smoothly with azimuth. At the near-lateral direction $\theta = 90°$, the low-frequency ILD is about 4 dB, a result attributed to the finite source distance (1.4 m) in the KEMAR HRTF measurements. This distance yields an incident wave that is not a perfect plane wave. Except for the slight difference at low frequency, the ILD of KEMAR is qualitatively consistent with that of a spherical head model. ILD increases with frequency. At frequencies 1.4 and 2.8 kHz, the maximal ILD does not occur at $\theta = 90°$ because of the inherent head shadow effect. Assuming that the head is approximated as a sphere, multi-path diffracted waves that correspond to a lateral incidence from $\theta = 90°$ are summed in-phase at the contralateral ear because these waves propagate at the exact same path lengths. This identicalness leads to an increase, or bright spot, in pressure at the contralateral ear and a consequent reduction in ILD. Even for a real head, the bright spot at the contralateral pressure remains. With increasing frequency, the ILD of KEMAR varies in a complex manner with source direction and frequency. At 5.6 kHz (nearly $ka = 8$ with $a = 0.08$ m), the ILD of KEMAR considerably deviates from that of the spherical head in Figure 1.9. This deviation is attributed to the imperfect spherical shape of the KEMAR head, along with diffractions from the pinnae.

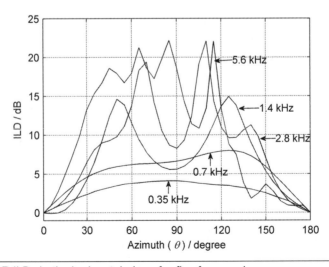

Figure 3.9 KEMAR ILDs in the horizontal plane for five frequencies

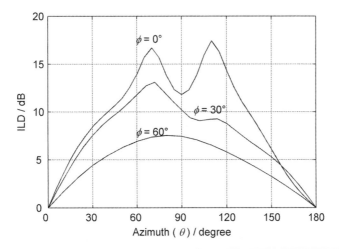

Figure 3.10 Wideband ILDs of KEMAR at three elevations

Regardless of distance, the far-field ILD is a multivariate function of frequency, azimuth, and elevation, making it a complex ILD. Some researchers calculated the mean ILD across a certain frequency range $f_L \le f \le f_H$, with the form:

$$ILD(\theta,\phi) = 10 \log_{10} \left[\frac{\int_{f_L}^{f_H} |H_R(\theta,\phi,f)|^2 \, df}{\int_{f_L}^{f_H} |H_L(\theta,\phi,f)|^2 \, df} \right]. \quad (3.23)$$

Wideband ILD can be obtained if the integral is implemented over the entire audio frequency range (Gardner, 1997). Although the wideband ILD may not be directly used as a localization cue, it is helpful in analyzing HRTF features. Figure 3.10 plots the wideband ILD of KEMAR, varying with azimuth at three elevations. The variations in wideband ILD with azimuth are less than that in high-frequency narrow-band ILD (Figure 3.9); the variation range of ILD maximizes in the horizontal plane, and decreases when the sound source deviates from the horizontal plane. A front-back asymmetry in the wideband ILD is also observed.

Using the concept of auditory filter introduced in Section 1.2.2, the ILD within each critical bandwidth (CB) or equivalent rectangular bandwidth (ERB) can be calculated as a function of the central frequency of the auditory filter, and as a function of source direction. This calculation method is similar to the auditory signal processing procedure for localization information. Because of the differences in anatomical structures and dimensions (particularly of pinnae), individual ILDs rapidly change with frequency and in a unique manner at frequencies above 5–6 kHz. At a given frequency, considerable deviations in ILD among individuals are expected. Therefore, calculating the mean ILD across individuals at each specified frequency is an unsuitable approach. Instead, HRTFs should be somewhat smoothed along frequencies prior to calculating ILDs. Here, calculating ILDs within each CB or ERB represents a certain type of HRTF smoothing.

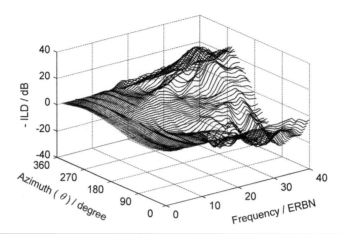

Figure 3.11 Mean ILD across 52 human subjects in the horizontal plane (ordinate specified as –ILD for clarity)

For a given subject and source direction, the energy of the HRTFs in each ERB is first evaluated by integration over each ERB. Then, the ILD in each ERB can be calculated as a function of source direction and central frequency in ERBN scale [Equation (1.7)]:

$$ILD(\theta,\phi,ERBN) = 10\log_{10} \frac{\int_{ERB} |H_R(\theta,\phi,f)|^2 \, df}{\int_{ERB} |H_L(\theta,\phi,f)|^2 \, df} \quad (dB). \qquad (3.24)$$

Finally, the mean ILD (i.e., ILD_{ave}) is obtained by averaging across subjects.

Figure 3.11 shows the ILD_{ave} in the horizontal plane, $\phi = 0°$, as a function of azimuth and frequency (in ERBN units). This ILD is calculated from the average of the 52 Chinese subjects. In the front and rear ($\theta = 0°$ and $180°$), the absolute magnitude of ILD_{ave}, $|ILD_{ave}|$, is close to 0 dB (≤ 1.0 dB) for $ERBN \leq 31$ (≤ 6.2 kHz). For $ERBN \geq 32$, $|ILD_{ave}|$ slightly increases but is less than 4.5 dB. In nature, for each subject, HRTFs are not perfectly left-right symmetric, and the ILD generated by the source in the front and rear is not exactly zero. However, the effects of left-right asymmetry can be effectively eliminated by averaging over all 52 subjects, so that the ILD_{ave} is close to 0 dB. An increase in $|ILD_{ave}|$ at high frequencies is caused by asymmetric anatomical structures or measurement errors. Similar results can be observed at other directions in the median plane.

When the sound source in the horizontal plane deviates from the front or rear, the $-ILD_{ave}$ demonstrates a complex dependence on frequency. Moreover, the $|ILD_{ave}|$ increases when the sound source approaches the lateral direction because of the head shadow effect. However, at low frequencies with $ERBN \leq 8$ ($f \leq 0.3$ kHz), $|ILD_{ave}|$ is less than 4.7 dB.

3.4. Spectral Features of HRTFs

3.4.1. Pinna-related Spectral Notches

The discussion in Section 1.4.4 indicated that the spectral features of pressure caused by pinnae reflections and diffractions are vital localization cues above 5–6 kHz. These spectral features can be analyzed through HRTFs (Figures 3.3 and 3.4). Although numerous studies on this issue have been conducted, completely consistent results have not been achieved because pinna-related spectral features are complicated and individualized.

As stated in Section 1.4.4, the first (lowest) frequency notch caused by the pinna (called the pinna notch) is the most compelling pinna-related spectral feature. The elevation dependence of the central frequency of the pinna notch (hereafter pinna notch frequency) is regarded as an important vertical localization cue. Shaw and Teranishi (1968) first reported the elevation dependence of pinna notch frequency. Similar results were also found by follow-up studies on artificial heads and human subjects (Hebrank and Wright, 1974; Butler and Belendiuk, 1977; Lopez-Poveda and Meddis, 1996; Kulkarni, 1997; Gardner, 1997). The general description accepted by many researchers is that the pinna notch frequency changes from 5 or 6 kHz to 12 or 13 kHz, when elevation varies from −40° to 60°.

Figure 3.12 shows the HRTF magnitude spectra of Chinese subject number 25 in the median plane with $\theta = 0°$ and elevation $\phi = -30°, 0°$, and $30°$ (Xie et al., 2007a). The pinna notch at 6–9 kHz is observed in the spectra. The pinna notch frequency increases with elevation and is roughly equal for both ears, with a difference of less than 0.3 kHz. For the right ear, the pinna notch frequency at $\phi = -30°, 0°$, and $30°$ are 6.5, 8.1, and 8.8 kHz, respectively. At high elevations with $\phi \geq 60°$, the pinna notch gradually vanishes. The observations depicted in Figure 3.12 agree with the results of these studies.

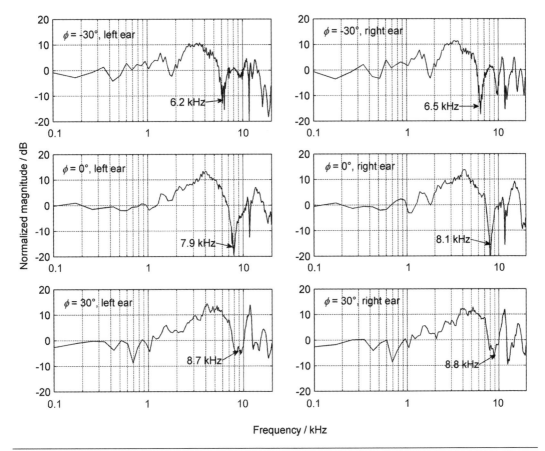

Figure 3.12 HRTF magnitude spectra for subject no. 25 at elevations −30°, 0°, 30° in the median plane

Considerable intersubject differences in spectral features are observed in the Chinese subject HRTF database. The results for some subjects are similar to those in Figure 3.12, but with quantitative differences in central frequency and pinna notch depth. Moreover, the pinna notch frequencies of two ears differ for some subjects. For subject number 4, for example, the pinna notch frequencies of the left and right ears at ($\phi = 0°$, $\theta = 0°$) are 8.6 and 9.3 kHz, respectively. This difference is caused by left-right asymmetry in the details of pinna anatomy. For subjects similar to this case, a left-right symmetric model of pinnae is invalid. For a few subjects, the pinna notches are shallow, or even completely disappearing. Figure 3.13 presents the HRTF magnitude spectra of subject number 29 at ($\phi = 0°$, $\theta = 0°$). A significant deviation between the left and right ear HRTF magnitude spectra is observed above 5–6 kHz. In particular, the pinna notch of the right ear diminishes and is replaced by a wideband roll-off between 6 and 12 kHz. This phenomenon is caused by the difference in pinna shape. If the pinna fails to focus the reflected sounds to the entrance of the ear canal, out-of-phase interference between the direct and reflected sounds diminishes.

No identical spectral features of HRTFs have been observed for any two or more subjects from the total 52, nor is an explicit rule found. Algazi et al. (2001b) reported the statistical results for 54 human subjects (43 subjects from the CIPIC database and 11 from another source), and indicated that the mean pinna notch frequency at ($\phi = 0°$, $\theta = 0°$) is 7.6 kHz, with a standard deviation of 1050 Hz. Given such a large deviation, the mean pinna notch frequency is less statistically significant.

These findings indicate that the pinna notch frequency may be an important localization cue for most individuals. However, pinna-related spectral features vary considerably among individuals. More efforts are required to further understand how the auditory system uses the individualized information introduced by pinnae for localization. Individualized pinna-related spectral features also explain the deterioration or individual dependence of perceived performance in a virtual auditory display (VAD) when nonindividualized HRTFs are used. The pinna notch frequencies denoted in Figure 3.12 are directly evaluated from a discrete Fourier transform-based analysis of HRTFs. To improve accuracy in extracting the central frequencies of pinna-related notches in HRTF magnitude spectra, Raykar et al. (2005) proposed an approach based on the combination of some signal processing techniques, including a linear prediction model, windowed autocorrelation functions, group-delay function, and all-pole modeling. Figure 3.14 shows the right ear HRTF magnitude spectra in the median plane for subject number 10 from the CIPIC database. In Figure 3.14, the abscissa pertains to the polar elevation angles in the interaural polar coordinate system described in Figure 1.3, and the ordinate is frequency. In Figure 3.14, the HRTF magnitude is represented in bright image in dB with bright images that correspond to large magnitudes, and vice

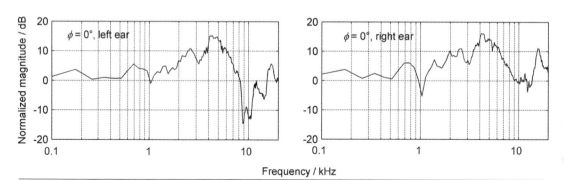

Figure 3.13 HRTF magnitude spectra for subject no. 29 at the front position with $\theta = 0°$ and $\phi = 0°$

versa. The images that correspond to the specified abscissa represent the HRTF magnitude spectra at a given elevation. The dark stripe in the figure represents the variations in the central frequencies of several pinna spectral notches with elevation; the dotted line represents the notches determined by Raykar's approach. The central frequency of the first pinna notch increases with rising elevation, consistent with the observations depicted in Figure 3.12.

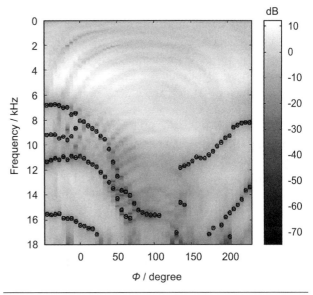

Figure 3.14 HRTF magnitude spectra in the median plane with pinna notch marked (from Raykar et al., 2005, by permission of J. Acoust. Soc. Am.)

3.4.2. Torso-related Spectral Cues

As mentioned in Section 1.4.4, the change in ipsilateral pressure spectra below 3 kHz caused by the scattering and reflection of the torso/shoulder is a potential vertical localization cue (Algazi et al., 2001a). Figure 3.15 is the bright image representation of ipsilateral HRTF magnitudes, as well as the corresponding HRIRs for KEMAR without pinna, at $\Theta = 25°$ (Algazi et al., 1999). The interaural pole coordinate system specified in Figure 1.3 is adopted. The out-of-phase interference between direct and torso-reflected sounds at the entrance to the ear canal results in notches in HRTF magnitude spectra. These notches are observed as a series of arc-shaped dark stripes in Figure 3.15(a). The central frequency of a notch varies with elevation Φ and extends to below 1 kHz. The lowest notch frequency occurs at about $\Phi = 80°$ (near the top), corresponding to a maximal path difference between direct and torso-reflected sounds. In Figure 3.15(b), the roughly horizontal bright line that represents the onset of HRIRs reveals that the arrival time of direct sound is approximately independent of Φ in the cone of confusion, with Θ constant (Section 1.4.3). The two V-shaped lines located in the frontal range,

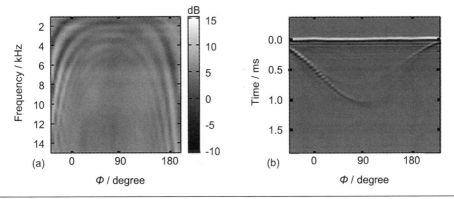

Figure 3.15 Bright image representation of ipsilateral HRTF magnitudes and corresponding HRIRs for KEMAR without pinnae at $\Theta = 25°$ (from Algazi et al., 2001d, by permission of IEEE): (a) HRTF; (b) HRIR

with $\Phi < 90°$, and in the rear range, with $\Phi > 90°$, represent the reflections of different torso parts. At $\Phi = 90°$, the arrival time difference between the direct and torso-reflected sounds is on the order of 0.8 ms, which corresponds to a path difference in the order of 0.27 m.

Because people frequently change clothes, different sound absorption and shoulder reflection, or scattering, occur. However, human sound localization is seldom influenced by clothes. To explain this phenomenon, Zhong and Xie (2006a) analyzed the HRTFs of human subjects and KEMAR, with and without pinnae, and then explored the comprehensive influence of clothes and pinnae on shoulder reflections and HRTFs. The results suggest that shoulder reflections are nearly irrelevant to clothes at frequencies below 3 kHz. Above 5 kHz, different clothes affect high-frequency shoulder reflections, as well as HRTF magnitude spectra when pinnae are removed. This influence is effectively concealed in the presence of pinnae, enabling the shoulder reflections below 3 kHz to function as robust localization cues. Determining to what extent torso- or shoulder-related reflections influence localization necessitates further investigation by psychoacoustic experiments.

3.5. Spatial Symmetry in HRTFs

In studies related to the analysis and application of HRTFs, left-right symmetric HRTF models are often adopted for simplicity. The head and torso, for example, are often approximated as spatially symmetric shapes to simplify HRTF calculation (see Chapter 4), resulting in symmetric HRTFs. However, symmetric models are merely approximations. In nature, the human head is not a perfect sphere, and pinnae further destroy the front-back symmetry of anatomical structures. Moreover, even two pinnae are not absolutely left-right symmetric for some individuals. Given that HRTFs reflect the overall acoustic filtering effect of anatomical structures, including the head, torso, and pinnae, the spatial symmetry of HRTFs is closely related to that of anatomical structures. Using physical measurements, Searle et al. (1975) confirmed the left-right asymmetry in HRTFs in the median plane, and claimed that the elevation-related left-right HRTF asymmetry introduced by pinnae is an important localization cue in the median plane. This finding nonetheless remains disputed. Figure 3.9 shows that wideband ILD exhibits a front-back asymmetry, but, for some human subjects, high-frequency HRTFs are left-right asymmetric. This section analyzes the spatially symmetric features of measured HRTFs (Zhong and Xie, 2007c), the influence of the symmetry of anatomical structures on that of HRTFs, and the applicable frequency range of some symmetric models.

3.5.1. Front-back Symmetry

At a given elevation, $\phi = \phi_0$, far-field HRTFs are a function of azimuth, θ, and frequency, f. To quantify the front-back symmetry of HRTFs, a front-back asymmetry ratio, R_{FB}, at an azimuth, $\theta = \theta_0$, is defined as:

$$R_{FB}(\theta_0, f) = \frac{|H(\theta_0, f) - H(\theta_1, f)|^2}{2[|H(\theta_0, f)|^2 + |H(\theta_1, f)|^2]}, \qquad (3.25)$$

where θ_1 is a front-back mirror azimuth with respect to θ_0. R_{FB} reflects the energy of the difference between two HRTFs, which is normalized by the doubled value of total energy. By definition, $0 \leq R_{FB}(\theta_0, f) \leq 1$ with $R_{FB}(\theta_0, f) = 0$ representing a perfect front-back symmetry.

The horizontal HRTFs of KEMAR, with and without pinnae, along with those of a human subject (measured by our laboratory; see Section 2.3) are used to investigate the influence of pinnae on the front-back symmetry of HRTFs. The HRTFs of both ears can be used in analyses, but only the results for the left ear are presented here. Note that $\theta_1 = (540° - \theta_0)$, for sources ipsilateral to the left ear (in the left semicircular space), and $\theta_1 = (180° - \theta_0)$, for sound sources contralateral to the left ear (in the right semicircular space). Figure 3.16 shows the calculated $R_{FB}(\theta_0, f)$ at some azimuths, θ_0, where $\theta_0 = 360°, 345°, 330°, 315°, 300°$, and $285°$ represent the source directions ipsilateral to the left ear, and $\theta_0 = 0°, 15°, 30°, 45°, 60°$, and $75°$ represent the source directions contralateral to the left ear.

Figures 3.16(a) and 3.16(b) show the $R_{FB}(\theta_0, f)$ of the left ear of KEMAR without pinnae, with the sources in the left and right semicircular spaces, respectively. The R_{FB} of the ipsilateral source directions is generally less than that of the contralateral source directions. This result can be attributed to the path differences in diffraction or scattering for sources at different spatial directions. For ipsilateral directions, sound waves from a front left and a mirror back left source directly reach the left ear, so that they are less influenced by the front-back asymmetry in the head. For contralateral directions, sound waves from a front right and a mirror back right source have to propagate along different paths around the head before reaching the left ear. In this case, the front-back asymmetry in the head significantly affects diffracted waves and, therefore, HRTFs. Thus, the contralateral HRTFs effectively reflect the front-back asymmetry in the head. Moreover, the head shadow diminishes the signal-to-noise ratio in contralateral HRTF measurement, thereby reducing the resultant R_{FB} value and causing many small fluctuations [Figure 3.16(b)]. The R_{FB} also

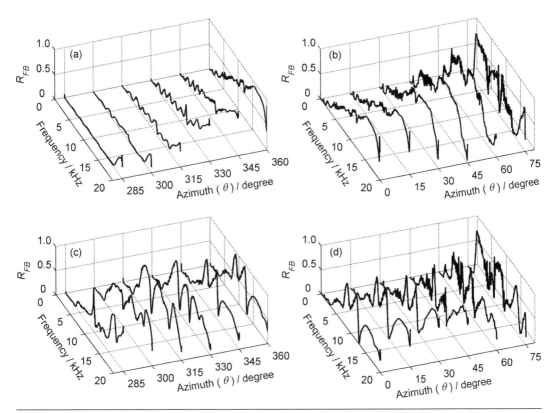

Figure 3.16 $R_{FB}(\theta_0, f)$ of KEMAR in the horizontal plane: (a) ipsilateral to the left ear without pinnae; (b) contralateral to the left ear without pinnae; (c) ipsilateral to the left ear with pinnae; (d) contralateral to the left ear with pinnae

increases when the ipsilateral source direction deviates from the lateral direction ($\theta_0 = 270°$). At 5.5 kHz, for example, the R_{FB} values are 0.03 for $\theta_0 = 285°$ and 0.24 for $\theta_0 = 360°$, a result also caused by the head's front-back asymmetry.

Figures 3.16(c) and 3.16(d) show the $R_{FB}(\theta_0, f)$ of the left ear of KEMAR, with pinnae, and with the source in the left and right semicircular space, respectively. The R_{FB} is less than 0.05 for all the azimuths, but only below 1.1 kHz (below 1.6 kHz, R_{FB} is less than 0.1). Under this condition, the HRTF is approximately front-back symmetric, and the front-back symmetric anatomical model yields a good approximation. The comparison of Figures 3.16(a) to 3.16(d) shows that the pinnae significantly destroy the front-back symmetry of the anatomical structure, and consequently generate a larger R_{FB} for KEMAR with pinnae than for KEMAR without pinnae. This phenomenon is more prominent at ipsilateral source directions, especially at $\theta_0 = 285°$. For example, the R_{FB} (285°, 5.5 kHz) is 0.03 in Figure 3(a), but 0.20 (285°, 5.5 kHz) in Figure 3(c). The existence of pinnae also eliminates the large difference between the ipsilateral and contralateral R_{FB}. Even with pinnae, however, fluctuations in the contralateral R_{FB} are still more obvious than those in the ipsilateral R_{FB} because of the head shadow [Figures 3.16(c) and 3.16(d)]. For a human subject, the tendencies of R_{FB} are similar to those of KEMAR [Figures 3.16(c) and 3.16(d)]. Because of complicated anatomical structures and relatively larger measurement errors, however, the R_{FB} of the human subject is slightly larger than that of KEMAR. For Chinese subject number 25, the frequency range within which the HRTFs are approximately front-back symmetric is $f \leq 0.7$ kHz under criterion $R_{FB} \leq 0.05$, or $f \leq 1.2$ kHz under criterion $R_{FB} \leq 0.1$.

3.5.2. Left-right Symmetry

Because the anatomical structures (especially the head) are approximately left-right symmetric in nature, HRTFs were often assumed to be left-right symmetric in many previous studies and expressed as follows:

$$H_R(\theta_0, f) = H_L(360° - \theta_0, f), \tag{3.26}$$

where H_L is the HRTF of the left ear and H_R is the HRTF of the right ear in a given elevation plane. To test this hypothesis, a criterion $R_{LR}(\theta_0, f)$ is defined to describe the left-right asymmetry in HRTFs at a given azimuth:

$$R_{LR}(\theta_0, f) = \frac{|H_R(\theta_0, f) - H_L(360° - \theta_0, f)|^2}{2[|H_R(\theta_0, f)|^2 + |H_L(360° - \theta_0, f)|^2]}. \tag{3.27}$$

Equation (3.27) is the normalized energy of the difference between the HRTFs of the right and left ears at a left-right mirror azimuth. The value of R_{LR} lies between [0, 1], with 0 representing perfect left-right symmetry.

Figure 3.17 shows the $R_{LR}(\theta_0, f)$ of two Chinese human subjects for azimuth θ_0 from 0° to 180°, at an interval of 5° in the horizontal plane. The R_{LR} does not exceed 0.05, thus, the HRTFs can be regarded as left-right symmetric at low and mid-frequency ranges ($f \leq 2.2$ kHz for subject no. 25, and $f \leq 5.1$ kHz for subject no. 10). As frequency increases, the R_{LR} increases and left-right asymmetry becomes obvious because of the left-right asymmetries in fine anatomical structures (such as head shape, hairstyle, and pinnae) and possible measurement errors at high frequencies. For subject number 25, the left-right asymmetry becomes prominent above 8.3 kHz, with the R_{LR} reaching a maximal value of 0.96. By contrast, the left-right symmetry for subject number 10 is relatively better, and the R_{LR} is generally less than that of subject number 25.

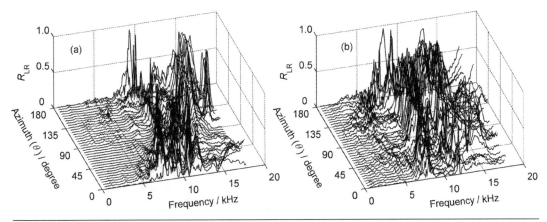

Figure 3.17 R_{LR} (θ_0, f) in the horizontal plane for (a) subject no. 10 and (b) subject no. 25

To investigate the left-right symmetry of the HRTFs of Chinese subjects, these analysis methods can be applied to each individual by turns. However, this procedure is tedious and complicated. Instead, the mean difference in the left-right energy level of the HRTFs of each subject is evaluated in each ERB (see Section 1.2.2) as follows:

$$\Delta(ERBN) = \frac{1}{72} \sum_{\theta=0°}^{355°} \left| 10 \log_{10} \frac{\int_{ERB} |H_R(\theta, 0°, f)|^2 \, df}{\int_{ERB} |H_L(360° - \theta, 0°, f)|^2 \, df} \right| \ (dB), \tag{3.28}$$

where the numerator is the HRTF energy in each ERB, with the central frequency in ERBN units for the right ear and source at horizontal azimuth θ; the denominator is that for the left ear and source in the mirror horizontal azimuth $(360° - \theta)$. The ratio of energy is transferred to the difference in energy level in decibel units. Summation and division by 72 represent the average over 72 azimuths in the horizontal plane. $\Delta = 0$ dB represents a perfect left-right symmetry; $\Delta > 0$ dB indicates left-right asymmetry. The symmetry in HRTF energy (magnitude) is included in the calculation of Δ, whereas that in phase is excluded. Because slight movements of subjects during measurement mainly cause errors in phase, the calculation of Equation (3.28) enables ruling out these phase errors.

Figure 3.18 shows the $\Delta_{ave}(ERBN)$, obtained by averaging individual $\Delta(ERBN)$ over 52 subjects. For each subject, $\Delta \geq 0$ dB. Hence, the contribution of each subject is not canceled out by averaging. At low frequencies $N \leq 16$ ($f \leq 1$ kHz), left-right symmetry is a good approximation, with $\Delta_{ave}(ERBN) \leq 1$ dB. As frequency increases, $\Delta_{ave}(ERBN)$ significantly rises, particularly for $N \geq 30$ ($f \geq 5.5$ kHz). This increase is caused by the contribution of left-right asymmetry in anatomical structures, which occur with increasing frequency. Measurement errors also increase with rising frequency.

In brief, the asymmetric details of human anatomical structures results in asymmetry in high-frequency HRTFs. According to the mean results for 52 subjects at 72 horizontal directions, the left-right symmetry in HRTFs is destroyed above 5.5 kHz, although the exact frequency and extent to which asymmetry occurs depends on the individual. For example, the HRTFs of subject number 25 appear to be left-right asymmetric above 2.2 kHz. Even for a given subject, the results vary at different azimuths. The conclusion drawn here can be used as guidance for determining the

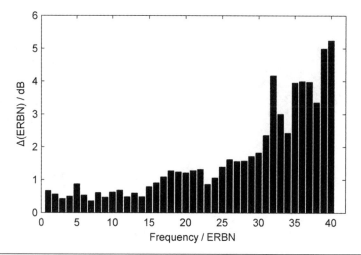

Figure 3.18 Mean Δ(*ERBN*) across 52 subjects

acceptable frequency range of left-right symmetric HRTF models. Naturally, the audibility of left-right asymmetry in HRTFs requires further psychoacoustic validation.

3.5.3. Symmetry of ITD

The symmetry of ITDs is closely related to that of HRTFs. As a dominant localization cue at low frequencies, ITD and its symmetry are meaningful in studies on binaural hearing. Moreover, left-right and front-back symmetric ITD models, such as the Woodworth model introduced in Section 1.4.1, is often employed in the signal processing of VADs, as well as in other practical applications. However, the assumption of symmetric ITD requires validation, which has been conducted using the Chinese subject HRTF database (Xie, 2006a).

The ITDs of $S = 52$ human subjects were evaluated by leading-edge detection with threshold $\eta = 10\%$. The mean value across $N = 52$ subjects is shown in Figure 3.8(c). Because the maximal variation range of ITD occurs in the horizontal plane, the horizontal ITD is chosen as an example in the subsequent analysis. In the horizontal plane, the ITD of subject s and azimuth θ is denoted by $ITD(\theta, s)$. The criterion for evaluating the left-right asymmetry in ITD is defined as:

$$\Delta_1(\theta) = \frac{1}{S} \sum_{s=1}^{S} |[|\, ITD(\theta,s)| - |ITD(360° - \theta,s)|]|| \qquad 0 \leq \theta \leq 180°, \tag{3.29}$$

or

$$\Delta_2(\theta) = \frac{1}{S} \sum_{s=1}^{S} [|\, ITD(\theta,s)| - |ITD(360° - \theta,s)|] \qquad 0 \leq \theta \leq 180°. \tag{3.30}$$

Although either Equations (3.29) or (3.30) can be used to assess the difference in ITD between azimuth θ and its left-right mirrored azimuth ($360° - \theta$), Equation (3.29) may be preferred because the absolute value of the difference in ITD is calculated prior to averaging across subjects, so that cancelling out the random left-right asymmetry in the ITD of each subject is avoided.

Similarly, the criterion for evaluating front-back asymmetry in ITD is defined as:

$$\Delta_3(\theta) = \frac{1}{S} \sum_{s=1}^{N} [ITD(\theta,s) - ITD(180° - \theta,s)] \qquad 0 \leq \theta \leq 90°, \tag{3.31}$$

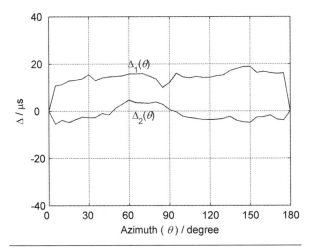

Figure 3.19 Calculated $\Delta_1(\theta)$ and $\Delta_2(\theta)$ for 52 subjects

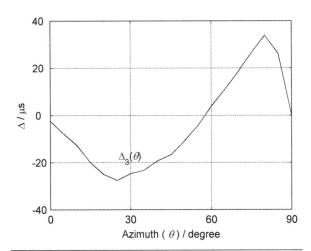

Figure 3.20 Calculated $\Delta_3(\theta)$ for 52 subjects

which reflects the difference in ITD at azimuth θ and its front-back mirror azimuth $(180° - \theta)$ in the right semicircular plane. The front-back difference in ITD in the left semicircular plane can be calculated in a similar manner.

The $\Delta_1(\theta)$ and $\Delta_2(\theta)$, calculated from 52 subjects and 72 horizontal azimuths, are plotted in Figure 3.19. The value of $\Delta_1(\theta)$ is not equal to zero but less than 19 µs. Therefore, the ITD of each subject may be slightly left-right asymmetric because of slight asymmetry in anatomical appearance as well as measurement error. Each subject exhibits a random and slight left-right asymmetry in ITD, suggesting that a relatively large |ITD| may occur at either of a pair of mirror azimuths. This randomness can be canceled out by the averaging in Equation (3.30), yielding a nearly zero $\Delta_2(\theta)$, with a value less than ± 6 µs. Figure 3.20 shows $\Delta_3(\theta)$. $\Delta_3(\theta) \neq 0$ indicates the front-back asymmetry in ITD, in which a positive value translates to a greater front ITD than rear ITD, and vice versa. A prominent ITD asymmetry occurs at azimuths 25° and 80° with $\Delta_3(25°) = -28$ µs and $\Delta_3(80°) = 34$ µs. This result is due to the overall influence of head shape, pinna position, and the front-back asymmetric nature of pinnae. Similar results are observed for the left semicircular plane with $270° \leq \theta \leq 360°$.

In conclusion, slight left-right asymmetry in ITD is observed for each subject; however, the left-right symmetric ITD model is valid for statistical averages. Subjected to front-back asymmetry in anatomical appearance, ITD is front-back asymmetric. Hence, a front-back symmetric ITD model causes certain deviations.

3.6. Near-field HRTFs and Distance Perception Cues

The previous sections analyze some issues related to far-field HRTFs, with source distance $r \geq 1.0$ m. Most of the measured HRTF data are far-field HRTFs. According to the definition of HRTFs, near-field HRTFs vary with distance. To explore the distance dependence of HRTFs and the associated perception cues, Brungart and Rabinowitz (1999a) measured the near-field HRTFs of KEMAR at a few source directions and distances. Figure 3.21 shows the left ear HRTF magnitudes at azimuths of 90° and 270° in the horizontal plane with $r = 0.125, 0.25, 0.50$, and 1.00 m.

The following tendencies were observed:

1. At a given source distance, the magnitude of the ipsilateral HRTF at $\theta = 270°$ demonstrates an overall increase above 3–4 kHz, intermingling some high-frequency notches and fluctuations caused by pinna-related diffraction and reflection. The magnitude of the contralateral HRTF at $\theta = 90°$ decreases above 3–4 kHz because of the head shadow effect. The spectral peak at 2–3 kHz, caused by the resonance of the occluded-ear simulator, is independent of source distance and direction. These observations are qualitatively consistent with the far-field results ($r = 1.4$ m) discussed in Section 3.1.2.

2. Within the near-field distance, $r < 1.0$ m, the ipsilateral HRTF magnitude increases with decreasing r when a direct propagation path from source to concerned ear exists; the contralateral HRTF magnitude decreases with r because of the enhancement of the head shadow when a direct propagation path is missing. The variations in HRTF magnitude with r increase the ILD associated with decreasing r. This phenomenon is particularly prominent at low frequencies. For example, the ILD at 100 Hz exceeds 20 dB with $r = 0.125$ m. This characteristic of near-field ILD differs from the case of far-field ILD.

3. When a direct propagation path from source to concerned ear exists, the increase in low-frequency magnitude with decreasing r is greater than that in high-frequency magnitude. This phenomenon relatively increases low-frequency magnitude and therefore causes a perceptible change in timbre.

4. The variations in HRTF magnitude spectrum with distance are more obvious in the range of $r < 0.5$ m than at 0.5 m $< r < 1.0$ m; the variations are negligible when $r > 1.0$ m. Thus, the HRTF with $r > 1.0 – 1.2$ m is often regarded as distance independent, and is referred to as a far-field HRTF in numerous studies. This issue is further discussed in Section 4.1.4.

The analysis of ITD shows that, although ITD increases with decreasing r, this tendency is less than 12% and is therefore not obvious, when compared with the ILD tendency. The ITD is evaluated

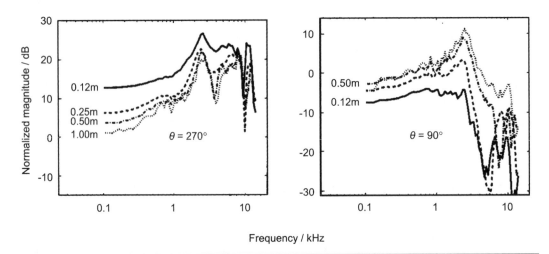

Figure 3.21 Left ear HRTF magnitudes of KEMAR at two horizontal azimuths and four source distances (from Brungart and Rabinowitz, 1999a, by permission of J. Acoust. Soc. Am.)

following the procedure indicated in Definition 7 (Section 3.2.1), but does consider the overall phase function of HRTFs from 100 Hz to 6.5 kHz.

These results are at least qualitatively applicable to lateral sources in the horizontal plane. The increase in ILD, as well as the relative proportion of low-frequency HRTF magnitude for $r < 1.0$ m, is especially compelling. Brungart et al. (1999a) argues that it is a cue for distance perception.

Aside from physical factors common to far-field HRTFs, such as the mirror reflection effect of the head on ipsilateral sources and the head shadow for contralateral sources, some other factors contribute to the near-field HRTF features just mentioned:

1. The head shadow is extraordinarily prominent in the near field. For a spherical head and infinitely distant source (or incident plane wave), a direct path from the source to the ipsilateral hemisphere exists, but the contralateral hemisphere lies in the shadowed region. For a nearby source, the shadowed region may extend to some part of the ipsilateral hemisphere because of the head's convex shape. As a result, the size of the head-shadowed region increases with decreasing source distance. However, the influence of the head shadow also depends on frequency. Thus, the behaviors of near-field HRTFs cannot be fully interpreted in terms of geometric acoustics. Some effects that are related to wave acoustics, such as scattering and diffraction, should be considered.

2. The head approximately acts as a mirror reflection plane for high-frequency sounds. In theory, pressure at the surface of an infinite rigid plane increases to 6 dB over that of free-field pressure. As source distance decreases, the frequency at which this mirror reflection occurs may decrease.

3. The ratio of the source distance relative to the left and right ears changes for a nearby source. As shown in Figure 1.7(a), the curved surface of the head is disregarded and the two ears are simplified as two points in free space separated by $2a$. Assuming that the source is positioned at $\theta = 90°$ with a distance r relative to the head center, the distances to the left and right ears are $r + a$ and $r - a$, respectively. According to the inverse relationship between radiant pressure and point source distance, the difference in left and right ear pressure levels (i.e., ILD) is:

$$\Delta P = 20 \log_{10} \frac{r+a}{r-a} \quad (dB).\tag{3.32}$$

For a distant source with $r \gg a$, $\Delta P \approx 0$ dB. The result for a nearby source is different: for example, $\Delta P = 15$ dB when $r = 0.125$ m and $a = 0.0875$ m. Hence, the change in the ratio of the source distance relative to left and right ears partially accounts for the observation that ILD increases with decreasing distance. Although this analysis uses a simplified head and model without pinnae, physical essence can be derived, at least qualitatively.

4. Figure 3.22 shows that, with the sound source in the horizontal plane approaching the head, source azimuth θ_1 relative to the ipsilateral ear changes—even if source azimuth θ relative to the head center is unaltered. This phenomenon is known as acoustic parallax (Suzuki *et al.*, 1998; Brungart, 1999d), which changes the effect of pinnae on incident sounds and therefore changes high-frequency HRTFs.

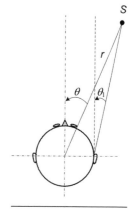

Figure 3.22 Acoustic parallax effect for nearby sound sources

3.7. HRTFs and Other Issues Related to Binaural Hearing

This book focuses on spatial hearing and VADs. Some basic physical features of HRTFs and various localization cues have been discussed in previous sections. These features and cues are also applicable to other issues related to binaural hearing.

First, the issue of directional loudness in the free field is presented. Psychoacoustic results show that, if the sound pressure at the head center with the head absent is fixed, then the free-field subjective loudness is related to frequency and direction of sound sources; for sounds with specific frequencies (such as pure tones or band-passed sounds, whose bandwidths are narrower than those of auditory filters), the subjective loudness depends merely on sound source direction. In the international standard related to loudness (ISO, 2003), the sound pressure level of a free progressive plane wave is measured at the position of the head center with the listener absent. The free-field equal-loudness level contours describe the subjective loudness (average over some populations) of a sound source directly in front of the listener.

Directional loudness is an issue related to binaural hearing. The direction-dependent sound pressures at the eardrums are formed by sounds that propagate from a sound source to both ears with scattering and diffractions of the head, torso, and pinnae. Binaural loudness can be calculated as follows. First, direction and frequency-dependent sound pressures at the two ears are evaluated through HRTFs; then, loudness models are used to convert the sound pressure to monaural specific loudness for each ear; finally, a combination of the monaural specific loudness from the two ears yields the binaural specific loudness as a function of sound source direction and frequency (Zacharov et al., 2001; Sivonen and Ellermeier, 2006, 2008; Moore et al., 1997, 2007). Sivonen and Ellermeier (2006) showed that most variations in subjective loudness with sound source direction can be analyzed in terms of HRTFs.

On the basis of Moore's loudness model, the American National Standard (ANSI S 3.4, 2007) specifies a procedure for calculating the monaural and binaural loudness of steady sounds from a directly frontal source, as perceived by listeners with normal hearing. The free field and diffuse field, to the eardrum, transfer functions for frontal incidence are obtained by averaging over a series of measurements and provided in each 1/3 octave band. The magnitudes of free-field transfer characteristic are listed in Table 3.3. The magnitude at an arbitrary frequency can be obtained by interpolation.

The second issue associated with binaural hearing is spatial unmasking. Masking refers to the phenomenon in which the detection threshold of a sound (target) may increase in the presence of another sound (masker). This increase often occurs when the target and masker are close to each other in the frequency spectrum. Correspondingly, the masking threshold is defined as the minimum audible sound pressure level of the target measured at the head center with the head absent. The masking threshold varies under different conditions. For given free-field frequency spectra of masker and target, as well as the sound pressure level of the masker measured at the head center position with the head absent, the masking threshold of a spatially separated masker and target is lower than that of a spatially coincident masker and target. The spatial separation of masker and target, in terms of direction and distance, decreases the masking threshold—a phenomenon often termed spatial unmasking.

Spatial unmasking can be partially interpreted by the source position tendency of HRTFs (Kopco and Shinn-Cunningham, 2003; Xie et al., 2006; Xie and Jin, 2008a, 2008b). The bandwidth of the masker is assumed less than that of an auditory filter, and the target is assumed a pure tone, whose frequency is within the bandwidth of the masker. At a specific frequency, when the positions of the

Table 3.3 Magnitudes of the free-field to eardrum transfer functions for frontal incidence, specified by (ANSI 3.4, 2007).

Frequency(Hz)	20	25	31.5	40	50	63	80	100
Magnitude(dB)	0.0	0.0	0.0	0.0	0.0	0.0	0.0	0.0
Frequency(Hz)	125	160	200	250	315	400	500	630
Magnitude(dB)	0.1	0.3	0.5	0.9	1.4	1.6	1.7	2.5
Frequency(Hz)	750	800	1000	1250	1500	1600	2000	2500
Magnitude(dB)	2.7	2.6	2.6	3.2	5.2	6.6	12.0	16.8
Frequency(Hz)	3000	3150	4000	5000	6000	6300	8000	9000
Magnitude(dB)	15.3	15.2	14.2	10.7	7.1	6.4	1.8	−0.9
Frequency(Hz)	10000	11200	12500	14000	15000	16000	20000	
Magnitude(dB)	−1.6	1.9	4.9	2.0	−2.0	2.5	2.5	

masker and target are spatially coincident, the diffractions that anatomical structures (such as the head) cause to the masker and target sound are the same. Accordingly, a certain target-to-masker sound pressure ratio (target-to-masker ratio) exists for each ear. When the masker and target are spatially separated, the diffractions imposed on their sounds differ from each other, thereby potentially increasing the target-to-masker ratio for one ear (called the better ear). The auditory system can detect the target with the information provided by the better ear, and therefore decrease the masking threshold. The target-to-masker ratio for each ear, which is related to the conditions of the target and masker (i.e., intensity, frequency, and spatial position), can be evaluated in terms of HRTFs. Details on quantitative analyses and experimental results can be found in the aforementioned literature. Some binaural cues (such as ITD) also change with spatial separation of target and masker, thereby contributing to spatial unmasking, apart from increasing the target-to-masker ratio for the better ear. Thus, spatial unmasking is a comprehensive effect that results from the combined binaural cues processed by the high-level neural system.

As with spatial unmasking of a pure tone by a narrowband masker, HRTFs are also helpful for partially interpreting the cocktail party effect (Shinn-Cunningham, 2001; Brungart and Simpson, 2002). The cocktail party effect, which is related to energy masking and information masking, essentially concerns the interfering source that masks the target source. When the interfering source and target are spatially separated in terms of direction and/or distance, the resultant binaural cues, other than the better ear effect, are likely to change. In this manner, the cocktail party effect presumably results from the combination of binaural cues processed by the high-level neural system of hearing.

In the research of noise immission, sound exposure was traditionally measured through the free-field or diffuse-field pressure level at the field point with the head absent, where the free-field pressure level is P_0 (converted to dB) given in Equation (1.15). The at-ear sound pressure level is also measured as sound exposure. For a sound source close to the ear, such as a headphone, the coupling between sound source and the external ear is non-negligible, and an artificial head or human subject is required for measuring binaural sound exposure. The International Organization for Standardization (ISO) has specified corresponding standards (ISO 11904-1, 2002; ISO 11904-2, 2004) for such measurements. For comparisons with traditional measurements, the at-ear sound pressure level can be converted to free-field or diffuse-field pressure level by HRTFs, or vice versa (Hammershøi and Møller, 2008). Although only some representative examples are presented here, they demonstrate the importance of HRTFs in binaural research, aside from their significance in sound localization.

3.8. Summary

Although far-field HRTFs are individual dependent, they share some common features. In the time domain, the main body of HRIRs lasts about 1.1–1.4 ms. HRTF magnitudes in the frequency domain appear to vary with sound source direction and frequency in a complex manner. For contralateral incidence, the head shadow acts as a low-pass filter and significantly decreases HRTF magnitudes above 4 kHz. For ipsilateral incidence, the head functions as a mirror reflection plane at high frequencies, leading to an increase in high-frequency magnitudes but not in low-frequency magnitudes. HRTFs can also be decomposed into minimum-phase, all-pass phase, and linear-phase functions. Studies indicate that the all-pass phase function is negligible under certain conditions, depending on source direction and frequency. Thus, the minimum-phase approximation of HRTFs is reasonable, although other studies provide different conclusions.

HRTFs can be used to analyze a variety of localization cues. Various interaural time difference (ITD) definitions and evaluation methods are presented in the literature. Among them, ITD_p derived from interaural phase delay difference can be directly applied as a localization cue at low frequencies. However, it is frequency dependent. Although the ITD obtained from other methods are not directly related to localization, they are closely associated with interaural phase delay difference and envelope delay difference. This relationship endows these ITD with practical significance. Because values of ITD may vary with different evaluation methods, comparing different values of ITD is meaningful only under the same conditions (e.g., using the same evaluation method).

The ITD evaluated by different methods and HRTFs exhibit similar variations with source direction. In the median plane, the ITD are nearly zero. As a sound source approaches lateral directions, the absolute value of ITD increases, reaching its maximum around $\theta = 90°$ and $\theta = 270°$ in the horizontal plane. The variation range of ITD maximizes in the horizontal plane, as compared with other elevations.

The maximal ITD derived from the Chinese subject HRTF database shows a statistically significant difference in gender, with the maximal ITD of males greater than that of females. Therefore, establishing separate models for males and females, or at least constructing a model based on equal numbers of male and female statistics, is beneficial. Compared with the maximal ITD derived using the CIPIC HRTF database, that derived for the Chinese subjects significantly differs from the maximal ITD of Western subjects. The Chinese maximal ITD is lower than that of the Western maximal ITD. Constructing an HRTF database for Chinese subjects is necessary; otherwise, biased results may be obtained when HRTF data on Western subjects are directly adopted in studies on Chinese subjects.

The analysis of interaural level difference (ILD) shows that low-frequency ILD is small and smoothly changes with source direction. As frequency increases, however, the relationship between ILD and direction, as well as the association between ILD and frequency, becomes complex. Calculating wideband interaural level difference can simplify analysis. The variation range of wideband ILD maximizes in the horizontal plane, as compared with other elevations. According to the concept of auditory filter, ILD can also be calculated in each equivalent rectangular bandwidth (ERB), which approximately matches the actual auditory localization process and is preferred for calculating the average interaural level difference across multiple subjects.

The spectral features of HRTFs, especially pinna-related spectral features, are highly individualized. An analysis of the Chinese subject HRTF database shows that the results for some subjects exhibit the general tendencies presented in existing studies; that is, the pinna notch frequency in the median plane rises as elevation increases, but gradually diminishes at elevations greater than 60°. However, the pinna notch frequencies of some subjects exhibit left-right asymmetry. For a few subjects, the notches are shallow, or even completely disappearing.

The symmetry in HRTFs is closely related to that in anatomical structures. The actual geometry of the head, ear location, and the presence of pinnae diminish the front-back symmetry of HRTFs; the left-right asymmetry in the fine structure of pinnae also degrades the left-right symmetry of HRTFs. Moreover, the occurring frequency and degree of asymmetry in HRTFs vary with individual. HRTFs are approximately front-back symmetric below 1–2 kHz. The average results for human subjects show that the left-right symmetry in HRTFs is destroyed at frequencies above 5.5 kHz. For each subject, the ITD may be slightly left-right asymmetric, and the left-right symmetric ITD model is statistically reasonable. The ITD is substantially front-back asymmetric because of the front-back asymmetry in anatomical structures. Thus, front-back symmetric interaural time difference models may cause errors.

When the source distance is greater than 1.0 m, HRTFs are almost independent of distance; these are called far-field HRTFs. At a distance of less than 1.0 m, however, the magnitude of HRTFs depends upon source distance. As source distance decreases, the ILD (especially at low frequencies) outside the median plane increases. Increase in ILD and the relative increase in low-frequency magnitude is regarded as an auditory distance perception cue in near fields.

In addition to being beneficial for source localization, HRTFs are helpful in research on directional loudness, spatial unmasking, cocktail party effect, sound exposure, and other issues related to binaural hearing.

4

Calculation of HRTFs

This chapter discusses the calculation methods for HRTFs, that is, the analytical and numerical methods for solving the acoustic scattering problem caused by the head, torso, and pinnae. The strict solutions of the HRTFs from a spherical head model, as well as the snowman model, are discussed. Numerical methods for calculating HRTFs from complicated head and pinnae models are also reviewed, with emphasis on the boundary element method.

Although measurement is the most accurate and common approach to obtaining head-related transfer functions (HRTFs), it necessitates complex apparatuses and is time-consuming. HRTFs also vary by the individual, and performing measurements for each individual is an impractical method. An alternative is calculation. From the mathematical and physical perspectives, calculating HRTFs pertains to solving the scattering problem caused by the head, torso, and pinnae; that is, solving the wave or Helmholtz equation, subject to certain boundary conditions. This chapter presents the calculation methods for HRTFs. The analytical solution to the HRTFs of a spherical head model (i.e., the simplest head model) is discussed in Section 4.1, including the cases of infinitely distant (incident plane wave) and close (near-field) source distances. A simplified head and torso model (the snowman model) that takes the torso into account is introduced in Section 4.2, and the corresponding solution is obtained by multi-scattering or multipole re-expansion. However, the strict solutions of HRTFs can be obtained only for rare, simplified models with symmetrical and regular head and torso shapes. To account for the complex geometry of a real head in HRTF calculation, researchers developed numerical methods, which are discussed in Section 4.3, with emphasis on the boundary element method (BEM).

4.1. Spherical Head Model for HRTF Calculation

4.1.1. Determining Far-field HRTFs and their Characteristics on the Basis of a Spherical Head Model

The spherical head model is the simplest model for HRTF calculation, based on which Lord Rayleigh first analyzed the sound pressures at two ears. As shown in Figure 4.1, the head is approximated as a rigid sphere, with radius a, and the pinnae and torso are omitted. For an incident plane wave or a sinusoidal point source that is infinitely distant from the sphere center, the pressure at the

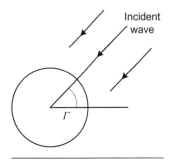

Incident wave

Γ

Figure 4.1 Scattering of incident plane wave by a sphere

sphere surface can be calculated using the analytical solutions to the scattering problem of spheres (Morse and Ingrad, 1968):

$$P(\Gamma, f) = -\frac{P_0}{(ka)^2} \sum_{l=0}^{\infty} \frac{(2l+1)j^{l+1}P_l(\cos\Gamma)}{dh_l(ka)/d(ka)}, \qquad (4.1)$$

where Γ is the angle between incident direction (the ray from sphere center to source) and the ray from the sphere center (origin) to the field (observed or received) point on the sphere surface; $k = 2\pi f/c$ is the wave number; P_0 denotes the pressure magnitude of the incident plane wave, that is, the free-field pressure magnitude at the origin with the head absent; $P_l(\cos\Gamma)$ is the Legendre polynomial of degree, l; $h_l(ka)$ represents the lth-order spherical Hankel function of the second kind.

The series in Equation (4.1) comprises infinite Legendre polynomial terms and the derivative of the spherical Hankel function. In practical calculation, the series is usually truncated, up to a certain order L, and the recursion relationships of special functions are applied. Cooper (1982) developed a calculation program and indicated that for three-decimal accuracy, the numbers of terms required are 3, 4, 6, 9, 19, and 43 for $ka = 0.1$, 0.3, 1.0, 3.0, 10, and 30, respectively. The truncation order problem will be comprehensively discussed in Section 6.3.4.

In Figure 4.2, the calculation results of Equation (4.1) are plotted as normalized logarithmic HRTF magnitudes $20\log_{10}|P(\Gamma, f)/P_0|$ against ka for several different angles, Γ. All the curves asymptotically approach 0 dB (free-field pressure) at low frequencies with $ka \leq 0.5$–0.6, which approximately corresponds to frequencies below 300–400 Hz for a standard head radius $a = 0.0875$ m. At ipsilateral incident directions, HRTF magnitude increases with frequency. For example, an approximately 6 dB increase in magnitude is observed at $\Gamma = 0°$ and $ka > 3$. As stated in Section 3.1.2, this increase is due to the approximate mirror-reflection effect of the head at high frequencies. Meanwhile, because of the head shadow effect, the high-frequency magnitude roughly rolls off at contralateral incident directions, with maximal attenuation occurring around $\Gamma = 150°$. Some interference ripples in the magnitude are also visible. By contrast, the magnitude at opposing

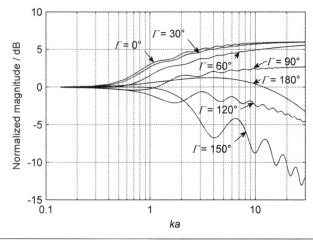

Figure 4.2 Normalized logarithmic HRTF magnitudes of the rigid spherical head model for an infinitely distant source

incident directions ($\Gamma = 180°$) is relatively flat because of the effect of bright spots that are caused by the in-phase interference of multipath diffracted waves around the spherical head (refer to Section 3.3).

We assume that two ears are located diametrically across the spherical head and Γ is the incident direction with respect to the right ear, as defined in Figure 1.3. The pressures at the two ears, $P_L = P(180° - \Gamma, f)$ and $P_R = P(\Gamma, f)$, are invariable on the cone of confusion with a given Γ (or equally, a given Θ in Figure 1.3) because of the symmetry in the spherical head model.

Transforming the positions into the default coordinate system adopted in this book is convenient. The locations of the two ears are specified by horizontal azimuths $\theta_L = 270°$ and $\theta_R = 90°$. For an incident plane wave or infinitely distant source at azimuth θ in the horizontal plane, the angles Γ with respect to the left and right ears are respectively replaced by:

$$\Gamma_L = 90° + \theta \qquad \Gamma_R = 90° - \theta. \tag{4.2}$$

According to the definition from Equation (1.15), binaural HRTFs can be calculated by substituting Equation (4.2) into Equation (4.1) as follows:

$$\begin{aligned} H_L(\theta, f) &= -\frac{1}{(ka)^2} \sum_{l=0}^{\infty} \frac{(2l+1) j^{l+1} (-1)^l P_l(\sin\theta)}{dh_l(ka)/d(ka)} \\ H_R(\theta, f) &= -\frac{1}{(ka)^2} \sum_{l=0}^{\infty} \frac{(2l+1) j^{l+1} P_l(\sin\theta)}{dh_l(ka)/d(ka)} \end{aligned} \tag{4.3}$$

Figure 4.3 plots the binaural HRTF magnitudes for incident direction $\theta = 90°$ in the horizontal plane, with head radius $a = 0.0875$ m. Below 300 Hz, the magnitudes of both ears are approximately equal. This result holds for an incident plane wave or an infinitely distant source, but not for the finite source distance (1.4 m) shown in Figure 3.3. In contrast to Figure 3.3, Figure 4.3 does not exhibit corresponding peaks and notches because the pinnae and torso have been omitted in the spherical head model.

Figure 4.4 shows the time-domain head-related impulse responses (HRIRs) that correspond to the HRTFs in Figure 4.3 (at a sampling frequency of 44.1 kHz), in which a suitable delay is complemented to the HRTFs before inverse Fourier transform is applied. This complementarity

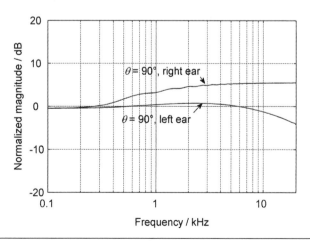

Figure 4.3 Normalized logarithmic HRTF magnitudes of the rigid spherical head for an infinitely distant source at horizontal azimuth $\theta = 90°$

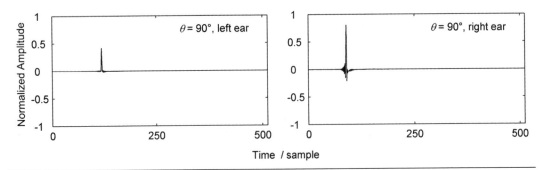

Figure 4.4 Time-domain HRIRs that correspond to Figure 4.3

guarantees the causality of HRIRs. Compared with the HRIRs of KEMAR (Figure 3.1), those of the rigid spherical head model are relatively simple and of a short duration. This result is also due to the omission of the pinnae and torso reflections and diffractions from the spherical head model.

The calculation also indicates that the HRTFs of a rigid spherical head are approximately minimum-phase, satisfying Equation (3.6) (Duda and Martens, 1998). This result holds for both far- and near-field HRTFs. Removing the reflections/diffractions/scattering caused by the pinnae and the torso eliminates certain factors that diminish the minimum-phase characteristics of HRTFs.

4.1.2. Analysis of Interaural Localization Cues

The interaural localization cues in the HRTFs of a rigid spherical head can be evaluated using Equation (4.3). Following the method provided by Kuhn (1977), we plot interaural phase delay difference (ITD$_p$) as a function of frequency for the incident plane wave from three different horizontal azimuths (Figure 4.5). Here, a standard head radius of $a = 0.0875$ m is chosen. The ITD$_p$ at a given incident direction is approximately independent of frequency below 0.4 kHz and above 3 kHz. In addition, the low-frequency ITD$_p$ is greater than its high-frequency counterpart. For example, the ITD$_p$ for $\theta = 90°$ incidence is 767 μs at 0.1 kHz and 676 μs at 3.0 kHz. Within a mid-frequency of 0.5–3 kHz, the ITD$_p$ is frequency-dependent and smoothly transitions from an asymptotic result at low frequency to that at high frequency. A similar behavior is also observed in the ITD$_p$ of KEMAR, as suggested by Kuhn (1977) and shown in Figure 3.7.

At low frequencies with $ka \ll 1$, only the terms $l = 0$ and 1 are retained, and higher order terms can be omitted in Equation (4.3). The HRTFs of the spherical head can then be approximated by:

$$H_L \approx 1 - j\frac{3}{2}ka\sin\theta \qquad H_R \approx 1 + j\frac{3}{2}ka\sin\theta. \qquad (4.4)$$

In this case, HRTF magnitude tends to unity (0 dB) and ITD$_p$ can be evaluated using the definition from Equation (3.10):

$$ITD_p = \frac{2\arctan\left(\frac{3}{2}ka\sin\theta\right)}{2\pi f} \approx \frac{3a}{c}\sin\theta. \qquad (4.5)$$

This low-frequency ITD$_p$ is 1.5 times that given by Equation (1.8). It is independent of frequency and only direction-dependent. Equation (1.8) is derived from a shadowless head model, while Equation (4.5) is the asymptotic result of wave acoustics. The head shadow effect extends the

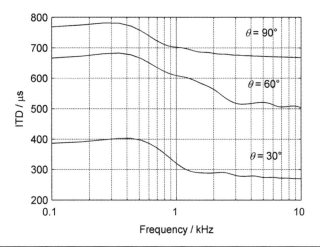

Figure 4.5 ITD$_p$ of the rigid spherical head model varying as a function of frequency for the incident plane wave at three different horizontal azimuths

effective path difference between two ears. Accordingly, an accurate low-frequency ITD$_p$ can be obtained by substituting a magnified head radius $a' = 1.5a$ into Equation (1.8).

The series in Equation (4.3) converges more slowly as ka increases. Kuhn (1977) expressed the waves on the surface of a rigid sphere as a series of creeping waves that travel around the sphere. At high frequencies with $ka \gg 1$, the higher order creeping wave attenuates more rapidly, such that only a few creeping waves significantly contribute to overall pressure. Then, ITD$_p$ can be approximately evaluated from the ratio of interaural path difference to the phase speed of the lower order creeping wave, leading to a result identical to Equation (1.9). The high-frequency asymptotic value of ITD$_p$ is therefore equivalent to the result of the Woodworth's formula derived from geometrical acoustics (ray tracing). It is also roughly equal to the interaural group delay difference (ITD$_g$) defined by Equation (3.15) and the interaural onset delay difference (ITD$_{lead}$) defined by Equation (3.19). These results further demonstrate the importance of Woodworth's formula. According to the analysis in Section 3.2.1, the HRTF phase consists of three components: the minimum-phase, all-pass phase, and linear-phase components. The results indicated here show that the high-frequency ITD$_p$ is mainly contributed to by the linear-phase component. The minimum- and all-pass phase components increase the low-frequency ITD$_p$, but their contributions to the high-frequency ITD$_p$ are negligible. The interaural level difference (ILD) for a spherical head, as a function of incident azimuth for different ka, can also be evaluated using Equation (4.3), as shown in Figure 1.9.

4.1.3. Influence of Ear Location

Some authors suggested that two ears are set backward about 10°, that is, at azimuths of 100° and 260°, rather than exactly opposite one another on a spherical head (Blauert, 1997; Cooper, 1987). Thus, Equation (4.2) becomes:

$$\Gamma_L = 100° + \theta \qquad \Gamma_R = 100° - \theta. \tag{4.6}$$

The HRTFs of a rigid spherical head with offset ear locations can also be calculated by substituting Equation (4.6) into Equation (4.1), after which interaural localization cues can be evaluated. Figure 4.6 shows the interaural cross-correlation delay difference (ITD$_{corre}$) defined by Equation

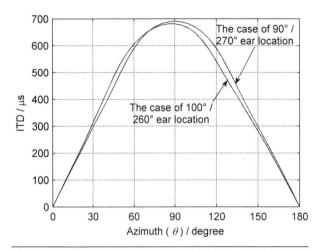

Figure 4.6 ITD_{corre} of the rigid spherical head model with two different ear locations

(3.16) as a function of incident azimuth θ in the horizontal plane with head radius $a = 0.0875$ m. Prior to calculating the interaural correlation to derive ITD_{corre}, the HRTFs are subjected to low-pass filtering at a cutoff frequency of 2.0 kHz. The ITD_{corre} values of the ears that are diametrically located (90°/270°) across the spherical head are also shown for comparison. As shown in the figure, the maximal ITD_{corre} occurs at azimuth $\theta = 90°$ with a value of 691 μs for the 90°/270° ear location. The ITD_{corre} is also front-back symmetric. For the 100°/260° ear location, the maximal ITD_{corre} occurs at azimuth $\theta = 85°$ with a value of 681 μs. Therefore, the backward ear location causes the maximal ITD_{corre} to occur at a forward azimuth with a slightly reduced value, and then results in the front-back asymmetry in ITD_{corre}. For example, the ITD_{corre} is 606 μs at $\theta = 60°$ and 541 μs at $\theta = 120°$ for the 100°/260° ear location. For human subjects, the anatomical structure of pinnae tends to yield the maximal ITD_{corre} at the azimuth backward from the two sides (while pinnae contribute less to HRTF magnitudes below 2–3 kHz, they do influence ITD to some extent). As a result, the effect of pinnae on interaural time difference (ITD) cancels out that caused by the offset ear location, so that the maximal ITD_{corre} occurs adjacent to an azimuth of 90° (or 270°), as indicated in Section 3.2.2.

Additionally, the variations in ILD with azimuth θ for a rigid spherical head model at the 100°/260° ear location is similar to those at the 90°/270° ear location (Figure 1.9). However, front-back asymmetry in ILD is observed. For example, the ILD at $ka = 8.0$ is 12.9 dB for $\theta = 60°$, and 11.1 dB for $\theta = 120°$, because of the front-back asymmetry in ear location.

4.1.4. Effect of Source Distance

This analysis is limited to an incident plane wave or an infinitely distant source. Duda and Martens (1998) extended their analysis to an arbitrary source distance and investigated the distance dependence of spherical head HRTFs. As an extension to Equation (4.1), the pressure at the surface of a rigid sphere generated by a sinusoidal point source at a distance r is given by (Morse and Ingrad, 1968):

$$P(r,\Gamma,f) = -j\frac{\rho_0 c Q_0}{4\pi a^2}\sum_{l=0}^{\infty}\frac{(2l+1)h_l(kr)P_l(\cos\Gamma)}{dh_l(ka)/d(ka)}, \tag{4.7}$$

where ρ_0 is the density of air; c is the speed of sound; Q_0 denotes the intensity of the point sound source; Γ represents the angle between the source direction and the ray from the sphere center (origin) to the field point on the sphere surface. The other symbols are identical to those used in Equation (4.1). With the head moved away, the pressure at the sphere center generated by the same sound source is:

$$P_0(r,f) = j\frac{k\rho_0 c Q_0}{4\pi r}\exp(-jkr). \tag{4.8}$$

According to the definition in Equation (1.15), the HRTFs at an arbitrary source distance can be calculated as the ratio of Equation (4.7) to Equation (4.8):

$$H(r,\Gamma,f)=-\frac{r\exp(jkr)}{ka^2}\sum_{l=0}^{\infty}\frac{(2l+1)h_l(kr)P_l(\cos\Gamma)}{dh_l(ka)/d(ka)}.\tag{4.9}$$

For convenience, source distance is normalized with head radius a (normalized distance ρ should not be confused with density ρ_0):

$$\rho=\frac{r}{a}.\tag{4.10}$$

Then, Equation (4.9) is rewritten as:

$$H(\rho,\Gamma,ka)=-\frac{\rho\exp(jka\rho)}{ka}\sum_{l=0}^{\infty}\frac{(2l+1)h_l(ka\rho)P_l(\cos\Gamma)}{dh_l(ka)/d(ka)}.\tag{4.11}$$

Figure 4.7 plots the magnitude $|H(\rho,\Gamma,ka)|$ in Equation (4.11) varying as a function of ka for different distances ρ and two angles $\Gamma=0°$ and 150°. The choice of $\Gamma=150°$ rather than $\Gamma=180°$ in the calculation is intended to avoid the occurrence of bright spots in the spherical head model (Section 3.3). Duda and Martens (1998) also provided the results for $ka\le100$. In the current work, we plot only the $ka\le30$ results, which approximately correspond to a frequency range below 18.6 kHz with head radius $a=0.0875$ m.

For both ipsilateral ($\Gamma=0°$) and contralateral ($\Gamma=150°$) source directions, the variations in magnitude $|H(\rho,\Gamma,ka)|$ with normalized source distance ρ and ka are qualitatively consistent with the results for KEMAR (Section 3.6). The primary tendencies are as follows:

1. The magnitude increases with a reduction in ρ when a direct path from the source to the concerned ear exists, and decreases with a reduction in ρ when the ear is located in the shadow of the head. The calculations reveal that, for head radius $a=0.0875$ m, magnitude visibly varies with source distance within a range of 1.0 m, especially within 0.5 m. Beyond the range of 1.0 m, the far-field HRTF magnitude is approximately independent of

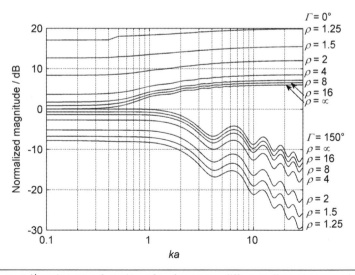

Figure 4.7 $|H(\rho,\Gamma,ka)|$ varies as a function of ka for some different distances ρ and angles $\Gamma=0°$ and 150°

distance. For example, the variations in magnitude are less than 1 dB as distance increases from 1.0–10 m.

2. As ρ decreases, the extent of increase in magnitude for the ipsilateral HRTF is relatively more prominent at low frequencies than at high frequencies. The high-frequency magnitude only rises 2 dB at $\Gamma = 0°$ and $\rho = 1.25$, in contrast to 6 dB for an infinitely distant source or incident plane wave.

3. Because ear canals and pinnae are disregarded in the spherical head model, the peak caused by ear canal resonance and the notches caused by pinna reflection and diffraction vanish from the resultant magnitudes.

The distance dependence of ILD can also be estimated using Equation (4.11). Assuming that two ears are located at $\theta = 100°$ and $260°$ in the horizontal plane, ILD can be calculated from the HRTF magnitudes evaluated by substituting Equation (4.6) into Equation (4.11). Figure 4.8 shows the variations in ILD with source azimuth θ in the horizontal plane for $ka = 0.5$ and several different source distances. The default coordinate system of this book is used in the figure. The ILD approaches 0 dB (less than 1 dB) for $\rho = \infty$. As ρ decreases, the ILD at lateral directions increases. For example, the ILD exceeds 26 dB for $\rho = 1.25$ and $\theta = 100°$ in the horizontal plane. Note that $ka = 0.5$ corresponds to a low frequency of 309 Hz with $a = 0.0875$ m. As stated in Section 3.6, an increase in low-frequency ILD at a nearby source distance is considered as a distance localization cue. In addition, when kr is greater than 10–12, the ILD slowly varies with ρ. However, even for $\rho = 16$ that corresponds to $r = 1.4$ m with $a = 0.0875$ m, the maximal ILD near the lateral direction remains 2 dB. This result is due to finite source distance, as explained in Section 3.1.2.

The calculation also demonstrates that bringing the source close to the spherical head slightly increases ITD, which is evaluated from the interaural phase delay difference or onset delay difference of HRIRs. This finding is consistent with the results presented in Section 3.6.

4.1.5. Further Discussion on the Spherical Head Model

The spherical head model is a type of rare head model, whose analytical solution can be obtained by solving the wave or Helmholtz equation, subject to certain boundary conditions. Although the

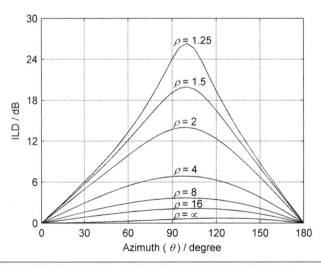

Figure 4.8 Variations in ILD with θ for $ka = 0.5$ and seven source distances

discussions thus far demonstrate that the HRTFs of a spherical head model exhibit some basic features of human HRTFs, the spherical head model is characterized by certain limitations.

First, the spherical head HRTFs are front-back symmetric for the 90°/270° ear location. According to the discussion in Section 3.5.1, however, the human HRTFs are roughly front-back symmetric only below a frequency of 1–2 kHz. A spherical head model with offset ear locations can, to a certain extent, account for the front-back asymmetric nature of human HRTFs. Moreover, evaluations show that the ear locations in the improved spherical head model are important (Zhong and Xie, 2007c). An appropriate ear location extends the applicable frequency range of the model to some extent, whereas inappropriate ear locations cause the opposite. As will be shown in Chapter 7, some individual discrepancies in ear locations have been observed in the measured anatomical parameters of Chinese subjects. Thus, carefully choosing ear locations is necessary for calculating HRTFs when using spherical head models.

Second, a real human head is not a perfect sphere. The scattering and diffraction by pinnae and the torso should also be taken into account. As frequency increases, the effects of head shape and pinnae become obvious. As stated in Chapter 3, pinna diffraction clearly distinguishes human HRTFs from those of spherical head models above 5–6 kHz. Moreover, the scattering caused by the torso may begin to affect HRTFs at frequencies below 1–2 kHz.

Overall, even if the effect of the torso is negligible, the spherical head model provides roughly reasonable results up to $ka \leq 5$; these results correspond to about 3 kHz with $a = 0.0875$ m (Cooper, 1987). A spheroidal/ellipsoidal head model, which is closer to the shape of a human head than a sphere, can yield more accurate HRTFs in calculations. Sugiyama et al. (1995) calculated the HRTF of a spheroidal head model. Jo et al. (2008) extended the calculation to near-field HRTFs. Thus far, pinnae remain disregarded in spheroidal/ellipsoidal models, and no details about the strict solution for ellipsoidal head HRTFs have been reported in the literature.

A completely rigid spherical head surface is assumed in the discussions just given. This hypothesis is approximately valid, if the acoustic impedance on the head surface is substantially greater than that of air. Treeby et al. (2007a) investigated the effect of a non-rigid and locally reacting surface on spherical head HRTFs. The results suggest that a uniformly distributed impedance boundary condition results in some changes in horizontal localization cues, ILD and ITD, especially for lateral source directions. Using a spherical head model with a hemispherical cover of hair and two diametrically located ears, Treeby et al. (2007b) further analyzed the contribution of hair to HRTFs. The results show that, unlike the rigid spherical head model, their model exhibits changes to both the ipsilateral and contralateral HRTFs at lateral source directions. These changes are introduced by hair, particularly degrade the front-back symmetry in HRTFs adjacent to the bright spot, and introduce asymmetric perturbations to ITD and ILD in the order of 25 μs and 4 dB (up to a frequency of 3 kHz), respectively.

Application of the spherical head model helps to correct the low-frequency response of measured HRTFs. As stated in Section 2.2.5, constrained by the low-frequency performance of a sound source and anechoic chamber, measured HRTFs are less accurate at low frequencies (i.e., below 100–200 Hz). Given that an accurate low-frequency HRTF response is important for a practical virtual auditory display (VAD), correcting the low-frequency response of measured HRTFs is necessary. By contrast, the HRTFs calculated for a rigid spherical head model are accurate at low frequencies, and are therefore suitable for correcting measured HRTFs. As indicated by Figures 4.2 and 4.7, as well as Equation (4.5), both far- and near-field HRTFs exhibit a nearly invariable magnitude and linear phase against frequency for $ka \leq 0.5$–0.6 (or below 300–400 Hz). Thus, provided that the HRTFs above 400 Hz are accurately measured, the inaccurate low-frequency HRTFs can be corrected in a straightforward manner through the calculation of results. That is, the low-frequency magnitude of measured HRTFs is flattened, and a linear phase is complemented (Xie, 2005a, 2009).

4.2. Snowman Model for HRTF Calculation

4.2.1. Basic Concept of the Snowman Model

To investigate the effect of the torso on HRTFs, Algazi et al. (2002a) proposed a simplified head-and-torso model, called the snowman model, for HRTF calculation. The model consists of a spherical head atop a spherical torso. Gumerov et al. (2002a, 2002b) applied multi-scattering or multipole re-expansion to calculate the HRTFs of the snowman model.

As shown in Figure 4.9, the head and torso are approximated by sphere A (SA) with radius a_A and sphere B (SB) with radius a_B, respectively. At field point r', we solve the wave or Helmholtz equation to identify pressure $P(r', r, f)$, which is generated by a sinusoidal point source located at r with intensity Q_0. According to acoustic theory (Morse and Ingrad, 1968), $P(r', r, f)$ satisfies the following equation:

$$\nabla'^2 P(r',r,f) + k^2 P(r',r,f) = -jk\rho_0 c Q_0 \delta(r'-r), \qquad (4.12)$$

where k is the wave number; ρ_0 is the density of air and c is the speed of sound; $\delta(r' - r)$ denotes the Dirac delta function; ∇'^2 is Laplacian with respect to field point r'. The Sommerfeld radiation condition requires:

$$\lim_{r'\to\infty} r' \left[\frac{\partial P(r',r,f)}{\partial r'} + jkP(r',r,f) \right] = 0, \qquad (4.13)$$

where $r' = |r'|$.

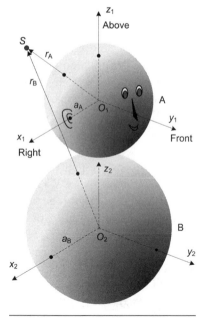

Figure 4.9 Snowman model and coordinate

Only the pressures at the two field points that specify two ears on the head surface are of interest in HRTF calculation. Calculating the pressure at either ear is adequate because of the symmetry in the snowman model. According to the acoustic principle of reciprocity, exchanging the positions of the source and field point yields identical pressures, that is, $P(r', r, f) = P(r, r', f)$. In HRTF calculation, therefore, the pressure at the ear can be equivalently obtained by calculating the pressure $P(r, r', f)$ at an arbitrary spatial position r, while the source is located on head surface r'. $P(r, r', f)$ satisfies Equations (4.14) and (4.15), which are obtained by exchanging the variables r' and r in Equations (4.12) and (4.13):

$$\nabla^2 P(r,r',f) + k^2 P(r,r',f) = -jk\rho_0 c Q_0 \delta(r-r'), \quad (4.14)$$

$$\lim_{r\to\infty} r \left[\frac{\partial P(r,r',f)}{\partial r} + jkP(r,r',f) \right] = 0, \quad (4.15)$$

where ∇^2 is Laplacian with respect to field point r.

For simplicity, a rigid head and torso surface is assumed (a non-rigid and locally reacting surface can also be assumed).

Consequently, the normal derivative of $P(r, r', f)$ should be zero on surface SA of the spherical head and surface SB of the spherical torso:

$$\left.\frac{\partial P(r,r',f)}{\partial n}\right|_{SA} = 0, \tag{4.16}$$

$$\left.\frac{\partial P(r,r',f)}{\partial n}\right|_{SB} = 0. \tag{4.17}$$

Therefore, the HRTFs of the snowman model can be obtained by solving Equation (4.14), subject to these boundary conditions.

From a physical perspective, overall pressure $P(r, r', f)$ consists of three parts: the free-field pressure $P_0(r, r', f)$ generated by a point source, as well as the scattered pressure $P_A(r, r', f)$ and $P_B(r, r', f)$ caused by spherical head A and spherical torso B, respectively. That is:

$$P(r,r',f) = P_0(r,r',f) + P_A(r,r',f) + P_B(r,r',f). \tag{4.18}$$

Similarly to Equation (4.8), $P_0(r, r', f)$ can be written as:

$$P_0(r,r',f) = j\frac{k\rho_0 cQ_0}{4\pi|r-r'|}\exp(-jk|r-r'|). \tag{4.19}$$

Equations (4.18) to (4.19) are valid in arbitrary coordinates. With the local coordinate centered at spherical head A, position or vector $r = r_A$ is specified by (r_A, α_A, β_A), where $r_A = |r_A|$; $0° \le \alpha_A \le 180°$ and $0° \le \beta_A < 360°$ (or $-90° < \beta_A \le 270°$) represent the direction of vector r_A. In the interaural polar coordinate system shown in Figure 1.3, we have $\alpha_A = \Gamma_A$ and $\beta_A = \Phi_A$. In the default coordinate system shown in Figure 1.1, we have $\beta_A = \theta$ and $\alpha_A = 90° - \phi$. The mathematical derivations of the HRTF calculation in two coordinate systems are similar. Let Ω_A denote the direction (α_A, β_A), that is, $\Omega_A = (\alpha_A, \beta_A)$. The scattered pressure $P_A(r, r', f)$ caused by spherical head A can be expanded as a series of complex-value spherical harmonic (SH) functions $Y_{lm}(\Omega_A)$. (Two equivalent forms of SH functions exist; i.e., real-value and complex-value SH functions. The derivation here is based on the paper of Gumerov et al., 2002a; 2002b.) Equivalently, $P_A(r, r', f)$ can be expanded as a series of complex-value SH functions (Appendix A):

$$P_A(r,r',f) = P_A(r_A,r'_A,f) = \sum_{l=0}^{\infty}\sum_{m=-l}^{l} A_{lm}^A h_l(kr_A)Y_{lm}(\Omega_A), \tag{4.20}$$

where $h_l(kr_A)$ is the lth-order spherical Hankel function of the second kind; A_{lm}^A is a set of SH coefficients to be determined. Equation (4.20) satisfies the condition given by Equation (4.15).

Similarly, in the local coordinate centered at spherical torso B, position or vector $r = r_B$ is specified by (r_B, α_B, β_B), where $r_B = |r_B|$; $0° \le \alpha_B \le 180°$ and $0° \le \beta_B < 360°$ represent the direction of vector r_B. Let $\Omega_B = (\alpha_B, \beta_B)$. Then the scattered pressure $P_B(r, r', f)$ caused by spherical torso B can also be expanded as a series of complex-value SH functions $Y_{lm}(\Omega_B)$:

$$P_B(r,r',f) = P_B(r_B,r'_B,f) = \sum_{l'=0}^{\infty}\sum_{m=-l'}^{l'} A_{l'm'}^B h_{l'}(kr_B)Y_{l'm'}(\Omega_B), \tag{4.21}$$

where $A_{l'm'}^B$ is a set of SH coefficients to be determined.

Substituting Equations (4.19), (4.20), and (4.21) into Equation (4.18) while applying the boundary conditions in Equations (4.16) and (4.17), SH coefficients A_{lm}^A and, $A_{l'm'}^B$ and, subsequently, pressure $P(r, r', f)$, can be obtained by truncating the series up to a certain finite order. HRTFs can then be evaluated by the ratio of $P(\mathbf{r}, \mathbf{r}', f)$ to the free-field pressure given by Equation (4.19). This is the main procedure for calculating the HRTFs of the snowman model by multipole re-expansion. More details can be found in Appendix B.

4.2.2. Results for the HRTFs of the Snowman Model

Algazi et al. (2002a) calculated the HRTFs of a snowman model, whose head and torso were approximated by two tangential rigid spheres with radii a_A = 0.0875 m and a_B = 0.23 m, respectively. Figure 4.10 is a bright image representation of the far-field and right ear HRTF magnitudes of the snowman model in the lateral plane. For convenience, a new angle $-90° < \Phi_1 \le 270°$ is introduced to specify the source direction in the lateral plane, and Φ_1 = −90° (or 270°), 0°, 90°, and 180° represent the bottom, right, top, and left directions, respectively.

Three arc-shaped dark stripes at ipsilateral directions are distinctly visible in Figure 4.10. They represent the spectral notches in HRTF magnitude caused by the out-of-phase interference of direct sound and torso reflection. The lowest frequency notch occurs around 1 kHz at Φ_1 = 80°, corresponding to a maximal path difference between direct and torso-reflected sounds. As source elevation deviates from Φ_1 = 80°, the central frequency of notch increases. This result roughly agrees with those for KEMAR presented in Section 3.4.2. The spectral feature caused by torso (especially shoulder) reflection may be an elevation localization cue.

When the source is located below certain low elevations, torso reflections are replaced by a torso shadow. In this case, the source at ipsilateral directions is shadowed by the torso alone. By contrast, the source at contralateral directions is shadowed by both the torso and the head, leading to an obvious magnitude attenuation above 1 kHz, as well as deeper and more closely-spaced interference stripes within the range 195° to 250° (Figure 4.10). The lowest magnitude does not occur directly below Φ_1 = 270° (or −90°). Similar to the bright spot caused by a spherical head (Section 3.3), this phenomenon is attributed to in-phase interference due to the symmetric spherical torso. In practice, torso bright spots disappear because of the asymmetry in a real torso.

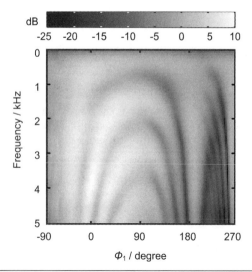

Figure 4.10 Far-field and right ear HRTF magnitudes of the snowman model in the lateral plane (from Algazi et al., 2002a, by permission of J. Acoust. Soc. Am.)

The snowman model is also one of the rare models, whose strict solutions for HRTFs can be obtained by solving the wave equation. When compared with the spherical head model, the torso introduces three new features in resultant HRTFs: torso reflections, torso shadows, and torso bright spots. Although solving the HRTFs of the snowman model is highly complicated (as discussed in Section 4.2.1 and Appendix B), it is still rough, and only valid below certain frequencies, because the effects of pinnae and head shape are disregarded.

As for spherical head HRTFs, HRTFs calculated for the snowman model can also be applied to correct the low-frequency response of measured HRTFs (Algazi et al., 2002b). Taking the effect of the torso into account, the snowman model provides more accurate low-frequency HRTFs than does the spherical head model. However, the low-frequency response of measured HRTFs can be reasonably corrected from the calculation results for the spherical head model because the effect of the torso on HRTFs is minimal below 400 Hz.

4.3. Numerical Methods for HRTF Calculation

The strict solutions of HRTFs can only be obtained for rare, simplified head/torso models, with the complex geometry of pinnae omitted. However, the geometry of a real human head and torso is complicated, and the contribution of pinnae to high-frequency HRTFs is significant. Therefore, the HRTFs calculated for a simplified head/torso model are only rough approximations of human HRTFs at low and mid-frequencies. To improve the accuracy of HRTF calculation, some numerical methods have been developed on the basis of the complex geometry of an artificial or human head model. Among these, the *boundary element method* (*BEM*) is the most commonly used in HRTF calculation (Kahana et al., 1998, 1999, 2007; Katz, 2001a, 2001b; Algazi et al., 2002a; Walsh et al., 2004; Otani and Ise, 2003, 2006). Differences exist among the methods used by various researchers, but the basic underlying ideas are similar.

4.3.1. Boundary Element Method for Acoustic Problems

BEM is an extensively used method for acoustic radiation and scattering problems (Ciskowski and Brebbia, 1991), in which the boundary problem of the wave or Helmholtz equation is first converted into a boundary surface integral. The boundary surface is then made discrete, into a mesh of elements, resulting in a set of linear algebra equations. Two BEM methods are used in acoustic computation: direct and indirect BEM (DBEM and IBEM). The former uses the pressure and normal velocity on the boundary surface as the unknowns, and results in an asymmetric matrix of equations; the latter uses the pressure jumps and normal velocity jumps on each side of the boundary as the unknowns, and results in a symmetric matrix of equations. DBEM and IBEM have been applied to HRTF calculation, depending on problem size and computational power. For brevity, only DBEM is addressed here. Some of the methods discussed below are applicable to IBEM.

Generally, for an acoustic radiation and scattering problem that regards a sinusoidal point source located at source point r with intensity Q_0, the pressure $P(r', r, f)$ at field point r' satisfies Equation (4.12). For an external problem, $P(r', r, f)$ should also satisfy Equation (4.13). Here, the acoustic principle of reciprocity is temporarily excluded from calculation. Within the volume V specified by a closed boundary surface S, the overall pressure in a field point is a combination of the free-field pressure generated by the point source and the pressure caused by boundary scattering and reflection. As a result, the solution to the boundary problem can be expressed as a Kirchhoff-Helmholtz integral equation (Morse and Ingrad, 1968):

$$C(\mathbf{r}')P(\mathbf{r}',\mathbf{r},f)= jk\rho_0 cQ_0 G(\mathbf{r}',\mathbf{r},f)$$
$$+ \iint_S \left[G(\mathbf{r}',\mathbf{r}'',f)\frac{\partial P(\mathbf{r}'',\mathbf{r},f)}{\partial n''} - P(\mathbf{r}'',\mathbf{r},f)\frac{\partial G(\mathbf{r}',\mathbf{r}'',f)}{\partial n''} \right] dS'', \tag{4.22}$$

$$C(\mathbf{r}')= \begin{cases} 1/2 & \mathbf{r}' \in S \\ 1 & \mathbf{r}' \in V, \\ 0 & other \end{cases} \tag{4.23}$$

where n'' is the outward normal direction; the integral is calculated over boundary surface S. $G(\mathbf{r}', \mathbf{r}, f)$ is the free-space Green function of a point source, which is defined as:

$$G(\mathbf{r}',\mathbf{r},f)=\frac{1}{4\pi|\mathbf{r}'-\mathbf{r}|}\exp(-jk|\mathbf{r}'-\mathbf{r}|). \tag{4.24}$$

Equation (4.22) indicates that, for a given source intensity and position, pressure at a field point $\mathbf{r}'$ within volume V is specified by the free-space Green function of a point source, along with the pressure and its normal derivative on boundary surface S. The normal derivative of pressure is zero on a rigid boundary surface. More generally, on a locally reacting boundary surface with a specific acoustic impedance of $Z(\mathbf{r}', f)$, or acoustic admittance $Y(\mathbf{r}', f) = 1/Z(\mathbf{r}', f)$, the boundary condition can be written as (Morse and Ingrad, 1968):

$$\left. \frac{\partial P(\mathbf{r},\mathbf{r},f)}{\partial n'} \right|_S =-j2\pi f\rho_0 v_{n'} =-jk\rho_0 cY(\mathbf{r},f)P(\mathbf{r},\mathbf{r},f). \tag{4.25}$$

Here, $v_{n'}$ is the normal velocity of media.

The basic idea of BEM is to make the boundary surface discrete, into a mesh of M small elements S_i, $i = 1,2 \dots M$. The central position of the ith element is specified by vector $\mathbf{r}'_i$, and the pressure and velocity at each element are supposed to be constants. When the pressure at each element is identified, the pressure at an arbitrary field point $\mathbf{r}'$ within volume V can be calculated. By substituting the boundary condition of Equation (4.25) into Equation (4.22), the surface integral in Equation (4.22) is approximated by a summation over M elements thus:

$$C(\mathbf{r}')P(\mathbf{r}',\mathbf{r},f)= jk\rho_0 cQ_0 G(\mathbf{r}',\mathbf{r},f)$$
$$-\sum_{i=1}^{M}\left[\iint_{Si} \frac{\partial G(\mathbf{r}',\mathbf{r}'',f)}{\partial n''}dS''+ jk\rho_0 cY(\mathbf{r}'_i,f)\iint_{Si} G(\mathbf{r}',\mathbf{r}'',f)dS'' \right]P(\mathbf{r}'_i,\mathbf{r},f). \tag{4.26}$$

At the M field points consistent with the centers of M elements on the boundary surface, we can apply position $\mathbf{r}' = \mathbf{r}'_q$ and $C(\mathbf{r}') = 1/2$, where $\mathbf{r}'_q$ with $q = 1,2 \dots M$ specifies the position of each element. Then, Equation (4.26) is converted into M algebra equations, which determine the pressures at M elements on the boundary surface:

$$\frac{1}{2}P(\mathbf{r}'_q,\mathbf{r},f)= jk\rho_0 cQ_0 G(\mathbf{r}'_q,\mathbf{r},f)-\sum_{i=1}^{M}[G^n_{qi}(f)+ jk\rho_0 cY(\mathbf{r}'_i,f)G_{qi}(f)]P(\mathbf{r}'_i,\mathbf{r},f)$$

$$G^n_{qi}(f)=\iint_{Si}\frac{\partial G(\mathbf{r}'_q,\mathbf{r}'',f)}{\partial n''}dS'' \qquad G_{qi}(f)=\iint_{Si}G(\mathbf{r}'_q,\mathbf{r}'',f)dS'' \tag{4.27}$$

$$q,i =1,2, \dots M.$$

Let $\boldsymbol{P}_M = [P(\boldsymbol{r'}_1, \boldsymbol{r}, f), P(\boldsymbol{r'}_2, \boldsymbol{r}, f), \dots P(\boldsymbol{r'}_M, \boldsymbol{r}, f)]^T$ represent an $M \times 1$ matrix (column vector) that comprises the pressures of M elements, where superscript T denotes the matrix transpose; $\boldsymbol{G}_0 = [G(\boldsymbol{r'}_1, \boldsymbol{r}, f), G(\boldsymbol{r'}_2, \boldsymbol{r}, f), \dots G(\boldsymbol{r'}_M, \boldsymbol{r}, f)]^T$ represents an $M \times 1$ matrix (column vector) that comprises the values of the free-space Green function at the position of M elements when a point source is located at $\boldsymbol{r}$; $[Y] = $ diagonal $[Y(\boldsymbol{r'}_1, f), Y(\boldsymbol{r'}_2, f), \dots Y(\boldsymbol{r'}_M, f)]$, which represents an $M \times M$ diagonal matrix that consists of the values of acoustic admittance $Y(\boldsymbol{r'}, f)$ at M discrete elements; $[G^n]$ and $[G]$ represent two $M \times M$ matrices with elements G^n_{qi} and G_{qi}, respectively; and $[I]$ represents an $M \times M$ identity matrix. With the aforementioned notations, Equation (4.27) can be written as a matrix equation:

$$\left\{ \frac{1}{2}[I] + [G^n] + jk\rho_0 c[G][Y] \right\} \boldsymbol{P}_M = jk\rho_0 cQ_0 \boldsymbol{G}_0. \tag{4.28}$$

For each specified frequency f, the pressure vector $\boldsymbol{P}_M$ of M elements can be solved from Equation (4.27) or Equation (4.28). Then, substituting the resultant $\boldsymbol{P}_M$ into Equation (4.26), the pressure $P(\boldsymbol{r'}, \boldsymbol{r}, f)$ at an arbitrary field point $\boldsymbol{r'}$ within volume V is found by:

$$C(\boldsymbol{r'})P(\boldsymbol{r'}, \boldsymbol{r}, f) = jk\rho_0 cQ_0 G(\boldsymbol{r'}, \boldsymbol{r}, f) - \left\{ \boldsymbol{G}^n_{r'} + jk\rho_0 c\boldsymbol{G}_{r'}[Y] \right\} \boldsymbol{P}_M \tag{4.29}$$

where $\boldsymbol{G}^n_{r'}$ and $\boldsymbol{G}_{r'}$ are $1 \times M$ row matrices (vectors), and their elements are respectively given by:

$$G^n_{r',i}(f) = \iint\limits_{Si} \frac{\partial G(\boldsymbol{r'}, \boldsymbol{r''}, f)}{\partial n''} dS'' \qquad G_{r',i}(f) = \iint\limits_{Si} G(\boldsymbol{r'}, \boldsymbol{r''}, f) dS'' \tag{4.30}$$

$$i = 1, 2 \dots M.$$

The first term on the right side of Equation (4.29) is the free-field pressure at field point $\boldsymbol{r'}$ generated by the point source. It is, therefore, relevant to source position $\boldsymbol{r}$. The second term on the right side is the pressure at field point $\boldsymbol{r'}$ caused by the scattering and reflections of the boundary surface. Vector $\boldsymbol{P}_M$ can be solved from Equation (4.28), which represents the pressures at M boundary elements and is relevant to source position $\boldsymbol{r}$. $\{\boldsymbol{G}^n_{r'} + jk\rho_0 c\boldsymbol{G}_{r'}[Y]\}$ can be regarded as the transfer matrix (or row vector) from M boundary elements to field point $\boldsymbol{r'}$, and is independent of source position $\boldsymbol{r}$. Therefore, the second term on the right side of Equation (4.29) indicates that the total scattered and reflected pressure at field point $\boldsymbol{r'}$ is a combination of those caused by M boundary elements. This idea is the basic principle that underlies the DBEM-based solution to the acoustic radiation and scattering problem.

4.3.2. Calculation of HRTFs by BEM

HRTF calculation is a typical exterior radiation and scattering problem of the wave equation, in which the source is located at the arbitrary spatial position $\boldsymbol{r}$ outside the human body, and the pressures at only two field points (two ears) are considered. In this case, one boundary surface S consists of the geometrical surface of the head, torso, and pinnae. Another boundary is infinitely distant, in which pressure satisfies Equation (4.13).

The geometrical surfaces of a human or artificial head (such as head and pinnae) are irregular. To accurately obtain the boundary surface, a laser three-dimensional (3-D) scanner [or other scanning device, such as a computed tomography scanner (CT) or nuclear magnetic resonance imaging (NMRI) unit] is often used to acquire data on human geometrical surfaces. A previously used laser 3-D scanner is equipped with a vertical linear transmitter and sensor array, so that geometrical surface data are acquired by rotating the array around a subject. Constrained by array size, however, the scanner only obtains geometrical surface data on the head and neck; those on the torso are often excluded. Given the difficulty in acquiring the geometrical surface data perpendicular to the array, the result is a cylindrical shell with both ends (the top of the head and bottom of the neck) open. The shell should be artificially closed. The ear canal is also excluded from the model because of the inaccessibility of a typical laser 3-D scanner, resulting in a blocked ear canal model for HRTF calculation. Some other methods should be supplemented to acquire geometrical data on the ear canal. Because a previously used 3-D laser scanner cannot accurately acquire both the fine structure and data on the back of pinnae, some other supplementary methods are required to compensate for this drawback. Certain current handheld laser scanners have exhibited improved performance, making these especially applicable to scanning pinnae. Figure 4.11(a) illustrates the geometrical model of a B & K 4128C artificial head obtained by Otani and Ise (2006), and Figure 4.11(b) shows the features of the external ears in the model. The back of the left ear is constructed from laser-scanned data and is, therefore, less accurate. The back of the right ear, on the other hand, is constructed from a set of supplementary micro CT-scanned data with high precision, so that the details behind the right ear are visible.

When the geometrical surface model is obtained, it is made discrete, into a mesh of M triangular elements. The maximal frequency to be analyzed depends upon element size, and the largest length of the elements should not exceed 1/4 to 1/6 of the shortest wavelength concerned. Figure 4.11(a) also shows a mesh of triangular elements. The largest length of the elements is 5.64 mm, which corresponds to a maximal frequency of 10 kHz (1/6 wavelength) or 15 kHz (1/4 wavelength). The torso has a simpler shape than does the head, but its larger surface prompts requirements for a considerably greater number of elements. Even if the torso is excluded from the model, the number of elements M required for head and neck geometry is already substantial. For example, 28,000 elements are included in Figure 4.11(a).

A problem with BEM-based HRTF calculation is its high computational cost. At each specified frequency, Equation (4.27) or Equation (4.28) is a set of M linear algebra equations. The computational cost for solving the equations rapidly increases with the number of M. To obtain HRTFs as

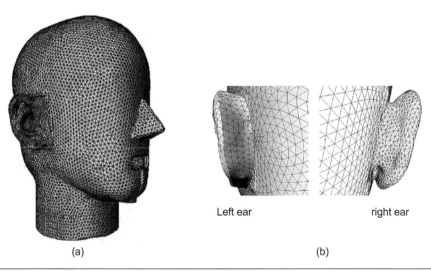

Left ear   right ear

(a)   (b)

Figure 4.11 Geometrical model of a B & K 4128C artificial head (from Otani and Ise, 2006, by permission of J. Acoust. Soc. Am.): (a) Entire model; (b) External ears

a function of frequency, the equations should be repeatedly solved for the surface pressure P_M at each discrete frequency. Because the column vector G_0 in Equation (4.28) is relevant to source position r, solution P_M depends on r. To obtain HRTFs at various source positions (spatially-sampled HRTFs), these calculations should be repeated for each source position r. Repeated calculations entail extensive computational time. For a maximal calculated frequency from several kHz to nearly 20 kHz with a frequency resolution of 100 Hz or finer, dozens to hundreds of hours are spent by a typical personal computer to calculate a set of HRTFs at various source directions by conventional BEM (depending on computational power, the number M of elements, frequency, and spatial resolution, etc.). High computational costs make calculations difficult.

In the conventional BEM-based HRTF calculation provided by Equations (4.28) and (4.29), considerable computational cost is involved in repeatedly solving Equation (4.28) for the surface pressure P_M at various source positions r. Although multi-source positions should be considered in HRTF calculation, only the two field points r' at the left and right ears are of interest. According to the acoustic principle of reciprocity, exchanging the source position and field point results in identical pressures. In HRTF calculation, therefore, source position can be fixed at the entrance to the ear canal and field points are selected at various spatial positions outside the body. Given that the matrix $\left\{\frac{1}{2}[I]+[G'']+ jk\rho_0 c[G][Y]\right\}$ on the left side of Equation (4.28) is independent of the field point, solving Equation (4.28) is required for only the source positions at the left and right ears. Furthermore, if left-right symmetry holds, the solution for the source located at either ear is sufficient. Eventually, the HRTFs for multiple source positions can be obtained by reusing the resultant surface pressure P_M in the calculation of Equation (4.29) with various field points. Thus, incorporating the acoustic principle of reciprocity into BEM-based HRTF calculation reduces computational cost, as confirmed by some researchers (Kahana, 1999; Katz, 2001a, 2001b).

Even if the acoustic principle of reciprocity is incorporated, the actual cost of BEM-based HRTF calculation remains high. Although the torso contributes to HRTFs and has a relatively simple geometrical shape, its large area requires a much higher number of elements, which in turn entails high computational cost. Constrained by computational power, the torso is often, if not always, excluded from actual calculation models. As a trade-off between accuracy and computational cost, the maximal frequency of HRTFs is also frequently limited to reduce the required number M of elements in reciprocal BEM calculation. Reciprocal BEM calculation also raises a new problem. That is, because source position is close to the boundary surface of the head, the accuracy of solving P_M from Equation (4.28) decreases.

To simultaneously reduce computational cost and avoid the potential error caused by reciprocal BEM calculation, Otani and Ise (2003) introduced a new algorithm for solving Equation (4.28). The $M \times M$ matrix $\left\{\frac{1}{2}[I]+[G'']+ jk\rho_0 c[G][Y]\right\}$ on the left side of Equation (4.28) is independent of source position. Its inverse matrix can therefore be calculated in advance, and then the surface pressure P_M for various source positions is rapidly calculated by multiplying the inverse matrix by G_0 as follows:

$$P_M = jk\rho_0 c Q_0 \left\{\frac{1}{2}[I]+[G'']+ jk\rho_0 c[G][Y]\right\}^{-1} G_0 . \tag{4.31}$$

Nonetheless, calculating the inversion of an $M \times M$ matrix is considerably more complex and time-consuming than directly solving a set of M linear algebra equations. To speed up the calculation, Otani and Ise (2006) proposed an improvement to this algorithm and incorporated a time-domain operation into the post-process calculation in Equation (4.29). The results indicate that, once the pressure P_M at the boundary surface is identified, the HRTF for an arbitrary source position can be calculated within 1 s.

An increasing number of M elements are required to extend the maximal frequency in BEM calculation. The computational cost and storage required for calculating the pressures at M elements by conventional BEM are of the order M^3 and M^2, respectively. Using the iterative calculation cannot reduce the required storage but can decrease the cost to $(M^2 N_{iter})$, where N_{iter} is the number of iterative calculations. The number of M is usually large; thus, an additional increase in M results in enormous computational cost that is beyond the computational power of a typical personal computer. To further reduce computational cost, some researchers proposed a fast multipole accelerated boundary element method (FMM BEM) for HRTF calculation (Gumerov et al., 2009, 2010; Kreuzer et al., 2009). The acoustic principle of reciprocity was also incorporated in the calculation. The positional dependence of HRTFs was first represented by a weighted combination of SH functions, so that data dimensionality is reduced (to be discussed in detail in Section 6.3.4). Then, BEM was used to calculate the SH coefficients (SH spectrum) to reduce computational cost and accelerate calculation. HRTFs up to a frequency of 20 kHz, as a continuous function of source direction and distance, can be calculated using FMM BEM. The computational cost and storage required for FMM BEM are (MN_{iter}) and M, respectively.

Another problem with BEM-based HRTF calculation is uniqueness. The unique and accurate solution P_M for Equation (4.28) exists only when matrix $\left\{\dfrac{1}{2}[I]+[G'']+jk\rho_0 c[G][Y]\right\}$ is regular. However, this matrix is frequency dependent. It is irregular at some frequencies, and Equation (4.28) has no unique solution for such cases. For an internal problem of the wave equation, the irregularity corresponds to the eigenfrequencies or resonance models inside a closed volume. For an exterior problem, however, such correspondence is illogical.

To address the problem of irregularity, M' field points located at $r'_{M+1}, r'_{M+2}, \dots r'_{M+M'}$ inside the boundary S of the external problem are added (Ise and Otani, 2002). Let the pressures at these M' internal field points be zero, that is, $P(r'_q, r, f) = 0$, $q = M + 1, M + 2, \dots M + M'$; the results are substituted into Equations (4.22) and (4.23). This substitution yields M'—supplemented equations similar to Equation (4.27):

$$0 = jk\rho_0 cQ_0 G(r'_q, r, f) - \sum_{i=1}^{M}[G''_{qi}(f) + jk\rho_0 cY(r'_i, f)G_{qi}(f)]P(r'_i, r, f)$$

(4.32)

$$i = 1, 2, \dots M, \quad q = M + 1, M + 2, \dots M + M'.$$

Combining Equation (4.32) with Equation (4.27) yields:

$$\left\{\frac{1}{2}[\tilde{I}]+[\tilde{G}'']+ jk\rho_0 c[\tilde{G}][Y]\right\}P_M = jk\rho_0 cQ_0\tilde{G}_0,$$

(4.33)

where P_M is an $M \times 1$ matrix (column vector); $[Y]$ is an $M \times M$ diagonal matrix, similar to that in Equation (4.28); $\tilde{G}_0 = [G(r'_1, r, f), G(r'_2, r, f), \dots G(r'_M, r, f), G(r'_{M+1}, r, f), \dots G(r'_{M+M}, r, f)]^T$ is an $(M + M)' \times 1$ matrix (column vector); both $[\tilde{G}'']$ and $[\tilde{G}]$ are $(M + M') \times M$ matrices, and their elements are, respectively, given by:

$$\tilde{G}''_{qi}(f) = \iint_{Si}\frac{\partial G(r'_q, r'', f)}{\partial n''}dS'' \qquad \tilde{G}_{qi}(f) = \iint_{Si}G(r'_q, r'', f)dS''$$

(4.34)

$$q = 1, 2 \dots M, M+1, \dots M + M', \qquad i = 1, 2, \dots M,$$

and $[\tilde{I}]$ is an $(M + M') \times M$ matrix. Except for the M diagonal elements in the left of the matrix, which take a value of 1, the other elements take a value of 0.

Equation (4.33) is an over-determined problem, with $(M + M')$ linear equations and M unknowns. Therefore, it has no precise solution. From the perspective of least square error, however, an approximate solution can be obtained by the pseudoinverse method:

$$\boldsymbol{P}_M = jk\rho_0 cQ_0 \{[U]^+ [U]\}^{-1} [U]^+ \tilde{\boldsymbol{G}}_0,$$

$$[U] = \left\{ \frac{1}{2}[\tilde{I}] + [\tilde{G}''] + jk\rho_0 c[\tilde{G}][Y]) \right\}, \tag{4.35}$$

where superscript "+" denotes the transpose and conjugation of the matrix. $P(\boldsymbol{r}', \boldsymbol{r}, f)$ or the HRTF can be obtained by substituting the resultant $\boldsymbol{P}_M$ into Equation (4.29). Equation (4.35), however, is characterized by two problems. First, the solution that it provides is only an approximation in terms of least square error. Second, calculating the pseudoinverse of the matrix is time consuming.

4.3.3. Results for BEM-based HRTF Calculation

Katz (2001a, 2001b) used BEM to calculate the HRTFs at the entrance of a blocked ear canal for a human subject. Constrained by computational power and time, the maximal frequency in the calculation was 5.4 kHz. The geometry of the head surface was first acquired using a laser 3-D scanner. The HRTFs were then calculated under the rigid boundary condition, and subsequently compared with the results calculated for a head model without pinnae modified from the complete head model by a computer, and for a rigid spherical head model. The results reveal that the pinna effect begins at a frequency of 1.5–2 kHz for the front and rear source directions, and above 3.5 kHz for the ipsilateral source direction. The results for the rigid spherical head model are similar to those for the head model without pinnae. These results are roughly consistent with the measurements discussed in Chapter 3. Katz extended the calculations to include the actual acoustic properties of the human head surface. The human skin is approximately rigid, but human hair is acoustically absorptive and should, therefore, be incorporated into the model (Katz, 2000). Accordingly, the measured impedance of hair was assigned into part of the mesh in the BEM model, but the effects of sound propagation through hair was not taken into account. The calculation indicates that hair results in a 0.5–6 dB change in HRTF magnitudes above 3 kHz. The change in magnitude is source direction dependent and obvious at contralateral directions. This result can be interpreted as the attenuation caused by hair absorption when sound diffracts around the head.

In earlier work, Kahana et al. (1998) used BEM to calculate the HRTFs of the KEMAR artificial head up to 6 kHz. The results are consistent with measurements. In subsequent work, the authors (Kahana et al., 1999) calculated the HRTFs of several models, including a spherical head, an ellipsoidal head, and artificial heads with and without pinnae (excluding the torso). The results demonstrate that the HRTFs of an ellipsoidal head are similar to those of a spherical head, up to 1.5 kHz, with a magnitude difference of less than 3 dB. Moreover, the maximal frequency of the KEMAR calculation was extended to 10–15 kHz. In later work, the transfer functions of human head and baffled pinnae were calculated (Kahana and Nelson, 2007), with the maximal frequency of the latter being 20 kHz. Because of the fine structures and individual features of pinnae, the transfer characteristics of an isolated pinna, called *pinna-related transfer functions* (*PRTFs*), are significant for hearing research. Kahana and Nelson (2000, 2006) also used BEM to analyze the relationship between the resonance mode of the pinna/ear canal (Section 1.4.4) and source directions.

Algazi et al. (2002a) used BEM to calculate the HRTFs of a modified snowman model. Distinguished from the original snowman model, which has a spherical head and torso (Section 4.2.2), the modified model includes a spherical head and an ellipsoidal torso fitted from the geometry of the KEMAR artificial head. A comparison among the calculated HRTFs of the original snowman model, the modified snowman model, and the measured HRTFs of KEMAR without pinnae show similar features at low frequencies.

To evaluate the overall effect of the neck and torso, Chen et al. (2012) used BEM to calculate the HRTFs of a head-neck-torso (HNT) model, which comprises a spherical head, a spherical torso, and a cylindrical neck. The results (up to 5 kHz) indicate that the HRTF magnitudes of the HNT model differ from those of the spherical head-and-torso (HAT) model at frequencies above 0.5 kHz, especially at the near-field source distance and source direction contralateral to the ear. The discrepancy in HRTF magnitudes causes a discrepancy in ILD between the HNT and HAT models. This discrepancy reaches a level of ±12 dB at the lateral direction and a source distance of 0.2 m. As source distance increases, the discrepancy in the results for the HNT and HAT models decreases. Therefore, the neck influences HRTFs and should be included in HRTF calculation, especially for near-field HRTFs.

Otani et al. (2006, 2009) used BEM to calculate HRTFs for a B & K 4128C artificial head (Figure 4.11), with various source distances ranging from 0.1–3.0 m at an interval of 0.01 m (including near and far fields). The results confirm that the acoustic parallax of the pinna (Section 3.6) causes near-field HRTF magnitudes to vary with source distance at high frequencies.

All these calculations excluded the ear canal; thus, only blocked ear canal HRTFs were obtained. Walsh et al. (2004) considered the ear canal and used BEM to calculate HRTFs, with the eardrum selected as the reference point. The anatomical geometry of the head and external ear was obtained by combining the experimental data on the human ear canal with those on the head model acquired by a CT scanner. The results indicate that a significant difference between the HRTFs of head models that include and exclude the ear canal exists, but that the HRTFs of both models encode the same directional information on sound source. That is, the blocked ear canal HRTFs include complete directional information for the sound source. The calculation results for the model with the ear canal also exhibits the resonance mode of the ear canal and concha (Section 1.4.4).

Aiming to investigate the variations in HRTF with the growth of children and to prepare for the potential requirement for hearing aids, Fels et al. (2004) acquired the anatomical geometry of children aged 4 to 6 years using a digital photogrammetric system, and then calculated the HRTFs via BEM. The frequency range for the calculation was up to 6 kHz. Given the difficulty in directly measuring the HRTFs of children, numerical calculation is especially important.

Corresponding to the round-robin to compare the measured HRTFs discussed in Section 2.2.9, some authors initiated a round-robin to compare the calculated HRTFs of the Neumann KU 100 artificial head (Greff and Katz, 2007). To reduce the potential variations caused by different geometric models of an artificial head, a single geometric model and mesh were created using a medical NMRI scanner, which allows for a maximal frequency of 20 kHz (1/4 wavelength) in calculation. Two research groups contributed HRTFs up through 2007. The preliminary results demonstrate that the HRTFs calculated by the two groups are similar, but some deviations from measured data are observed at frequencies above 5 kHz.

Gumerov et al. (2010) used the FMM-accelerated BEM mentioned in Section 4.3.2 to calculate the HRTFs of a spherical head, a Neumann KU-100 model, KEMAR artificial heads, and a simplified head/torso model fitted from KEMAR. The maximal frequency in calculation is 20 kHz, and the calculation results were compared with measurements.

4.3.4 Simplification of Head Shape

The complex nature of the human head and pinna geometry prompts the use of BEM-based HRTF calculation, but BEM presents large computational costs. To investigate the effect of a simplified head shape on HRTFs, Tao et al. (2002, 2003a, 2003b) proposed a parameterized model of head geometry, in which the head shape is represented by SH functions and simplified by spatial low-pass filtering. Suppose that the geometrical surface of the head is obtained by a scanner and described by the following function:

$$R(\Omega') = R(\alpha', \beta'),\tag{4.36}$$

where R is the distance of the head surface relative to the head center. $\Omega' = (\alpha', \beta')$, $0° \le \alpha' \le 180°$ and $0° \le \beta' < 360°$ are the elevation and azimuth, respectively; these are related to the default coordinate (θ, ϕ) of this book as $\beta' = \theta$, $\alpha' = 90° - \phi$. Thus, $\alpha' = 0°$ and $90°$ represent the top and horizontal directions, respectively. (Except for differences in notation, the coordinate chosen here is consistent with that in the literature.)

As explained in Appendix A, $R(\Omega')$ can be represented by a combination of complex or real-valued SH functions (real-valued SH functions were used in the literature):

$$R(\Omega') = \sum_{l=0}^{\infty}\sum_{m=-l}^{l} d_{lm} Y_{lm}(\Omega') = a_{00} Y_{00}^1(\Omega') + \sum_{l=1}^{\infty}\sum_{m=0}^{l}[a_{lm}Y_{lm}^1(\Omega') + b_{lm}Y_{lm}^2(\Omega')]\tag{4.37}$$

where $Y_{l0}^2(\Omega') = 0$, which is preserved in Equation (4.37) for convenience in writing. The SH coefficients in Equation (4.37) are calculated by:

$$d_{lm} = \int R(\Omega')Y_{lm}^*(\Omega')d\Omega' = \int_{\beta'=0}^{2\pi}\int_{\alpha'=0}^{\pi} R(\alpha',\beta')Y_{lm}^*(\alpha',\beta')\sin\alpha'\,d\alpha'\,d\beta',\tag{4.38}$$

$$a_{l0} = d_{l0} \quad a_{lm} = \frac{\sqrt{2}}{2}[d_{lm} + d_{l,-m}] \quad b_{lm} = \frac{\sqrt{2}}{2}j[d_{lm} - d_{l,-m}] \quad 1 \le m \le l.\tag{4.39}$$

The integral in Equation (4.38) is calculated over the surface of unit sphere S. In practical calculation, M appropriate directions on the surface of the unit sphere are selected, and the value of the function $R(\Omega')$ at each direction, that is, $R(\Omega'_i)$, $i = 0,1,2 \ldots (M - 1)$, is determined. The integral in Equation (4.38) can then be approximately calculated by the weighted summation over M directions. Here, the Gauss–Legendre sampling scheme provided in Appendix A is used to select $M = 2L^2$ directions. For a given L, SH coefficients up to $l \le (L - 1)$ can be evaluated. Substituting the resultant SH coefficients in Equation (4.37) and truncating (low-pass filtering) the summation to $l \le (L - 1)$ yields the SH representation of $R(\Omega')$ as:

$$R(\Omega') = \sum_{l=0}^{L-1}\sum_{m=-l}^{l} d_{lm} Y_{lm}(\Omega') = a_{00} Y_{00}^1(\Omega') + \sum_{l=1}^{L-1}\sum_{m=0}^{l}[a_{lm}Y_{lm}^1(\Omega') + b_{lm}Y_{lm}^2(\Omega')].\tag{4.40}$$

The truncated order in Equation (4.40) determines the accuracy of the parameterized model of head geometry. Given that $Y_{00}^1(\Omega') = 1/\sqrt{4\pi}$, the first term on the right side of Equation (4.40) represents a spherical head model with radius $R(\Omega') = a = a_{00}/\sqrt{4\pi}$. The low-order terms roughly represent the basic deviation between the shapes of a real head and a spherical head, and the high-order terms describe the fine structures of the head. Therefore, the higher the order of the terms retained in Equation (4.40), the more accurate—but complicated—the resultant head model.

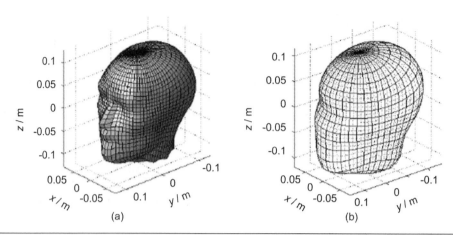

Figure 4.12 KEMAR without pinnae model (from Tao et al., 2003a, 2003b, by permission of AES): (a) $(L - 1) = 34$; (b) $(L - 1) = 17$

As an example, Figure 4.12(a) shows a model of a KEMAR artificial head without pinnae with truncated order $(L - 1) = 34$. Figure 4.12(b) illustrates the same model but with truncated order $(L - 1) = 17$. Some fine structures, such as the protruding nose, have been smoothed in Figure 4.12(b).

Suppose that the $R(\Omega')$ of a reference head shape is represented by SH up to the $(L - 1)$ order, as shown by Equation (4.40), and that the SH coefficients are evaluated from the value of $R(\Omega'_i)$ at $M = 2L^2$ directions. A simplified head shape $R'(\Omega')$ is provided by truncating the SH representation up to the $(L' - 1)$ order. Relative to the reference shape, the percentage root mean square (RMS) error introduced by head simplification is given by:

$$\varepsilon_{R,\%} = \sqrt{\frac{\sum_{i=0}^{M-1}\left[R(\Omega'_i) - R'(\Omega'_i)\right]^2}{\sum_{i=0}^{M-1}\left[R(\Omega'_i)\right]^2}} \times 100\% . \tag{4.41}$$

The calculation results indicate that the accuracy of the head shape representation monotonically improves with the increasing SH terms included in Equation (4.40). The percentage RMS error steadily decreases when truncation order $(L' - 1)$ increases, from 2 to approximately 9. Thereafter, the percentage RMS error slowly decreases with increasing $(L' - 1)$. A similar phenomenon is observed from shape error at the ear position. When truncation order $(L' - 1)$ increases from 2 to 10, the shape error at the ear position visibly drops; thereafter, it decreases slightly. By contrast, shape error at the nose position converges relatively slowly with increasing truncation order. Therefore, the protruding parts of the head surface require a higher order of SH representation.

The pressures at the simplified head surface can be calculated using BEM. The pressure error caused by a simplified head shape can subsequently be evaluated. Let $P(\Omega', r, f)$ denote the pressure at the surface of a reference head model with SH representation up to $l \leq (L - 1)$, and $P'(\Omega', r, f)$ denote the pressure at the simplified head surface with SH representation up to $l \leq (L' - 1)$, where Ω and r represent the head surface and source position, respectively. The percentage RMS pressure error caused by simplification in head shape is defined by:

$$\varepsilon_{P,\%} = \sqrt{\frac{\sum_{i=0}^{M-1}\left[P(\Omega'_i,r,f)-P'(\Omega'_i,r,f)\right]^2}{\sum_{i=0}^{M-1}\left[P(\Omega'_i,r,f)\right]^2}} \times 100\% . \tag{4.42}$$

The calculation results for frequency f = 250 Hz, 500 Hz, 1 kHz, 2 kHz, 3 kHz and the incident plane wave at six horizontal directions indicate that the percentage RMS pressure error steadily decreases when truncation order $(L' - 1)$ increases, from 2 to approximately 10. Similarly, when truncation order $(L' - 1)$ increases from 2 to 10, the pressure error at the ear position visibly drops and decreases less thereafter.

These results indicate that below a frequency of 3 kHz, an SH representation up to an 11 order is sufficient to describe the shape of KEMAR without pinnae. When the wavelength is short, and the contribution of the fine structures of the head increases at high frequencies above 3 kHz, a higher order SH representation of the head shape is required. The SH representation of the head shape can further be used to simplify the calculation of the pressures at the head surface. According to Equation (4.37), $R(\Omega')$ is completely determined by SH coefficients e_{lm}^σ :

$$R(\Omega') = R(e_{lm}^\sigma) \qquad e_{lm}^\sigma = \begin{cases} a_{lm} & \sigma = 1 \\ b_{lm} & \sigma = 2 \end{cases} . \tag{4.43}$$

For a free-field point source at a given position r, the pressure $P'(\Omega', r, e_{lm}^\sigma, f)$ at an arbitrary position $\Omega' = (\alpha', \beta')$ on the head surface is determined by the head shape and the acoustic impedance on the head surface (for simplicity, a rigid surface is assumed here). Let the pressure at the surface of a given template head be $P_{00} = P_{00}(\Omega', r, f)$. The difference between the concerned and template head shapes is determined by the difference Δe_{lm}^σ in corresponding SH coefficients. For a small change Δe_{lm}^σ in head shape, the pressure at the surface of the concerned head can be approximated by a Taylor expansion up to the first order:

$$P(\Omega',r,e_{lm}^\sigma,f) \approx P_{00} + \sum_{\sigma=1}^{2}\sum_{l=0}^{L-1}\sum_{m=0}^{l}\frac{\partial P(\Omega',r,e_{lm}^\sigma,f)}{\partial e_{lm}^\sigma}\Delta e_{lm}^\sigma , \tag{4.44}$$

and the partial differential in Equation (4.44) can be approximately calculated using the forward-difference formula:

$$\frac{\partial P(\Omega',r,e_{lm}^\sigma,f)}{\partial e_{lm}^\sigma} \approx \frac{P(\Omega',r,e_{lm}^\sigma,f)-P_{00}}{\Delta e_{lm}^\sigma} . \tag{4.45}$$

P_{00} can be pre-calculated, and the first-order derivatives of all shape SHs are calculated using Equation (4.45) and stored for later use. When the differences between the concerned and template head shapes are identified, pressure $P'(\Omega', r, e_{lm}^\sigma, f)$ can be easily calculated from Equation (4.45). In particular, if a rigid sphere with radius $a = a_{00}/\sqrt{4\pi}$ is chosen as the template head, the analytical solution of P_{00} can be obtained using the method discussed in Section 4.1. The difference between the concerned and template head shapes is described by.

$$\Delta e_{lm}^\sigma = e_{lm}^\sigma, \quad \sigma = 1, 2, \quad l = 1, 2....(L-1), \quad m = 0, 1...l \tag{4.46}$$

This method for calculating pressure at the head surface is called differential pressure synthesis (DPS) because the difference in pressure at the surface of the concerned and template heads is calculated. As an example, a rigid sphere is used as the template head and an SH representation up to

$(L' - 1) = 17$ for KEMAR without pinnae is chosen as the concerned head. The HRTFs were calculated by DPS and compared with those calculated by conventional BEM. The results indicate that, below a frequency of 1 kHz, the percentage RMS pressure error of the DPS calculation is roughly 8%, whereas that of the rigid spherical head model is as large as 29%. The percentage RMS pressure error of DPS calculation visibly increases with frequency, on the order of 23% at 2 kHz and 43% at 3 kHz. The approximation in DPS calculation is valid only when the difference between the concerned and template head shapes is much smaller than the wavelength. As frequency increases, the wavelength is shortened and the fine structures of the head impose a more considerable effect. The approximation then gradually becomes invalid.

The pinnae are excluded in this calculation because a high order SH representation is required to describe fine pinna shape and structures. Hetherington and Tew (2003) proposed a method for parameterizing the two-dimensional cross-sectional contours of the human head and pinnae by elliptic Fourier function, and then calculated HRTFs, with pinnae included.

4.3.5. Other Numerical Methods for HRTF Calculation

In addition to BEM, other numerical methods for calculating HRTFs have been proposed. In an early work, Kahana et al. (1999) used the infinite-finite element method (IFEM) to calculate the HRTFs of a KEMAR artificial head.

The finite-difference time-domain (FDTD) method is another numerical method for HRTF calculation, in which the wave equation and boundary are both spatially and temporally made discrete, resulting in a finite-difference algorithm. In essence, the problem of solving the partial differential equation is converted into an issue of iteration calculation.

Xiao and Liu (2003) applied the FDTD method to calculate the HRTFs of a KEMAR artificial head, and introduced a perfectly matched layer (PML) to absorb the outgoing waves at the truncated boundary of an unbounded medium. Mokhtari et al. (2007) used NMRI to acquire the head geometry of two human subjects and then used the FDTD method to calculate the HRTFs at 133 directional distributions in the front hemispherical space. The comparison of measurements demonstrates that the mean spectral distortion [Equation (5.19)] of the HRTFs calculated across 133 directions for two subjects are 4.7 and 3.8 dB, respectively. The HRTFs were measured at a far-field distance $r = 1.3$ m. Constrained by the number of cells (thus, computational cost) in the FDTD method, however, the authors calculated only the HRTFs at near-field distance $r = 0.3$ m. According to the discussion in Section 3.6, the spectral features of near-field HRTFs are distinguished from those of far-field HRTFs; therefore, the results of near-field calculation and far-field measurement may be incomparable.

Kirkeby et al. (2007) proposed the ultra-weak variational formulation (UWVF), a numerical method similar to the FDTD method, to calculate the HRTF of a B & K 4128C artificial head. To investigate the effect of the torso, the HRTFs calculated for three different models (a rigid head alone, a rigid head with a rigid torso, and a rigid head with an absorbing torso) were compared. Turku et al. (2008) conducted a psychoacoustic experiment to evaluate the UWVF-based HRTFs and measured the HRTFs of B & K 4128C from a perceptual point of view. The results indicate that two sets of HRTFs are equally well exhibited in terms of localization performance.

Takemoto et al. (2012) used FDTD to calculate HRTFs of the whole head and PRTFs of an isolated pinna. By comparison with the calculated results, mechanisms for generating peaks and notches in media-plane HRTFs were identified.

4.4. Summary

HRTFs vary with the individual, and measuring HRTFs for every individual is complicated and time-consuming. Calculation is another method for obtaining HRTFs, in which the wave or Helmholtz equation is solved, subject to the boundary conditions presented by the head, pinnae, and torso. The strict solutions of HRTFs can be obtained for rare simplified head/torso models.

The spherical head (without pinnae and a torso) is the simplest model for calculating HRTFs. The spherical head model is roughly valid below a frequency of 3 kHz (at most), and demonstrates some basic features of HRTFs, including ITD, ILD, the effect of ear location, and the distance dependence of near-field HRTFs.

The snowman model, which consists of a spherical head and torso with different radii, is another simplified model for HRTF calculation. The strict solutions of snowman HRTFs can be obtained by multi-scattering or multipole re-expansion. The snowman model can account for the reflection and shadow of the torso, at least qualitatively.

The geometry of a real head (and torso) is complicated and irregular. An accurate head model should include pinnae, which are vital to high-frequency HRTFs. Because obtaining the strict solutions of the HRTFs of a head model with complicated geometry is difficult, some numerical methods for HRTF calculation have been developed, among which the BEM is the most commonly used. In the BEM, the boundary problem of the wave or Helmholtz equation is first converted into a boundary surface integral. Then, the boundary surface is made discrete, as a mesh of small elements, resulting in a set of linear algebra equations. The head geometry for BEM-based HRTF calculation is usually acquired by laser, or some other type of scanners. More accurate HRTFs are obtained with BEM calculation.

A drawback to BEM-based HRTF calculation is the enormous number of elements required for the head geometry, resulting in high computational costs. Incorporating the acoustic principle of reciprocity into calculation visibly reduces computational cost, but may introduce errors in resultant HRTFs. Some algorithms for accelerating BEM-based HRTF calculation have recently been developed. Uniqueness is another problem with BEM-based HRTF calculation.

Prior to calculating HRTFs, the head geometry can be simplified by truncating the spherical harmonics representation of the head shape to some extent; such an approach includes low-pass filtering of head shape in the spatial domain. Differential pressure synthesis can also be used to simplify BEM-based HRTF calculation, but this method is valid only below 3 kHz. Aside from the BEM, other numerical methods, such as the FDTD method, have been used to calculate HRTFs.

5

HRTF Filter Models and Implementation

> *This chapter presents HRTF filter models and their implementation. Mathematical analysis methods that are used to assess HRTF models and approximation are outlined. The classification of HRTF filter models, some physical and perceptual considerations on HRTF filter design, as well as various designs and implementations of HRTF filters, are discussed.*

Chapters 2 and 4 presented approaches to obtaining head-related transfer functions (HRTFs); that is, empirical measurements and theoretical calculations, respectively. An important application of HRTFs is binaural synthesis processing for virtual auditory displays (VADs). As stated in Section 1.8.3, a straightforward approach to synthesizing binaural signals for a free-field virtual sound source is to multiply the mono stimuli by a pair of corresponding HRTFs in the frequency domain, or to equally convolve the mono stimuli with a pair of head-related impulse responses (HRIRs) in the time domain. In practice, however, this approach suffers from inefficiency. Given that an HRTF is a system transfer function of a linear-time-invariant (LTI) system, its characteristics can be simulated by a variety of LTI system models and then effectively implemented by corresponding filters. These implementations indicate that HRTFs can be represented by various kinds of filters, whose frequency response is equivalent, or approximate, to that of the HRTFs. Early studies implemented HRTF-based signal processing with analog filters. With the development of computers and digital signal processors (DSPs), HRTFs are often implemented by digital filter models, an approach that has received considerable attention in research on HRTFs and VADs. This chapter focuses on issues related to digital filter models for HRTFs and their implementation. Section 5.1 reviews the mathematical analysis methods used to assess HRTF approximation, forming the analytic fundamentals for the follow-up discussion. The basic concepts and classification of HRTF filter models, as well as some physical and perceptual considerations related to HRTF filter design, are presented in Section 5.2. Section 5.3 comparatively discusses various methods for HRTF filter design. The corresponding filter structures and implementation are provided in Section 5.4. Finally, Section 5.5 introduces the frequency warping transform and related applications to HRTF filter design.

5.1. Error Criteria for HRTF Approximation

The ultimate purpose of HRTF modeling or filter design is to identify a model or filter, whose response $\hat{H}(\theta, \phi, f)$ approximately matches the response $H(\theta, \phi, f)$ of a given original, or target

HRTF in terms of a given error criterion. Prior to constituting a model or designing a filter, specifying mathematical measures to quantify the errors in various HRTF approximations is necessary. For brevity, θ and ϕ, which denote source direction, are excluded in the succeeding discussion, unless otherwise stated.

The frequency-domain square error at a specific frequency is defined as:

$$\varepsilon_s(f) = |H(f) - \hat{H}(f)|^2. \tag{5.1}$$

An error measure commonly used in HRTF approximation is the summation of Equation (5.1) over a certain frequency range:

$$\varepsilon_{\Sigma S} = \int |H(f) - \hat{H}(f)|^2 df, \tag{5.2}$$

which is conventionally called a *square error*. The *mean square error* across a certain frequency range is also used:

$$\varepsilon_{MS} = \frac{1}{f_h - f_l} \int |H(f) - \hat{H}(f)|^2 df, \tag{5.3}$$

where f_h and f_l are the upper and lower frequency bounds of interest, respectively; $(f_h - f_l)$ determines the integral frequency range.

In digital signal processing, $H(f)$ is represented by its N samples $H(f_k)$ at N discrete frequencies $f_k(k = 0, 1 \ldots N - 1)$, or $H(\omega_k)$ at N discrete digital angular frequencies ω_k. Equation (2.20) indicates the relationship among ω_k, f_k, and k as:

$$\omega_k = \frac{2\pi k}{N} = \frac{2\pi f_k}{f_s} \qquad f_k = \frac{k}{N} f_s \qquad k = 0, 1 \ldots N - 1, \tag{5.4}$$

where f_s denotes the sampling frequency of the digital system. Hence, the discrete frequency can be specified by either ω_k, f_k, or k, and the corresponding $H(f)$ can be written as either $H(f_k)$, $H(\omega_k)$, or $H(k)$, depending on the specific issues discussed. Equations (5.2) and (5.3) can be respectively expressed in the discrete frequency domain as follows:

$$\varepsilon_{\Sigma S} = \sum_{f_k} |H(f_k) - \hat{H}(f_k)|^2 = \sum_{\omega_k} |H(\omega_k) - \hat{H}(\omega_k)|^2 = \sum_{k=0}^{N-1} |H(k) - \hat{H}(k)|^2, \tag{5.5}$$

$$\varepsilon_{MS} = \frac{1}{N} \sum_{f_k} |H(f_k) - \hat{H}(f_k)|^2 = \frac{1}{N} \sum_{\omega_k} |H(\omega_k) - \hat{H}(\omega_k)|^2 = \frac{1}{N} \sum_{k=0}^{N-1} |\;(\;) - \hat{}\;(\;)|^2. \tag{5.6}$$

The N-point frequency-domain HRTF $H(f_k)$ is related to the N-point time-domain HRIR $h(n)$ by the N-point discrete Fourier transform (DFT) [see Equation (2.22)]. Correspondingly, the square error and mean square error of time-domain HRIRs are:

$$\varepsilon_{\Sigma S} = \sum_{n=0}^{N-1} |h(n) - \hat{h}(n)|^2, \tag{5.7}$$

$$\varepsilon_{MS} = \frac{1}{N} \sum_{n=0}^{N-1} |h(n) - \hat{h}(n)|^2, \tag{5.8}$$

where $h(n)$ denotes the original HRIR, and $\hat{h}(n)$ represents the approximated HRIR. The time- and frequency-domain expressions of the square error and mean square error are equivalent, according to Parseval's theorem (Oppenheim et al., 1999).

With the concept of vector space, each N-point HRTF can be represented by an $N \times 1$ complex-valued vector, $\boldsymbol{H}(f) = [H(f_0), H(f_1),\dots H(f_{N-1})]^\mathrm{T}$, where each term is the value of the HRTFs at a specific discrete frequency, and superscript "T" denotes a transposition operation. Equally, each N-point HRIR can be represented by an $N \times 1$ real-valued vector, $\boldsymbol{h}(n) = [h(0), h(1), \dots h(N-1)]^\mathrm{T}$, where each term is the value of the HRIRs at a specific discrete time. Correspondingly, the square norm of $\boldsymbol{H}(f)$ and $\boldsymbol{h}(n)$ can be written as:

$$\|\boldsymbol{H}\|_2 = \boldsymbol{H}^+ \boldsymbol{H} \qquad \|\boldsymbol{h}\|_2 = \boldsymbol{h}^\mathrm{T} \boldsymbol{h}, \tag{5.9}$$

where superscript "+" denotes transposed conjugation. Thus, the square error defined in Equations (5.5) and (5.7) can be interpreted as the square norm of the error vector:

$$\varepsilon_{\Sigma S} = \|\boldsymbol{H} - \hat{\boldsymbol{H}}\|_2 = \|\boldsymbol{h} - \hat{\boldsymbol{h}}\|_2 . \tag{5.10}$$

In addition to the commonly used square error and mean square error, some other error criteria for evaluating HRTF approximations have been suggested.

1. The *relative energy error* is defined as:

$$\varepsilon_R(f) = \frac{|H(f) - \hat{H}(f)|^2}{|H(f)|^2} (\times 100\%). \tag{5.11}$$

The *mean relative energy error* in terms of discrete frequency is:

$$\varepsilon_R = \frac{\displaystyle\sum_{f_k} |H(f_k) - \hat{H}(f_k)|^2}{\displaystyle\sum_{f_k} |H(f_k)|^2} (\times 100\%), \tag{5.12}$$

whose corresponding time-domain expression is:

$$\varepsilon_R = \frac{\displaystyle\sum_{n} |h(n) - \hat{h}(n)|^2}{\displaystyle\sum_{n} |h(n)|^2} (\times 100\%). \tag{5.13}$$

In practice, the relative energy error is often scaled in dB thus:

$$\varepsilon_R = 10 \log_{10} \left[\frac{\displaystyle\sum_{f_k} |H(f_k) - \hat{H}(f_k)|^2}{\displaystyle\sum_{f_k} |H(f_k)|^2} \right] (dB). \tag{5.14}$$

2. The *signal-to-distortion ratio (SDR)* is defined as:

$$SDR(f) = 10 \log_{10} \frac{|H(f)|^2}{|H(f) - \hat{H}(f)|^2} (dB). \tag{5.15}$$

Similar to signal-to-noise ratio, a large SDR means few errors. The *mean SDR* over frequencies is:

$$SDR_M = 10 \log_{10} \frac{\displaystyle\sum_{f_k} |H(f_k)|^2}{\displaystyle\sum_{f_k} |H(f_k) - \hat{H}(f_k)|^2} (dB). \tag{5.16}$$

In the time domain, Equation (5.16) is expressed as:

$$SDR_M = 10\log_{10} \frac{\sum_n |h(n)|^2}{\sum_n |h(n) - \hat{h}(n)|^2} \quad (dB).$$ (5.17)

3. *Spectral distortion (SD)* is defined as:

$$SD(f) = \left| 20\log_{10} \frac{|H(f)|}{|\hat{H}(f)|} \right| \quad (dB).$$ (5.18)

SD specifies the deviation between the original and approximate HRTF magnitudes in a logarithmic scale. An SD close to 0 dB means less deviation. The *mean SD* across a certain frequency range is:

$$SD_M = \sqrt{\frac{1}{N}\sum_{f_k}\left(20\log_{10}\frac{|H(f_k)|}{|\hat{H}(f_k)|}\right)^2} \quad (dB).$$ (5.19)

4. A *normalized cross-correlation coefficient* can be used to quantify the similarity between two functions. Similar to Equation (3.7) for $\tau = 0$, the normalized cross-correlation coefficient for discrete-time HRIRs is:

$$\Phi(\theta,\phi) = \frac{\sum_n h(\theta,\phi,n)\hat{h}(\theta,\phi,n)}{\sqrt{\left[\sum_n h^2(\theta,\phi,n)\right]\left[\sum_n \hat{h}^2(\theta,\phi,n)\right]}},$$ (5.20)

where $0 \leq |\Phi(\theta, \phi)| \leq 1$ denotes the similarity of the original and approximated HRTFs with $\Phi(\theta, \phi) = 1$ indicating complete identicality.

5.2. HRTF Filter Design: Model and Considerations

5.2.1. Filter Model for Discrete-time Linear-time-invariant System

According to signal processing theory (Oppenheim et al., 1999), the general form of a differential equation for a discrete-time LTI system with input $x(n)$ and output $y(n)$ is:

$$y(n) + a_1 y(n-1) + \dots + a_P y(n-P) = b_0 x(n) + b_1 x(n-1) + \dots + b_Q x(n-Q)$$ (5.21a)

or:

$$y(n) + \sum_{p=1}^{P} a_p y(n-p) = \sum_{q=0}^{Q} b_q x(n-q).$$ (5.21b)

That is, the output at discrete time n is related to the weighted sum of the preceding $(n - P)$ outputs, aside from input $x(n)$ and the weighted sum of the preceding $(n - Q)$ inputs. This system model is called the *autoregressive moving-average model* (ARMA).

Applying Z-transform to Equation (5.21b) yields:

$$Y(z)\left(1 + \sum_{p=1}^{P} a_p z^{-p}\right) = \left(\sum_{q=0}^{Q} b_q z^{-q}\right)X(z)$$

Then, the system function is expressed as:

$$H(z) = \frac{Y(z)}{X(z)} = \frac{\sum\limits_{q=0}^{Q} b_q z^{-q}}{1 + \sum\limits_{p=1}^{P} a_p z^{-p}} = \frac{b_0 + b_1 z^{-1} + .. \quad b_Q z^{-Q}}{1 \quad a_1 z^{-1} \quad a_p z^{-P}} \tag{5.22}$$

+

That is, $H(z)$ can be written as a rational function of z. $H(z)$ can also be expressed in pole-zero form:

$$H(z) = K z^{(P-Q)} \frac{\prod\limits_{q=1}^{Q}(z - z_q)}{\prod\limits_{p=1}^{P}(z - z_p)}, \tag{5.23}$$

where K is a constant related to system gain. Equation (5.23) contains Q zeros and P poles, except those at the origin in the Z-plane.

Alternatively, Equation (5.22) can be expressed in cascade form. When P and Q are even numbers, the second-order cascade form of the system function is:

$$H(z) = \frac{\prod\limits_{q=1}^{Q/2}(b_{0q} + b_{1q} z^{-1} + b_{2q} z^{-2})}{\prod\limits_{p=1}^{P/2}(1 + a_{1p} z^{-1} + a_{2p} z^{-2})} = K \frac{\prod\limits_{q=1}^{Q/2} 1 + b_{1q} z^{-1} + b_{2q} z^{-2}}{\prod\limits_{p=1}^{P/2}\left(1 + a_{1p} z^{-1} + a_{2p} z^{-2}\right)}, \tag{5.24}$$

with $= \prod\limits_{q=1}^{Q/2}{}_{0q}, \quad '_{1q} = {}_{1q}/{}_{0q}), \quad '_{2q} = b_{2q}/b_{0q}.$ $a \quad z$

If $b_q = 0$ with $q = 1, 2 \ldots Q$, except b_0, Equation (5.21b) is called the *autoregressive (AR) model*, with input $x(n)$ and output $y(n)$ satisfying:

$$y(n) + \sum\limits_{p=1}^{P} a_p y(n - p) = b_0 x(n). \tag{5.25}$$

Correspondingly, the system function is written as:

$$H(z) = \frac{b_0}{1 + \sum\limits_{p=1}^{P} a_p z^{-p}} = z^P \frac{K}{\prod\limits_{p=1}^{P}(z - z_p)}. \tag{5.26}$$

The system function contains P poles without zeros outside the origin of the Z-plane, and is therefore called the *all-pole model*, a specific case of the ARMA model.

Applying inverse Z-transform to Equation (5.22) or (5.26) reveals that the length of impulse response $h(n)$ for the ARMA- or AR-modeled system is infinite. Hence, these systems are referred to as *infinite impulse response (IIR)* systems, whose characteristics can be represented by IIR filters with orders (Q, P). Specifically, the IIR filter of an AR system has $Q = 0$.

If $a_p = 0$ with $p = 1, 2 \ldots P$, Equation (5.21b) is called the *moving average (MA) model*, with input $x(n)$ and output $y(n)$ satisfying:

$$y(n) = b_0 x(n) + b_1 x(n-1) + \ldots + b_Q x(n-Q) = \sum\limits_{q=0}^{Q} b_q x(n-q) \tag{5.27}$$

That is, the output at discrete time n is the weighted sum of these inputs at discrete time n and the preceding $(n - Q)$. The system function in Equation (5.22) is then simplified as:

$$H(z) = \frac{Y(z)}{X(z)} = b_0 + b_1 z^{-1} + ... + b_Q z^{-Q} = \sum_{q=0}^{Q} b_q z^{-q} = K z^{-Q} \prod_{q=1}^{Q} (- _q). \qquad (5.28)$$

This system contains only zeros, except the poles at the origin of the Z-plane. Applying inverse Z-transform to Equation (5.28), or simply comparing it with Equation (2.19), yields the impulse response of the MA system:

$$h(n) = \begin{cases} b_n & 0 \le n \le Q \\ 0 & others \end{cases}, \qquad (5.29)$$

The impulse response has a finite length of $N = Q + 1$ points (or taps), and is therefore called a *finite impulse response (FIR)* system, whose characteristics can be represented by an FIR filter with order Q.

5.2.2. Basic Principles and Model Selection in HRTF Filter Design

Being the transfer function of an LTI system, an HRTF can be approximated by the models presented in Section 5.2.1, after which digital filters can be realized. From the perspective of signal processing, HRTFs can be implemented by either the FIR or IIR filter model. Although differences in filter design methods exist for the two kinds of filter models, they share some preliminary considerations in the filter design stage (Jot et al., 1995; Huopaniemi and Karjalainen, 1997a; Huopaniemi and Zacharov, 1999b).

The essential goal of modeling HRTFs by FIR or IIR filters is to enable the transfer function of filter $\hat{H}(\omega)$ [i.e., the value of system function $\hat{H}(z)$ on the unit circle with $z = \exp(j\omega)$] to match or approximate target HRTFs. As stated in Section 5.1, various measures are available for evaluating the approximation errors of HRTFs, among which the frequency- or time-domain square error defined in Equation (5.5) or (5.7) is commonly used. With the square error criterion, designing an HRTF filter involves determining the coefficients of $\hat{H}(\omega)$ in a least square error sense by:

$$\min \varepsilon_{\Sigma S} = \min \left[\sum_{\omega_k} |H(\omega_k) - \hat{H}(\omega_k)|^2 \right] = \min \left[\sum_{f_k} |H(f_k) - \hat{H}(f_k)|^2 \right], \qquad (5.30)$$

where ω_k is related to f_k by Equation (5.4). Some researchers have used other error criteria in HRTF filter design, as will be introduced later in this chapter.

As stated in Section 5.2.1, an FIR system is represented by an MA model, whereas an IIR system is represented by an ARMA or AR model. Although the MA, ARMA, and AR models are generally accepted and can be converted into any of the three types from a signal processing perspective, an appropriate system model can simplify the procedures in HRTF filter design and, consequently, reduce filter order. According to the discussions in Sections 1.4.4, 3.1.2, and 3.4, the notches and peaks in HRTF magnitude spectra correspond to the zeros and poles in the system function. In this sense, the ARMA model in Equation (5.22) is reasonable. However, the peaks in HRTF magnitude spectra are relatively smooth, and can therefore be appropriately modeled by the MA model (i.e., an FIR filter).

Generally, the order of an IIR filter required to approximate HRTFs in terms of the ARMA model is lower than that of an FIR filter in terms of the MA model. Efficiency therefore dictates a preference for the IIR filter. Conversely, the design of an IIR filter is more complicated than that of an FIR filter, and the former tends to be unstable when inappropriately designed. Given that each kind of filter has advantages and shortcomings, both FIR and IIR filters have been used in HRTF modeling.

The IIR filter based on the AR model is unable to fit the pinna notches in HRTF magnitude spectra efficiently because it lacks zeros, except at the origin in the Z-plane. Because some important information on directional localization is encoded in pinna notches, the AR model is seldom used independently in HRTF filter design.

5.2.3. Length and Simplification of Head-related Impulse Responses

Whether an FIR or IIR model is used, a filter is designed on the basis of given time-domain HRIRs or frequency-domain HRTFs. As indicated in Table 2.1, the length of measured HRIRs usually ranges from 128 to 4096 points at a sampling frequency of 44.1–50 kHz, that is, 2.5–80 ms. A short HRIR corresponds to a simple filter. The discussion in Section 3.1.1 indicated that the main energy of HRIRs is concentrated at 1.1–1.4 ms. Hence, truncating HRIRs by using a time-domain window shortens the duration of HRIRs while preserving their essence. Such truncation, however, results in the frequency-domain smoothing of HRTFs, and a brief window reduces the spectral resolution of HRTFs, according to the uncertainty principle of time and frequency [see Equation (2.25)]. In this sense, time window truncation may influence the accuracy of low-frequency HRTFs. For example, the frequency resolution that corresponds to a 2 ms time window is merely 0.5 kHz. As indicated in Section 1.2.2, the frequency resolution of the auditory system is relatively high at low frequencies. Fortunately, the magnitude of low-frequency HRTFs is nearly invariable (see Section 3.1.2). Hence, an appropriate time window, or equivalent frequency-domain smoothing, has little influence on low-frequency HRTFs—provided that raw HRTFs are sufficiently accurate at low frequencies.

Using psychoacoustic experiments, several researchers have investigated the effects of HRIR duration on generating free-field virtual sound sources (Zahorik et al., 1995; Sevona et al., 2002). Localization experiments, associated with Gaussian noise and individualized HRIRs at 354 directions, revealed that an equivalent localization performance is achieved for free-field real and virtual sources at an HRIR duration of either 10.24 or 20.48 ms (Sevona et al., 2002); localization performance diminishes, as HRIR duration ranges from 0.32–5.12 ms. However, if a slight loss of fidelity is tolerable, then HRIRs with a duration of 1.28 ms are sufficient because the concentrated energy of HRIRs occurs mainly in this time range.

As discussed in Section 3.1.1, some preceding samples of HRIR are approximately zero, roughly corresponding to the propagation delay, from sound source to ear (truncated by a time-domain window). Section 3.1.3 showed that an HRTF can be represented by a product of its minimum-phase function $H_{min}(\theta, \phi, f)$ by its excess-phase function $\exp[j\psi_{cess}(\theta, \phi, f)]$, which can be further decomposed into an all-pass function $\exp[j\psi_{all}(\theta, \phi, f)]$ and a linear-phase function $\exp[-j2\pi f T(\theta, \phi)]$. The propagation delay from sound source to ear (contained in the linear-phase function) is equal to the pure delay in filter design, and can be implemented as a pure delay line z^{-m}. Thus, only minimum-phase and all-pass functions require consideration in HRTF filter design. Under minimum-phase approximation, when the all-pass function is negligible, the energy of the minimum-phase impulse response $h_{min}(t)$ is optimally concentrated at the beginning, allowing for short filter lengths for HRIRs. Therefore, extracting the pure delay in HRIRs, or approximating HRIRs as minimum-phase impulse responses, simplifies filter design (Jot et al., 1995). In this case, the system function of an IIR filter that corresponds to Equations (5.22) and (5.24) is:

$$\hat{H}(z) = z^{-m} \frac{\sum\limits_{q=0}^{Q} b_q z^{-q}}{1 + \sum\limits_{p=1}^{P} a_p z^{-p}} = z^{-m} \frac{\prod\limits_{q=1}^{Q/2}(b_{0q} + b_{1q}z^{-1} + b_{2q}{}^{-2})}{\prod\limits_{p=1}^{P/2}(1 + {}_{1p}z^{-1} + \hat{b}^p{}_{b}{}^{-2}z)}. \tag{5.31}$$

Similarly, the system function of an FIR filter that corresponds to Equation (5.28) can be written as:

$$\hat{H}(z) = z^{-m} \sum\limits_{q=0}^{Q} b_q z^{-q}. \tag{5.32}$$

Equations (5.31) and (5.32) differ from Equations (5.22) and (5.28) insofar as a pure delay z^{-m} has been extracted in Equations (5.31) and (5.32). The procedures for filter design are similar in both cases, except that the minimum phase component of HRIRs (HRTFs) is used for filter design in minimum-phase approximation. The simplification introduced by minimum-phase approximation can be generalized to other types of filter design.

The digital delay order m is selected as:

$$m = round\,[T(\theta,\phi)f_s], \tag{5.33}$$

where $T(\theta, \phi)$ is in units of s (seconds); f_s is the sampling frequency of the system in units of Hz; and "round" refers to rounding to the nearest unit-delay multiple. If f_s = 44100 Hz, rounding the delay length to the nearest sample leads to a worst-case error of approximately 10 μs. The delay error from rounding can be avoided by using fractional delay filtering (Laakso et al., 1996).

5.2.4. HRTF Filter Design Incorporating Auditory Properties

These HRTF filter models are mainly designed from the physical or mathematical point of view. That is, attempts are made to ensure that the physical response of a designed filter matches target HRTFs (or HRIRs), or minimizes the deviation between the designed filter and the target HRTFs (or HRIRs) in a specified mathematical sense, such as that implemented for least square errors. In applications for VADs, on the other hand, more concern is placed on errors (distortions) in auditory perception. Because of the specific properties of human hearing, an optimal HRTF filter design, in a physical or mathematical sense, may not be preferred, from the perspective of hearing. Therefore, incorporating human auditory properties into HRTF filter design is necessary. This incorporation enables the construction of a simple filter structure that has a tolerable perceptible error, or a specified filter structure that has less perceptible error. Nevertheless, auditory perception is a complex and multidimensional process, for which a complete description has thus far been elusive. In current studies, measures that only partially reflect auditory properties are included, such as nonuniform frequency resolution and nonlinear loudness perception. More insight into auditory perception guarantees that more appropriate and comprehensive auditory models will be constructed and subsequently incorporated into HRTF filter design. This backdrop indicates that auditory-based HRTF filter design is a flexible and continually developing field.

Human auditory properties are incorporated into HRTF filter design through various means, such as auditory filter smoothing, auditory weighting, and logarithmic magnitude error.

(1) Smoothing HRTF

An HRTF is a complex function of frequency, but not all details are significant to auditory perception. Therefore, raw HRTF magnitudes can be smoothed, to some extent, prior to filter design. The smoothing process involves discarding high-order variations in HRTF magnitude, thereby simplifying HRTF filter design.

As mentioned in Section 5.2.3, windowing HRIRs is equivalent to smoothing HRTFs in the frequency domain. A time window with a duration of Δt results in a frequency-domain resolution of $\Delta f = 1/\Delta t$. Kulkarni and Colburn (1998) investigated the audibility of fine structures in HRTF magnitude, and proposed a smoothing method that was based on the truncation of coefficients in the Fourier series expansion of the log-magnitude spectrum of HRTFs. A measured N-point HRIR can be expressed in the frequency domain as:

$$H(k) = H(\omega_k) = H(z)\big|_{z=\exp(j\omega_k)} \qquad \omega_k = \frac{2\pi k}{N} \quad k = 0,1..N-1. \tag{5.34}$$

Its log-magnitude spectrum can also be expressed by the Fourier series, or cosine harmonics:

$$H_{\log}(k) = \log_{10}|H(k)| = \sum_{n=0}^{N/2} C(n)\cos\left(\frac{2\pi nk}{N}\right), \tag{5.35}$$

where $C(n)$ is the nth coefficient. Truncating the Fourier series in Equation (5.35)—that is, reconstructing $H_{\log}(k)$ up to the preceding $N' < N/2$ degree—yields a smoothed HRTF magnitude spectrum:

$$\hat{H}_{\log}(k) = \sum_{n=0}^{N'} C(n)\cos\left(\frac{2\pi nk}{N}\right). \tag{5.36}$$

Psychoacoustic experiments showed that no audible difference was detected between measured HRIRs with $N = 1024$ ($N/2 = 512$) and smoothed HRIRs with $N' \geq 32$.

Qian and Eddins (2008) analyzed the linear magnitudes of HRTFs in the spectral modulation frequency (SMF) domain. The SMF domain representation of HRTFs is obtained by applying Fourier transform to HRTFs in the frequency domain, and scaling in the SMF domain in units of cycles per octave. Low-pass filtering in the SMF domain is equivalent to smoothing processing in the frequency domain. Both the results of theoretical and psychoacoustic experiments indicate that low-pass filtering at 2 cycles/octave does not significantly influence localization accuracy. Time-domain HRIRs are related to frequency-domain HRTFs by Fourier transform. Hence, low-pass filtering in the frequency domain is equivalent to smoothing in the time domain because the HRIR components that rapidly vary with time are discarded. A similar relationship holds between frequency-domain and SMF-domain HRTFs. In this sense, the smoothing method proposed by Kulkarni and Colburn (1998) is similar to that put forward by Qian and Eddins. In the former, however, smoothing is performed in the frequency domain, whereas the latter features smoothing in the SMF domain.

In these two studies, smoothing was carried out on a linear frequency scale; that is, smoothing with a uniform frequency resolution, which inherently characterizes both raw HRTF data and the time window. A uniform frequency resolution is appropriate from the perspective of signal processing, such as Fourier transform. In nature, however, the frequency resolution of human hearing is nonuniform and decreases with increasing frequency (see Section 1.2.2); therefore, uniform frequency resolution does not match the properties of human hearing. Instead, either a critical bandwidth (CB) or an equivalent rectangular bandwidth (ERB) filter model would specify human hearing properties. The desired smoothing processing should consider human hearing and retain the spectral features relevant to auditory perception.

The most straightforward way to accomplish frequency-dependent smoothing on HRTF magnitudes is to use moving averaging in a variable-size frequency window (Koring and Schmitz, 1993):

$$|H_s(f)| = \sqrt{\frac{1}{f_h - f_l} \int_{f_l}^{f_h} |H(f')|^2 \, df'}, \qquad (5.37)$$

where $f_h - f_l$ is the window width. For discrete frequency-domain HRTFs, the integral in Equation (5.37) is replaced by summation. Different choices within the upper and lower boundaries of the frequency window are available. In defining:

$$f_l = \frac{f}{\sqrt{K}} \qquad f_h = f\sqrt{K}. \qquad (5.38)$$

$K = 2$ corresponds to a one-octave width, and $K = 5/4$ is equivalent to the one-third-octave width that roughly matches the properties of human hearing. If the window width in Equation (5.37) is chosen according to auditory filters, such as the Δf_{CB} in Equation (1.4) or the ERB in Equation (1.6), the corresponding results are assumed to be optimal from a psychoacoustic point of view. The performance of the CB and ERB models is similarly accurate above 500 Hz, whereas that of the ERB model is relatively accurate below 500 Hz. Some other researchers attempted to use GammaTone filter banks in auditory smoothing (see Section 13.6; Breebaart and Kohlrausch, 2001).

This HRTF smoothing was carried out within an audible frequency range. No additional efforts were made to evaluate the audibility of high-frequency spectral details in HRTFs. Above 5 kHz, in which the wavelength is short and the effect of fine anatomical structures (such as pinnae) are obvious, HRTFs vary with frequency in a complex manner. The errors caused by various approximations of HRTFs, such as those executed in HRTF filter design, the spatial interpolation, or the principal components analysis and reconstruction (to be discussed in Chapter 6), are visible at high frequencies. Measured HRTFs also exhibit considerable errors at high frequencies.

In previous studies, researchers smoothed binaural HRTFs to the same degree, without quantitatively comparing the results for ipsilateral and contralateral HRTFs. Because sound waves diffract around the head before reaching the contralateral ear, multipath interference from diffracted sound waves causes contralateral HRTFs to vary with frequency in a more complicated manner than do ipsilateral HRTFs. The measurement and errors in various approximations of HRTFs are therefore greater at the contralateral ear than at the ipsilateral ear. Given the shadow effect of the head, however, sound pressure at the contralateral ear is considerably attenuated at high frequencies. Consequently, the complex spectral details of contralateral HRTFs at high frequencies may be inaudible. As stated in Section 1.4.4, the contribution of contralateral HRTF spectra to localization is less than that of ipsilateral HRTF spectra. Thus, in practical signal processing, some simple filter models can be used to approximate contralateral HRTFs at high frequencies.

In psychoacoustic experiments, Xie and Zhang (2010) investigated the audibility of spectral details of HRTFs at high frequencies. The magnitudes of individualized HRTFs at frequencies above 5 kHz were smoothed by a moving frequency window with various bandwidths, and subsequently used to create a virtual sound source in headphone presentation. The primary results for six subjects demonstrate that, in the horizontal and lateral planes, the ipsilateral and contralateral HRTF magnitudes above 5 kHz can be smoothed only with a bandwidth of 2.0 and 3.5 ERB, respectively—without the introduction of audible artifacts. The binaural HRTF magnitudes above 5 kHz can also be simultaneously smoothed with a bandwidth of 2.0 ERB for the ipsilateral ear and 3.5 ERB for the contralateral ear. In the median plane, the binaural HRTFs above 5 kHz can be simultaneously smoothed with a bandwidth of 2.0 ERB, without the introduction of audible artifacts. In practical signal processing for VADs, therefore, it is unnecessary to use excessively complicated signal models to represent the high-frequency spectral details in HRTF magnitudes.

(2) Auditory weighting in HRTF filter design

The basic principle of the HRTF filter design is to minimize the deviation between approximation $\hat{H}(\omega)$ and target $H(\omega)$ in a mathematical sense. The least square error in Equation (5.30), for example, treats different frequencies in an equal manner, with attempts to minimize overall error across the entire frequency range. This treatment might be optimal in a mathematical sense, but not for real auditory perception.

To more accurately match human hearing, the auditory weighting function is introduced at each frequency point where approximation error is evaluated. Greater weighting is assigned to frequencies with high auditory resolutions, whereas less weighting is assigned to those with low auditory resolutions. The weighted least square error is defined as:

$$\min \varepsilon_{\Sigma W} = \min\left[\sum_{\omega_k} W(\omega_k)\,|H(\omega_k)-\hat{H}(\omega_k)|^2\right] = \min\left[\sum_{f_k} W(f_k)\,|H(f_k)-\hat{H}(f_k)|^2\right], \quad (5.39)$$

where weighting function $W(\omega)$ or $W(f)$ is frequency-dependent.

Various choices in weighting function $W(f)$ are available, among which the inverse of the Δf_{CB} in Equation (1.4) or the ERB in Equation (1.6) are commonly used:

$$W(f) = \frac{1}{\Delta f_{CB}} \quad or \quad W(f) = \frac{1}{ERB} \qquad (5.40)$$

Because both Δf_{CB} and ERB increase with increasing central frequency, $W(f)$ decreases with rising central frequency, suggesting that a relatively large high-frequency error is tolerable. This situation precisely reflects the properties of human auditory resolution.

5.3. Methods for HRTF Filter Design

Section 5.2 presented some basic principles of HRTF filter design. With the development of signal processing, various filter design methods have been employed in HRTF filter design. Outstanding studies have been carried out by various researchers; for those reviews, see Jot et al. (1995), Huopaniemi and Karjalainen (1997a), and Huopaniemi and Zacharov (1999a, 1999b).

5.3.1. Finite Impulse Response Representation

Approximating HRTFs by using a finite impulse response (FIR) filter is straightforward and simple. Windowing and direct frequency sampling methods are commonly used. In addition to the windowing method and frequency sampling method, there are other methods for FIR filter design, such as the Parks-McClellan method (Oppenheim et al., 1999).

(1) Windowing method

A time-domain HRIR $\hat{h}(n)$ with a finite length of $N = Q + 1$ can be regarded as the result of truncating the infinite response $h(n)$ by a window function $w(n)$:

$$\hat{h}(n) = h(n)w(n). \tag{5.41}$$

Applying Z-transform to Equation (5.41) yields:

$$\hat{H}(z) = w(0)\hat{h}(0) + w(1)\hat{h}(1)z^{-1} + \ldots + w(N-1)\hat{h}(N-1)z^{-(N-1)} = \sum_{n=0}^{N-1} w(n)\hat{h}(n)z^{-n} \tag{5.42}$$

Two issues related to the windowing method are worth noting. The first is window length. The length of $w(n)$ should be greater than the effective length of HRIRs, and window length can be effectively reduced under minimum-phase approximation (Section 5.2.3). In practical signal processing for free-field virtual source synthesis, further windowing HRIRs is possible, given that the length of raw HRIRs is greater than the minimum length acceptable in auditory perception. The second issue is window shape. Windowing HRIRs in the time domain is equivalent to convoluting the corresponding HRTFs with a frequency-domain representation of window functions. The simplest window is a rectangular window defined by:

$$w(n) = \begin{cases} 1 & 0 \le n \le N-1 \\ 0 & others \end{cases}. \tag{5.43}$$

Sandvad and Hammershøi (1994) studied the effect of various window functions on HRTF filter design. They found that rectangular windowing induces the Gibbs phenomenon to occur as ripples around magnitude response discontinuities. However, rectangular windowing remains a favorable technique because practical HRTF magnitudes are continual.

(2) Frequency sampling method

Uniformly sampling a known HRTF with N-point in the frequency range of $0 \le \omega < 2\pi$ yields:

$$\hat{H}(k) = \hat{H}(z)\Big|_{z = \exp(j\frac{2\pi k}{N})} = \hat{H}(\omega_k) \qquad \omega_k = \frac{2\pi k}{N} \quad k = 0,1..N-1 \tag{5.44}$$

The impulse response of an N-point FIR filter can then be obtained by applying inverse DFT to Equation (5.44):

$$\hat{h}(n) = IDFT[\hat{H}(k)] \qquad n = 0,1.....N-1. \tag{5.45}$$

The frequency sampling method is equivalent to the windowing method; thus, considerations in the windowing method are applicable to the N-point impulse response obtained from N-point uniform frequency-domain sampling.

(3) Interaural transfer function and Wiener filtering

The aforementioned methods for FIR filter design directly model left- and right-ear HRTFs. Some other studies focused on the ratio between contralateral and ipsilateral HRTFs. Assuming that $H_\alpha(z)$ and $H_\beta(z)$ represent ipsilateral and contralateral HRTFs, respectively, then the interaural transfer function (ITF) is defined as:

$$ITF(z) = \frac{H_\beta(z)}{H_\alpha(z)}. \tag{5.46}$$

Ou and Bai (2007) approximated the contralateral $H_\beta(z)$ via ITF and a stochastic Wiener filtering technique. According to Equation (5.46), $H_\beta(z)$ can be written as:

$$H_\beta(z) = ITF(z)H_\alpha(z),\tag{5.46a}$$

indicating that $H_\beta(z)$ can be represented by the $H_\alpha(z)$ and a cascaded $ITF(z)$. That is, the binaural HRTFs of a specific source position can be represented by $H_\alpha(z)$ and $ITF(z)$, where $H_\alpha(z)$ is the same for ipsilateral and contralateral HRTF approximations. According to the FIR representation in Equation (5.32), $ITF(z)$ is expressed as:

$$ITF(z) = \frac{\max(b_{q,\beta})}{\max(b_{q,\alpha})} z^{-(m_\beta - m_\alpha)} \frac{H'_\beta(z)}{H'_\alpha(z)} = \frac{m}{} \frac{b_{q,\beta}}{b_{q,\alpha}} z^{-(m_\beta - m_\alpha)} ITF \quad z$$

with

$$H_\beta \quad z = \sum_{q=0}^{Q} \frac{b_{q,\beta}}{\max(\quad_{q,\beta})}^{-q} \quad '(\ ) = \sum_{}^{Q} \frac{\max(\quad_{q,\beta}}{\max(\quad)_{q,\alpha})}^{-q} \quad ITF'(z) = \frac{H'_\beta(z)}{H'_\alpha(z)},\tag{5.47}$$

in which the subscripts "β" and "α" refer to the contralateral and ipsilateral ears, respectively; "max" refers to the maximal element in the bracketed object; and $ITF'(z)$ denotes the normalized interaural transfer function with initial delay extracted. This way, $ITF(z)$ is represented by $ITF'(z)$, with initial delays and normalized magnitude factor $\max(b_{q,\beta})/\max(b_{q,\alpha})$ added.

To obtain FIR-based models of $ITF'(z)$ by time-domain stochastic Wiener filtering, Equation (5.46a) is converted to the time domain by inverse Z-transform as:

$$h'_\beta(n) = \sum_i itf'(n-i)h'_\alpha(i),\tag{5.48}$$

where the variables in lowercase letters represent the time-domain counterparts (i.e., impulse responses) of those in the initial letters, and both n and i represent discrete time. If $h'_\beta(n)$ and $h'_\alpha(n)$ are known, then $itf'(n)$ can be derived by solving the Wiener–Hopf equation and realized using an FIR filter. Details on the derivation appear in works related to stochastic signal processing (e.g., Hayes, 1996).

$ITF(z)$ eliminates the direction-independent components of HRTFs, such as ear canal resonance. Unlike direct filter models of HRTFs, the FIR-based filter model of $ITF(z)$ is relatively short. Moreover, Ou and Bai (2007) adopted a perceptual criterion—the absolute threshold of hearing—to simplify the filter design of contralateral HRTFs. Analyses revealed that the $ITF'(z)$ above 15 kHz can be set to zero because the corresponding magnitude level is always below the absolute threshold of hearing. Finally, a 40-point FIR filter was designed to approximate $ITF'(z)$ without detectable perceptual quality degradation, when compared with a 256-point FIR filter.

5.3.2. Infinite Impulse Response Representation by Conventional Methods

Prony's and the Yule-Walker methods are conventional techniques for IIR filter design. Prony's method is implemented in the time domain; thus, the resultant model contains information for both magnitude and phase of HRTFs. Assuming that the time-domain HRIR $h(n)$ is known, the task is to design a causal IIR filter with a time-domain impulse response $\hat{h}(n)$, or, equivalently, a system function in the form of Equation (5.22):

$$\hat{H}(z) = \sum_{n=0}^{\infty} \hat{h}(n)z^{-n} = \frac{\sum_{q=0}^{Q} b_q z^{-q}}{1 + \sum_{p=1}^{P} a_p z^{-p}} = \frac{b_0 + b_1 z^- + \dots + b_Q z^{-Q}}{1 + a_1 z^-_1 + \dots + a_P z^{-P}}$$

The main goal is to determine $(P + Q + 1)$ coefficients a_p and b_q with $p = 1,2 \ldots P$ and $q = 0,1 \ldots Q$, to ensure that the square error between target $h(n)$ and approximated $\hat{h}(n)$ in Equation (5.7) is minimized:

$$\min \varepsilon_{\Sigma S} = \min \left[\sum_{n=0}^{U} |h(n) - \hat{h}(n)|^2 \right], \tag{5.49}$$

where U denotes the upper boundary of summation.

To derive a_p and b_q, Equation (5.22) is rewritten as:

$$(1 + a_1 z^{-1} + \ldots a_P z^{-P}) \hat{H}(z) = (b_0 + b_1 z^{-1} + \ldots + b_Q z^{-Q})$$

in which the coefficients are assumed as:

$$a(n) = \begin{cases} 1 & n = 0 \\ a_n & 1 \leq n \leq P \\ 0 & others \end{cases}, \qquad b(n) = \begin{cases} b_n & 0 \leq n \leq Q \\ 0 & others \end{cases}. \tag{5.50}$$

In this way, $a(n)$ and $b(n)$ can be regarded as finite length sequences with lengths of $(P + 1)$ and $(Q + 1)$ points, respectively. Applying inverse Z-transform to Equation (5.50) yields:

$$a(n) * \hat{h}(n) = b(n), \tag{5.51}$$

where '*' denotes convolution. Equation (5.51) can also be written as follows:

$$\hat{h}(n) + \sum_{p=1}^{P} a_p \hat{h}(n - p) = b_n \qquad 0 \leq n \leq Q,$$
$$\hat{h}(n) + \sum_{p=1}^{P} a_p \hat{h}(n - p) = 0 \qquad n \geq (Q+1). \tag{5.51a}$$

Directly substituting Equation (5.51a) into Equation (5.49) yields nonlinear equations of a_p and b_q that are difficult to resolve. Fortunately, approximated results can be derived as follows. First, a_p is identified to minimize the following error:

$$\varepsilon_1 = \sum_{n=Q+1}^{\infty} |h(n) - \hat{h}(n)|^2 = \sum_{n=Q+1}^{\infty} \left| h(n) + \sum_{p=1}^{P} a_p \hat{h}(n - p) \right|^2 \approx \sum_{n=Q+1}^{\infty} \left| h(n) + \sum_{p=1}^{P} a_p h(n - p) \right|^2,$$

for which the second equation in Equation (5.51a) is used to derive the second equality, and $\hat{h}(n - p)$ is replaced by $h(n - p)$ in the last approximation. The condition used to ensure the minimization of error ε_1 is:

$$\frac{\partial \varepsilon_1}{\partial a_p} = 0 \qquad p = 1,2 \ldots P. \tag{5.52}$$

Then, P linear equations regarding a_p are obtained, and these have the matrix form:

$$\begin{bmatrix} r_{11} & r_{12} & \cdots & r_{1P} \\ r_{21} & r_{22} & \cdots & r_{2P} \\ \vdots & \vdots & \vdots & \vdots \\ r_{P1} & r_{P2} & \cdots & r_{PP} \end{bmatrix} \begin{bmatrix} a_1 \\ a_2 \\ \vdots \\ a_P \end{bmatrix} = - \begin{bmatrix} r_{10} \\ r_{20} \\ \vdots \\ r_{P0} \end{bmatrix}, \text{ with } r_{lk} = \sum_{n=Q+1}^{\infty} h(n - l)h(n - k) \; l = 1,2 \ldots P, \; k = 0,1 \ldots P. \tag{5.53}$$

Here, r_{lk} is the autocorrelation function of $h(n)$, such that resolving Equation (5.53) yields P coefficients a_p. With approximations, $\hat{h}(n - p) = h(n - p)$ and $\hat{h}(n) = h(n)$, $(Q + 1)$ coefficients b_q are obtained by substituting coefficients a_p back into the first equation of Equation (5.51a) thus:

$$b_q = h(n) + \sum_{p=1}^{P} a_p h(n-p) \qquad q = 0,1...Q. \tag{5.54}$$

Finally, a desired IIR filter model of HRTFs is constructed by substituting a_p and b_q into Equation (5.22). Prony's method can be conveniently implemented by function "prony" in Matlab software.

Figure 5.1 compares the original HRIRs/HRTFs (solid lines) and the responses of the Prony-based IIR filter with orders $(Q, P) = (26, 26)$ (dashed lines) separated for the left and right ears. The original pair of HRIRs is from MIT-KEMAR data at horizontal azimuth 30°. As stated in Section 3.1.1, the HRIR of the left ear equipped with a DB-061 small pinna is used, and the right ear HRIR at azimuth 30° is taken from the left ear HRIR at azimuth 330° because of left-right symmetry. The raw or measured HRIR is 512 points at a sampling frequency of 44.1 kHz, with corrections made to the low-frequency response of the sound source for HRTF measurement. The HRIR is then truncated to 128 points to form an original HRIR, and then used as the basis for filter design. In Figure 5.1, moderate agreement is observed with relative energy errors [defined in Equation (5.14)] of about −9.8 and −19.1 dB for the left (contralateral) and right (ipsilateral) ears, respectively.

The Yule-Walker algorithm is another method for IIR filter design. Assume that the target HRTF at N discrete digital (angular) frequencies ω_k with $k = 0, 1 \dots N - 1$ is denoted by $H(\omega_k)$, and the transfer function of approximation is:

$$\hat{H}(z) = \frac{b_0 + b_1 z^{-1} + \dots + b_Q z^{-Q}}{1 + a_1 z^{-1} + \dots + a_p z^{-P}}. \tag{5.55}$$

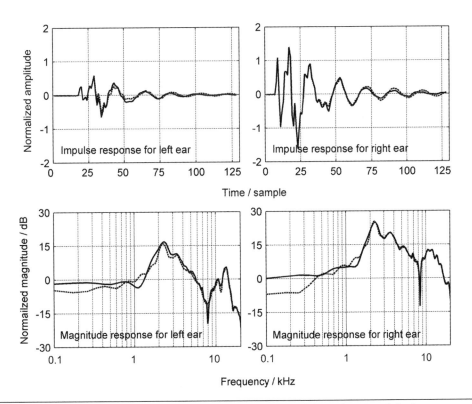

Figure 5.1 Examples for Prony-based HRTF filter design; solid lines denote original responses, and dashed lines represent approximations

The goal is to determine $(P + Q + 1)$ coefficients a_p, b_q with $p = 1,2 \ldots P$ and $q = 0,1 \ldots Q$, to ensure the square error of the magnitude response between the target and approximation is minimized:

$$\min \varepsilon_{\Sigma S} = \min \sum_{k=0}^{N-1} [|H(\omega_k)| - |\hat{H}(z)|_{z=\exp(j\omega_k)}]^2. \tag{5.56}$$

The reader is referred to signal processing books for details. The Yule-Walk algorithm is conveniently realized by function "yulewalk" in Matlab software.

Figure 5.2 shows the magnitudes of the original HRTFs (solid lines) and responses of the Yule-Walk-based IIR filter with orders $(Q, P) = (26, 26)$ (dashed lines) separated for the left and right ears. The 128-point original HRIR pair is identical to that shown in Figure 5.1. General agreement is observed in Figure 5.2, with relative energy errors in HRTF magnitudes [defined similar to Equation (5.14), with magnitudes instead of complex-valued HRTFs] of about −25.1 and −23.3 dB for the left (contralateral) and right (ipsilateral) ears, respectively. Because the Yule-Walk algorithm merely accounts for magnitude, the errors of the left and right ears are similar. In practical filter design of HRTFs, the Yule-Walk algorithm should be supplemented with the phase that has been neglected, in addition to the pure delay that corresponds to the linear-phase component of HRTFs.

5.3.3. Balanced Model Truncation for IIR Filter

The ARMA IIR filter is favorable, from the standpoint of efficiency in implementation. Empirical HRIRs have finite lengths, on the basis of which conventional IIR filters are designed. Balanced model truncation (BMT) is an attractive approach to HRTF filter design; in BMT, a set of state variables are used to represent an LTI system, and system state is determined by state space and output equations. The state variables that contribute less relevant information are discarded for simplification. In what follows, a brief outline of BMT-based HRTF filter design is presented (for more details, see Mackenzie et al., 1997; Beliczynski et al., 1992).

According to signal processing theory, a single-input and single-output system with Q unit delay branches can be represented by Q state variables, which vary as functions of discrete time n and form a $Q \times 1$ column matrix (vector) as:

$$W(n) = [w_1(n), w_2(n), \ldots, w_Q(n)]^T, \tag{5.57}$$

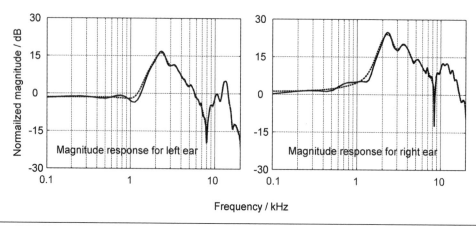

Figure 5.2 Examples for Yule-Walk-based HRTF filter design; solid lines denote original responses, and dashed lines represent approximations

where superscript "*T*" denotes the matrix transpose. The system is described by the state equation:

$$W(n+1)=[A]W(n)+Bx(n), \tag{5.58}$$

and output equation:

$$y(n)=C^T W(n)+D x(n), \tag{5.59}$$

where $x(n)$ and $y(n)$ are system input and output at time n, respectively. [A] is a $Q \times Q$ matrix:

$$[A]=\begin{bmatrix} a_{11} & a_{12} & ..... & a_{1Q} \\ a_{21} & a_{22} & ..... & a_{2Q} \\ \vdots & & & \\ a_{Q1} & a_{Q2} & ..... & a_{QQ} \end{bmatrix}, \tag{5.60}$$

B and C^T represent a $Q \times 1$ column matrix (vector) and $1 \times Q$ row matrix (vector), respectively,

$$B=[b_1,b_2.....b_Q]^T, \qquad C^T=[c_1,c_2,....,c_Q], \tag{5.61}$$

and D is a constant.

Applying Z-transform to Equations (5.58) and (5.59) yields system function $H(z)$:

$$H(z)=\frac{Y(z)}{X(z)}=C^T\{z[I]-[A]\}^{-1}B+D, \tag{5.62}$$

where [I] is a $Q \times Q$ identity matrix. In Equation (5.62), the transfer characteristics of the system are completely determined by [A], B, C^T, and D.

This method can be applied to HRTF filter design. If an N-point HRIR $h(n)$ with $n = 0,1... N-1$ is known, it can be written as $H(z)$ after Z-transform:

$$H(z)=h(0)+h(1)z^{-1}+h(2)z^{-2}+....+h(Q)z^{-Q}=h(0)+H_1(z \tag{5.63}$$

where:

$$H_1(z)=h(1)z^{-1}+h(2)z^{-2}+....+h(Q)z^{-Q}; Q=N-1. \tag{5.64}$$

The $H(z)$ in Equation (5.63) contains Q delayed branches, similar to the system described in Equations (5.58) and (5.59). Therefore, the matrices that correspond to those in Equations (5.60) and (5.61) are obtained by (Huopaniemi and Zacharov, 1999b):

$$[A]=\begin{bmatrix} 0 & 0 & 0 & ... & 0 & 0 \\ 1 & 0 & 0 & ... & 0 & 0 \\ 0 & 1 & 0 & ... & 0 & 0 \\ \vdots & \vdots & \vdots & \vdots & \vdots & \vdots \\ 0 & 0 & 0 & ... & 1 & 0 \end{bmatrix}, \tag{5.65}$$

$$B=[1,0,....0]^T, \quad C^T=[h(1), h(2),.....h(Q)], D=h(0).$$

Because $H_1(z)$ differs from $H(z)$ by lacking $h(0)$, [A], B, and C^T for $H_1(z)$ are the same as those for $H(z)$ in Equation (5.65), except that $D = 0$.

$H_1(z)$ is an FIR filter with corresponding Hankel matrix:

$$[h_{an}]=\begin{bmatrix} h(1) & h(2) & .... & h(Q) \\ h(2) & h(3) & .... & 0 \\ \vdots & \vdots & \vdots & \vdots \\ h(Q) & 0 & \vdots & 0 \end{bmatrix} \tag{5.66}$$

where $[h_{an}]$, which is determined by the coefficients of $H_1(z)$, is a $Q \times Q$ real and symmetric matrix. With the theory of matrix analysis, matrix $[h_{an}]$ is decomposed into:

$$[h_{an}] = [V][\Lambda][V]^T, \tag{5.67}$$

where $\Lambda = \text{diag}[\lambda_1, \lambda_2, \dots \lambda_Q]$, which is a diagonal matrix that consists of Q real eigenvalues in descending magnitude (i.e., $|\lambda_1| \geq |\lambda_2| \geq \dots \geq |\lambda_Q|$); and $[V]$ comprises normalized orthogonal eigenvectors of $[h_{an}]$ with $[V][V]^T = [I]$ (a $Q \times Q$ identity matrix):

$$[V] = [\mathbf{v}_1, \mathbf{v}_2 \dots \mathbf{v}_Q] \qquad [h_{an}]\mathbf{v}_q = \lambda_q \mathbf{v}_q \qquad q = 1, 2 \dots Q,$$

$$\mathbf{v}_{q'}^T \mathbf{v}_q = \delta_{q'q} = \begin{cases} 1 & q' = q \\ 0 & q' \neq q \end{cases}. \tag{5.68}$$

When matrix $[V]$ and $[\Lambda]$ are known, a truncated system of $H_1(z)$ (or, equivalently, $[A]$, $\mathbf{B}$, $\mathbf{C}^T$, and $D = 0$)—up to order $S < Q$ can be obtained with:

$$[A_S] = [V_S]^T [A][V_S] \qquad \mathbf{B}_S = [V_S]^T \mathbf{B} \qquad \mathbf{C}_S^T = \mathbf{C}^T [V_S] \qquad D_S = 0, \tag{5.69}$$

where $[V_S]$ is a $Q \times S$ matrix that consists of the preceding S columns in $[V]$; that is,

$$[V] = \{[V_S], [V_{Q-S}]\}. \tag{5.70}$$

Consequently, $[A_S]$, $\mathbf{B}_S$, and $\mathbf{C}_S^T$ are $S \times S$, $S \times 1$, and $1 \times S$ matrices, respectively. Together, they determine an S-order truncated system (i.e., a subsystem). The S-order truncated system has S delayed branches and S state variables, with the system function in the form:

$$\hat{H}_1(z) = \mathbf{C}_S^T \{z[I] - [A_S]\}^{-1} \mathbf{B}_S. \tag{5.71}$$

$[I]$ is an $S \times S$ identity matrix. According to Equation (5.62), supplementing $D = h(0)$ results in:

$$\hat{H}(z) = \mathbf{C}_S^T \{z[I] - [A_S]\}^{-1} \mathbf{B}_S + h(0). \tag{5.72}$$

As proved, $\hat{H}(z)$ approximates $H(z)$ with the error quantified by the Hankel norm (Beliczynski et al., 1992):

$$\varepsilon_H = \| H_1(z) - \hat{H}_1(z) \|_H \leq 2tr([\Sigma_2]) \tag{5.73}$$

where tr represents the matrix trace. Matrix Σ_2 is determined by matrix Σ:

$$[\Sigma] = \text{diag}[\sigma_1, \sigma_2, \dots \sigma_Q] = \begin{bmatrix} [\Sigma_1] & 0 \\ 0 & [\Sigma_2] \end{bmatrix}. \tag{5.74}$$

where $\sigma_1 = |\lambda_1|, \sigma_2 = |\lambda_2|, \dots, \sigma_Q = |\lambda_Q|$ are the Hankel singular values determined by Equation (5.68), $[\Sigma_1]$ and $[\Sigma_2]$ are $S \times S$ and $(Q - S) \times (Q - S)$ diagonal matrices, respectively. Given that $\sigma_1, \sigma_2 \dots \sigma_Q$ is arranged in descending order, only the state variables with minimal contributions are omitted. This is the basic principle of BMT.

In practical calculation, because of the specific properties of $[h_{an}]$, we can obtain:

$$[A_S] = [V(2:Q, 1:S)]^T [V(1:Q-1, 1:S)]$$
$$\mathbf{B}_S = [V(1, 1:S)]^T \qquad \mathbf{C}_S^T = \mathbf{C}_T[V(1:Q, 1:S)]. \tag{5.75}$$

This method simulates one HRTF at a time. In synthesizing multiple free-field virtual sources or a single source with multiple reflections (i.e., M HRTFs for M sources), each source (and thus, HRTF) should be processed separately. Some researchers generalized BMT for a single source to M simultaneous sources (Georgiou and Kyriakakis, 1999; Grantham et al., 2005; Adams and Wakefield, 2007, 2008, and 2009). That is, they replaced Equations (5.58) and (5.59) with the state equation and output equation of an M-input and one-output state system, where each input represents a source direction. The system was then truncated by the singular value decomposition (SVD) of the matrix, or the Hankel norm optimal approximation. For modeling a single source with multiple reflections, a single-input/multiple-output state system is implemented. For binaural rendering, an M-input and two-output state system can be used. In multiple-input (and output) systems, multiple HRTFs share direction-independent poles (resonance), thereby reducing the total number of state variables and the complexity of the system.

5.3.4. HRTF Filter Design Using the Logarithmic Error Criterion

In Section 5.3.2, HRTF filter design was based on the least square error, between the target and designed frequency responses for Equation (5.5). In the frequency range of relatively low HRTF magnitudes (or energy), a large deviation between target $H(f)$ and designed $\hat{H}(f)$ contributes little to the square error criterion in Equation (5.5). Therefore, the square error criterion emphasizes the errors that occur at spectral peaks, rather than the notches that also provide information important for directional localization (see Section 1.4.4).

The spectral distortion SD defined in Equation (5.18) represents the difference between target $H(f)$ and designed $\hat{H}(f)$ on a logarithmic basis from human hearing. Blommer et al. (1997) proposed an error criterion based on log-magnitude spectrum differences, rather than on magnitude or square-magnitude spectrum differences in ARMA filter design. Blommer assumed that the orders (P, Q) in an ARMA model are both even numbers. Then, the model can be realized with a second-order cascade form, along with a pure delay z^{-m} that corresponds to the linear-phase component of HRTFs. According to Equations (5.24) and (5.31), the system function of a (P, Q)-order filter can be represented as follows:

$$\hat{H}(z) = z^{-m} K \frac{\prod_{q=1}^{Q/2}(1 + b'_{1q}z^{-1} + b'_{2q}z^{-2})}{\prod_{p=1}^{P/2}(1 + a_{1p}z^{-1} + a_{2p}z^{-2})}. \tag{5.76}$$

Assuming that $z = \exp(j\omega)$, the corresponding filter transfer function is:

$$\hat{H}(\omega) = \hat{H}(z)\big|_{z=\exp(j\omega)}. \tag{5.77}$$

Further assuming that target $H(f)$ is presented as a function of discrete frequency or, equivalently, N samples $H(\omega_k)$ at N discrete angular frequencies $\omega_k = 2\pi k/N$ ($k = 0,1 \ldots N - 1$), the logarithmic error criterion is defined as:

$$\varepsilon_{\log} = \sum_{k=0}^{N-1} W(\omega_k) \left| \ln \frac{H(\omega_k)}{\hat{H}(\omega_k)} \right|^2 = \sum_{k=0}^{N-1} W(\omega_k) \left| \ln H(\omega_k) - \ln \hat{H}(\omega_k) \right|^2, \tag{5.78}$$

where $W(\omega_k)$ is a frequency-dependent weighting function. For convenience in calculation, the natural logarithm is adopted in Equation (5.78), and the SD in Equation (5.18) is replaced with $\varepsilon_{\log}$. With $\ln(H) = \ln|H| + j\arg(H)$, Equation (5.78) is equivalently written as:

$$\varepsilon_{\log} = \varepsilon_{\log,M} + \varepsilon_{\log,P},$$

with:

$$\varepsilon_{\log,M} = \sum_{k=0}^{N-1} W(\omega_k) \Big[\ln|H(\omega_k)| - \ln|\hat{H}(\omega_k)| \Big]^2,$$

$$\varepsilon_{\log,P} = \sum_{k=0}^{N-1} W(\omega_k) \Big[\arg H(\omega_k) - \arg \hat{H}(\omega_k) \Big]^2.$$
(5.79)

The logarithmic error $\varepsilon_{\log}$ consists of the log-magnitude error $\varepsilon_{\log,M}$ and phase error $\varepsilon_{\log,P}$. If only concerned with magnitude, then phase error is disregarded. Because the log-magnitude error $\varepsilon_{\log,M}$ is related to $[\ln|H(\omega_k)/\hat{H}(\omega_k)|]^2$, the contribution of errors to spectral peak and notch are equal. This feature is a characteristic specific to logarithmic error.

Designing an IIR filter involves determining $(Q + P + 2)$ unknown coefficients in Equation (5.76), including m, K, b'_{1q}, b'_{2q}, a_{1p}, and a_{2p}, to minimize the logarithmic error $\varepsilon_{\log}$ in Equations (5.78) or (5.79). A necessary condition for identifying minimum $\varepsilon_{\log}$ is to solve:

$$\frac{\partial \varepsilon_{\log}}{\partial m} = 0 \qquad \frac{\partial \varepsilon_{\log}}{\partial K} = 0,$$

$$\frac{\partial \varepsilon_{\log}}{\partial a_{1p}} = \frac{\partial \varepsilon_{\log}}{\partial a_{2p}} = \frac{\partial \varepsilon_{\log}}{\partial b'_{1q}} = \frac{\partial \varepsilon_{\log}}{\partial b'_q} = 0 \qquad p = 1,2....P/2; \qquad q = 1,2...Q/2.$$
(5.80)

Equation (5.80) yields a series of nonlinear equations, for which an exact solution is difficult to determine. Various approximation methods have been proposed. Blommer et al. (1997) used the quasi-Newtonian gradient search algorithm to minimize logarithmic error. In the gradient search algorithm, the optimization result generally depends upon the value of the initial established parameter. Tsujino et al. (2006) proposed a two-step design method for this optimization. The first step is to form a rough design based on the least square error measure, and, in the second step, the gradient search algorithm is used for precise optimization with the logarithmic error measure. On the basis of Blommer's work, Durant and Wakefield (2002) developed an IIR filter design method in which a genetic algorithm is used. The method features a computational speed higher than that achieved by conventional methods.

Kulkarni and Colburn (2004) proposed a new scheme for HRTF filter design. First, an HRTF is represented by multiplying a direction-dependent component (directional transfer function, DTF) by a direction-independent component. Second, the DTF is roughly approximated, using an AR model with coefficients derived by the linear prediction method. On the basis of the AR model, a general ARMA model for DTF is derived by iteration from the perspective of weighted square minimization with frequency weighting functions on a logarithmic scale. This scheme has the significant advantage of simplicity because only sets of linear equations need resolution. A four-interval, two-alternative forced choice (4I/2AFC, to be discussed in Section 13.2) subjective discrimination paradigm is further used to assess the perceptual performance of the modeled HRTFs. Results indicate that the DTF from the $(Q, P) = (6, 6)$ IIR filter is indistinguishable from the target DTF derived from the measured HRTFs.

5.3.5. Common-acoustical-pole and Zero Model of HRTFs

As discussed in Section 5.2.2, the notches and peaks in HRTF magnitude spectra correspond to the zeros and poles of the system function; thus, an ARMA model that uses Equation (5.22) is reasonable. According to Equations (5.22) and (5.23), the IIR filter properties that correspond to an ARMA model are determined by its zeros z_q and poles z_p, or, equivalently, filter coefficients a_p and b_q. In Sections 5.3.2, 5.3.3, and 5.3.4, a set of coefficients for an IIR filter with a preset order is used to represent the HRTF at a specific source direction. Consequently, the zeros and poles are direction dependent, making the resultant filters complex.

Haneda et al. (1999) proposed a common-acoustical-pole and zero (CAPZ) model for a group of HRTFs. In the CAPZ model, the direction-independent poles common to HRTFs at various source directions correspond to the direction-independent peaks in HRTF magnitude spectra that are caused by the physical resonances of an ear canal; the direction-dependent zeros represent spatial variations in HRTFs. For a group of HRTFs, the CAPZ model is simpler than conventional ARMA models of HRTFs because it extracts the direction-independent characteristics that are expressed by common poles with few parameters.

On the basis of Equation (5.22), the system function of an ARMA model for an HRTF at direction θ [strictly speaking, direction should be denoted by (θ, ϕ); only θ is used here for convenience] is expressed as:

$$\hat{H}(\theta,z) = \frac{b_0(\theta) + b_1(\theta)z^{-1} + \dots + b_Q(\theta)z^{-Q}}{1 + a_1 z^{-1} + \dots + a_p z^{-P}} = \frac{1}{A(z)} B(\theta,z), \tag{5.81}$$

$$A(z) = 1 + a_1 z^{-1} + \dots + a_p z^{-P}, \qquad B(\theta,z) = b_0(\theta) + b_1(\theta)z^{-1} + \dots + b_Q(\theta)z^{-Q}.$$

$\hat{H}(\theta, z)$ consists of two parts. One is the direction-independent denominator $A(z)$ with P coefficients $(a_1, a_2 \dots a_p)$ that determine the P common poles in $\hat{H}(\theta, z)$. The other is the direction-dependent numerator $B(\theta, z)$ with $Q + 1$ coefficients $b_0(\theta)$, $b_1(\theta)$, ... $b_Q(\theta)$ that determine the Q direction-dependent zeros in $\hat{H}(\theta, z)$. $1/A(z)$ represents the common transfer function that is irrelevant to direction, and $B(\theta, z)$ represents the directional transfer function.

A similar derivation procedure for Equation (5.51) can be used to estimate the coefficients of Equation (5.81) as follows:

$$\hat{h}(\theta,n) = -\sum_{p=1}^{P} a_p \hat{h}(\theta,n-p) + \sum_{q=0}^{Q} b_q(\theta)\delta(n-q), \tag{5.82}$$

where $\hat{h}(\theta, n)$ represents the impulse response of the filter defined by Equation (5.81), and $\delta(n - q)$ is defined by Equation (2.17).

Each of the original HRIRs at M source directions can be written as an N-point $h(\theta_i, n)$, with $i = 0,1 \dots M - 1$ and $n = 0,1 \dots N - 1$. For the ith direction, the square error between the original HRIR and the modeled HRIR (that is, the impulse response of the filter used for approximating the original HRIR) is defined by:

$$\varepsilon_{\Sigma S}(\theta_i) = \sum_{n=0}^{N+P-1} |h(\theta_i,n) - \hat{h}(\theta_i,n)|^2$$

$$= \sum_{n=0}^{N+P-1} |h(\theta_i,n) + \sum_{p=1}^{P} a_p \hat{h}(\theta_i,n-p) - \sum_{q=0}^{Q} b_q(\theta_i)\delta(n-q)|^2. \tag{5.83}$$

To avoid resolving nonlinear equations, the $\hat{h}(\theta, n)$ in Equation (5.83) is approximately replaced by $h(\theta_i, n)$. Thus, Equation (5.83) becomes:

$$\varepsilon_{\Sigma S}(\theta_i) = \sum_{n=0}^{N+P-1} |h(\theta_i,n) + \sum_{p=1}^{P} a_p h(\theta_i,n-p) - \sum_{q=0}^{Q} b_q(\theta_i)\delta(n-q)|^2. \tag{5.84}$$

The overall square error across directions (the cost function) is:

$$\varepsilon_{all} = \sum_{i=0}^{M-1} \varepsilon_{\Sigma S}(\theta_i) = \sum_{i=0}^{M-1} \sum_{n=0}^{N+P-1} |h(\theta_i,n) + \sum_{p=1}^{P} a_p h(\theta_i,n-p) - \sum_{q=0}^{Q} b_q(\theta_i)\delta(n-q)|^2. \tag{5.85}$$

P direction-independent coefficients a_p ($p = 1, 2...P$) and $M(Q + 1)$ direction-dependent coefficients $b_q(\theta_i)$ ($q = 0,1 ... Q$; $i = 0,1 ... M − 1$) exist in Equation (5.85). A total of $[P + M(Q + 1)]$ unknowns should be determined.

The foregoing analyses can be expressed in the form of a matrix for brevity. Assume an $[M(N + P)] \times 1$ column matrix (a vector) as:

$$e = \begin{bmatrix} e_0 \\ e_1 \\ \vdots \\ e_{M-1} \end{bmatrix} \qquad e_i = \begin{bmatrix} h(\theta_i,0) - \hat{h}(\theta_i,0) \\ h(\theta_i,1) - \hat{h}(\theta_i,1) \\ \vdots \\ h(\theta_i,N+P-1) - \hat{h}(\theta_i,N+P-1) \end{bmatrix}. \qquad (5.86)$$

The unknowns to be determined are written as a $[P + M(Q + 1)] \times 1$ column matrix x:

$$x = [a_1,a_2,...a_P,b_0(\theta_0),b_0(\theta_1)....b_0(\theta_{M-1}),.......b_Q(\theta_0),b_Q(\theta_1)...b_Q(\theta_{M-1})]^T. \qquad (5.87)$$

According to Equations (5.83) to (5.85), e is related to x by:

$$e = h_1 - [A]x, \qquad (5.88)$$

where h_1 is an $[M(N + P)] \times 1$ column matrix (a vector), $[A]$ is an $[M(N + P)] \times [P + M(Q + 1)]$ matrix constructed from $h(\theta_i, n)$ with $i = 0,1 ... M − 1$ and $n = 0,1 ... N − 1$. More details can be found in Haneda et al. (1994).

The overall square error defined in Equation (5.85) can then be written in matrix form:

$$\varepsilon_{all} = e^T e = \{h_1 - [A]x\}^T \{h_1 - [A]x\}, \qquad (5.89)$$

where "+" denotes transpose manipulation. Minimizing ε_{all} by resolving:

$$\frac{\partial \varepsilon_{all}}{\partial x_l} = 0 \qquad l = 1,2....[P + M(Q+1)], \qquad (5.90)$$

yields solutions for the $[P + M(Q + 1)]$ coefficients of the CAPZ model as:

$$x = \{[A]^T[A]\}^{-1}[A]^T h_1. \qquad (5.91)$$

Finally, the system function of the CAPZ model is obtained by substituting the coefficients into Equation (5.81).

In Haneda's work, the HRTFs measured from an artificial (B & K 4128) with length $N = 128$ points (i.e., samples) at a sampling frequency of 48 kHz were used. Then, the magnitude response of the common component of the $(Q, P) = (40, 20)$-order CAPZ model [i.e., $|1/A(z)|$ in Equation (5.81)] was derived from the horizontal HRTFs at every 30° from $\theta = 0°$ to 330° (Figure 5.3; Haneda et al., 1999). Three common resonance peaks, at 2.8, 9.0, and 12.2 kHz, were observed, and are displayed in Figure 5.3. These frequencies almost match three of the five resonance frequencies of the external ear shown by Shaw

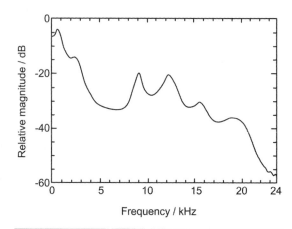

Figure 5.3 Magnitude response of transfer function with the estimated common poles in the CAPZ model (from Haneda et al., 1999, by permission of IEEE)

and Teranishi (1968). The slight difference in resonance frequencies from Shaw's work may be due to the differences in the ears used for the measurements. The direction-dependent zeros [i.e., $b_q(\theta)$] in the CAPZ model of HRTFs can also be used to interpret the relationship between the magnitude of two of the three resonance modes and the incident angle of sound waves. Other researchers attempted to construct a CAPZ model of a group of HRTFs using other techniques, such as the BMT discussed in Section 5.3.3 (Liu and Hsieh, 2001).

5.3.6. Comparison of Results of HRTF Filter Design

In the preceding sections, a variety of approaches to HRTF filter design and some examples have been presented, and some reviews that featured critiques of previously published research have been provided (Jot et al., 1995; Huopaniemi and Zacharov, 1999b). This section compares those approaches.

Through visual inspection, Posselt et al. (1986) concluded that an FIR filter, windowed by a 5 ms Hamming window, retains sufficient spectral information on HRTFs. Begault (1991) used an FIR-modeled HRTF in VADs, and found that the length of the FIR filter for a 10.24 ms HRIR (i.e., 512 points at a sampling frequency of 50 kHz) can be reduced to about 80 points—if only the HRTF magnitude is modeled—with HRTF phase represented by a pure time delay. For FIR-based HRTF filter design, Kulkarni and Colburn (1995) proposed a weighted least square error criterion based on the log-magnitude, and found that a 64-point FIR filter contains the majority of spatial information on HRTFs. Hartung and Raab (1996) concluded that a 48-point FIR filter that uses the techniques of log-magnitude weighting function and nonuniform frequency resolution may not influence localization.

In the early 1980s, the use of IIR models to approximate HRTFs was proposed (Kendall and Rodgers, 1982; Kendall and Martens, 1984). Asano et al. (1990) investigated sound localization in the median plane, and derived IIR models of different orders from individualized HRTFs. Psychoacoustic experiments showed that a (40, 40)-order IIR filter designed from a least square error criterion achieves equivalent localization performance, relative to a target 512-point HRIR, except that, in the former, front-back confusion increases for some subjects. Jot et al. (1995) realized localization in the entire horizontal plane by using IIR filters with orders of (16, 16) to (20, 20).

Sandvad and Hammershøi (1994) studied the effect of different-order FIR and IIR filters through psychoacoustic experiments that included 17 virtual source directions (that is, 17 different HRTFs). For FIR filter representation, a 256-point FIR filter (i.e., 5.33 ms at a sampling frequency of 48 kHz) was used as reference; in simplified FIR filter design, the raw HRIR was simplified by rectangular-window truncation, and the time delay at the beginning of the HRIR was represented by pure delay. 3I/3AFC (3 interval, 3 alternative forced choice) subjective discrimination experiments (see Section 13.2.1) indicate that a simplified 72-point FIR filter (1.5 ms) is indistinguishable from the reference, when white noise with a certain on- and off-ramp is used as a stimulus. For IIR filter representation, a 256-point FIR for the minimum-phase approximated HRTF was used as reference, and the test filter was designed based on minimum-phase HRTFs that were approximated using a modified Yule-Walk algorithm. 3I/3AFC subjective discrimination experiments indicate that a simplified (48, 48)-order IIR filter is perceptually equivalent to the reference, when white noise with a certain on- and off-ramp is used as a stimulus.

Mackenzie et al. (1997) proposed using the BMT (discussed in Section 5.3.3) as a means of designing low-order IIR filter models of HRTFs. In Mackenzie's work, 512-point MIT KEMAR HRIRs (Table 2.1) were employed with the following processes:

1. The initial time delay of the HRIRs was eliminated.

2. An HRTF was diffuse-field equalized [Equation (2.47)] to remove direction-independent features, such as ear canal resonance.
3. The magnitude response was smoothed, with bandwidth varying from 0.25 to 1 CB (see Section 5.2.4).
4. A minimum-phase version of the HRTF was constructed.

The resultant 128-point response was used as reference and basis for generating the BMT-based 10-order IIR filter. For comparison, the 10-order IIR filters generated by the Prony and Yule-Walker methods were also presented. Analyses of the SDRs [given by Equation (5.15)] of 38 filters for the left and right ears at 19 different horizontal directions indicate that the SDR of BMT ranges from 24–36 dB with an average of 29 dB, whereas those of the Prony and Yule-Walker methods are 19.3 and 19.0 dB, respectively. Therefore, the BMT method has an advantage of 10 dB over the other two.

Blommer et al. (1997) used a logarithmic error criterion in HRTF filter design, as presented in Section 5.3.4. The target HRTF has a length of 2048 points at a sampling frequency of 40 kHz. In the design processing, the authors considered both log-magnitude and phase errors, according to Equation (5.80), with frequency weighting function $W(\omega_k)$ as a constant. The results indicate that the (40, 40)-order IIR filter exhibit log-magnitude and phase errors less than 2.1 dB and 0.27 radians above 200 Hz, respectively, resulting in indistinguishable perception artifacts. The performance of a (40, 40)-order IIR filter designed from a logarithmic error criterion is equivalent to that of a (60, 60)-order IIR filter designed from a conventional least square error criterion. A (14, 14)-order IIR filter is derived when only the log-magnitude error is considered. All these results demonstrate that HRTF filters can be simplified using a logarithmic error criterion.

In Kulkarni's work (Section 5.3.4; Kulkarni and Colburn, 2004), a (6, 6)-order IIR filter sufficiently represents the direction-dependent component of HRTFs (i.e., DTF). Adding another (6, 6)-order IIR filter that represents the direction-independent component of HRTFs enables the derivation of a complete HRTF representation. Kulkarni and Colburn also pointed out that greater model error appears at source locations contralateral to the concerned ear. This result is due to the complex structure of the contralateral HRTF; such a structure is caused by the complicated interaction of sounds that reach the contralateral ear through different transmission paths. Moreover, the signal-to-noise ratio of the measured contralateral HRTF is inherently poor because of the head shadow effect, which is also an error source. With increasing elevation, HRTF variations become smoother, leading to fewer model errors—a phenomenon that occurs in various types of HRTF filter design.

The CAPZ model was introduced in Section 5.3.5. Haneda et al. (1999) found that, although both the $(Q, P) = (40, 20)$-order CAPZ model and conventional 60-order FIR filter model use 60 parameters to represent the zeros and poles of HRTFs, the CAPZ filter model remains superior because only 40 direction-dependent zeros (parameters) are used in the CAPZ filter, as opposed to 60 in the FIR filter. Conversely, when 36 HRTFs, spaced in the horizontal plane at intervals of 10°, were modeled with the conventional (40, 20)-order IIR filter, the average modeling error was about 4 dB better than that generated by the CAPZ model. However, the former required $36 \times 60 = 2160$ parameters (i.e., the total number of zeros and poles), whereas the latter needed only $20 + (36 \times 40)$ = 1460 parameters. When the total number of zeros and poles was set equally to 1460 [i.e., when the HRTFs were modeled by a conventional (30, 10)-order IIR filter], the average modeling error was 8 dB worse than that produced by the CAPZ model. Thus, the CAPZ model more efficiently reduces the total number of parameters with a fixed number of zeros and poles.

The foregoing analyses show that different design methods provide different results—even with the same kind of filter structure, such as IIRs. Such variances are partly due to differences in the original HRTF data. The most probable influence is the methods used in various studies. Some processing techniques, such as approximation in physics or auditory perception (e.g., minimum-phase approximation and smoothing) and introduction of perceptually relevant error criteria (e.g., auditory weighting and logarithmic error), can also effectively reduce the required filter order. In addition to objective evaluation, the performance indicated by these design results requires validation by psychoacoustic experiments.

The foregoing filters are implemented in time domain or frequency domain. Alternatively, HRTF filters can also be implemented in the time-frequency domain or the subband domain. Liang and Xie (2012) proposed a subband model for HRTF filters. An HRIR is first decomposed by a two-level wavelet packet (such as db4) and then represented by subband filters and reconstruction filters. The coefficients of the subband filters are the zero interpolation of the wavelet coefficients of the HRIR. The coefficients of the reconstruction filters can be calculated from the wavelet function. The model is simplified by applying a threshold method to reduce the wavelet coefficients. The calculated results indicate that, when compared with an original 128-point HRIR at a sampling frequency of 44.1 kHz, the error in reconstructed HRIR with 30 wavelet coefficients is about 1%. And the result of a psychoacoustic experiment shows that a model with 35 wavelet coefficients is perceptually indistinguishable from the original HRIR. Using the NOBEL identities in wavelet analysis, a common bank of reconstruction filters can be used for multiple virtual sources. This subband filter implementation visibly improves the computational efficiency for multiple virtual sources synthesis, as compared with traditional HRTF filters.

Finally, a prevalent practice in HRTF filter design is the aimless introduction of novel and complex mathematical methods and signal processing techniques, despite applicability issues and physical constraints. The most important issue in HRTF studies is meaningful conclusions concerning physics and auditory perceptual effects. Mathematics is only a tool for analysis. In some cases, complex mathematical tools may not solve problems, but increase problem complexity instead. Moreover, some studies excessively emphasize the accuracy of filter design in a mathematical sense, regardless of possible discrepancies in perceptual performance. Some mathematical errors in filter design may not lead to perceptible effects. Original HRTF data have also been contaminated by measurement errors (Section 2.2.9), not to mention the discrepancy caused by using non-individualized HRTFs in most practical applications. Thus, excessive emphasis on mathematical accuracy in HRTF filter design is meaningless.

5.4. Structure and Implementation of HRTF Filter

Binaural virtual source synthesis [see Equation (1.36)] requires a pair of HRTF filters; that is, $\hat{H}_L(z)$ for the left ear and $\hat{H}_R(z)$ for the right ear. As stated in Section 5.2.3, the linear-phase component of HRTFs is usually extracted and then realized by a delay line z^{-m}. For a given source direction (θ, ϕ), two different delay lines (one for each ear) are needed because the left-ear time delay $T_L(\theta, \phi)$ differs from the right-ear time delay $T_R(\theta, \phi)$ [Figure 5.4(a)]. Alternatively, the ITD (θ, ϕ) that is equal to $|T_L(\theta, \phi) - T_R(\theta, \phi)|$ is cascaded to the lagged ear. Thus, the time delays for the two ears can be addressed using a single delay line. For example, $|ITD(\theta, \phi)|$ should be cascaded to the left ear when azimuth ranges from 0° to 180° [Figure 5.4(b)]. If the pair of HRTFs at (θ, ϕ) is known, the corresponding ITD (θ, ϕ) can be estimated by the methods presented in Section 3.2, such as

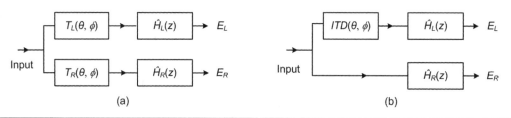

(a)   (b)

Figure 5.4 Implementation of HRTF processing: linear delay is realized by (a) processing the delay of the left and right ears separately, or by (b) cascading a delay |ITD (θ, ϕ)| to the lagged ear (here, the left ear)

the leading-edge detection method. Under the minimum-phase approximation of HRTFs, $\hat{H}_L(z)$ and $\hat{H}_R(z)$ can be realized with a pair of minimum-phase filters; ITD can be evaluated by various methods, among which the interaural cross-correlation delay (ITD$_{corre}$), estimated from cross-correlation calculation, is optimal from a psychoacoustic point of view (Busson et al., 2005). In practical use, some approximation methods [e.g., Equation (1.9), and those to be presented in Section 7.2] have been employed to approximately evaluate ITD (θ, ϕ).

Once the system function $\hat{H}(z)$ of HRTF filters is determined by the methods in Section 5.3, HRTFs can be realized by various filter structures (Oppenheim et al., 1999). Although various structures for FIR or IIR filters are, in principle, applicable to HRTF filters, a practical choice requires comprehensive consideration of the efficiency, accuracy, and stability of the implementation, on the basis of the conditions of system hardware (DSP) and software.

For an FIR filter of the system function given by Equation (5.42), the most direct realization is shown in Figure 5.5. The system for an FIR filter with a length $N = Q + 1$ requires Q delay units, and an output is obtained with N multiplication operations, as well as $N - 1$ addition operations. Alternatively, Equation (5.42) can be rewritten as a product of second-order factorization:

$$\hat{H}(z) = \prod_{q=1}^{[\frac{N}{2}]} (b_{0q} + b_{1q} z^{-1} + b_{2q} z^{-2}). \tag{5.92}$$

Here, $[N/2]$ denotes rounding. When N is even, one of the coefficients b_{2q} is zero. According to Equation (5.92), an FIR filter can also be realized by a second-order cascade structure (Figure 5.6).

With regard to an IIR filter whose system function is given by Equation (5.22), Figure 5.7(a) shows Type I direct realization with number of zeros Q equal to number of poles P. If $Q \neq P$, the gain of certain branches should be zero. Alternatively, an IIR filter can be realized by a Type II structure. When representing a transfer function with $P = Q$, Type II requires only half the number of delay units (i.e., Q) compared with Type I. A direct-realization IIR filter needs $(Q + P + 1)$ multiplication operators and $(Q + P)$ addition operators for each output signal sample.

Given that IIR filters contain a recursive structure, the most significant problems are the errors in filter coefficients and resultant output signals caused by finite word length, which may influence system stability, particularly for hardware implementation associated with fixed-point processing. Therefore, an IIR filter is usually realized in the cascade form provided by Equation (5.24); that is,

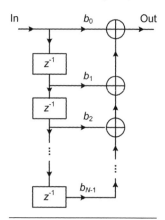

Figure 5.5 Direct realization of an FIR filter

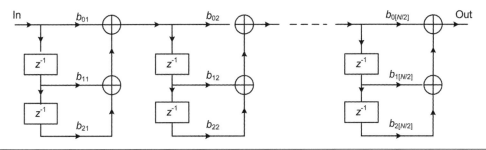

Figure 5.6 A second-order cascade structure of an FIR filter

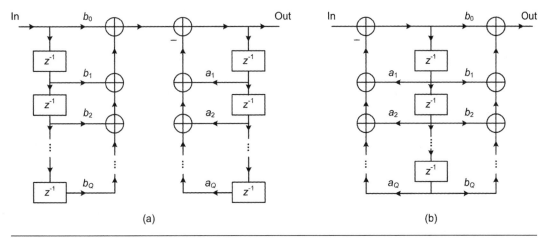

Figure 5.7 Direct realization of an IIR filter: (a) Type I; (b) Type II

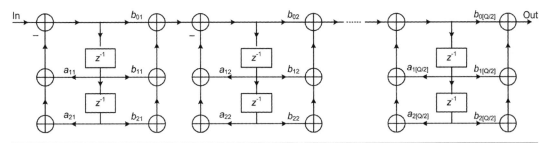

Figure 5.8 Cascade structure of an IIR filter using second-order sections

decomposing the IIR filter into a series of cascade sections. In particular, when $Q = P$ and both are even numbers, the system function of the IIR filter can be written as:

$$\hat{H}(z) = \prod_{q=1}^{Q/2} \hat{H}_q(z) \qquad \hat{H}_q(z) = \frac{b_{0q} + b_{1q}z^{-1} + b_{2q}z^{-2}}{1 + a_{1q}z^{-1} + a_{2q}z^{-2}}. \tag{5.93}$$

If $Q \neq P$, some coefficients (b_{1q}, b_{2q}) or (a_{1q}, a_{2q}) are zero. Figure 5.8 illustrates the cascade structure with second-order sub-filter sections. In the cascade structure, a zero-pole pairing method (i.e., a zero and a pole that are close or conjugated are arranged in the same second-order section) reduces the effect caused by finite word length. Exchanging cascade sequence also facilitates such reduction.

Finally, although HRTF filters can be implemented in a number of ways, a form that is simple, stable, less prone to error, and accords with actual software and hardware specifications is preferable for practical application.

5.5. Frequency-warped Filter for HRTFs

Almost every HRTF filter design method discussed thus far is implemented by conventional Z-transform or Fourier transform on a linear and uniform frequency scale. As pointed out in Section 5.2.4, however, the resolution of human hearing is inherently nonuniform. Therefore, incorporating nonuniform frequency resolution into HRTF filter design is justified. Section 5.2.4 presented some auditory-based filter design techniques, such as pre-smoothing HRTFs and introducing the frequency-dependent weighting function, but the ultimate HRTF filter is implemented on a linear frequency scale. Some studies introduced nonuniform sampling (Bai and Zeung, 2004) and wavelet decomposition to HRTF filter design (Fujinami 1998; Tan and Gan, 1999; Chong et al., 2001; Hacihabiboglu et al., 2002; Torres et al., 2004; Liang and Xie, 2012). The wavelet decomposition method enables the use of nonuniform frequency resolution, possibly improving the performance of the resultant HRTF filter. Despite the advantage of this approach, it may not represent the frequency resolution of human hearing completely. A promising way to decrease perceptible errors in HRTF filter design is to transform the conventional linear frequency scale to a special type of frequency scale that is close to human hearing. On this scale, the resultant HRTF filter is optimal in an auditory sense. This goal can be achieved by frequency warping, which will be discussed in this section. Compared with the indirect methods (e.g., smoothing and weighting) in Section 5.2.4, frequency warping directly describes the nonuniform frequency feature of the auditory system, presenting advantages, such as explicit signal processing models and easy implementation.

5.5.1. Frequency Warping

More details on frequency warping can be found in Harma et al. (2000). Here, only issues concerning HRTFs or HRIRs are outlined.

Let the original HRIR be expressed as a causal sequence $h(n)$. The counterpart of $h(n)$ with Z-transform is:

$$H(z) = Z[h(n)] = \sum_{n=0}^{\infty} h(n)z^{-n}. \tag{5.94}$$

Correspondingly, $h(n)$ can be recovered from $H(z)$ by inverse Z-transform:

$$h(n) = Z^{-1}[H(z)]. \tag{5.95}$$

As discussed in Section 2.1.2, the N-point DFT of $h(n)$ is obtained by sampling $H(z)$ on unit circle $z = \exp(j\omega)$ with N uniform points $\omega_k(k = 0,1 \ldots N-1)$. Thus, the traditional Fourier transform implies a linear frequency scale.

Warping can be applied to complex variable z. That is, z^{-1} in Equation (5.94) is substituted with $\bar{z}^{-1}$ as:

$$z^{-1} \leftarrow \bar{z}^{-1} = D(z) = \frac{z^{-1} - \lambda}{1 - \lambda z^{-1}}, \tag{5.96}$$

where λ is the warping coefficient. This equation indicates that frequency warping is realized by substituting unit delays in Equation (5.94) with the first-order all-pass sections $D(z)$. Equation (5.96) is an operation that maps a unit circle from the z domain to the $\bar{z}$ domain. Correspondingly, inverse frequency warping [i.e., the inverse transform of Equation (5.96)] is defined as:

$$\bar{z}^{-1} \leftarrow z^{-1} = \frac{\bar{z}^{-1} + \lambda}{1 + \lambda \bar{z}^{-1}}, \tag{5.97}$$

that is, $\bar{z}^{-1}$ is substituted with z^{-1}, or a unit circle from the $\bar{z}$ domain is mapped to the z domain.

After the mapping operation in Equation (5.96), the uniform distribution of frequencies $\omega_k (k = 0,1 \dots N-1)$ on the unit circle in the z domain is converted to nonuniform distribution on the unit circle in the $\bar{z}$ domain. The same is applied to the points $\bar{\omega}_k$ $(k = 0, 1 \dots N-1)$ that are uniformly distributed on the unit circle in the $\bar{z}$ domain. Warping therefore changes frequency scale.

Different coefficients λ correspond to different degrees of frequency warping. Except for the conventional linear frequency scale, some auditory-relevant frequency scales (such as the Bark and ERB scales mentioned in Section 1.2.2) are of interest. When 0.7364 is selected as the λ at a sampling frequency of $f_s = 48$ kHz, Equation (5.96) is proven to be approximately reflective of the transformation from linear frequency scale to Bark scale.

Applying Equation (5.96) to Equation (5.94) yields:

$$H(\bar{z}) = \sum_{n=0}^{\infty} h(n)\bar{z}^{-n} = \sum_{n=0}^{\infty} h(n)[D(z)]^n = \sum_{n=0}^{\infty} h(n)\left(\frac{z^{-1} - \lambda}{1 - \lambda z^{-1}}\right)^n, \tag{5.98}$$

which can be interpreted from two perspectives:

1. In the $\bar{z}$ domain, $H(\bar{z})$ is the system function of raw $h(n)$. Appling inverse $\bar{z}$ transform yields $h(n)$. Thus, variants $\bar{z}$ are similar to z, but with different symbols.
2. In the z domain, expanding the last term in Equation (5.98) into a power series of z^{-1} yields:

$$H(\bar{z}) = H_W(z) = \sum_{n=0}^{\infty} h_W(n)z^{-n}. \tag{5.99}$$

When Equations (5.99) and (5.94) are compared, $H(\bar{z})$ or $H_W(z)$ can be regarded as warped $H(z)$, for which inverse Z-transform is written as $h_w(n)$. $h_w(n)$ is actually warped $h(n)$.

Inversely, applying inverse warping to $H(\bar{z})$ in Equation (5.98) by substituting $\bar{z}^{-1}$ with z^{-1} provides:

$$H(z) = \sum_{n=0}^{\infty} h(n)z^{-n}. \tag{5.100}$$

Thus, $H(z)$ can be recovered from $H(\bar{z})$ by inverse warping. Reapplying the warping operation of Equation (5.96) to Equation (5.99)—that is, substituting z^{-1} with $\bar{z}^{-1}$—yields the original $H(z)$:

$$H_W(\bar{z}) = \sum_{n=0}^{\infty} h_W(n)\bar{z}^{-n} = H(z). \tag{5.101}$$

Equation (5.101) also indicates that $H(z)$ can be obtained by applying Z-transform with respect to variable $\bar{z}$ to warped $h_w(n)$.

5.5.2. Frequency-warped Filter for HRTFs

Huopaniemi and Zacharov (1999b) discussed the warped filter design of HRTFs. Similar to conventional HRTF filter design, warped filter design is divided into two categories: warped finite impulse response (WFIR) filters and warped infinite impulse response (WIIR) filters.

As for the traditional windowing method for FIR filters (Section 5.3.1), the design procedure for WFIR filters is as follows:

1. The frequency warping in Equations (5.98) and (5.99) is applied to the desired time-domain impulse response $h(n)$, resulting in $h_W(n)$.
2. $h_W(n)$ is truncated by a time window in Equation (5.41), yielding $\hat{h}_W(n)$.
3. According to Equation (5.99), applying Z-transform to $\hat{h}_W(n)$ yields:

$$\hat{H}_W(z) = \hat{h}_W(0) + \hat{h}_W(1)z^{-1} + ... + \hat{h}_W(N-1)z^{-(N-1)} = \sum_{n=0}^{N-1} \hat{h}_W(n)z^{-n}$$

4. According to Equation (5.101), substituting z^{-1} in the equation in step (3) with $\bar{z}^{-1}$ provides:

$$\hat{H}_W(\bar{z}) = \hat{H}(z) = \sum_{n=0}^{N-1} \hat{h}_W(n)\bar{z}^{-n}. \tag{5.102}$$

Equation (5.102) is the system function of the WFIR filter.

The frequency sampling method for FIR filter design (Section 5.3.1) can also be applied to WFIR design in a similar manner:

1. For a desired frequency-domain $H(z)$, applying:

$$z^{-1} = \frac{\bar{z}^{-1} + \lambda}{1 + \lambda \bar{z}^{-1}}$$

obtains $H_W(\bar{z}) = H(z)$ in Equation (5.101).
2. On the new frequency scale, sample $H_W(\bar{z})$ with N points that are uniformly spaced on unit circle $0 \leq \bar{\omega} < 2\pi$, and then the transfer function of the filter is determined as:

$$\hat{H}_W(k) = \hat{H}_W(\bar{z})\big|_{\bar{z}=\exp(j\bar{\omega}_k)} \quad where \quad \bar{\omega}_k = \frac{2\pi k}{N}, \quad k = 0,1...N-1.$$

3. Applying N-point inverse DFT to the equation in step (2) according to Equations (5.101) and (2.22b) yields:

$$\hat{h}_W(n) = IDFT[\hat{H}_W(k)] \quad n = 0,1...N-1$$

4. The system transfer function of the WFIR filter [i.e., Equation (5.102)] can then be obtained from $\hat{h}_W(n)$.

Similar to that for a WFIR filter, the design procedure for a WIIR filter is as follows:

1. The frequency warping in Equations (5.98) and (5.99) is applied to the desired time-domain impulse response $h(n)$, resulting in $h_W(n)$.
2. $h_W(n)$ is truncated by a time window in Equation (5.41), yielding $\hat{h}_W(n)$.
3. $\hat{h}_W(n)$ is approximated by a conventional IIR filter with Prony's method (Section 5.3.2), thus obtaining the system function, $\hat{H}_W(z)$.

4. Substituting z^{-1} in the equation in step (3) with $\overline{z}^{-1}$ according to Equation (5.101) yields the system transfer function of the WIIR filter as:

$$\hat{H}(z)=\hat{H}_W(\overline{z})=\frac{\sum_{q=0}^{Q}\overline{b}_q\overline{z}^{-q}}{1+\sum_{p=1}^{P}\overline{a}_p\overline{z}^{-p}}. \tag{5.103}$$

Two implementation methods can be used to realize WFIR and WIIR filters: direct implementation in the frequency-warping domain, or mapping back to the traditional linear frequency domain prior to implementation. Consider the WFIR filter in Equation (5.102) as an example. Direct implementation is realized by substituting every unit delay z^{-1} in Figure 5.5 with the all-pass unit $\overline{z}^{-1} = D(z)$ in Equation (5.96). The resultant block diagram is illustrated in Figure 5.9 with filter coefficients that differ from those in Figure 5.5. However, the WIIR filter in Equation (5.103) cannot be directly implemented because of delay-free recursive propagation through $D(z)$ units. This problem can be addressed by appropriately mapping coefficients (Huopaniemi and Zacharov, 1999b).

Alternatively, WFIR or WIIR can be implemented using a traditional FIR or IIR structure. First, WFIR or WIIR is mapped back to the original linear frequency domain; that is, WFIR or WIIR is expressed as a function of z^{-1} by substituting Equation (5.96) into Equations (5.102) or (5.103). Especially for the WIIR filter in Equation (5.103), the system function is written as:

$$\hat{H}(z)=\frac{\displaystyle\prod_{q=1}^{Q'/2}(b_{0q}+b_{1q}z^{-1}+b_{2q}z^{-2})}{\displaystyle\prod_{p=1}^{P'/2}(1+a_{1p}z^{-1}+a_{2p}z^{-2})}, \tag{5.104}$$

and implemented by using the second-order cascade structure in Figure 5.8.

Huopaniemi and Zacharov (1999b) compared the performance of warped filters and conventional FIR/IIR filters. The original blocked ear canal HRIRs were measured in the horizontal plane at azimuths of 0°, 40°, 90°, and 210° with a sampling frequency of 48 kHz. The FIR filter designed

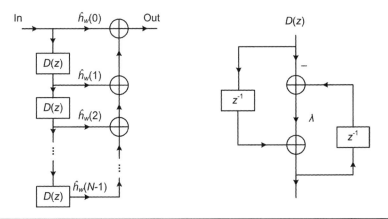

Figure 5.9. WFIR filter structure

on the basis of 257-point (256-order) HRTFs was used as the baseline. Traditional FIR filters with various orders, Prony-based IIR filters, and WIIR filters were compared, and all the filters were reconstructed according to minimum-phase approximation. The WIIR filters were designed using Prony's method with a warping coefficient $\lambda = 0.65$, a lower value than that of approximate Bark scale warping ($\lambda = 0.7364$). Because frequency warping is an auditorily motivated transform, the objective error criteria in Section 5.1 may not enable the comprehensive assessment of frequency warping performance. In Huopaniemi's work, therefore, an improved objective evaluation method, based on the binaural auditory model, was introduced (see Section 13.6 for details). Meanwhile, a listening experiment with a five-grade rating was conducted to verify the filter design results. Both objective and subjective results indicate that the $(Q, P) = (24, 24)$-order WIIR filter is perceptually equivalent to the baseline; a similar performance is achieved with a conventional 96-order (97-point) FIR filter or a $(Q, P) = (48, 48)$-order IIR filter. Thus, appropriate frequency warping improves the performance of HRTF filters.

5.6. Summary

HRTFs are transfer functions of a linear-time-invariant system, which can be accurately or approximately modeled by various linear-time-invariant system models. A variety of filter models for HRTFs have been proposed and used in binaural synthesis processing.

Generally, HRTFs can be approximated by the autoregressive moving-average model (ARMA), autoregressive (AR), and moving average (MA) models, among which the AR type is inferior because of poor efficiency. The impulse response lengths of the ARMA and AR models are infinite, whereas that of the MA model is finite. Therefore, the former are called IIR filter models, whereas the latter are referred to as FIR filter models.

The aim of HRTF filter design is to determine a specific finite impulse response (FIR) or infinite impulse response (IIR) filter, whose transfer function matches or approximates target HRTFs. A variety of mathematical criteria have been suggested to quantify the extent of approximation or errors in HRTF filter design. Among these, the least square error criterion is commonly used.

Some physical and auditory rules are usually incorporated to simplify the HRTF filters. For example, extracting the linear-phase component of HRTFs (which can be separately processed as a delay line) can reduce the order required for HRTF filters. Some auditory-based techniques (such as HRTF smoothing, auditory weighting, and a logarithmic error criterion) are also somewhat effective. Moreover, appropriate frequency warping transforms the conventional linear frequency scale to other frequency scales (such as Bark scale) that closely replicate auditory properties. Incorporating frequency warping into HRTF filter design reduces perceptible errors.

Various signal processing methods can be applied in HRTF filter design. For FIR filters, design methods include traditional windowing, frequency sampling, and Wiener filtering methods. For FIR filters, examples of design techniques are Prony's, Yule-Walk, and balanced model truncation methods. The CAPZ-based (common-acoustical-pole and zero) FIR filter also enables the representation of a group of HRTFs with few parameters.

For a certain type of filter, such as an FIR filter, the results obtained by different methodologies in various studies differ—possibly because of differences in methodology and original HRTF data. The physical or perceptual approximation of original HRTFs and the incorporation of auditory considerations into error criteria can simplify design results.

Designed HRTF filters can be realized using various structures. In practice, simple and stable structures with few errors are preferred.

6

Spatial Interpolation and Decomposition of HRTFs

This chapter focuses on the spatial characteristics of HRTFs. The basic principles and schemes for HRTF spatial interpolation are given. Two basic types of HRTF decomposition, spectral shape basis function decomposition and spatial basis function decomposition, are analyzed. The relationships between HRTF spatial interpolation, decomposition, and signal mixing in multi-channel sound are clarified, and some simplified algorithms for HRTF-based binaural synthesis are introduced. The microphone array beamforming models for synthesizing binaural signals and HRTFs are also discussed.

As discussed in Chapter 5, some (digital) filters have been designed to match the responses of head-related transfer functions (HRTFs) or head-related impulse responses (HRIRs) at a given source direction, resulting in filter models and the implementation of HRTF-based signal processing. However, an HRTF is also a complex function of source position. Synthesizing virtual sources at different spatial positions necessitates considering the spatial characteristics of HRTFs or HRIRs, as well as related signal processing techniques. Such spatial characteristics are the topic of this chapter. Section 6.1 presents the basic concepts and schemes for spatial interpolation, which enable the reconstruction of HRTFs at unmeasured directions. HRTFs can be efficiently represented by various linear decomposition techniques, which are closely related to the spatial and temporal, or frequency, characteristics of HRIRs/HRTFs. Section 6.2 discusses HRTF decomposition with spectral shape basis functions. The algorithms for conventional principal components analysis (PCA) and the associated compact representation of HRTFs are analyzed in detail. The subset selection of HRTFs as spectral shape basis functions is also outlined. Section 6.3 focuses on HRTF decomposition with spatial basis functions. This section analyzes two kinds of HRTF decomposition with predetermined spatial basis functions: the azimuthal Fourier series and spherical harmonic (SH) functions. These types of decompositions are then used as bases for deriving the azimuthal or spatial sampling (Shannon-Nyquist) theorem of HRTFs. Spatial PCA, discussed in Section 6.3, is used to derive a spatially compact basis function representation of HRTFs. The spatial characteristics and decomposition of HRTFs are applicable to signal processing in virtual auditory displays (VADs). The analogy between HRTF spatial interpolation, decomposition, and signal mixing (panning) in multi-channel sound are discussed in Section 6.4. Some simplified algorithms for HRTF-based binaural synthesis are introduced in Section 6.5. Finally, Section 6.6 describes the microphone array beamforming models for synthesizing binaural signals and HRTFs.

6.1. Directional Interpolation of HRTFs

6.1.1. Basic Concept of HRTF Directional Interpolation

Far-field HRTFs are continuous functions of source direction (θ, ϕ). As stated in Chapter 2, HRTFs are usually measured at discrete and finite directions—that is, sampled at directions around a spatial spherical surface. Under certain conditions, the HRTFs at unmeasured directions can be reconstructed or estimated from measured data by various interpolation schemes.

At a constant source distance $r = r_0$, for example, the HRTFs at the arbitrary unmeasured azimuth θ can be estimated from the HRTFs measured at M horizontal azimuths [i.e., $H(\theta_i, f)$ with $i = 0, 1, \dots, M - 1$] by the linear interpolation scheme:

$$\hat{H}(\theta, f) \approx \sum_{i=0}^{M-1} A_i H(\theta_i, f), \qquad (6.1)$$

where $\hat{H}(\theta, f)$ is the interpolated HRTF and $A_i = A_i(\theta)$ is a set of weights related to the target azimuth. Different methods for selecting measured azimuths and weights lead to various interpolation schemes. At a given azimuth, the measured HRTFs are usually represented by samples at N discrete frequencies (or HRIRs at N discrete times). Accordingly, Equation (6.1) can be interpreted from two viewpoints. First, it is an azimuthal interpolation equation at each discrete frequency $f = f_k$ ($k = 0, 1 \dots N - 1$). Second, the HRTF at a given azimuth can be written as an $N \times 1$ complex-valued vector (matrix) $\mathbf{H}(\theta_i)$, with each element representing the HRTF sample at each discrete frequency. Then, Equation (6.1) is regarded as an equation for evaluating vector $\hat{\mathbf{H}}(\theta)$ at an arbitrary unmeasured azimuth from a linear combination of M known vectors $\mathbf{H}(\theta_i)$. In what follows, although the HRTF is written as a function of continuous frequency, it consists of samples at discrete frequencies (denotations with black characters represent the $N \times 1$ vector, to distinguish it from the continuous function of frequency or time).

Equation (6.1) can be extended to three-dimensional spatial directions as:

$$\hat{H}(\theta, \phi, f) \approx \sum_{i=0}^{M-1} A_i H(\theta_i, \phi_i, f), \qquad (6.2)$$

where $H(\theta_i, \phi_i, f)$ with (θ_i, ϕ_i) ($i = 0, 1, 2 \dots M - 1$) denotes the measured HRTFs at a constant source distance $r = r_0$ and M appropriate spatial directions.

Equations (6.1) and (6.2) are HRTF directional interpolation equations in the frequency domain. According to the linear characteristics of Fourier transform, these two equations are also applicable to HRIR interpolation in the time domain, in which HRTFs are substituted with corresponding HRIRs. For example, the time-domain version of Equation (6.2) is:

$$\hat{h}(\theta, \phi, t) \approx \sum_{i=0}^{M-1} A_i h(\theta_i, \phi_i, t). \qquad (6.3)$$

The directional interpolation of HRIRs is mathematically equivalent to that of complex-valued HRTFs.

As stated in Section 3.1.3, an HRTF can be represented by its minimum-phase function cascading with a linear phase or pure delay component under certain conditions. The directional interpolation scheme can, therefore, be separately applied to minimum-phase HRTFs (or HRIRs) and pure delay. Because a minimum-phase function is uniquely determined by its magnitude, the directional interpolation of minimum-phase HRTFs can be replaced by the directional interpolation of HRTF magnitudes. As will be discussed later in this Section as well as Sections 6.2 and 6.3, the interpolation of minimum-phase HRTFs or HRTF magnitudes exhibits improved performance. For the interpolation of HRTF magnitudes, Equation (6.2) becomes:

$$|\hat{H}(\theta,\phi,f)| \approx \sum_{i=0}^{M-1} A_i \, |H(\theta_i,\phi_i,f)|. \tag{6.4}$$

Some directional interpolation schemes are discussed in the following sections. For simplicity, all equations are initially meant for HRTFs in the frequency domain, but they are also suitable for HRTF magnitudes, HRIRs, and the minimum-phase representation of HRTFs or HRIRs. In practical calculation, the continuous frequency f or continuous time t are replaced by the discrete frequency f_k and discrete time n, respectively.

6.1.2. Some Common Schemes for HRTF Directional Interpolation

The simplest scheme for directional interpolation is *adjacent linear interpolation*. Within the azimuthal region $\theta_i < \theta < \theta_{i+1}$ in the horizontal plane $\phi = 0°$, Equation (6.1) is approximated by the first-order term of its Taylor expansion of θ:

$$\hat{H}(\theta,f) \approx H(\theta_i,f) + \frac{\partial H(\theta,f)}{\partial \theta}\bigg|_{\theta=\theta_i} (\theta - \theta_i)$$

$$\approx H(\theta_i,f) + \frac{H(\theta_{i+1},f) - H(\theta_i,f)}{\theta_{i+1} - \theta_i}(\theta - \theta_i)$$

Then,

$$\hat{H}(\theta,f) \approx A_{i+1} H(\theta_{i+1},f) + A_i H(\theta_i,f), \tag{6.5a}$$

where:

$$A_{i+1} = \frac{\theta - \theta_i}{\theta_{i+1} - \theta_i} \qquad A_i = 1 - \frac{\theta - \theta_i}{\theta_{i+1} - \theta_i}. \tag{6.5b}$$

Equation (6.5) is the equation of adjacent linear interpolation, which indicates that the HRTF at an arbitrary target azimuth θ can be interpolated from the samples at two neighboring azimuths θ_i and θ_{i+1} with frequency-independent weights A_i and A_{i+1}. The smaller the interval between azimuths θ_i and θ_{i+1} and the smoother the $H(\theta,f)$ azimuthal variation, the more accurate the HRTF interpolation. Moreover, when the target azimuth θ lies between neighboring azimuths θ_i and θ_{i+1}, i.e., $\theta_i < \theta < \theta_{i+1}$, two weights always satisfy $0 < A_i, A_{i+1} < 1$, so that they are in phase.

Equation (6.5) can be extended to the case of $\theta_i < \theta_{i+1} < \theta$ or $\theta < \theta_i < \theta_{i+1}$—that is, predicting the HRTFs' outside region $[\theta_i, \theta_{i+1}]$ using the HRTFs at azimuths θ_i and θ_{i+1}. In this case, weights A_i and A_{i+1} are out of phase. Equation (6.5) is also applicable to HRTF azimuthal interpolation at elevations outside the horizontal plane. A similar equation for HRTF interpolation in the median plane can be derived by substituting the azimuth θ in Equation (6.5) with elevation ϕ.

Bilinear interpolation is an extension of adjacent linear interpolation (Wightman et al., 1992b). Given that HRTFs are measured at a constant source distance $r = r_0$ with respect to the head center, the spherical surface (upon which the source is located) is sampled along both azimuthal and elevation directions, resulting in a measurement grid, with its vertices representing the source directions for measurement.

Figure 6.1 shows that the intervals of the grid along θ and ϕ are denoted by θ_{grid} and ϕ_{grid}, respectively; the four measured source directions nearest the target direction are numbered 1, 2, 3, and 4. The associated HRTFs are denoted by:

$$H(1,f) = H(\theta_1, \phi_1, f) \qquad\qquad H(2,f) = H(\theta_1 + \theta_{grid}, \phi_1, f)$$
$$H(3,f) = H(\theta_1 + \theta_{grid}, \phi_1 + \phi_{grid}, f) \qquad H(4,f) = H(\theta_1, \phi_1 + \phi_{grid}, f)$$

With the Taylor expansion of a multivariate function, the HRTF at a target direction $(\theta, \phi) = (\theta_1 + \Delta\theta, \phi_1 + \Delta\phi)$ within the grid can be approximated as a weighted sum of the HRTFs associated with the four nearest directions:

$$\hat{H}(\theta,\phi,f) \approx A_1 H(1,f) + A_2 H(2,f) + A_3 H(3,f) + A_4 H(4,f), \tag{6.6a}$$

where:

$$
\begin{aligned}
A_1 &= (1 - A_\theta)(1 - A_\phi) \quad A_2 = A_\theta(1 - A_\phi) \\
A_3 &= A_\theta A_\phi \qquad\quad A_4 = (1 - A_\theta)A_\phi \\
A_\theta &= \frac{\Delta\theta}{\theta_{grid}} \qquad A_\phi = \frac{\Delta\phi}{\phi_{grid}}
\end{aligned}
\tag{6.6b}
$$

Equation (6.6b) is the bilinear interpolation equation of HRTFs.

Alternatively, some researchers chose to use normalized linear inverse distances, between the target direction and the four neighboring measured directions, as the four associated weights in Equation (6.6a) (Hartung et al., 1999a). The weights obtained in this manner are approximately equal to that given by Equation (6.6b), provided that the intervals between the adjacent measured directions are small.

Spherical triangular interpolation is similar to bilinear interpolation. The HRTFs at the four neighboring directions (numbered 1, 2, 3, and 4) are measured (Figure 6.2). The target P at direction (θ, ϕ) lies within the grid. Suppose that one of the four directions (e.g., direction 1 in the figure) is the nearest to target P. Then, the HRTF at the target direction can be estimated by a weighted sum of the HRTFs associated with the nearest direction and the two other directions (2 and 4 in the figure) adjacent to it:

$$H(\theta,\phi,f) = A_1 H(1,f) + A_2 H(2,f) + A_4 H(4,f), \tag{6.7a}$$

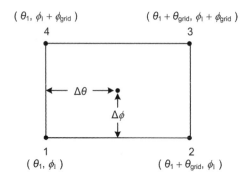

Figure 6.1 Illustration of bilinear interpolation

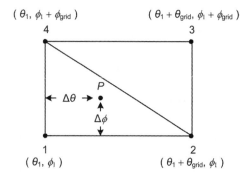

Figure 6.2 Illustration of spherical triangular interpolation

where:

$$A_1 = 1 - A_\theta - A_\phi \quad A_2 = A_\theta \quad A_4 = A_\phi$$

$$A_\theta = \frac{\Delta\theta}{\theta_{grid}} = \frac{|\theta - \theta_1|}{\theta_{grid}} \qquad A_\phi = \frac{\Delta\phi}{\phi_{grid}} = \frac{|\phi - \phi_1|}{\phi_{grid}}. \tag{6.7b}$$

Equation (6.7b) is the spherical triangular interpolation equation of HRTFs, in which (θ_1, ϕ_1) denotes the neighboring direction nearest the target. The interpolation equations for calculating HRTFs at other target directions within the grid are similar to Equations (6.7a) and (6.7b).

Cubic spline interpolation is a scheme in which segmented cubic polynomials are used to fit HRTFs. Let $H(\theta_0, f)$, $H(\theta_1, f)$, ... $H(\theta_M, f)$ denote the measured HRTFs at $M + 1$ horizontal azimuths $\theta_0 < \theta_1 < ... < \theta_M$. Azimuth periodicity yields $\theta_M = \theta_0 + 2\pi$ and $H(\theta_M, f) = H(\theta_0, f)$. At each specific frequency f, the periodic cubic spline function $S_f(\theta)$ should satisfy the following requirements:

1. Within each subregion $[\theta_i, \theta_{i+1}]$ $(i = 0,1, ... M - 1)$, function $S_f(\theta)$ can be represented by a cubic polynomial of azimuth θ.
2. Within region $[\theta_0, \theta_M]$, function $S_f(\theta)$ and its first- and second-order derivatives are continuous.
3. At azimuth $\theta = \theta_i$ $(i = 0,1 ... M)$, $S_f(\theta_i) = H(\theta_i, f)$.
4. The first- and second-order derivatives of $S_f(\theta)$ are periodic functions of azimuth θ.

The algorithm for deriving function $S_f(\theta)$ from measured HRTFs at $M + 1$ azimuths can be found in mathematical textbooks (Richard et al., 2005). When the function $S_f(\theta)$ in each subregion is obtained, the interpolated HRTFs can be estimated. The cubic spline interpolation scheme can also be used to interpolate the HRTFs in the median plane. Finally, this scheme is easily implemented using Matlab.

6.1.3. Performance Analysis of HRTF Directional Interpolation

The performance of HRTF interpolation is evaluated by objective and subjective methods. The objective method involves analyzing the physical errors in interpolated HRTFs or analyzing the similarity between actual and interpolated HRTFs. The various error or similarity criteria presented in Section 5.1 are applicable to evaluating the performance of HRTF interpolation. The subjective method entails analyzing the perceived performance of interpolated HRTFs by various psychoacoustical experiments, as comprehensively discussed in Chapter 13.

HRTF interpolation performance depends upon interpolation schemes (Runkle et al., 1995; Hartung et al., 1999a), but denser directional samples generally yield better interpolation performance. Interpolation performance is also related to frequency range, target direction, and HRTFs used.

Christensen et al. (1999) used the measured HRIRs from a VALDEMAR artificial head with a directional interval of 2° (Table 2.1, database no. 7) to evaluate the effectiveness of applying adjacent linear interpolation on minimum-phase HRIRs in the horizontal, median, lateral, and $\phi = 45°$ elevation planes. Interpolation error was evaluated by the spectral distortion (SD) in HRTF magnitudes [Equation (5.18)]. The results showed that interpolation error increases with increasing frequency. Interpolation error is low for ipsilateral source directions and high for contralateral source directions, a difference that stems from the complicated nature of contralateral HRTFs at high frequencies. Such complexity is caused by multi-path interference from diffracted sound waves around the head. Additionally, the signal-to-noise ratio of measured contralateral HRTFs is inherently poor because of the attenuation caused by the head shadow effect. Nevertheless, large errors in interpolated contralateral HRTF magnitudes at high frequencies may be imperceivable because of their small contribution to auditory perception. As elevation increases, the variations of HRTF with azimuth become smooth, thereby reducing interpolation error. The calculation results indicate that, below 7 kHz and with an azimuthal interval $\Delta\theta = 36°$, the error SD are within 3 dB for most azimuths in the horizontal plane, except for the contralateral directions between azimuths of 45° and 135°. For most source directions in the median and lateral planes, the errors with an angle interval $\Delta\phi = 20°$ are within 3 dB. At an elevation $\phi = 45°$, errors with an azimuthal interval $\Delta\theta = 36°$ are within 3 dB up to 10 kHz. Furthermore, three-interval, three-alternative forced-choice (3I/3AFC) subjective discrimination experiments (see Section 13.2.1) were conducted, to evaluate the directional interval or resolution required in HRTF measurements (Plogsties, 2000b; Minnaar et al., 2005). The results for the horizontal, median, and lateral planes indicate that an interval of 4° is required for some directions, to avoid distinguishing adjacent linear-interpolated minimum-phase HRIRs from measured ones. Hoffmann et al. (2008a) also used the measured HRIRs from the VALDEMAR artificial head and carried out a 3I/2AFC subjective discrimination experiment to investigate the audibility of the difference in adjacent HRTF magnitude spectra. The results showed that when the directional interval exceeds a direction-dependent threshold of 2.8° to 17.2°, differences among adjacent HRTF magnitude spectra are detectable. Therefore, interpolation is required for binaural signal synthesis processing, when the directional interval of measured HRTFs exceeds the detection threshold. However, the results of another experiment by Hoffmann et al. (2008b) suggest that the mean detectable threshold for changes in ITDs range from 87.8 to 163 μs, a value that is considerably larger than that in previous results from other researches.

Using data measured from the artificial head, Nishino et al. (1999) evaluated the effectiveness of applying adjacent linear interpolation and cubic spline interpolation on median-plane HRTF magnitudes. The performance of the two schemes are comparable at an elevation interval of 30° or less, with the mean spectral distortion SD_M across an audible frequency range [Equation (5.19)] at about 2 dB. However, cubic spline interpolation is effective at large and even intervals, whereas adjacent linear interpolation exhibits improved performance for a set of appropriately selected uneven intervals.

Using individualized HRTFs and the bilinear interpolation scheme, Wightman et al. (1992b) carried out a psychoacoustical experiment to evaluate the localization accuracy of interpolated complex-valued HRTFs and the interpolated minimum-phase HRTFs that were obtained solely from interpolated magnitudes under minimum-phase approximation. The interpolated minimum-phase HRTFs and measured HRTFs yield comparable localization performance at an azimuth interval $\Delta\theta = 15°$ and elevation interval $\Delta\phi = 12°$. The localization accuracy of the interpolated minimum-phase HRTFs slightly decreases, when azimuth and elevation interval increase to $\Delta\theta = 30°$ and $\Delta\phi = 24°$, respectively. The interpolated complex-valued HRTFs degrade localization accuracy (manifested primarily as an increase in front-back confusion), even with $\Delta\theta = 15°$ and $\Delta\phi = 12°$. Wenzel and Foster (1993b) used nonindividualized HRTFs to evaluate the localization performance of the minimum-phase HRTFs obtained from bilinear interpolation of HRTF magnitudes. At an azimuth interval $\Delta\theta = 15°$ and elevation interval $\Delta\phi = 18°$, although the nonindividualized HRTFs produce considerable localization errors (front-back and up-down confusion), the interpolated complex-valued HRTFs and interpolated minimum-phase HRTFs exhibit similar localization performance.

6.1.4. Problems and Improvements of HRTF Directional Interpolation

At a constant source distance r relative to the head center, source-to-ear distance varies with source direction. Thus, the arrival time of an HRIR for a given ear (e.g., the right ear) is direction dependent (Section 3.1.1). Directly interpolating HRIRs may cause spectral discrepancies. Consider a simple case of a pair of horizontal HRIRs (simplified as Dirac delta functions) at two neighboring azimuths θ_i and θ_{i+1} with arrival times τ_i and τ_{i+1}, respectively—that is, $h(\theta_i, t) = \delta(t - \tau_i)$ and $h(\theta_{i+1}, t) = \delta(t - \tau_{i+1})$. When a target azimuth θ lies between θ_i and θ_{i+1}, the desired HRIR should be $h(\theta, t) = \delta(t-\tau)$ with $\tau \approx (\tau_i + \tau_{i+1})/2$. However, the adjacent linear-interpolated HRIR estimated from Equation (6.5) is $h(\theta, t) = [\delta(t - \tau_i) + \delta(t-\tau_{i+1})]/2$, which is different from the desired HRIR. Although an actual HRIR is not a Dirac delta function, this case illustrates the problem. Equation (6.5) is derived by approximating the Taylor series representation of an HRTF or HRIR to its first-order term. When an HRIR rapidly varies with θ, the approximation is invalid. A similar problem exists in the direct interpolation of HRTFs in the frequency domain.

Section 6.1.1 showed that the interpolation of minimum-phase HRIRs or HRTF magnitudes alone exhibits improved performance. As stated in Section 3.1.3, however, the minimum-phase approximation of HRTFs does not always hold. Moreover, the nature of linear interpolation lies in estimating the HRTFs or HRIRs at unmeasured directions from a linear combination of measured ones. According to signal processing theory, the linear combination of minimum-phase functions does not always result in a minimum-phase function. To avoid the problem caused by interpolating minimum-phase HRTFs, some researchers suggested imposing arrival time correction on HRIR interpolation (Matsumoto et al., 2004). That is, prior to interpolation, the arrival time of the HRIR for each source direction is made synchronous by shifting the onset time of each HRIR. If all the arrival times are corrected to $t = 0$, the resultant HRIRs or HRTFs include the minimum-phase and all-pass phase components, with the linear-phase component excluded [Equation (3.1)]. The arrival time, which is useful in interaural time difference (ITD) calculation, can be evaluated by leading-edge detection or cross-correlation calculation, similar to that implemented in Section 3.2. The arrival times can then be separately interpolated. For KEMAR horizontal HRIRs with an azimuth interval $\Delta\theta = 15°$, the mean signal-to-distortion ratio (SDR_M) values [Equation (5.17)] for adjacent linear interpolation are 14.3 and 7.7 dB for cases with and without arrival time correction, respectively. Therefore, arrival time correction reduces interpolation errors.

A similarly improved interpolation scheme can be applied to HRTFs in the frequency domain. A complex-valued HRTF can be represented by its magnitude and phase as:

$$H(\theta,\phi,f)=|H(\theta,\phi,f)|\exp[j\psi(\theta,\phi,f)], \tag{6.8}$$

where $\psi(\theta, \phi, f)$ is the unwrapped phase. The magnitude and unwrapped phase can be separately interpolated, as opposed to separately interpolating the real and complex parts of the HRTFs in Equation (6.2). Some researchers also proposed interpolating the HRIRs in the wavelet domain (Torres and Petraglia, 2009), in which the wavelet coefficients of the wavelet-decomposed HRIRs are interpolated.

The directional interpolation algorithm and associated weights A_i are identical for each (discrete) frequency in the linear interpolation schemes given by Equation (6.2). On the basis of the spatial frequency response surfaces (SFRSs) discussed in Section 3.1.2, Cheng et al. (1999) proposed an alternative HRTF interpolation scheme. At each given frequency, far-field HRTF magnitude is only a function of direction (θ, ϕ) and can be represented by an SFRS. Similar to the spherical triangular interpolation stated above, the SFRS is divided by a triangular grid with its vertices representing measured directions. The HRTF magnitude at an arbitrary target direction can be estimated by a weighted sum (mean) of the HRTF magnitudes associated with the three vertex directions. Distinguished from spherical triangular interpolation, the three vertices are not always selected as the nearest to the target (Figure 6.2). Instead, they are selected so that the HRTF magnitude varies smoothly within the region closed by them. At different frequencies, HRTF magnitude varies with direction in different ways; accordingly, the selection of vertices also differs. That is, the SFRS-based interpolation scheme enables the interpolation of HRTF magnitudes at different frequencies with various grids and weights to optimize interpolation performance.

Some researchers used the neural network method to interpolate horizontal KEMAR HRIRs (Nishino et al., 1996). The overall performance of the method is inferior to that of the linear interpolation of HRIRs with arrival time correction. This case reveals that aimlessly using some novel mathematical or signal-processing tools, regardless of applicability or physical nature, may not produce desirable results.

Keyrouz and Diepold (2008) described the directional variations of HRTFs by state space (the temporal variations of HRIRs were described by state space in Section 5.3.3), and proposed an interpolation scheme based on the solution for the rational state-space interpolation problem.

To render a moving sound source in HRTF-based VADs, a large set of HRTFs should be interpolated in real time, which requires reducing the computational cost of interpolation. On the basis of the spherical triangular interpolation in Section 6.1.2, Freeland et al. (2004) introduced the interpositional transfer function (IPTF) and proposed an IPTF-based interpolation scheme. In Figure 6.2, suppose that, among the four vertices, vertex 1 is the nearest to the target direction P. Then, Equation (6.7a) can be written as:

$$H(\theta,\phi,f) = H(1,f)[A_1 + A_2 IPTF(2,1,f) + A_4 IPTF(4,1,f)], \tag{6.9}$$

where:

$$IPTF(2,1,f) = \frac{H(2,f)}{H(1,f)} \quad IPTF(4,1,f) = \frac{H(4,f)}{H(1,f)}. \tag{6.10}$$

The IPTF in Equation (6.10) denotes the ratio of HRTFs at different directions. Signal processing is simplified because the IPTF can be implemented by a low-order filter.

Computational complexity can also be reduced by directly interpolating the zeros of an HRTF-based filter (Hacihabiboglu, 2005). According to Equation (5.42), an HRTF can be modeled by an FIR filter:

$$H(\theta,\phi,z) = \sum_{n=0}^{N-1} h(\theta,\phi,n)z^{-n}, \tag{6.11}$$

where N denotes the filter length. Therefore, Equation (6.3) can be interpreted as the directional interpolation of the coefficients of an FIR filter. As discussed in Section 5.2.1, an N-point FIR filter has $Q = N - 1$ zeros (as well as $N - 1$ coincident poles at $z = 0$), including L pairs of complex-conjugate zeros and S real zeros with $2L + S = Q = N - 1$. Then, the system function of the filter can be written in the following form:

$$H(z) = KH_1(z)H_2(z),$$
$$H_1(z) = \prod_{l=1}^{L}(1 - z_l z^{-1})(1 - z_l^* z^{-1}) \quad H_2(z) = \prod_{s=1}^{S}(1 - z_s z^{-1}), \tag{6.12}$$

where z_l and z_l^* are a pair of complex-conjugate zeros, and z_s is a real zero. Given the HRTFs at neighboring azimuths θ_i and θ_{i+1}, their complex-conjugate zeros and real zeros are denoted by

$(z_{l,i}, z_{l,i}^*, l = 1,2 \ldots L, z_{s,i}, s = 1,2 \ldots S)$, $(z_{l,i+1}, z_{l,i+1}^*, l = 1,2 \ldots L, z_{s,i+1}, s = 1,2 \ldots S)$, respectively. The complex-conjugate zero pairs of the filter $H(z)$ associated with an in-between azimuth $\theta_i < \theta < \theta_{i+1}$ can be obtained by the adjacent linear interpolation scheme:

$$z_{l,\theta} = \lambda z_{l,i+1} + (1 - \lambda) z_{l,i} \qquad z_{l,\theta}^* = \lambda z_{l,i+1}^* + (1 - \lambda) z_{l,i}^* \quad 0 < \lambda < 1. \tag{6.13}$$

The real zeros of filter $H(z)$ can also be obtained by interpolation, but with a complicated procedure. Finally, the interpolated HRTFs are obtained by substituting the interpolated zeros into Equation (6.12).

Watanabe et al. (2005) proposed an HRTF interpolation scheme on the basis of the CAPZ model discussed in Section 5.3.5. The CAPZ model represents the HRTF by a filter with direction-independent poles and direction-dependent zeros. Directional interpolation is applied only to the coefficients associated with the zeros, thereby simplifying the computation. In practice, the HRTFs at different directions are represented by the ARMA model given by Equation (5.81). The P coefficients $a_1, a_2 \ldots a_P$ in the denominator of Equation (5.81) are related to the common poles of the ARMA model and are therefore independent of source direction. The $Q + 1$ coefficients $b_0(\theta), b_1(\theta) \ldots b_Q(\theta)$ in the numerator of Equation (5.81) are related to the zeros of the ARMA model and are thus direction dependent. Directional interpolation is separately applied to the $Q + 1$ coefficients $b_0(\theta), b_1(\theta) \ldots b_Q(\theta)$. Various conventional interpolation schemes, such as adjacent linear interpolation, are applicable to coefficient interpolation. Similar to other interpolation schemes, arrival time correction for HRIRs prior to CAPZ-based interpolation improves interpolation performance. The results for KEMAR horizontal HRTFs indicate that CAPZ-based interpolation yields the same accuracy as direct HRTFs interpolation (e.g., adjacent linear interpolation), but reduces data to about 30%. Watanabe et al. (2003) also represented HRTFs as partial-fraction expansions, in which the pole of each fraction is direction independent, but the coefficients of each fraction and residue are direction dependent. Conventional interpolation methods are separately applied to coefficients and residues.

A problem related to HRTF interpolation is the directional extrapolation, or prediction, of HRTFs. As stated in Section 2.2.4, measuring HRTFs at low elevations $\phi < -40°$ is difficult, and most available HRTF databases exclude data on low elevations. In practical use, however, low-elevation HRTFs are sometimes required. Therefore, extrapolating low-elevation HRTFs from the HRTFs measured at other elevations is necessary. Because the performance of the traditional extrapolation scheme is unsatisfactory, Zhong and Xie (2005a) proposed a directional extrapolation scheme on the basis of neural networks.

6.2. Spectral Shape Basis Function Decomposition of HRTFs

6.2.1. Basic Concept of Spectral Shape Basis Function Decomposition

Even in the far field, HRTFs are complex-valued functions of source direction (θ, ϕ) and frequency f. They are also individual dependent. This multivariable–dependent characteristic yields substantial dimensionality of entire HRTF data, so that the analysis and representation of HRTFs are complicated. Accordingly, the variety of HRTFs with certain variables is usually analyzed with other variables fixed. For example, the frequency responses of HRTFs are analyzed at each given source direction; or the directional dependence of HRTF magnitudes is analyzed at each given frequency (band) using the SFRS discussed in Section 6.1.4. Alternatively, the efficient or low-dimensional representation of HRTFs can be achieved by decomposing HRTFs into a weighted sum of appropriate basis functions, where the dependencies of HRTFs on different variables are separately represented by the variations in basis functions and weights.

HRTF linear decomposition is categorized into two basic types: *spectral shape basis function decomposition* and *spatial basis function decomposition*. The former represents HRTFs by a linear combination of spectral shape basis functions and is discussed in this section. The latter represents HRTFs by a linear combination of spatial basis functions and is discussed in Section 6.3.

When decomposed by spectral shape basis functions, the HRTF of a given ear is generally represented by:

$$H(\theta,\phi,f,s) = \sum_q w_q(\theta,\phi,s)d_q(f),$$ (6.14)

where s is the identity (ID) of the subject; $d_q(f)$ is a series of *spectral shape basis functions* that depend on frequency; $w_q(\theta, \phi, s)$ are source direction- and individual-dependent weights. When the basis functions $d_q(f)$ are selected, $H(\theta, \phi, f, s)$ are completely determined by weights $w_q(\theta, \phi, s)$. Note that $d_q(f)$ may be identical or different for left and right HRTFs; they may also be individual independent or dependent, contingent upon the decomposition algorithm used. By contrast, $w_q(\theta, \phi, s)$ always differs for left and right HRTFs. When unnecessary, the notation for the left or right ear is excluded from the denotation of HRTFs.

Various methods for selecting or deriving the basis functions $d_q(f)$ are available. A set of orthonormal basis functions is often preferred. If the correlations among the HRTFs at different directions are completely removed, so that the HRTFs can be represented by a small set of spectral shape basis functions, data dimensionality is efficiently reduced. The appropriate selection of basis functions depends upon the situation. Two important algorithms for deriving spectral shape basis functions—PCA and subset selection—are discussed here.

Corresponding to frequency-domain HRTF decomposition, equivalent time-domain HRIR decomposition is achieved by applying inverse Fourier transform to Equation (6.14) as:

$$h(\theta,\phi,t,s) = \sum_q w_q(\theta,\phi,s)g_q(t),$$ (6.15)

where $g_q(t)$—the inverse Fourier transform of $d_q(f)$—is the *time-domain basis function*.

Measured HRTFs are usually represented by their samples at discrete frequencies (or HRIRs at discrete times) and source directions. Accordingly, Equations (6.14) and (6.15) are analyzed with respect to discrete variables. In this case, both equations can be simply written as vector equations. Let $i = 0,1 \ldots (M-1)$ denote M discrete source directions (θ_i, ϕ_i); $k = 0,1 \ldots (N-1)$ denote N discrete frequencies; the $N \times 1$ column vector (or matrix) $\mathbf{H}_i$ denote the samples of HRTFs at N discrete frequencies and a given source direction (θ_i, ϕ_i); and the $N \times 1$ column vectors (or matrix) $\mathbf{d}_q$, called *spectral shape basis vectors*, denote the samples of $d_q(f)$ at N discrete frequencies. Then, Equation (6.14) can be written as:

$$\mathbf{H}_i = \sum_q w_q(\theta_i,\phi_i)\mathbf{d}_q = \sum_q w_q(i)\mathbf{d}_q \qquad i=0,1\ldots(M-1),$$

$$\mathbf{H}_i = \mathbf{H}(\theta_i,\phi_i) = [H(\theta_i,\phi_i,0),H(\theta_i,\phi_i,1),\ldots H(\theta_i,\phi_i,N-1)]^T,$$ (6.16)

$$\mathbf{d}_q = [d_q(0),d_q(1)\ldots,d_q(N-1)]^T,$$

where superscript "T" denotes the matrix transpose. For simplicity, the notation s (subject ID) is omitted here.

Similarly, let $n = 0,1...(N-1)$ denote N discrete times; the $N \times 1$ column vector (or matrix) $\boldsymbol{h}_i$ denote the samples of HRIRs at N discrete times and a given source direction (θ_i, ϕ_i); and the $N \times 1$ column vectors (or matrix) $\boldsymbol{g}_q$, called *time-domain basis vectors*, denote the samples of $g_q(t)$ at N discrete times. Then, Equation (6.15) can be written as:

$$\boldsymbol{h}_i = \sum_q w_q(\theta_i, \phi_i)\boldsymbol{g}_q = \sum_q w_q(i)\boldsymbol{g}_q \qquad i = 0,1....(M-1)$$

$$\boldsymbol{h}_i = \boldsymbol{h}(\theta_i, \phi_i) = [h(\theta_i, \phi_i, 0), h(\theta_i, \phi_i, 1), ...h(\theta_i, \phi_i, N-1)]^T, \qquad (6.17)$$

$$\boldsymbol{g}_q = [g_q(0), g_q(1)....., g_q(N-1)]^T.$$

6.2.2. Principal Components Analysis of HRTFs

Principal Components Analysis (PCA) is a statistical algorithm for deriving spectral shape basis functions and decomposing HRTFs. It effectively eliminates correlations among HRTFs, thereby reducing dimensionality. Accordingly, HRTFs can be simply represented by a small set of spectral shape basis functions. Martens (1987) first introduced PCA into the analysis of HRTFs. Since the work of Kistler and Wightman (1992), many researchers have also applied PCA on HRTFs (Middlebrooks and Green, 1992b; Chen et al., 1995; Wu et al., 1997). Although the algorithms and terms used in different studies may vary (for example, PCA is termed *Karhunen-Loeve expansion* by Chen), the basic principle is the same.

Prior to decomposition, the mean HRTF across source directions is subtracted from each HRTF to effectively eliminate the correlation among HRTFs. Equation (6.14) then becomes:

$$H(\theta, \phi, f, s) - H_{av}(f) = \sum_q w_q(\theta, \phi, s)d_q(f), \qquad (6.18)$$

where $H_{av}(f)$ is a *mean spectral function* that is frequency dependent (it may also be individual dependent, contingent upon the algorithm used). Equation (6.18) can be written as:

$$H(\theta, \phi, f, s) = \sum_q w_q(\theta, \phi, s)d_q(f) + H_{av}(f). \qquad (6.19)$$

A simple case with the specific ear of a given individual is first discussed, in which the notations for the left and right ears, as well as for the individual, are temporarily excluded from the equations. Let the HRTF at each direction be represented by its samples at N discrete frequencies; that is, $H(\theta_i, \phi_i, f_k) = H(i, k)$, with i and k denoting discrete direction and frequency, respectively. Then, Equation (6.19) can be written in the form of discrete variables:

$$H(i,k) = \sum_q w_q(i)d_q(k) + H_{av}(k) \quad i = 0,1..(M-1), \quad k = 0,1..(N-1). \qquad (6.20)$$

In Equation (6.20), $H_{av}(k)$ is the kth sample of the weighted mean of the HRTFs (or mean spectral function) across M source directions:

$$H_{av}(k) = \frac{1}{M}\sum_{i=0}^{M-1} \lambda_i H(i,k), \qquad (6.21)$$

where λ_i is the weight coefficient at direction i. If a weight coefficient $\lambda_i = 1$ is chosen for all M directions, Equation (6.21) is the arithmetic mean that represents the direction-independent component of HRTFs.

To eliminate the correlation among HRTFs at different directions and derive $d_q(k)$, the weighted mean of HRTFs is subtracted from each weighted HRTF. The resultant data form an $N \times M$ matrix $[H_\Delta]$ with elements:

$$H_{\Delta,k,i} = \lambda_i H(i,k) - H_{av}(k) \quad i = 0,1....(M-1), \quad k = 0,1....(N-1). \tag{6.22}$$

Thus, each row and column of the matrix corresponds to a specified frequency and direction, respectively. An $N \times N$ matrix $[R]$ can be constructed from $[H_\Delta]$ as:

$$[R] = \frac{1}{M}[H_\Delta][H_\Delta]^+. \tag{6.23}$$

Here, superscript "+" denotes the transpose and conjugation of the matrix. If $\lambda_i = 1$ is chosen for all directions, $[H_\Delta]$ is a matrix with a mean of zero across the columns at each row, and $[R]$ is the covariance matrix against N discrete frequencies (average over M directions), whose elements represent the similarity between the HRTFs in each pair of frequencies.

Because $[R]$ is an $N \times N$ Hermitian matrix, its eigenvalues are nonnegative and real numbers. The spectral shape basis vectors d_q shown in Equation (6.20), which represent the samples of $d_q(f)$ at N discrete frequencies, can be obtained from the eigenvectors of matrix $[R]$ for all $Q' \leq N$ positive eigenvalues γ_q^2:

$$[R]\, d_q = \gamma_q^2\, d_q \quad q = 1,2...Q' \quad \gamma_1^2 > \gamma_2^2 > ....\gamma_{Q'}^2 > 0. \tag{6.24}$$

The resultant spectral shape basis vectors are orthonormal to each other:

$$d_q^+ d_q = \begin{cases} 1 & q = q' \\ 0 & q \neq q' \end{cases}. \tag{6.25}$$

When the spectral shape basis vectors d_q are obtained, the weights $w_q(i)$ for direction i can be evaluated using the orthonormality in Equation (6.25) thus:

$$w_q(i) = \sum_{k=0}^{N-1} [H(i,k) - H_{av}(k)]d_q^*(k), \tag{6.26}$$

where superscript "*" denotes complex conjugation. Finally, the spectral shape basis function decomposition of HRTFs is obtained by substituting the $H_{av}(k)$ in Equation (6.21), $d_q(k)$ in Equation (6.24), and $w_q(i)$ in Equation (6.26) into Equation (6.20).

Strictly, the term "principal component" refers to the entire set of weights $w_q(i)$ that is associated with a particular basis vector. In the literature, however, it sometimes refers to a set of weighted basis vectors, and sometimes to weights or basis vectors alone. To avoid confusion, as suggested by Kistler and Wightman (1992), we use the term "*principal component*" or PC to describe the set of weighted basis vectors obtained by PCA; the term "*PC-vectors*" to describe (spectral shape) basis vectors alone; and the term "*PC-weights*" to describe weights alone. PC-weights determine the contribution of the corresponding PC-vectors to the HRTFs at different source directions, so that they are also termed "*spatial characteristic functions.*"

When HRTFs are reconstructed by all Q' PCs, Equation (6.20) is the exact representation of original HRTFs. If the HRTFs are reconstructed by the $Q < Q'$ PCs associated with the preceding Q largest and positive eigenvalues $\gamma_q^2, q = 1, 2 \dots Q$, Equation (6.20) is an approximate representation of original HRTFs:

$$\hat{H}(i,k) = \sum_{q=1}^{Q} w_q(i)d_q(k) + H_{av}(k) \quad i = 0,1..(M-1), \quad k = 0,1..(N-1). \tag{6.27}$$

Corresponding to Equation (6.16), Equation (6.27) can be written as a vector equation:

$$\hat{H}_i = \sum_{i=1}^{Q} w_q(i) d_q + H_{av},\qquad(6.28)$$

where:

$$H_{av} = [H_{av}(0), H_{av}(1)....., H_{av}(N-1)]^T.\qquad(6.29)$$

It is an $N \times 1$ vector (matrix) that consists of the N samples of the weighted mean of the HRTFs across directions.

The preceding $Q < Q'$ PC-vectors $d_1, d_2 \ldots d_Q$ are orthonormal but incomplete. Accordingly, the more PCs included in Equation (6.28), the more exact, but also larger, the dimensionality of reconstructed HRTFs. The first PC contributes the most to $H(i, k)$, while the contributions of the other PCs decline. The preceding Q PCs are the most important or representative, and they minimize square errors in reconstruction. In this sense, PCA can also be regarded as a smoothing algorithm for HRTFs, retaining primary features while discarding minor details. These features make up the basic principle of the PCA decomposition and reconstruction of HRTFs. The cumulative percentage variance of the energy represented by the preceding Q PCs in Equation (6.27) is evaluated by:

$$\eta = \frac{\sum_{i=0}^{M-1}\sum_{k=0}^{N-1}|\hat{H}(i,k)-H_{av}(k)|^2}{\sum_{i=0}^{M-1}\sum_{k=0}^{N-1}|H(i,k)-H_{av}(k)|^2}\times 100\% = \frac{\sum_{q=1}^{Q}\gamma_q^2}{\sum_{q=1}^{Q'}\gamma_q^2}\times 100\%.\qquad(6.30)$$

In addition, the dimensionality of original HRTFs is N (frequencies) $\times M$ (directions). As indicated by Equation (6.27), the dimensionality of PCA-reconstructed HRTFs is $N \times Q + Q \times M + N = (N + M) \times Q + N$. With sufficient reconstruction accuracy, data dimensionality is reduced if Q satisfies the following condition:

$$Q < \frac{N(M-1)}{N+M}.\qquad(6.31)$$

The PCA decomposition of HRTFs can also be written in matrix form. Let $[H]$ be an $N \times M$ matrix that comprises the original HRTF samples at N frequencies and M directions, with each row and column of the matrix corresponding to a specified frequency and direction, respectively. The element of the matrix is:

$$H_{k,i} = H(i,k) \qquad k=0,1...(N-1) \quad i=0,1....(M-1).\qquad(6.32)$$

Similarly, let $[\hat{H}]$ be an $N \times M$ matrix that contains the reconstructed HRTFs at N frequencies and M directions given by Equation (6.28). The elements of the matrix are:

$$\hat{H}_{k,i} = \hat{H}(i,k) \qquad k=0,1...(N-1) \quad i=0,1....(M-1).\qquad(6.33)$$

Let $[D']$ be an $N \times Q'$ matrix that comprises all the Q' PC-vectors d_q associated with positive eigenvalues and in descending order, with each row and column of the matrix corresponding to a specified frequency and PC-vector, respectively:

$$[D] = [d_1, d_2, ...d_{Q'}].\qquad(6.34)$$

Let $[W']$ be a $Q' \times M$ matrix that contains all PC-weights, with each row and column of the matrix corresponding to a specified PC-vector and direction, respectively. The elements of the matrix are:

$$W'_{q,i} = w_q(i) \qquad q = 1,2....Q', \quad i = 0,1....(M-1). \tag{6.35}$$

Finally, let $[H_{av}]$ be an $N \times M$ matrix related to $H_{av}(k)$ and with identical elements in its each row. The elements of the kth row are:

$$H_{av;k,i} = H_{av}(k) \qquad k = 0,1....(N-1), \quad i = 0,1....(M-1). \tag{6.36}$$

With the matrix notation above, Equation (6.26) can be written as:

$$[W'] = [D']^{+}\{[H]-[H_{av}]\}. \tag{6.37}$$

Further, Equation (6.27) or (6.28) becomes:

$$[\hat{H}] = [D][W] + [H_{av}]. \tag{6.38}$$

Here, the $N \times Q$ matrix $[D]$ and $Q \times M$ matrix $[W]$ are constructed by selecting the preceding Q columns from matrix $[D']$ and preceding Q rows from matrix $[W']$.

The foregoing analysis can also be carried out on the basis of the SVD (singular value decomposition) of the matrix (Larcher et al., 2000; Deif, 1982). For simplicity, a weight coefficient of $\lambda_i = 1$ is chosen for all the M directions ($i = 0,1 \dots M - 1$) in Equations (6.21) and (6.22). Assume that the rank of the $N \times M$ complex-valued matrix $[H_\Delta]$ defined by Equation (6.22) is Q'. Then, $[H_\Delta][H_\Delta]^{+}$ and $[H_\Delta]^{+}[H_\Delta]$ are $N \times N$ and $M \times M$ Hermitian matrices, respectively. They share an identical set of real eigenvalues, in which the Q' eigenvalues are positive, with $\delta_1^2 \geq \delta_2^2 \geq \delta_{Q'}^2 > 0$, and the others are zero:

$$\{[H_\Delta][H_\Delta]^{+}\}d_q = \delta_q^2 d_q \qquad \{[H_\Delta]^{+}[H_\Delta]\}v_q = \delta_q^2 v_q \quad q = 1,2....Q', \tag{6.39}$$

where d_q and v_q are $N \times 1$ and $M \times 1$ eigenvectors, respectively. Except for a scaling factor $1/M$, matrix $\{[H_\Delta][H_\Delta]^{+}\}$ is equivalent to matrix $[R]$ in Equation (6.23). Accordingly, matrix $\{[H_\Delta][H_\Delta]^{+}\}$ and $[R]$ share the same set of eigenvectors d_q, except for a scaling difference between respective eigenvalues, namely, $\delta_q^2 = M\gamma_q^2$. According to SVD, matrix $[H_\Delta]$ can be decomposed as:

$$[H_\Delta] = [U][\Gamma'][V]^{+}, \tag{6.40}$$

where $[\Gamma']$ is an $N \times M$ matrix, with its Q' non-zero left-diagonal elements being the singular values of matrix $[H_\Delta]$ in descending order—that is, $\delta_1 = M^{1/2}|\gamma_1| \geq \delta_2 = M^{1/2}|\gamma_2| \dots \geq \delta_{Q'} = M^{1/2}|\gamma_{Q'}| > 0$. $[U]$ and $[V]$ are $N \times N$ and $M \times M$ unitarity matrices, respectively, with $[U]^{-1} = [U]^{+}$ and $[V]^{-1} = [V]^{+}$. The preceding Q' columns of matrix $[U]$ are also constructed from the Q' orthonormal eigenvectors d_q, whereas the preceding Q' columns of matrix $[V]$ are constructed from the Q' orthonormal eigenvectors v_q.

$[H_\Delta]$ can be approximated by retaining the preceding $Q < Q'$ eigenvectors and disregarding the contributions of the other eigenvectors in Equation (6.40) thus:

$$[\hat{H}_\Delta] = [D][\Gamma][V]^{+}, \tag{6.41}$$

where $[D]$ is an $N \times Q$ matrix that is constructed by selecting the preceding Q columns from the $N \times N$ matrix $[U]$—that is, Equation (6.34); $[\Gamma]$ denotes a $Q \times M$ matrix constructed by selecting the preceding Q rows from the $N \times M$ matrix $[\Gamma']$. Comparing Equations (6.38) and (6.22), the PCA- or SVD-reconstructed HRTFs can be obtained by:

$$[\hat{H}] = [\hat{H}_\Delta] + [H_{av}] = [D][W] + [H_{av}] \qquad [W] = [\Gamma][V]^{+}. \tag{6.42}$$

6.2.3. Discussion of Applying PCA to HRTFs

In the aforementioned analysis, PCA is applied to complex-valued HRTFs in the frequency domain; so that both magnitude information and phase information are retained in the reconstructed HRTFs. PCA can also be applied to linear or logarithmic HRTF magnitudes alone. The only requirement is to replace the complex-valued HRTFs in the aforementioned equations with linear or logarithmic HRTF magnitudes. In this case, phase information is lost in the reconstructed HRTFs. Under the minimum-phase approximation in Section 3.1.3, complex-valued HRTFs can be recovered from HRTF magnitudes and a pure delay.

PCA can also be applied to HRIRs in the time domain. Similar to that executed with Equation (6.27), the HRIR at a given source direction i can be approximately reconstructed as:

$$\hat{h}(i,n) = \sum_{q=1}^{Q} w_q(i)g_q(n) + h_{av}(n) \quad i=0,1..(M-1), \quad n=0,1..(N-1), \tag{6.43}$$

where i and n denote the discrete direction and time, respectively; $g_q(n)$ ($q = 1 \dots Q$, $n = 0,1 \dots N - 1$) are the Q time-domain basis functions, or PC-vectors; $w_q(i)$ represent the PC-weights; and $h_{av}(n)$ is the weighted mean of the HRIRs across source directions. The algorithms for applying PCA on HRIRs are similar to those applied on HRTFs. The only requirements are to replace all HRTFs with HRIRs and to replace the discrete frequency k by the discrete time n in the derivations from Equations (6.20) to (6.31).

The PCA of time-domain HRIRs is closely related to that of frequency-domain HRTFs in the following ways:

1. If Equation (6.27) represents the PCA-based decomposition of complex-valued HRTFs, then $d_q(k)$ is related to $g_q(n)$ in Equation (6.43) by N-point discrete Fourier transform (DFT), and so is $H_{av}(k)$ and $h_{av}(n)$. In this case, the PC-weights $w_q(i)$ in Equation (6.43) are identical to those in Equation (6.27). In other words, the PCA of the time-domain HRIRs and that of the complex frequency-domain HRTFs are mutually equivalent. Therefore, applying PCA on time-domain HRIRs to obtain additional results that cannot be derived from the PCA of complex frequency-domain HRTFs, or vice versa, is impossible (Rao and Xie, 2010). This obstacle has been neglected in some studies (Hwang and Park, 2008a).
2. If Equation (6.27) represents the PCA-based decomposition of linear or logarithmic HRTF magnitudes, then no concise relationship between $d_q(k)$ and $g_q(n)$ exists in Equation (6.43). Accordingly, the PC-weights $w_q(i)$ in Equation (6.43) differ from those in Equation (6.27).
3. The PCA of complex-valued HRTFs involves complex-valued computation and storage. By contrast, the PCA of HRIRs deals with real-valued computation and storage. The PCA of HRIRs appears simpler than that of complex-valued HRTFs. However, complex-valued HRTFs, which are related to HRIRs by Fourier transform, are conjugate-symmetrical around the Nyquist frequency (Oppenheim et al., 1999). Taking advantage of conjugate symmetry, the dimensionality of complex-valued HRTF data can be reduced by about half, so that the computational efficiency of complex-valued HRTFs is comparable to that derived in the PCA of HRIRs.
4. If the arrival times of HRIRs are corrected prior to PCA, the number of PCs required is reduced for the same accuracy in reconstruction.
5. In practice, PCA can be applied to complex-valued HRTFs, as well as to linear or logarithmic HRTF magnitudes and HRIRs, depending on the problem of interest.

Reconstructed HRTFs, including complex-valued HRTFs, as well as linear HRTF magnitudes and HRIRs, usually exhibit increased error at high frequencies, particularly at contralateral directions and low elevations. This increase is due to the complicated nature of contralateral HRTFs at high frequencies; the complexity is caused by multi-path interference from diffracted sound waves around the head. Moreover, because of the attenuation caused by the head shadow effect, the signal-to-noise ratio of measured contralateral HRTFs is inherently poor. The reconstruction errors of logarithmic HRTF magnitudes tend to be more normally distributed among directions than those of linear-magnitude or complex-valued HRTFs. Theoretically, the effectiveness of the reconstruction of logarithmic HRTF magnitudes at contralateral directions can be partially improved, but at the cost of reducing accuracy at ipsilateral directions. In practice, however, this improvement is sometimes minimal because of random errors mixed into the original (measured) contralateral HRTFs.

The PCA algorithm can be extended to near-field HRTFs or HRIRs. Near-field HRTFs consist of data on various source directions and distances. In this case, the discrete variable i in Equations (6.20) to (6.27) denotes source position in terms of direction and distance.

Some correlations also exist among the HRTFs of different individuals and ears. To eliminate these correlations, the PCA algorithm discussed in Section 6.2.2 is extended to the binaural HRTFs of multiple individuals. Given the far-field HRTFs at M directions for both ears of S subjects, the samples of each HRTF at N discrete frequencies are denoted by $H(i, k, s, ear)$, where $i = 0,1 \ldots (M-1)$ denotes the discrete direction; $k = 0, 1 \ldots (N-1)$ is the discrete frequency; $s = 1,2 \ldots S$ denotes different subjects; and $ear = L$ or R represents the left or right ear, respectively. To extend Equation (6.27), HRTFs can be approximately reconstructed by Q PCs as:

$$\hat{H}(i,k,s,ear) = \sum_{q=1}^{Q} w_q(i,s,ear)d_q(k) + H_{av}(k),$$

$$i = 0,1..(M-1), \quad k = 0,1..(N-1), \ s = 1,2....S, \quad ear = L,R$$

(6.44)

where $H_{av}(k)$ is the kth sample of the mean HRTF across M directions, S subjects, and two ears (for simplicity, $\lambda_i = 1$ is chosen as the coefficient for all the calculations of weighted mean):

$$H_{av}(k) = \frac{1}{2SM}\sum_{ear}\sum_{s=1}^{S}\sum_{i=0}^{M-1} H(i,k,s,ear).$$

(6.45)

The procedures for deriving the spectral shape basis functions $d_q(k)$ or PC-vectors d_q are similar to those in Equations (6.22) to (6.25). The differences are that the dimensions of matrix $[H_\Delta]$ become $N \times (2SM)$, with each matrix row corresponding to a specified frequency and each column corresponding to a specified direction, subject, and ear. The scaling factor $1/M$ on the right side of Equation (6.23) should be replaced by $1/(2SM)$. Corresponding to Equation (6.26), PC-weights are determined by:

$$w_q(i,s,ear) = \sum_{k=0}^{N-1}[H(i,k,s,ear) - H_{av}(k)]d_q^*(k)$$

(6.46)

The HRTFs of different subjects, as well as those of left and right ears, are decomposed by the same set of PC-vectors. The individual and left-right differences of HRTFs are encoded in PC-weights $w_q(i,s,ear)$.

In some works, the mean HRTF across directions is separately calculated for each subject and ear, rather than across all directions, subjects, and ears. That is,

$$H_{av}(k,s,ear) = \frac{1}{M}\sum_{i=0}^{M-1} H(i,k,s,ear).$$

(6.47)

Then, Equation (6.44) becomes:

$$\hat{H}(i,k,s,ear)=\sum_{q=1}^{Q}w_q(i,s,ear)d_q(k)+H_{av}(k,s,ear).$$

$$i=0,1..(M-1),\quad k=0,1..(N-1),\quad s=1,2....S,\quad ear=L\,or\,R \tag{6.48}$$

The dimensions of matrix $[H_\Delta]$ are still $N\times(2SM)$, and the elements of the matrix are given by:

$$H_\Delta(i,k,s,ear)=H(i,k,s,ear)-H_{av}(k,s,ear).$$

$$i=0,1..(M-1),\quad k=0,1..(N-1),\quad s=1,2....S,\quad ear=L\,or\,R \tag{6.49}$$

Equation (6.46) becomes:

$$w_q(i,s,ear)=\sum_{k=0}^{N-1}[H(i,k,s,ear)-H_{av}(k,s,ear)]d_q^*(k). \tag{6.50}$$

Except for the differences, from Equation (6.47) to Equation (6.50), the PCA procedures are similar to those in the aforementioned discussions. Unlike those in Equation (6.44), the individual and left-right differences of the HRTFs in Equation (6.48) are reflected by both PC-weights $w_q(i,s,ear)$ and the mean spectral function $H_{av}(k,s,ear)$.

Kistler and Wightman (1992) used the PCA procedures in Equations (6.47) to (6.50) to decompose logarithmic HRTF magnitudes, and reconstruct complex-valued HRTFs by minimum-phase approximation. With the HRTFs from Equations (6.47) to (6.50) being replaced by their logarithmic magnitudes, the decomposition in Equation (6.48) becomes:

$$20\log_{10}|\hat{H}(i,k,s,ear)|=\sum_{q=1}^{Q}w_q(i,s,ear)d_q(k)+\frac{20}{M}\sum_{i'=0}^{M-1}\log_{10}|H(i',k,s,ear)|. \tag{6.51}$$

6.2.4. PCA Results for HRTFs

As previously stated, many researchers applied PCA to decompose and reconstruct HRTFs. In early work by Martens (1987), the 24 critical band-filtered HRTFs at 35 horizontal source azimuths were represented by four PCs.

Chen et al. (1995) measured the HRIRs at 2188 directions from a KEMAR artificial head, including 29 elevations, ranging from $\phi=-36°$ to $90°$ at an interval of $\Delta\phi=4.5°$, and 80 azimuths, at an interval of $\Delta\theta=4.5°$ at each elevation (given the limitations of the measurement apparatus, the HRTFs at some azimuths at low elevations were not measured). After post-processing, the resultant 128-point HRTFs at a sampling frequency of 51.2 kHz represent a frequency range of up to 25.6 kHz. Then, the PCA algorithm from Equations (6.20) to (6.30) was applied to the resultant complex-valued HRTFs.

Each elevation has 80 azimuthal HRTF samples. When the source direction deviates from the horizontal plane, the distance between neighboring azimuthal samples decreases. To prevent the HRTFs at high (or low) elevations from dominating the covariance matrix and the mean HRTFs across directions, Chen et al. incorporated the following directional weighting coefficients in the calculations of Equations (6.21) and (6.22):

$$\lambda_i=\lambda(\theta_i,\phi_i)=1-|\sin\phi_i|. \tag{6.52}$$

λ_i decreases as source direction deviates from the horizontal plane, thereby compensating for the dense grid in HRTF measurements at high elevations; that is, over sampling near the pole region is avoided (the choice of λ_i depends on the interval of measured directions). The PCA results for the HRTFs at $M = 2188$ directions ($\Delta\phi = 4.5°$, $\Delta\theta = 4.5°$) and those at $M = 561$ directions ($\Delta\phi = 9.0°$, $\Delta\theta = 9.0°$) are similar. The preceding PCs capture the significant variations of HRTFs. The cumulative percentage variance of energy in reconstruction [Equation (6.30)] increases with the number of PCs included. When the HRTFs were reconstructed from $Q = 12$ PCs, the cumulative percentage variance of energy exceeds 99.9%, which is adequate to represent the variations in the major structures of HRTFs. In addition, the mean reconstruction error calculated by Equation (5.12) is in the order of 0.1%, which also validates the ideal performance of PCA reconstruction. As expected, reconstruction error occurs primarily at high frequencies, particularly at contralateral directions and low elevations. For example, the mean reconstruction error [the mean of Equation (5.12) across directions] for right-ear HRTFs is 0.5% (-23.0 dB) within regions $0° \leq \phi \leq 30°$ and $45° \leq \theta \leq 90°$, but 7.6% (-11.2 dB) within regions $-36° \leq \phi \leq 0°$ and $180° \leq \theta \leq 225°$.

Kistler and Wightman (1992) measured the HRTFs of both the ears of $S = 10$ human subjects at $M = 265$ source directions (including 11 elevations that range from $\phi = -48°$ to $72°$ at an interval of $\Delta\phi = 12°$, 24 azimuths at an interval of $\Delta\theta = 15°$ in each elevation, as well as a top direction $\phi = 90°$). The sampling frequency was 50 kHz. The logarithmic HRTF magnitudes at $N = 150$ discrete frequencies that range from 0.2 to 15 kHz were included in the analysis. The dimensions of matrix $[H_\Delta]$ in Equation (6.23) were 150×5300. The preceding five spectral shape basis functions are shown in Figure 6.3, in which the discrete basis functions (PC-vectors) are plotted as continuous functions. These five PCs account for 74.3%, 6.6%, 4.5%, 2.7%, and 2.2% variance in logarithmic HRTF magnitudes; that is, they account for about 90% cumulative percentage variance.

In Figure 6.3, all five spectral shape basis functions are approximately constant, with a value of zero below 2 to 3 kHz. This result reveals that the logarithmic HRTF magnitudes in Equation (6.51) are dominated by the direction-independent component in this frequency range. All five spectral shape basis functions deviate from zero above 3 kHz, reflecting variations in high-frequency HRTF magnitudes. The directional and individual dependencies of HRTF magnitudes are encoded in PC-weights—the spatial characteristic functions $w_q(i,s,ear)$. The results of Kistler and Wightman also indicate individual differences in PC-weights $w_q(i,s,ear)$, but similar directional tendencies of the preceding PC-weights across all the subjects. Accordingly, the combination of PC-weights and spectral shape basis functions (or PC-vectors) accounts for some fundamental characteristics of HRTFs. For example, the first PC-weight exhibits positive values at contralateral source directions and negative values at ipsilateral source directions. The first spectral shape basis function, which accounts for a 74.3% variance in logarithmic HRTF magnitudes, tends toward a negative value as frequency increases. Therefore, combining the first PC-weight and spectral shape basis function demonstrates an attenuation of contralateral logarithmic magnitudes and an increase of ipsilateral logarithmic magnitudes at high frequencies. This result is consistent with the basic features of the HRTF magnitudes and resultant ILD (interaural level difference) presented in Section 3.3. Although the variations in the remaining four PC-weights and spectral shape basis functions are somewhat complicated, their combination reflects, to some extent, the spectral features of HRTFs.

Kistler and Wightman (1992) also conducted a psychoacoustical experiment to compare localization performance of the headphone-presented binaural signals synthesized from individual (measured) and reconstructed HRTFs (Section 13.3). The logarithmic magnitudes of reconstructed HRTFs were obtained from Equation (6.51) with a variety of PCs included. The phase of the reconstructed HRTFs was recovered by minimum-phase approximation, cascading with an interaural pure delay. The results indicate that the localization performance of the reconstructed HRTFs with five PCs is nearly identical to that of the measured HRTFs. For HRTF reconstruction with less than five PCs, localization performance is degraded and front-back or up-down confusion rate rises because of high-frequency spectral distortions in the reconstructed HRTFs.

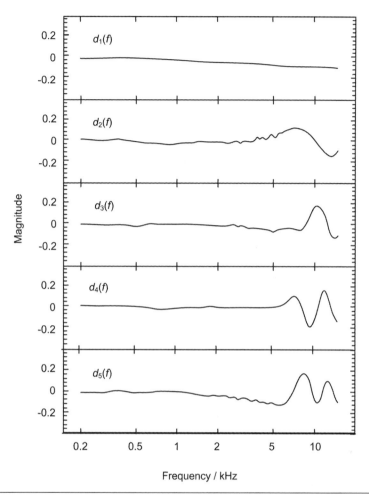

Figure 6.3 Preceding five spectral shape basis functions (from Kistler and Wightman, 1992, by permission of J. Acoust. Soc. Am.)

Sodnik et al. (2006) applied PCA to the linear HRTF magnitudes of KEMAR at five elevations: $\phi = -20°, -10°, 0°, 10°, 20°$. The preceding two PCs adequately describe the azimuthal variations of HRTFs (0.3–8.5 kHz), with the mean correlation (across azimuths) between the original and reconstructed HRTF magnitudes being 0.99. Sodnik et al. also proposed an equation to fit the azimuthal variations of two PC-weights.

Xie and Zhang (2012a) applied PCA to the near-field HRIRs of KEMAR at 493 source directions and 9 distances from 0.2 to 1.0 m at an equal interval of 0.1 m (Section 2.4.2). The results indicate that minimum-phase HRIRs can be decomposed by 15 PCs along with a time-domain mean function, and the cumulative percentage variance of energy [Equation (6.30)] is 97.4%.

Because PCA eliminates the correlation among HRTFs and, therefore, reduces data dimensionality, it is applicable to simplifying a large number of HRTFs prior to performing subsequent physical analyses. Riederer (2000) proposed a method for examining the errors in a large number of measured HRTFs. The method involves verifying whether the HRTFs that are reconstructed from the first PCs deviate from the structure common to all HRTFs.

Hwang and Park (2007, 2008) applied PCA to the median-plane HRIRs of 45 subjects taken from the CIPIC database (including 43 human subjects, and KEMAR with either DB-060/061 small pinnae or DB-065/066 large pinnae). The arrival times of the HRIRs were corrected prior to PCA. The results for the left ear indicate that median-plane HRIRs can be approximately reconstructed by 12 PCs with an energy error of less than 5%. The directional and individual dependencies of the HRIRs were encoded in PC-weights $w_q(i,s,ear)$, where $q = 1,2 \dots 12$ denotes the PC, $i = 0,1 \dots 48$ denotes the elevation, $s = 1,2 \dots 45$ represents the subject, and $ear = $ left refers to the left ear. Hwang and Park also calculated the mean and standard deviation of $w_q(i,s,ear)$ across subjects with different q and i to analyze the statistical distribution of the individual characteristics of the HRIRs.

In the study by Qian and Eddins [(2008); see Section 5.2.4)], PCA was first applied to the HRTF magnitudes of 26 human subjects at 360 source directions, and the results were transformed to the spectral modulation frequency domain for further analyses.

To evaluate the influence of pinnae on HRTFs, Xie (2008a) applied PCA to HRTF magnitudes at 72 horizontal azimuths for KEMAR with (DB-060/061) and without pinnae. The results indicate that the preceding 1, 2, and 3 PCs account for 80.1%, 91.8%, and 94.9% cumulative percentage variance of energy [Equation (6.30)], respectively, in HRTF magnitudes for KEMAR with pinnae. The PCs account for 91.6%, 97.9%, 98.4% cumulative percentage variance of energy, respectively, in HRTF magnitudes for KEMAR without pinnae. Including pinnae complicates HRTF magnitude spectra, so that more PCs are required to represent their variations. Pinnae also exert greater influence on ipsilateral HRTF magnitudes than on contralateral HRTF magnitudes in terms of energy variation.

Rao (2008a) applied PCA to the logarithmic HRTF magnitudes of 52 Chinese subjects, and reconstructed them with 10 PCs (with a cumulative percentage variance of $\eta > 90\%$). The difference between the PC-weights of male and female HRTF representations was evaluated by multivariate hypothesis testing. The results indicate that at a 0.05 significance level, the logarithmic HRTF magnitudes for males and females are significantly different at 59 out of 72 horizontal source directions. This result may be attributed to gender differences in anatomical structures, such as head dimensions. Moreover, the differences in logarithmic HRTF magnitudes are more obvious at the back than at the front because differences in hairstyle more strongly influence sound wave incident from the back.

6.2.5. Directional Interpolation under PCA Decomposition of HRTFs

In Section 6.1, directional interpolation was separately calculated for the HRTF samples at each discrete frequency $f = f_k$ or for the HRIR samples at each discrete time $t = t_n$. When HRTFs are decomposed into the weighted sum of Q PC-vectors by Equation (6.20) or (6.28), the spatial dependence of HRTFs is represented by Q PC-weights or the spatial characteristic functions $w_q(i)$ ($q = 1,2 \dots Q$, $i = 1,2 \dots M - 1$). For each PC-vector $\boldsymbol{d}_q$, the values $w_q(i) = w_q(\theta_i, \phi_i)$ of corresponding PC-weights at M measured directions are known. The HRTFs at these M measured directions can therefore be reconstructed from Q PC-vectors and PC-weights, along with the mean spectral function $H_{av}(k)$.

Similar to that done in Equation (6.20), suppose that the HRTFs at the arbitrary (unmeasured) directions (θ, ϕ) can also be represented by a weighted sum of Q spectral shape basis functions or PC-vectors:

$$\hat{H}(\theta,\phi,k) = \sum_{q=1}^{Q} w_q(\theta,\phi)d_q(k) + H_{av}(k) \qquad k = 0,1 \dots (N-1). \qquad (6.53)$$

Equation (6.53) includes the Q functions $w_q(\theta, \phi)$ with $q = 1,2 \dots Q$, which satisfy $w_q(\theta_i, \phi_i) = w_q(i)$ at M measured directions (θ_i, ϕ_i), $i = 0,1 \dots M - 1$. Then, the problem of HRTF interpolation becomes the problem of individual PC-weight interpolation. The Q PC-weights $w_q(\theta, \phi)$ are separately interpolated. Provided that the number of Q is less than the length N of HRTFs, interpolation on PC-weights is simpler than direct interpolation on HRTFs.

The HRTF interpolation schemes presented in Section 6.1.2 (bilinear interpolation, spherical triangular interpolation, cubic spline interpolation) can also be used to interpolate PC-weights. The continuous function representation of $w_q(\theta, \phi)$ can also be recovered by fitting its M directional samples. Such fitting can be achieved by azimuthal Fourier or SH decomposition and reconstruction to be discussed in the next section (Zhang and Xie, 2008a, 2008b). Chen et al. (1993) proposed a thin-plate spline model for estimating the continuous function representation of $w_q(\theta, \phi)$, which they accomplished by solving the problem:

$$\min_{\hat{w}_q(\theta,\phi)} \sum_{i=0}^{M-1} [\hat{w}_q(\theta_i,\phi_i) - w_q(\theta_i,\phi_i)]^2 + \kappa \, \| S[\hat{w}_q(\theta,\phi)] \|^2 , \tag{6.54}$$

where $\hat{w}_q(\theta,\phi)$ is the continuous PC-weight or spatial characteristic functions to be estimated; κ is a regularization parameter used to balance the smoothness and accuracy of $\hat{w}_q(\theta,\phi)$; and:

$$\| S[\hat{w}_q(\theta,\phi)] \|^2 = \iint d\theta d\phi \left\{ \left[\frac{\partial^2 \hat{w}(\theta,\phi)}{\partial^2 \theta} \right]^2 + 2 \left[\frac{\partial^2 \hat{w}(\theta,\phi)}{\partial\theta\partial\phi} \right]^2 + \left[\frac{\partial^2 \hat{w}(\theta,\phi)}{\partial^2 \phi} \right]^2 \right\}. \tag{6.55}$$

is the roughness measure. Carlile et al. (2000) compared the performance of spherical thin-plate spline interpolation and adjacent linear interpolation on PC-weights and concluded that the former is superior to the latter.

6.2.6. Subset Selection of HRTFs

Any vector in three-dimensional space can be represented by a linear combination of three orthonormal unit vectors (e.g., the unit vectors along the three orthonormal axes x, y, and z of a Cartesian coordinate). Accordingly, these three unit vectors form a complete set of orthonormal basis in three-dimensional vector space. Alternatively, an arbitrary vector in three-dimensional space can be represented by a linear combination of any three linearly independent vectors. As a result, any three linearly independent vectors also form a complete set of basis. The basis functions (vectors) decomposition of HRTFs is analogous to the basis vector representation of arbitrary vectors in three-dimensional space.

As previously stated, the HRTF at an arbitrary direction can be represented by a linear combination of spectral shape basis vectors. A set of Q' orthonormal and complete basis vectors can be derived using the statistical algorithm in Section 6.2.2. The basis vectors obtained in this manner are analogous to the orthonormal unit vectors in three-dimensional space. The essence of PCA lies in appropriately selecting a small set of $Q < Q'$ basis vectors to represent the primary variations of HRTFs and minimize the mean square error in HRTF reconstruction.

Equation (6.2) indicates that under certain conditions (Shannon-Nyquist theorem, Section 6.3), the HRTF at an arbitrary direction (θ, ϕ) can be obtained by interpolating a finite set of complete basis vectors that consist of measured HRTFs. In other words, the HRTF at an arbitrary direction can be represented by a linear combination of actual HRTFs. As a result, the finite set of actual HRTFs can be regarded as a complete set of basis vectors, which are analogous to the three linearly independent vectors in three-dimensional space. The issue here is how to appropriately choose a subset of K representative basis vectors—that is, choosing K privileged HRTFs from a set of M known HRTFs—so that they can represent the primary variations of all M HRTFs or arbitrary HRTFs. Various subset selection methods provide a solution to this issue.

Gardner (1999) proposed the QR subset method for choosing a subset of HRTFs on the basis of the QR decomposition of HRTF matrices. A procedure slightly different from, but mathematically equivalent to, Gardner's original derivation is provided in the current discussion. The HRIRs (or HRTFs) for M directions are denoted by M vectors h_i with $i = 0,1 \ldots M - 1$, and the dimensions of each vector are $N \times 1$. The energy of each HRIR is evaluated by the L_2-norm of vector h_i, that is, $E_i = (h_i^T h_i)$, where superscript "T" denotes the transpose of the vector (for complex-valued HRTFs, it should be replaced by the transpose and conjugation of the vector). The subset selection involves appropriately choosing K HRIRs ($u_0, u_1, \ldots, u_{K-1}$) from M known HRIRs, by which the remaining HRIRs are represented. The procedures for QR subset selection are as follows:

1. The HRIR vector with maximum L_2-norm is selected as u_0 from M known HRIR vectors, and the corresponding normalized (unit) vector is given by:

$$e_0 = \frac{u_0}{(u_0^T u_0)^{1/2}}$$

2. The $M - 1$ remaining HRIR vectors are projected to the orthonormal subspace of e_0, that is, $(e_0^T h_i)e_0$ is subtracted from each of the $M - 1$ remaining HRIR vectors:

$$h'_i = h_i - (e_0^T h_i)e_0$$

The L_2-norm of each $M - 1$ vector h_i' is calculated, and the vector h_i associated with the maximum L_2-norm of h_i' is selected as u_1.

3. The two orthonormal and unit basis vectors for the subspace that is spanned by vectors u_0 and u_1 are derived. One can be selected as e_0, and the other can be expressed as a linear combination of vectors u_0 and u_1 thus:

$$e_1 = b_{10}u_0 + b_{11}u_1$$

Weights b_{10} and b_{11} and, therefore, vector e_1 can be obtained by the following orthonormal and normalized conditions:

$$(e_0^T e_1) = 0 \qquad (e_1^T e_1) = 1$$

4. The $M - 2$ remaining HRIR vectors are projected to the orthonormal subspace spanned by (u_0, u_1) or (e_0, e_1), that is, $(e_0^T h_i)e_0$ and $(e_1^T h_i)e_1$ are subtracted from each $M - 2$ remaining HRIR vectors:

$$h''_i = h_i - (e_0^T h_i)e_0 - (e_1^T h_i)e_1$$

The L_2-norm of each $M - 2$ vector h_i'' is calculated, and the vector h_i associated with the maximum L_2-norm of h_i'' is selected as u_2.

5. The K privileged HRIR vectors $u_0, u_1 \ldots u_{K-1}$ and corresponding directions can be selected following the preceding steps in sequence. The resultant HRIR vectors constitute a subset of the basis for original HRIR vectors.

6. The HRIR at the arbitrary direction (θ, ϕ) are approximated by a linear combination of basis vectors $u_0, u_1 \ldots u_{K-1}$:

$$\hat{h}(\theta,\phi) \approx \sum_{i=0}^{K-1} A_i(\theta,\phi) u_i. \tag{6.56}$$

The basis vectors obtained by the QR subset method minimize the mean square error between $\hat{h}(\theta)$ and original $h(\theta)$ for a given number K of subset basis vectors.

7. These procedures are applicable to selecting the basis vector for HRIRs and HRTFs.

Subset selection can also be implemented by a combination of SVD (or equal PCA) and QR decomposition of HRIR or HRTF matrices. This approach is called SVD subset selection. Similar to that executed in Equations (6.41) and (6.42), the zero-mean HRIRs (or zero-mean HRTFs) at M directions can be represented by the product of the $N \times Q$ matrix $[D]$ and $Q \times M$ matrix $[W]$, where matrix $[D]$ consists of Q time-domain basis vectors; and matrix $[W]$ consists of weights with each of column corresponding to a specified direction associated with HRTFs. Each column of matrix $[W]$ constitutes a $Q \times 1$ column vector w_i with $i = 0,1 \ldots M - 1$, denoting the weighting vector associated with a specified direction. In steps (1) to (6), replacing the HRIR vectors h_i by weighting vectors w_i also yields K privileged directions. Then, the HRIR vectors h_i associated with the K privileged directions are chosen as the basis vectors $u_0, u_1, \ldots , u_{K-1}$, by which the HRIR at an arbitrary direction is approximately represented. These steps account for the procedure of SVD subset selection.

Gardner (1999) compared the results of QR subset selection, SVD subset selection, and PCA. The minimum-phase HRIRs of KEMAR and five human subjects were used in the analysis. The KEMAR-HRIRs included 72 azimuthal measurements with an interval of 5° in the horizontal plane. The human HRIRs were selected from the IRCAM database (Table 2.1), including 24 azimuthal measurements with an interval of 15° in the horizontal plane for each subject. All the HRIRs were diffuse field equalized. The results indicate that SVD subset selection and PCA require $K = 6$ and $K = 5$ basis vectors, respectively, to achieve a mean error [see Equation (5.14)] of −20 dB. With the same number of basis vectors, the smallest mean error is generated by PCA, followed by the SVD method; the QR method exhibits the largest mean error. However, the results for the five human subjects and KEMAR indicate that the privileged directions chosen by the SVD subset selection method vary across individuals. The six privileged directions for the right ear HRTFs of KEMAR are (listed in the order of contribution) 40°, 15°, 140°, 70°, 110°, and 270°. By contrast, the privileged directions chosen by the QR subset selection method are consistent across subjects. The overall results for the five human subjects are 90°, 30°, 60°, 105°, 0°, and 150°.

The ipsilateral HRIRs tend to dominate subset basis vectors because ipsilateral HRIRs contain more energy than contralateral HRIRs. This difference in energy stems from the diffraction and shadow effect of the head. Gardner's results indicate that contralateral minimum-phase HRIRs can be approximated by a linear combination of ipsilateral HRIRs. Contralateral HRIRs contain less energy, but include information on head shadow and diffraction. Whether such information can be accurately recovered from ipsilateral HRIRs alone is debatable, requiring more validation by psychoacoustic experiments.

Although PCA is superior to subset selection in terms of data reduction, an advantage of the latter over PCA is that its basis vectors come directly from actual HRIRs or HRTFs. This feature makes subset selection beneficial to binaural signal processing (in particular, the virtual loudspeaker-based algorithms, to be discussed in Section 6.5.1).

In addition to the QR subset selection and SVD subset selection methods, genetic algorithms were used (Cheung et al., 1998) to select privileged HRTF subsets and optimize reconstruction on the whole. Some researchers used spatial clustering to choose basis vectors (Shimada et al., 1994; Fahn and Lo, 2003). The HRTFs for different directions are first grouped into clusters according to certain similarity or distance criteria. A subset of representative HRTFs, one for each cluster, is selected as the center of the clusters so that the mean distance from each center to all the HRTFs contained in the cluster is minimized. Each center represents the main features of the corresponding cluster. Finally, the HRTFs at other or arbitrary directions can be approximately reconstructed by a linear combination of the representative HRTF subsets or cluster centers. In practice, various error criteria (including those discussed in Section 5.1.1) and cepstrum distances are applicable to measuring the similarity or distance among HRTFs. Nicol et al. (2006) compared the similarity criteria for HRTF clustering to reduce reconstruction error. A neural network-based method was also proposed for selecting representative HRTF subsets (Lemaire et al., 2005). Various interpolation schemes can be used to reconstruct HRTFs from representative subsets.

6.3. Spatial Basis Function Decomposition of HRTFs

6.3.1. Basic Concept of Spatial Basis Function Decomposition

As stated in Section 6.2.1, HRTFs can also be decomposed into a weighted combination of spatial basis functions. In this case, the HRTF of a given ear is generally represented by:

$$H(\theta,\phi,f,s)=\sum_{q}d_{q}(f,s)w_{q}(\theta,\phi). \tag{6.57}$$

In contrast to the $w_q(\theta, \phi)$ in Equation (6.14), that in Equation (6.57) is a series of spatial basis functions rather than direction-dependent weights. Here, $w_q(\theta, \phi)$ depends only on source direction. $d_q(f, s)$ is a set of frequency- and individual-dependent weights. When the spatial basis functions $w_q(\theta, \phi)$ are specified, $H(\theta, \phi, f, s)$ are completely determined by weights $d_q(f, s)$. Note that $d_q(f, s)$ always differs for left and right HRTFs, whereas $w_q(\theta, \phi)$ may be identical or different for left and right HRTFs, depending on the decomposition algorithms used. Moreover, the notation for the left or right ear is excluded from the denotation of HRTFs for simplicity; such a notation is supplemented when necessary. Corresponding to the decomposition of HRTFs in the frequency domain, the equivalent decomposition of HRIRs in the time domain is achieved by applying inverse Fourier transform to Equation (6.57) thus:

$$h(\theta,\phi,t,s)=\sum_{q}g_{q}(t,s)w_{q}(\theta,\phi), \tag{6.58}$$

where $g_q(t, s)$ are the time-domain weights and inverse Fourier transform of $d_q(f, s)$.

Various methods are used to select or derive the spatial basis functions $w_q(\theta, \phi)$. A set of orthonormal basis functions is often desired, but appropriate selection often depends on the situation. This section discusses three kinds of important spatial basis functions: azimuthal Fourier series (harmonics), spatial SH functions, and those derived from spatial PCA (SPCA).

The spatial basis function decomposition of HRTFs was used for the HRTF calculation in Chapter 4, in which HRTF calculation involved solving the wave, or Helmholtz, equation within certain boundary conditions. The separation of variables is a common algorithm for solving the wave equation, whose solution is decomposed into a linear combination of orthonormal spatial basis functions. For example, the HRTF in Equation (4.3) is decomposed into a linear combination of Legendre polynomials (orthonormal basis functions) with various degree l, and the coefficients associated with each Legendre polynomial are merely frequency-dependent weights.

Suppose that the basis functions $w_q(\theta, \phi)$ in Equation (6.57) or (6.58) are specified, and the summarization can be truncated up to order Q. Substituting the measured HRTFs at M directions into Equation (6.57) yields:

$$H(\theta_{i,},\phi_{i},f,s)=\sum_{q=1}^{Q}d_{q}(f,s)w_{q}(\theta_{i},\phi_{i}) \qquad i=0,1,2...(M-1). \tag{6.59}$$

At each frequency f, Equation (6.59) is a set of M linear equations, with the number of unknown $d_q(f, s)$ equal to the number Q of basis functions. Under certain conditions, the exact or approximate solution of $d_q(f, s)$ can be obtained from Equation (6.59). The spatial basis functions representation of $H(\theta, \phi, f, s)$ can then be realized by substituting the resultant $d_q(f, s)$ into Equation (6.57). Therefore, spatial basis functions decomposition of HRTFs can also be regarded as a spatial interpolation, or fitting algorithm, for HRTFs. Moreover, if HRTFs can be exactly or approximately represented by a small set of basis functions (sometimes called a sparse set of basis functions), HRTF dimensionality is significantly reduced.

6.3.2. Azimuthal Fourier Analysis and Sampling Theorem of HRTFs

As an example of spatial basis function decomposition, Zhong and Xie (2005b, 2009a) decomposed the HRTFs at each elevation using the azimuthal Fourier series (harmonics) and then derived the azimuthal sampling theorem of HRTFs. For a specified individual and ear, a far-field HRTF is a continuous function of azimuth with a period of 2π at a given elevation. It can therefore be expanded as a real- or complex-valued azimuthal Fourier series thus:

$$H(\theta,\phi_0,f)=\sum_{q=-\infty}^{+\infty}d_q(f)\exp(jq\theta)=a_0(f)+\sum_{q=1}^{+\infty}[a_q(f)\cos q\theta+b_q(f)\sin q\theta]. \tag{6.60}$$

In Equation 6.60, $H(\theta, \phi_0, f)$ denotes the azimuthal HRTF at a given elevation $\phi = \phi_0$; $\exp(jq\theta)$ or $(\cos q\theta, \sin q\theta)$ is the azimuthal harmonic, a set of predetermined orthonormal basis functions that depend on azimuth rather than on individual and ear; and $d_q(f)$ or $\{a_0(f), a_q(f), b_q(f)\}$ are frequency-dependent weights or azimuthal Fourier coefficients that can be evaluated by:

$$d_q(f)=\frac{1}{2\pi}\int_0^{2\pi}H(\theta,\phi_0,f)e^{-jq\theta}d\theta \qquad q\neq 0,$$

$$d_0(f)=a_0(f)=\frac{1}{2\pi}\int_0^{2\pi}H(\theta,\phi_0,f)d\theta, \tag{6.61}$$

$$a_q(f)=d_q(f)+d_{-q}(f) \qquad b_q(f)=j[d_q(f)-d_{-q}(f)] \qquad q=1,2,3.....$$

In addition to frequency dependence, $d_q(f)$ or $\{a_0(f), a_q(f), b_q(f)\}$ are also relevant to elevation, individual, and ear. For simplicity, however, these variables are excluded in the following discussion.

Equations (6.60) and (6.61) indicate that azimuthal HRTFs can be decomposed into a weighted sum of infinite azimuthal harmonics, and weights $d_q(f)$ or $\{a_0(f), a_q(f), b_q(f)\}$ can be evaluated from continuous $H(\theta, \phi_0, f)$. If the azimuthal harmonic expansion in Equation (6.60) is convergent, Equation (6.60) can be truncated as:

$$H(\theta,\phi_0,f)=\sum_{q=-Q}^{Q}d_q(f)\exp(jq\theta)=a_0(f)+\sum_{q=1}^{Q}[a_q(f)\cos q\theta+b_q(f)\sin q\theta], \tag{6.62}$$

where Q is the truncation order, above which higher-order harmonics are disregarded because their contribution is insignificant. The azimuthal HRTF then consists of $(2Q + 1)$ weighted azimuthal harmonics, and is completely determined by its $(2Q + 1)$ weights or azimuthal Fourier coefficients. In this situation, the $(2Q + 1)$ weights can be evaluated from the discrete azimuthal samples of $H(\theta, \phi_0, f)$. Sampling the azimuthal HRTF from azimuth 0 to 2π at M positions with an even interval $2\pi/M$ yields:

$$H(\theta_i,\phi_0,f)=\sum_{q=-Q}^{+Q}d_q(f)\exp(jq\theta_i) \qquad \theta_i=\frac{2\pi i}{M}, \quad i=0,1,....,M-1,$$

where θ_i is the measured azimuth, and $H(\theta_i, \phi_0, f)$ is the measured HRTF at azimuth θ_i. If the number of azimuthal measurements satisfies $M \geq (2Q + 1)$, then the $(2Q + 1)$ weights can be obtained by resolving these M linear equations. The solution involves using the orthogonality of $\exp(jq\theta_i)$:

$$d_q(f) = \frac{1}{M}\sum_{i=0}^{M-1} H(\theta_i, \phi_0, f)exp(-jq\theta_i) \qquad |q| \leq Q$$

$$d_q(f) = 0 \qquad \begin{array}{l} Q < |q| \leq (M-1)/2 \quad \text{if } M \text{ is odd} \\ -(M/2) \leq q < -Q, \text{ or } Q < q \leq (M/2)-1 \quad \text{if } M \text{ is even} \end{array} \tag{6.63}$$

Substituting Equation (6.63) into Equation (6.62) yields the following azimuthal interpolation equation of HRTFs, which enables the HRTFs at unmeasured azimuths and the azimuthal-continuous HRTFs to be reconstructed from M measured HRTFs:

$$H(\theta, \phi_0, f) = \frac{1}{M}\sum_{i=0}^{M-1} H(\theta_i, \phi_0, f) \frac{\sin\left[\left(Q+\frac{1}{2}\right)(\theta-\theta_i)\right]}{\sin\left(\frac{\theta-\theta_i}{2}\right)}. \tag{6.64}$$

At each given elevation, the azimuthal HRTFs can be decomposed as a weighted sum of azimuthal harmonics. Accordingly, the problems of HRTF azimuthal sampling rate and interpolation are closely related to the highest order of azimuthal harmonics. If the azimuthal HRTF can be represented by the azimuthal harmonics up to order Q, the azimuthal-continuous HRTF can be reconstructed from the $M \geq (2Q + 1)$ azimuthal measurements uniformly distributed in the $0 \leq \theta < 2\pi$ region. In other words, the azimuthal sampling rate should be at least twice that of the azimuthal Fourier harmonic bandwidth of HRTFs. This statement is the azimuthal sampling theorem of HRTF, which is similar to the Shannon-Nyquist theorem for time.

6.3.3. Analysis of Required Azimuthal Measurements of HRTFs

As seen in Section 6.1, the errors in various HRTF interpolation schemes differ but follow a general tendency. That is, the denser the directional grid in HRTF measurement, the fewer the errors that occur in interpolated HRTFs. In achieving accurate interpolation performance, measurements with high directional resolution are preferable. Because dense measurement is time-consuming and tiresome, however, measuring HRTFs with excessively fine directional resolution for each individual is inconvenient. The problem then lies in determining an appropriate directional resolution for HRTF measurement. Some preliminary results on this issue have been introduced in Section 6.1. These results are based on the objective analysis of physical errors or the subjective examination of audible errors in interpolated HRTFs, rather than restricting efforts to theoretical analysis. The directional resolution needed to reconstruct a continuous spatial function is determined by the complexity of the function itself. The azimuthal sampling theorem of HRTFs requires that the minimal number of azimuthal measurements be $M_{min} = (2Q + 1)$, where Q is the highest order of azimuthal harmonics. Consequently, the azimuthal resolution needed is represented by the maximum allowable azimuthal interval $\Delta\theta = 360° / (2Q + 1)$ in measurement. In what follows, the azimuthal resolution needed in HRTF measurement is analyzed in detail, on the basis of the azimuthal sampling theorem (Zhong and Xie, 2005b, 2009a).

Evaluating the contribution of each azimuthal harmonic and then determining Q necessitates defining the relative power of the qth-order azimuthal harmonic term against total power by:

$$\eta_q(f) = \frac{|d_q(f)|^2 + |d_{-q}(f)|^2}{\displaystyle\sum_{q'=-\infty}^{+\infty} |d_{q'}(f)|^2} \qquad q = 1,2,3.... \tag{6.65}$$

In Equation (6.65), the denominator is an infinite sum over weights, which should be calculated from a continuous azimuthal HRTF [see Equation (6.61)]. Unfortunately, the continuous azimuthal HRTF within the audible frequency range is unavailable because only the HRTFs at discrete azimuthal directions are usually measured. However, given an HRTF database that contains a sufficient number of measured HRTFs and satisfies the azimuthal sampling theorem, the weights can be calculated by Equation (6.63). Equation (6.65) can then be rewritten as:

$$\eta_q(f) = \frac{|d_q(f)|^2 + |d_{-q}(f)|^2}{\displaystyle\sum_{q'=-Q_1}^{+Q_2} |d_{q'}(f)|^2} \qquad \begin{array}{l} q = 1,2,3...\min(Q_1,Q_2) \\ Q_1 = Q_2 = (M-1)/2, \quad M = odd \\ Q_1 = M/2, \quad Q_2 = (M/2)-1, \quad M = even \end{array} \tag{6.66}$$

where $(Q_1 + Q_2 + 1) = M$ is the number of azimuthal measurements. To evaluate the contribution of the preceding Q'-order azimuthal harmonic terms, the relative power ratio η is defined as:

$$\eta = \frac{\displaystyle\sum_{q=-Q'}^{+Q'} |d_q(f)|^2}{\displaystyle\sum_{q=-Q_1}^{+Q_2} |d_q(f)|^2}. \tag{6.67}$$

The steps for determining Q and $M_{min} = (2Q + 1)$ for each elevation are as follows.

1. At each discrete frequency point, whose value is determined by sampling frequency, η for $Q' = 1,2, \ldots$, is calculated until $Q' = Q'' < \min(Q_1, Q_2)$ when η reaches a power criterion, such as 0.95 or 0.99. This criterion is chosen according to the desired reconstruction accuracy. Then, Q'' is regarded as the truncation order Q, above which the contribution of higher-order terms is negligible. According to the azimuthal sampling theorem of HRTFs, $(2Q + 1) = (2Q'' + 1)$ is M_{min}.

2. If $Q'' < \min(Q_1, Q_2)$, then $(2Q'' + 1) < 2 \times \min(Q_1, Q_2) + 1 \leq (Q_1 + Q_2 + 1)$, where the first term is M_{min} and the last term is the number of azimuthal measurements. In this case, the number of azimuthal measurements is sufficient, with some redundancy under the chosen power criterion, because M_{min} is less than the number of actual azimuthal measurements. If $Q'' = \min(Q_1, Q_2)$, which indicates that truncation rests on the harmonic term with the highest order that the employed HRTFs can provide, then evaluating the sufficiency of the number of azimuthal measurements is difficult. In this case, the number of azimuthal measurements analyzed is regarded as insufficient.

3. The M_{min} calculated by the preceding steps is intended for each discrete frequency point. For a given frequency range, the maximum M_{min} among the discrete frequency points included in the frequency range is chosen as the M_{min} of the frequency range. In what follows, the $M_{min\text{-}audio}$ for the audible frequency range ($f \leq 20$ kHz) is calculated.

The MIT-KEMAR HRIRs of the left ear are used first in the analysis. The number M of azimuthal measurement and corresponding azimuthal interval $\Delta\theta$ at each elevation are shown in Table 2.2,

Section 2.3. Each dataset contains a 512-point HRIR with a sampling frequency of 44.1 kHz, from which a 512-point HRTF can be obtained by DFT.

Calculated from the 72 measured azimuthal HRTFs in the horizontal plane $\phi = 0°$, the relative contribution of each azimuthal harmonic is shown in Figure 6.4. The relative contribution of lower-order azimuthal harmonics, which correspond to the smooth azimuthal variation of the HRTF, is significant at low frequencies and extends to high frequencies. The relative contribution of higher-order azimuthal harmonics, which corresponds to the rapid azimuthal variation of the HRTF, is negligible at low frequencies and becomes significant with increasing frequency. For example, the relative contributions of the 20th-order azimuthal harmonic are 3.2×10^{-7} at 172 Hz and 0.12 at 15.4 kHz. Moreover, the relative contribution of the azimuthal harmonics demonstrates a clear downward trend with increasing expansion order, verifying the hypothesis that the azimuthal harmonic expansion of HRTFs is convergent. As a result, the truncation order Q is low and M_{min} is small at low frequencies, and these variables visibly increase with frequency. Chapter 4 demonstrated that a weighted sum of low-degree Legendre polynomials, which is equivalent to a combination of low-order directional harmonics, sufficiently represents spherical head HRTFs below 2–3 kHz. Because the irregularities of head shape and pinnae influence higher order azimuthal harmonics at high frequencies, analyzing the variations in high-order azimuthal harmonics is a means to exploring the effect of fine anatomical structures on high-frequency HRTFs (Zhong and Xie, 2007c). The results for other elevation planes are similar, but the relative contribution of high-order azimuthal harmonics gradually declines as elevation deviates from the horizontal plane.

Figure 6.5 shows the M_{min} of KEMAR as a function of frequency for some elevations; the M_{min} is calculated from Equation (6.67) with power criteria of 0.95 and 0.99. Table 2.2 showed that the azimuthal interval of elevation ±40° is $\Delta\theta = 6.43°$ in MIT-KEMAR HRTF measurement. The azimuthal interval in actual measurements is only an approximation of this value and is, therefore, at most nearly uniform. Thus, the results for these two elevations are not included in the figure. Because the M_{min} curves for $\phi = -10°$, $-20°$, and $-30°$ are nearly identical to those for $\phi = 10°$, $20°$, and 30°, respectively, they are also omitted in the figure. Figure 6.5 shows some general tendencies:

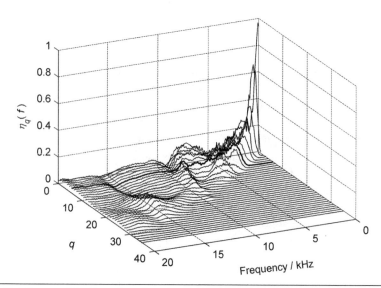

Figure 6.4 Relative contribution of each azimuthal harmonic at different frequencies

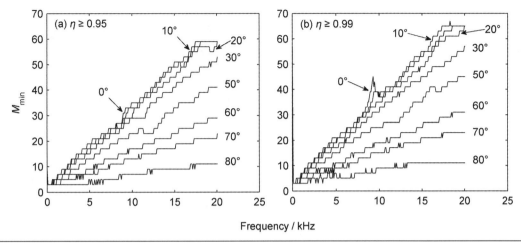

Figure 6.5 Minimal number M_{min} of azimuthal measurements for KEMAR HRTFs: (a) $\eta \geq 0.95$; (b) $\eta \geq 0.99$

1. Given an elevation plane, the M_{min} of each frequency point increases stepwise with increasing frequency, although a few exceptions may exist.
2. Given a frequency sampling point, M_{min} gradually decreases as elevation deviates from the horizontal plane.
3. Given an elevation plane and a frequency sampling point, the M_{min} with criterion 0.99 is always larger than that with criterion 0.95.

The calculated $M_{min\text{-}audio}$ for the audible frequency range and actual number of azimuthal measurements for MIT-KEMAR HRTFs at various elevations are listed in Table 6.1. Comparison among data in rows 2–4 in Table 6.1 indicates that the actual number of azimuthal measurements for the MIT data are sufficient for most elevations, except for elevations 70° and 80° with a power criterion of 0.95, or elevations 50°, 70°, and 80° with a power criterion of 0.99.

As indicated by the results in Table 6.1, the azimuthal interval in each elevation listed in Table 2.3 was chosen in the measurements for the Chinese HRTF database. After the measurement, the $M_{min\text{-}audio}$ of each subject and elevation was verified (Zhong and Xie, 2009a). The results indicate

Table 6.1 Calculated $M_{min\text{-}audio}$ for the audible frequency range and actual number of azimuthal measurements for MIT-KAMER HRTFs at various elevations

ϕ	−30°	−20°	−10°	0°	10°	20°	30°	50°	60°	70°	80°
Actual M	60	72	72	72	72	72	60	45	36	24	12
$M_{min\text{-}audio}$ ($\eta \geq 0.99$)	57	63	63	65	67	65	57	45	31	23	11
$M_{min\text{-}audio}$ ($\eta \geq 0.95$)	53	57	57	59	59	59	53	41	29	23	11
$M_{min\text{-}audio}$ ($\eta \geq 0.99$, with arrival time correction)	49	35	31	43	59	57	39	31	19	17	11
$M_{min\text{-}audio}$ ($\eta \geq 0.95$, with arrival time correction)	21	17	17	21	25	23	19	21	17	15	11
$M_{min\text{-}audio}$ ($\eta \geq 0.99$, magnitude alone)	15	15	17	23	25	19	15	9	7	5	5
$M_{min\text{-}audio}$ ($\eta \geq 0.95$, magnitude alone)	9	9	9	11	9	11	9	7	5	3	3

that in the horizontal plane and up to 20 kHz, the $M_{min\text{-}audio}$ of 51 out of 52 subjects varies in a range from 49 to 69 with a power criterion $\eta \geq 0.95$; the $M_{min\text{-}audio}$ of one subject is 71 (for this subject, actual azimuthal measurement $M = 72$ is sufficient up to 16 kHz). The mean $M_{min\text{-}audio}$ across 52 subjects is 64. The third row in Table 6.2 lists the mean $M_{min\text{-}audio}$ and standard deviation across 52 subjects for various elevations. All mean $M_{min\text{-}audio}$ are less than the actual number M of azimuthal measurements listed in the second row of Table 6.2. These results verify that the azimuthal interval in each elevation listed in Table 2.3 was reasonable, that is, the azimuthal measurements for the database were sufficient.

As stated in Section 6.1.4, if the arrival time of HRIRs is corrected prior to interpolation—that is, if the arrival time for each source position is made synchronous by shifting the onset time of HRIRs—the interpolation performance of HRIRs or HRTFs improves. Interpolation to HRTF magnitudes alone also exhibits better interpolation performance. This result motivates the arrival time correction in the original HRTFs, or the consideration of HRTF magnitudes alone in the azimuthal Fourier analysis (Zhong and Xie, 2008, 2009a). Rows 5–8 of Table 6.1 show the resultant $M_{min\text{-}audio}$ of the MIT-KEMAR HRTFs. The results for the 52 Chinese subjects are shown in the fourth and fifth rows of Table 6.2.

$M_{min\text{-}audio}$ visibly decreases after arrival time correction. The arrival time correction representation of HRTFs cascaded with pure delay retains all the information from the original HRTFs. Such a correction is therefore especially suitable for simplifying HRTF measurement. Tables 6.1 and 6.2 also show that the actual azimuthal measurement for the arrival time correction representation of the HRTFs of KEMAR and Chinese subjects is sufficient. In addition, the $M_{min\text{-}audio}$ of HRTF magnitudes alone at each elevation further decreases, which is important in practical application.

According to the results listed in Table 6.2, the 95% confidence range of $M_{min\text{-}audio}$ can be evaluated as mean $M_{min\text{-}audio} \pm 2\sigma$. To ensure convenient practical measurement and statistical satisfaction of the azimuthal sampling theorem, the suggested number of azimuthal measurements M for the arrival time correction representation of HRTFs is an integer equal to, or slightly larger than, mean $M_{min\text{-}audio} + 2\sigma$, which is the upper limit of the 95% confidence range. The results are listed in the second row of Table 6.3, with the corresponding azimuthal interval $\Delta\theta = 360°/M$ bracketed. Because conventional HRTF measurement is usually performed at azimuthal intervals of 5°, an alternative suggestion is provided in the third row of Table 6.3.

The third row of Table 6.1 also implies that the relationship between the $M_{min\text{-}audio}$ of elevation ϕ and that of the horizontal plane can be approximated by Equation (6.68), with a correlation coefficient of 0.996:

$$M_{min\text{-}audio}(\phi) = M_{min\text{-}audio}(0)\cos\phi, \tag{6.68}$$

Table 6.2 Calculated $M_{min\text{-}audio}$ for the audio frequency range and actual number of azimuthal measurements for 52 Chinese subjects at various elevations

ϕ	−30°	−15°	0°	15°	30°	45°	60°	75°	90°
Actual M	72	72	72	72	72	72	36	24	1
Mean $M_{min\text{-}audio}$ ($\eta \geq 0.95$) and standard deviation σ	60	63	64	62	58	52	34	23	1
	±4.8	±3.9	±3.9	±4.8	±4.9	±7.4	±1.3	±1.4	
Mean $M_{min\text{-}audio}$ ($\eta \geq 0.95$, with arrival time correction) and standard deviation σ.	37	29	36	29	28	24	16	14	1
	±11.3	±10.0	±10.3	±5.7	±7.7	±8.0	±3.2	±3.6	
Mean $M_{min\text{-}audio}$ ($\eta \geq 0.95$, magnitude alone) and standard deviation σ	15	16	19	16	13	11	8	8	1
	±6.0	±5.9	±7.9	±5.4	±5.0	±4.7	±2.3	±3.2	

Note: The calculated frequency in Table 6.2 is up to 20 kHz.

Table 6.3 Suggested azimuthal measurement numbers for human HRTF measurement (with azimuthal resolutions in parentheses, which are determined with arrival time correction).

ϕ	−30°	−15°	0°	15°	30°	45°	60°	75°	90°
Suggested M and $\Delta\theta$ derived from the 95% confidence range of the mean $M_{min\text{-}audio}$	60 (6°)	60 (6°)	60 (6°)	40 (9°)	45 (8°)	40 (9°)	24 (15°)	24 (15°)	1
Suggested M and $\Delta\theta$ chosen from multiples of 5°	72 (5°)	72 (5°)	72 (5°)	72 (5°)	72 (5°)	72 (5°)	24 (15°)	24 (15°)	1

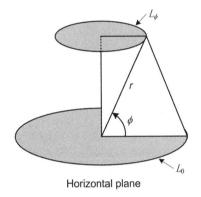

Horizontal plane

Figure 6.6 Perimeter at different elevations

Therefore, for HRTFs without arrival time correction, the azimuthal measurement needed decreases as the source position deviates from the horizontal plane. Bovbjerg et al. (2000) obtained a similar result from an empirical equation, rather than from the azimuthal sampling theorem of HRTFs. Coincidently, the same relationship exists in the perimeter between the elevation plane ϕ and the horizontal plane of a sphere, shown in Figure 6.6—$L_\phi = L_0 \cos\phi$. Thus, the $M_{min\text{-}audio}$ of each elevation is directly proportional to the perimeter ratio of the elevation to the horizontal plane.

Ajdler et al. (2005) analyzed the required azimuthal measurement of HRTFs using the acoustic principle of reciprocity. The authors assumed that the free sound field is sampled along a horizontal circle with radius r, and a source is located within the circle. The azimuthal sample theorem was introduced by analyzing the two-dimensional temporal-azimuthal frequency spectrum and its sound field bandwidth. The results indicate that the required azimuthal measurement of HRTFs in the horizontal plane is $M = 72$, or, equally, an even interval of 5°, which is similar to the results of azimuthal measurement without arrival time correction in Tables 6.1 and 6.2. However, Ajdler used a free-field model and did not consider the scattering and diffraction of anatomical structures (such as head, torso, and pinnae). The HRTFs of the spherical head and KEMAR were included in further analysis in 2008, and a similar conclusion was drawn (Ajdler et al., 2008). An algorithm for interpolating HRTFs in the subband domain was also proposed by Ajdler et al. In the subbands where spatial aliasing occurs on traditional interpolation schemes, the subband component of an HRTF is first decomposed into its complex envelope and carrier. The spatial interpolation is carried out in the complex temporal envelope, and the carriers of the different azimuthal positions are aligned to the carrier of a reference azimuthal position, which improves interpolation performance.

The azimuthal interpolation equation (6.64) is derived when the azimuthal sampling theorem is analyzed in Section 6.3.2. The azimuthal periodicity of the HRTF at each elevation is used in Equation (6.64), so that all the HRTF samples at M measured azimuths contribute to the interpolated HRTFs with various (but frequency-independent) weights. This difference is prominent between Equation (6.64) and traditional adjacent interpolation schemes.

To evaluate the interpolation performance of Equation (6.64), the horizontal HRTFs of the left ear of KEMAR are interpolated. $M = 72$ azimuthal measurements, with an even interval of $\Delta\theta = 5°$, exist in the horizontal HRTFs provided by MIT. For comparison, the $M = 36$ measured HRTFs, with azimuths ranging from $\theta = 0°\text{–}350°$ in 10° increments, are substituted into Equation (6.64) to recover continuous HRTFs. The remaining HRTFs, with azimuths ranging from 5°–355° at 10° increments, are used as references for evaluation.

Interpolation performance can be quantificationally evaluated by the mean SDR defined in Equation (5.16). As indicated in Figure 6.5, $M = 36$ azimuthal measurements are sufficient for recovering azimuthally continuous HRTFs in the horizontal plane up to 10 kHz (without arrival time correction). Accordingly, the calculation in Equation (5.16) is limited to a frequency range up to 10 kHz. The results are shown in Figure 6.7, which illustrates that SDR_M is high for ipsilateral source directions (θ = 180°–360°) and low for contralateral source directions (θ = 0°–180°). Similar to the case of PCA, this difference is due to the complicated

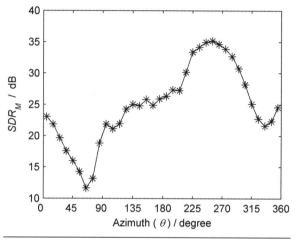

Figure 6.7 Mean signal-to-distortion ratio (SDR_M) of azimuthal interpolation in the horizontal plane

nature of contralateral HRTFs at high frequencies. The attenuation caused by the head shadow effect yields an inherently poor signal-to-noise ratio for measured contralateral HRTFs. Overall, the average SDR_M over 36 reference directions is about 25 dB, indicating good interpolation performance on average.

6.3.4. Spherical Harmonic Function Decomposition of HRTFs

In the preceding discussion, the HRTF for each elevation is decomposed by a set of orthonormal azimuthal basis $\exp(jq\theta)$. The far-field HRTF is a function of both azimuth θ and elevation ϕ, and can therefore be decomposed by the orthonormal basis functions of azimuth and elevation. Evans et al. (1998a) proposed the use of SH functions in decomposing HRTFs. As stated in Appendix A, a source direction is specified by $\Omega = (\alpha, \beta)$ with $0° \leq \alpha \leq 180°$ and $0° \leq \beta < 360°$. This source direction is related to the default coordinate in Figure 1.1 by $\beta = \theta$, $\alpha = 90° - \phi = \phi'$. Thus, the far-field HRTFs of a given individual and ear are denoted by $H(\alpha, \beta, f) = H(\Omega, f)$, and can be decomposed by complex- or real-valued SH functions (SHFs) as (real-valued SHFs were used in Evans's original paper).

$$H(\Omega, f) = \sum_{l=0}^{\infty} \sum_{m=-l}^{l} d_{lm}(f) Y_{lm}(\Omega)$$
$$= a_{00}(f) Y_{00}^1(\Omega) + \sum_{l=1}^{\infty} \sum_{m=0}^{l} [a_{lm}(f) Y_{lm}^1(\Omega) + b_{lm}(f) Y_{lm}^2(\Omega)],$$

(6.69)

where $Y_{l0}^2 = 0$, which is preserved in Equation (6.69) for convenience in writing. Spherical harmonics are a set of predetermined orthonormal basis functions of direction $\Omega = (\alpha, \beta)$, and are independent of individual and ear. The weights or SH coefficients for the complex-valued SHFs decomposition are evaluated as:

$$d_{lm}(f) = \int H(\Omega, f) Y_{lm}^*(\Omega) d\Omega = \int_{\beta=0}^{2\pi} \int_{\alpha=0}^{\pi} H(\alpha, \beta, f) Y_{lm}^*(\alpha, \beta) \sin\alpha \, d\alpha \, d\beta,$$
$$l = 0, 1, 2 \dots, \quad |m| \leq l$$

(6.70)

and the SH coefficients $\{a_{00}(f), a_{lm}(f), b_{lm}(f)\}$ for the real-valued SHFs decomposition are related to $d_{lm}(f)$ by Equations (A.15) and (A.16) in Appendix A. In addition to frequency dependence, individual and ear dependence characterize SH coefficients. For simplicity, however, the notation for the left or right ear is omitted from the denotation of SH coefficients.

Similar to the azimuthal Fourier or harmonic decomposition of HRTFs, a directionally continuous $H(\Omega, f)$ is required for calculating SH coefficients by Equation (6.70). The available HRTFs are measured at discrete and finite M directions, and the integral in Equation (6.70) should be calculated by a weighted sum at all M directions. Some calculation methods are outlined in Appendix A. Given that the HRTFs at M directions are sampled by appropriate schemes—that is, $H(\Omega_i, f)$ with $i = 0, 1, \ldots (M - 1)$—the integral in Equation (6.70) at each specified frequency can be evaluated by the summation:

$$d_{lm}(f) = \sum_{i=0}^{M-1} \lambda_i \, H(\Omega_i, f) Y_{lm}^*(\Omega_i). \tag{6.71}$$

Evans used the Gauss-Legendre sampling scheme (Appendix A) to select $M = 2L^2$ directions. For a given L, the SH coefficients up to the $l \le (L - 1)$ order can be evaluated. Substituting the resultant SH coefficients into Equation (6.69) and truncating the summation to order $l \le (L - 1)$ yield:

$$H(\Omega, f) = \sum_{l=0}^{L-1} \sum_{m=-l}^{l} d_{lm}(f) Y_{lm}(\Omega)$$

$$= a_{00}(f) Y_{00}^1(\Omega) + \sum_{l=1}^{L-1} \sum_{m=0}^{l} [a_{lm}(f) Y_{lm}^1(\Omega) + b_{lm}(f) Y_{lm}^2(\Omega)]. \tag{6.72}$$

If the contribution of SHFs with order $l \ge L$ is insignificant, so that Equation (6.69) is convergent, then Equation (6.72) is a reasonable representation of HRTFs. Compared with Equation (6.57), Equation (6.72) features the selection of a set of $Q = L^2$ SHFs, $\{Y_{lm}(\Omega)\}$ or $\{Y_{lm}^1(\Omega), Y_{lm}^2(\Omega)\}$, as the spatial basis functions $w_q(\alpha, \beta) = w_q(\Omega)$, and the corresponding $Q = L^2$ weights are $\{d_{lm}(f)\}$ or $\{a_{lm}(f), b_{lm}(f)\}$. In Equations (6.71) and (6.72), let:

$$A_i(\Omega) = \sum_{l=0}^{L-1} \sum_{m=-l}^{l} Y_{lm}^*(\Omega_i) Y_{lm}(\Omega), \tag{6.73}$$

then:

$$H(\Omega, f) = \sum_{i=0}^{M-1} \lambda_i A_i(\Omega) H(\Omega_i, f). \tag{6.74}$$

Equation (6.74) indicates that the HRTF at the arbitrary direction Ω can be obtained by a linear combination of measured HRTFs at M appropriate directions Ω_i. Therefore, Equation (6.72) or (6.74) can be regarded as an HRTF directional interpolation, or fitting, equation. Equation (6.72) can be converted to the expression in the default coordinate (θ, ϕ) by using $\alpha = (90° - \phi)$, $\beta = \theta$.

Evans measured the HRIRs of a B & K 4127 artificial head at $L = 18$ elevations, including $\phi = \pm5.0°, \pm14.5°, \pm24.0°, \pm34.0°, \pm44.0°, \pm53.5°, \pm63.0°, \pm73.0°$, and $\pm82.5°$, as well as $2L = 36$ azimuths (with an even azimuthal interval $\Delta\theta = 10°$ at each elevation). After processing, each dataset contains a 96-point HRIR with a sampling frequency of 20.20 kHz. The resultant 96-point HRTFs are obtained by DFT (discrete Fourier transform). Influenced by the structure of the apparatus, the HRTFs measured at the lowest four elevations, of $\phi < -50°$, are considerably distorted. To reduce the error in calculating SH coefficients, these data are excluded and simply replaced by the measured data at $\phi = -44.0°$. Naturally, this processing method may introduce some errors into the resultant SH coefficients.

SH decomposition can be applied to HRTFs in the frequency domain or to HRIRs in the time domain. According to Equation (6.8), the complex-valued HRTFs in the frequency domain can be represented by their magnitude and unwrapped phase responses. Then, each response can be separately decomposed by SHFs. According to Equation (6.71), SH coefficients below 10 kHz and up to order $l = 17$ for both magnitude and unwrapped phase components can be evaluated on the basis of measured HRTFs. Then, the l-order SH spectrum is defined as:

$$S_l(f) = \sqrt{\frac{\sum_{m=-l}^{l} |d_{lm}(f)|^2}{2l+1}} = \sqrt{\frac{\sum_{m=0}^{l} (|a_{lm}(f)|^2 + |b_{lm}(f)|^2)}{2l+1}}, \tag{6.75}$$

where $S_l(f)$ represents the relative contribution of an l-order SH component to the HRTF magnitude or phase response. Figure 6.8 shows the relationship between $\max[|S_l(f)|]$ (the maximum absolute value of function $S_l(f)$ for an HRTF magnitude over a frequency range of $f \le 10$ kHz) and the order l of SH components. The relative contribution of lower-order SH components is significant; this contribution corresponds to the smooth directional variation of HRTFs. The $\max[|S_l(f)|]$ decays with increasing order l. Similar to the azimuthal harmonic representation of HRTFs, the irregularities of the head shape and pinnae influence higher-order SH components at high frequencies. These irregularities are reflected by the rapid directional variations of HRTFs.

The HRTF magnitude results calculated by Evans indicate that the highest order ($l = 17$) SH components contribute only a proportion of 5.2×10^{-3} to the total energy of all $l \le 17$-order SH components. The preceding $l \le 7$ SH components contribute a proportion of more than 0.9 (90%) of total energy. When represented by the SH components up to order $l = 17$, the averages of mean relative energy error [Equation (5.12)] over directions for the interpolated HRTF magnitudes in Equation (6.72) are:

1. above the horizontal plane, 2.08% to 7.14%;
2. below the horizontal plane, 2.37% to 20.92%.

Evans et al. also decomposed the HRIRs using SHFs and revealed that the arrival time correction of HRIRs prior to decomposition improves reconstruction and interpolation performance.

With the SHF representation of HRTFs, the azimuthal sampling theorem of HRTFs can be extended to full spatial sampling. The truncation order $(L - 1)$ in Equation (6.72) is chosen according to average inter-element spacings of half a wavelength over the surface area of the sphere (Duraiswami et al., 2004):

$$(L-1) = integer(ka), \tag{6.76a}$$

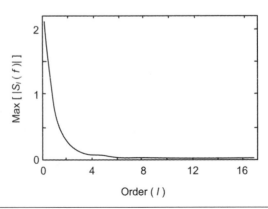

Figure 6.8 Relationship between max $[|S_l(f)|]$ for HRTF magnitude with $f \le 10$ kHz and order l (from Evans et al., 1998a, by permission of J. Acoust. Soc. Am.)

where k is the wave number; a represents the radius of the smallest sphere that encloses all significant scattering sources; and function "integer" denotes rounding up to the nearest integer. Or, alternatively, a slightly larger truncation order is chosen according to the characteristics of fast and monotonic decay to zero of the spherical Bessel function when its order is no less than $(eka/2)$ (Zhang and Abhayapala et al., 2010); that is,

$$(L-1) = integer\left(\frac{eka}{2}\right) \tag{6.76b}$$

where $e = 2.7183$. Accordingly, the required number of SH components (or spatial basis functions) in Equation (6.72) is provided by:

$$L^2 = [integer(ka)+1]^2, \tag{6.77a}$$

or by:

$$L^2 = [integer\left(\frac{eka}{2}\right)+1]^2. \tag{6.77b}$$

The truncation order and required number of SH components increase with k or frequency. According to the results of linear algebra, when HRTFs are represented as a weighted sum of L^2 spatial basis functions, L^2 directional samples are at least required for calculating the SH coefficients by Equation (6.71). This restriction is imposed, by the Shannon-Nyquist theorem, to HRTF spatial sampling. For a spherical head with radius $a = 0.0875$ m ≈ 0.09 m and a frequency of up to 20 kHz, Equations (6.76a) and (6.77a) indicate that $(L-1) = 33$ and $L^2 = 1156$, or Equations (6.76b) and (6.77b) show that $(L-1) = 46$ and $L^2 = 2209$. Therefore, the total number of directional samples is large. For an artificial head or human with a torso, a large spherical radius a that encloses the torso should theoretically be selected in calculating truncation order. However, the torso mainly influences HRTFs below 3 kHz and truncation order increases almost linearly with frequency. Evaluating the maximal truncation order $(L-1)$ and minimum directional measurement of HRTFs on the basis of the results for a spherical head at 20 kHz is a reasonable approach.

The Shannon-Nyquist theorem determines the lower bound of directional samples for calculating L^2 SH coefficients. The practical number of directional samples is usually larger than the lower bound, depending on the directional sampling scheme applied. As described in Appendix A, the equiangle sampling and Gauss-Legendre sampling schemes require $M = 4L^2$ and $M = 2L^2$ directional samples, which are fourfold and twice the lower bound, respectively. The nearly uniform sampling scheme requires at least $M = L^2$ directional samples, but, in practice, the value is 1.3–1.5 times the lower bound.

These methods require that the HRTF samples cover the sphere of directions to accurately calculate SH coefficients. As previously stated, measuring HRTFs at low elevations is difficult, and most of the available HRTF data provides the measurements at an open grid with the bottom cut off. The HRTFs at low elevations should be rationally supplemented, to alleviate the spatial Gibbs phenomenon caused by abrupt data truncation. An alternative method is to calculate the SH coefficients by various approximate algorithms, such as the pseudoinverse or the regularization algorithm. Further analysis indicates that the regularization algorithm is especially suitable for calculating SH coefficients using the data from an open measurement grid (Zotkin et.al, 2009). Zhang (W.), Zhang (M.), and Kennedy, et al. (2012) applied a hierarchical sampling scheme called IGLOO, from astrophysics, for efficient HRTF measurements and SH decompositions.

So far, the SH decomposition of distance-independent far-field HRTFs has already been discussed. Duraiswami et al. (2004) extended the SH decomposition algorithm to distance-dependent near-field HRTFs, yielding a method for estimating near-field HRTFs on the basis of far-field measurements. According to the acoustic principle of reciprocity, pressure at an ear can be equally calculated by the pressure at spatial location r caused by a point source at the ears. With the wave equation provided by Equation (4.14) and the Sommerfeld radiation condition given by Equation (4.15), the pressure at spatial location $(r, \Omega) = (r, \alpha, \beta)$ can be decomposed by complex- or real-valued SHFs (complex-valued SHFs were used in Duraiswami's original paper). Thus,

$$
\begin{aligned}
P(r,\Omega,f) &= \sum_{l=0}^{\infty}\sum_{m=-l}^{l} d'_{lm}(f)h_l(kr)Y_{lm}(\Omega) \\
&= a'_{00}(f)h_0(kr)Y_{00}^1(\Omega) \\
&\quad + \sum_{l=1}^{\infty}\sum_{m=0}^{l}[a'_{lm}(f)h_l(kr)Y_{lm}^1(\Omega)+b'_{lm}(f)h_l(kr)Y_{lm}^2(\Omega)],
\end{aligned}
\tag{6.78}
$$

where $h_l(kr)$ is the lth-order spherical Hankel function of the second kind; $\{d'_{lm}(f)\}$ or $\{a'_{00}(f),$ $a'_{lm}(f), b'_{lm}(f)\}$ is a set of SH coefficients to be determined [$Y_{lm}^2(\Omega) = 0$ is preserved in Equation (6.78) for convenience in writing].

According to Equations (1.15) and (1.16), the free-field pressure of a point source can be expressed as $P_0(r, f) = A \exp(-jkr)/r$. Then, the HRTF at an arbitrary source direction and distance can be written as:

$$
\begin{aligned}
H(r,\Omega,f) &= \frac{P(r,\Omega,f)}{P_0(r,f)} = \frac{r\exp(jkr)}{A}\sum_{l=0}^{\infty}\sum_{m=-l}^{l} d'_{lm}(f)h_l(kr)Y_{lm}(\Omega) \\
&= \frac{r\exp(jkr)}{A}\{a'_{00}(f)h_0(kr)Y_{00}^1(\Omega) \\
&\quad + \sum_{l=1}^{\infty}\sum_{m=0}^{l}[a'_{lm}(f)h_l(kr)Y_{lm}^1(\Omega)+b'_{lm}(f)h_l(kr)Y_{lm}^2(\Omega)]\}
\end{aligned}
$$

The spherical Hankel function of the second kind at infinitely distant distance tends to:

$$
h_l(kr)\xrightarrow[r\to\infty]{}\frac{j^{l+1}}{kr}\exp(-jkr)
$$

A normalized radial function can be introduced as:

$$
\Xi_l(kr)=(-j)^{l+1}kr\exp(jkr)h_l(kr).
$$

Then, the HRTF at an arbitrary source direction and distance becomes:

$$
\begin{aligned}
H(r,\Omega,f) &= \sum_{l=0}^{\infty}\sum_{m=-l}^{l} d_{lm}(f)\Xi_l(kr)Y_{lm}(\Omega) \\
&= a_{00}(f)\Xi_0(kr)Y_{00}^1(\Omega) \\
&\quad + \sum_{l=1}^{\infty}\sum_{m=0}^{l}[a_{lm}(f)\Xi_l(kr)Y_{lm}^1(\Omega)+b_{lm}(f)\Xi_l(kr)Y_{lm}^2(\Omega)],
\end{aligned}
\tag{6.79}
$$
$$
d_{lm}(f)= j^{l+1}d'_{lm}(f)/Ak \quad a_{lm}(f)= j^{l+1}a'_{lm}(f)/Ak \quad b_{lm}(f)= j^{l+1}b'_{lm}(f)/Ak.
$$

Because $\Xi_l(kr)$ tends to unity at an infinitely distant distance, $\{d_{lm}(f)\}$ or $\{a_{lm}(f), b_{lm}(f)\}$ is the only SH coefficient of the far-field HRTF decomposition given by Equation (6.69).

Equation (6.79) includes an infinite number of SH components. Similar to the far-field case, the SH representation is truncated up to the $(L - 1)$ order, resulting in L^2 SH components and corresponding coefficients. The exponential growth of the spherical Hankel function $h_l(kr)$ for the small argument kr requires $(L-1) = integer\,(kr_{min})$, where r_{min} denotes a minimum source distance. The truncation order should be chosen by comprehensively considering the minimum source distance and Equation (6.76).

The M directional samples of HRTFs at source distance $r = r_0$ are substituted into Equation (6.79) and then the summation is truncated up to the order $(L-1)$:

$$
\begin{aligned}
H(r_0,\Omega_i,f) &= \sum_{l=0}^{L-1}\sum_{m=-l}^{l} d_{lm}(f)\Xi_l(kr_0)Y_{lm}(\Omega_i) \\
&= a_{00}(f)\Xi_0(kr_0)Y_{00}^1(\Omega_i) \\
&\quad + \sum_{l=1}^{L-1}\sum_{m=0}^{l}[a_{lm}(f)\Xi_l(kr_0)Y_{lm}^1(\Omega_i)+b_{lm}(f)\Xi_l(kr_0)Y_{lm}^2(\Omega_i)].
\end{aligned}
\tag{6.80}
$$

$$i = 0,1,.....M-1$$

Similar to the case of far-field HRTFs, at each specified frequency f, Equation (6.80) is a set of M linear equations with L^2 SH coefficients unknown. This equation can also be written as the matrix equation:

$$[R]A = H, \tag{6.81}$$

where A is an $L^2 \times 1$ column vector (matrix) that consists of L^2 unknown SH coefficients; H is an $M \times 1$ column vector (matrix) that comprises M directional samples of HRTFs; and $[R]$ denotes an $M \times L^2$ matrix that consists of the products of the radial function and directional samples of SHFs. If $M = L^2$ and matrix $[R]$ is invertible, the unique solution of Equation (6.81) is given by $A = [R]^{-1}H$. Because the HRTFs at low elevations are usually unavailable, $[R]$ is often ill conditioned and therefore non-invertible. Duraiswami et al. used the Tikhonov regularization to solve Equation (6.81) thus:

$$A = \{[R]^+[R]+\delta[D]\}^{-1}[R]^+H, \tag{6.82}$$

where δ is a small regularization parameter that balances the robustness and accuracy of the solution, and $\delta = 0$ corresponds to a common pseudoinverse solution; $[D]$ is an $L^2 \times L^2$ diagonal regularization matrix, with its diagonal element—$D_l = 1 + l(l+1)$—corresponding to an l-order SH coefficient.

The source-position continuous representation of HRTFs is obtained by substituting the resultant SH coefficients into Equation (6.79) with summation, up to order $l \leq (L-1)$. Equation (6.79) can also be converted into the expression in the default coordinate (θ, ϕ) by letting $\alpha = (90° - \phi)$, $\beta = \theta$.

Equations (6.79) and (6.82) not only enable the reconstruction of directional continuous HRTFs from M directional measurements at a given distance r_0, but also allow for estimating the HRTFs at an arbitrary source distance from the measurements at a given distance r_0. Therefore, Equation (6.79) or (6.82) is a distance or range extrapolation equation of HRTFs. As stated in Section 2.4, the algorithm for estimating near-field HRTFs on the basis of far-field measurements is attractive because it resolves some of the difficulties encountered in near-field HRTF measurement.

Similarly, Pollow et al. (2012) studied the problem of HRTF range extrapolation by SH decomposition, and confirmed that the results of range extrapolation are in good agreement with the measurements and HRTFs calculated from artificial heads that are based on the boundary element method (BEM). The results also showed that outward range extrapolation (i.e., from near-field to far-field measurements) is theoretically more stable than inward extrapolation.

The SH representation of HRTFs at different source directions and distances was also analyzed by Zhang and Abhayapala et al. (2010); the nature of this study is similar to that of the works already discussed. Moreover, their results are validated only by using the spherical head HRTFs and far-field measured HRTFs of KEMAR and CIPIC, rather than by using measured near-field HRTFs.

Aside from the SH decomposition-based algorithm, a virtual method of conventional and local wave field synthesis that stems from spatial sound reproduction was also proposed for directional interpolation and distance extrapolation of HRTFs (Spors et al., 2011). Actually, HRTF interpolation is closely related to multi-channel spatial sound reproduction, as will be shown in Section 6.4.

SH decomposition can be regarded as an extension of azimuthal Fourier decomposition. Both methods represent HRTFs by a set of predetermined spatial basis functions, which are continuous functions of azimuth or direction and are independent of frequency, individual, and ear. The frequency and individual dependencies, as well as the differences between left and right HRTFs, are encoded in the weights. Mathematically, the predetermined spatial basis function representation of HRTFs is closer to the analytical solutions of HRTFs for the wave or Helmholtz equation. Such solutions are obtained by separating the variables. However, calculating SH coefficients is complicated. Special directional sampling schemes for HRTFs are required for accurately calculating SH coefficients; this requirement results in complicated measurement. Most available HRTF databases do not satisfy the requirements of directional sampling schemes. These schemes also require measurement at low elevations, which is difficult to accomplish in practice. Various methods for supplementing samples at low elevations may result in errors. The pseudoinverse or the regularization algorithm can partially solve this problem, but the results are approximate solutions under some least error criteria. Therefore, SH decomposition is not always the optimal choice in terms of required directional measurements. These factors restrict the application of SH decomposition to HRTFs. By contrast, azimuthal Fourier decomposition is separately applied to each elevation, so that the HRTF samples at low elevations are not always necessary. Moreover, the number of azimuthal HRTF samples required for calculating azimuthal Fourier coefficients varies across elevations, and most available HRTF databases satisfy this requirement. Therefore, azimuthal Fourier decomposition has strengths that are complementary to SH decomposition.

6.3.5. Spatial Principal Components Analysis and Recovery of HRTFs from a Small Set of Measurements

In previous sections, two sets of predetermined spatial harmonics, namely, azimuthal Fourier harmonics and spatial SHFs, are chosen as orthonormal basis functions for HRTF decomposition and reconstruction, respectively. Accordingly, the spatial sampling (Shannon-Nyquist) theorem of HRTFs is derived. Because a large set of spatial harmonics is required to precisely represent HRTFs, the spatial harmonics are, thus far, a less than optimal solution to spatial basis functions in terms of efficiency (Section 6.3.1). By contrast, PCA eliminates the correlations among HRTFs, so that the HRTFs can be represented by the weighted sum of a small set of basis functions or PC-vectors. Therefore, PCA effectively derives a small set of basis functions. However, the traditional PCA (discussed in Section 6.2) is applied in the frequency or time domain. Consequently, HRTFs (or HRIRs) are represented by a weighted combination of a small set of spectral shape (or time-domain) basis functions, rather than by spatial basis functions.

The PCA of HRTFs can also be applied in the spatial domain. This approach is called *spatial principal component analysis* (*SPCA*) to distinguish it from traditional PCA in the frequency or time domain (Xie, 2012b). The algorithm of SPCA is similar to that of traditional PCA, but the result is an efficient spatial basis function (vector) representation of HRTFs.

Given the far-field HRTFs at M directions for both ears of S subjects, each HRTF is represented by its samples at N discrete frequencies. Although the HRTFs of each individual may be slightly left-right asymmetrical, they are approximately left-right symmetrical overall. With reference to the right ear, therefore, azimuthal reflection transform is applied to the data on the left ear, and the resultant data are processed, together with those on the right ear. The right ear HRTFs and reflected left ear HRTFs are denoted by $H(i,k,s,ear)$, where $i = 0,1 \ldots (M-1)$ represents the discrete directions; $k = 0, 1 \ldots (N-1)$ denotes the discrete frequencies; $s = 1,2 \ldots S$ refers to the different subjects; and $ear = L$ or R denotes the left or right ear. According to Equation (6.57) and corresponding to Equation (6.44), $H(i,k,s,ear)$ can be represented by:

$$H(i,k,s,ear) = \sum_q d_q(k,s,ear)w_q(i) + H_{av}(i)$$

$$i = 0,1....,(M-1), \quad k = 0,1....,(N-1), \quad s = 1,2....,S, \quad ear = L \text{ or } R, \tag{6.83}$$

where $w_q(i)$ is a set of universal or common spatial basis functions that are only direction dependent, whereas $d_q(k,s,ear)$ is a set of associated weights that are frequency and individual dependent. These weights also differ for the left and right ears. $H_{av}(i)$, called the mean spatial function, is the mean $H(i,k,s,ear)$ across frequencies, individuals, and two ears. This function, therefore, depends only on direction:

$$H_{av}(i) = \frac{1}{2SN} \sum_{ear} \sum_{s=1}^{S} \sum_{k=0}^{N-1} H(i,k,s,ear) \quad i = 0,1...(M-1). \tag{6.84}$$

Equation (6.83) directly represents the HRTFs of the right ear, but with an azimuthal reflection applied to $w_q(i)$ and $H_{av}(i)$, Equation (6.83) can also represent the HRTFs of the left ear. In addition, the same $w_q(i)$ and $H_{av}(i)$ are used in Equation (6.83) to represent the $H(i,k,s,ear)$ of both ears, which denotes the relationship of left-right reflection for spatial basis functions and mean spatial function. The left-right asymmetric features in the fine structures of HRTFs are encoded in the difference between the weights $d_q(k,s,ear)$ of the left and right ears.

To derive the spatial basis functions $w_q(i)$, the mean value $H_{av}(i)$ is subtracted from the HRTF at each direction and the resultant data form a $(2NS) \times M$ zero-mean matrix $[H_\Delta]$. Each column of matrix $[H_\Delta]$ corresponds to a specific source direction, and each row represents the data on a specific subject and a given ear at a discrete frequency. An $M \times M$ covariance matrix $[R]$ can be constructed from matrix $[H_\Delta]$ by:

$$[R] = \frac{1}{2NS}[H_\Delta]^+[H_\Delta], \tag{6.85}$$

where superscript "+" denotes the transpose and conjugation of the matrix. The spatial basis vectors, which represent the samples of spatial basis functions at M discrete directions, can be obtained from the eigenvectors $\mathbf{w}_q$ of matrix $[R]$ for all $Q' \leq M$ positive eigenvalues γ_q^2:

$$[R]\mathbf{w}_q = \gamma_q^2 \mathbf{w}_q \quad q = 1,2...Q' \quad \gamma_1^2 > \gamma_2^2 > ....\gamma_{Q'}^2 > 0, \tag{6.86}$$

where:

$$\mathbf{w}_q = [w_q(0), w_q(1)....., w_q(M-1)]^T. \tag{6.87}$$

The spatial basis vectors are orthonormal to each other:

$$\mathbf{w}_q^+ \mathbf{w}_q = \begin{cases} 1 & q = q' \\ 0 & q \neq q' \end{cases} \tag{6.88}$$

When the spatial basis vectors w_q are obtained, weights $d_q(k,s,ear)$ can be evaluated using the orthonormality in Equation (6.88):

$$d_q(k,s,ear) = \sum_{i=0}^{M-1} [H(i,k,s,ear) - H_{av}(i)] w_q^*(i). \tag{6.89}$$

Finally, the spatial basis function decomposition of HRTFs is obtained by substituting the $H_{av}(i)$ in Equation (6.84), $w_q(i)$ in Equation (6.86), and $d_q(k,s,ear)$ in Equation (6.89) into Equation (6.83).

To distinguish traditional PCA in the frequency or time domain from the subsequently used terms, *spatial principal component*, or *SPC*, is used to describe the set of weighted basis vectors obtained by SPCA; *SPC-vectors* is used to describe the spatial basis vectors w_q alone; and *SPC-weights* is adopted to describe weights $d_q(k,s,ear)$ alone.

If the preceding $Q < Q'$ SPCs are included in HRTF reconstruction, Equation (6.83) is an approximate representation of the original HRTF:

$$\hat{H}(i,k,s,ear) = \sum_{q=1}^{Q} d_q(k,s,ear) w_q(i) + H_{av}(i). \tag{6.90}$$

The cumulative percentage variance of energy represented by the preceding Q SPCs in Equation (6.90) is evaluated by:

$$\eta = \frac{\sum_{q=1}^{Q} \gamma_q^2}{\sum_{q=1}^{Q'} \gamma_q^2} \times 100\,\%. \tag{6.91}$$

Therefore, if a small set ($Q << M$) of w_q is adequate to account for a large percentage of η, then w_q is a set of efficient basis vectors, and Equation (6.90) is an efficient representation of HRTFs.

SPCA is applicable to complex-valued HRTFs and to linear or logarithmic HRTF magnitudes. It can also be applied to HRIRs in the time domain, by replacing the discrete frequencies with discrete times in Equations (6.83) to (6.91). Similar to that in traditional PCA, if the arrival time of HRIRs is corrected prior to SPCA, the number of SPCs required to retain the same accuracy in SPCA reconstruction is reduced.

SPCA is used to simplify HRTF measurement (Xie, 2012b). As stated in previous sections, HRTF measurement is usually carried out at discrete and limited spatial directions (i.e., sampling in three-dimensional space). The HRTFs at unmeasured directions should be estimated by spatial interpolation. Given that traditional interpolation schemes are subject to the spatial sampling (Shannon-Nyquist) theorem, the spatial sampling rate should be at least twice the spatial bandwidth of HRTFs for accurate recovery of the data at unmeasured directions. Accordingly, a dense measurement grid is required; otherwise, spatial aliasing occurs in the interpolated HRTFs. However, measuring HRTFs with high directional resolution for each individual is time-consuming and tiresome. The Shannon-Nyquist theorem is based on harmonic decomposition. It is merely a sufficient, rather than a necessary, condition for the accurate reconstruction of signals or spatial functions (Candes and Wakin, 2008). As mentioned in Section 6.3.1, if HRTFs can be decomposed as a weighted combination of orthonormal spatial basis functions, and if the contributions of high-order weighted coefficients are insignificant and therefore negligible, then the HRTFs can be approximated as the weighted combination of a small set of spatial basis functions in terms of least square error. This approximation reduces the dimensionality of HRTFs and enables the recovery of HRTFs with high directional resolution from a small set of directional measurements.

SPCA is applicable to deriving a small or efficient set of spatial basis functions. One of the problems with SPCA is that it requires prior knowledge of HRTFs (all the SPC-weights in decomposition) to evaluate the significance of each SPC-weight. However, satisfying this requirement is infeasible for measuring HRTFs. To address the issue, SPCA is applied to a premeasured baseline HRTF datasets with high directional resolution, from which prior knowledge of HRTFs, and, therefore, a small set of spatial basis functions, are acquired by statistical analysis.

In detail, SPCA is first applied to a baseline dataset with the HRTFs of the two ears of S subjects and at M source directions for each subject, so that a small set of universal or common spatial basis functions (or, more exactly, spatial basis vectors w_q) and the mean spatial function $H_{av}(i)$ are derived. Subsequently, the HRTFs of the two ears of subject s_1 outside the baseline dataset are measured at a few source directions (denoted by M_1) chosen from M directions. Similarly to Equation (6.83), the HRTFs of the new subject can be represented by spatial basis vectors. Specifically for the right ear and at M_1 measured directions, a set of the following equations is obtained:

$$H(i,k,s_1,ear) = \sum_{q=1}^{Q} d_q(k,s_1,ear)w_q(i) + H_{av}(i),$$

$$i = 0,1....,(M_1-1), \quad k = 0,1....,(N-1), \quad ear = R.$$

(6.92)

Similarly, applying an azimuthal reflection to $w_q(i)$ and $H_{av}(i)$ yields a set of equations from the M_1 measured HRTFs of the left ear.

At each discrete frequency k, Equation (6.92) is a set of M_1 linear equations with $w_q(i)$ and $H_{av}(i)$ derived from the baseline HRTF dataset with high directional resolution; the only unknowns are the Q individual SPC-weights $d_q(k,s_1,ear)$ of each ear. Ideally, the HRTFs of the new subject at $M_1 = Q$ measured directions are required to solve Equation (6.92). Under such a condition, however, the solution is usually unstable. Alternatively, if the HRTFs of the new subject at $Q < M_1 < M$ appropriate directions are measured, the Tikhonov regularization in Equation (6.82) can be used to solve Equation (6.92). Usually, number M_1 is selected as approximately twice Q, which enables deriving a robust solution to Equation (6.92). Therefore, the required number of measurements rises with an increasing number of spatial basis functions (vectors). This result confirms that efficient spatial basis function representation of HRTFs is important for measurement simplification.

As an illustrative case, the SPCA method is used to recover individualized HRTF magnitudes from a small set of measurements. A baseline dataset with the HRTFs of the two ears of 20 subjects (10 males and 10 females) and 493 source directions (Table 2.3) for each subject are chosen from the HRTF database on Chinese subjects. The results of SPCA for the baseline dataset demonstrate that $Q = 35$ SPCs are adequate for precisely representing the variations in HRTF magnitudes, with a 98.5% cumulative percentage variance of energy [Equation (6.91)]. The spatial basis vectors and mean spatial function (vector) derived from the baseline dataset are used to recover the HRTF magnitudes of six subjects outside the baseline dataset. The results show that the individualized HRTF magnitudes at 493 directions can be recovered from measurements at $M_1 = 73$ directions, with an average mean SDR [i.e., average of Equation (5.16) over directions] of about 19 dB. Recovery errors basically occur at the contralateral directions near the horizontal plane and at low elevations. The method is also applicable to recovering HRIRs in the time domain.

SPCA can be used to recover the HRTFs at discrete directions. The directional resolution of recovered HRTFs depends on that of the baseline dataset. On the basis of the baseline dataset with a high directional resolution (e.g., 1°), we can use the method described above to recover individualized HRTFs with the same directional resolution as the baseline dataset from a small set of measurements. The baseline dataset with a high directional resolution can be directly obtained by measurement. However, this approach is also tiresome. An alternative method is premeasuring an HRTF dataset at or above the Nyquist directional resolution. The baseline dataset is then obtained by applying traditional interpolation schemes on the premeasured dataset.

In addition to SPCA, some other algorithms were proposed to derive an efficient set of basis functions, from which individualized HRTFs were recovered from a small set of measurements. As mentioned at the end Section 6.2.6, Lemaire at el. (2005) proposed a neural network-based method for selecting a representative subset of HRTFs and confirmed that individualized HRTFs can be recovered from measured HRTFs at 50 source directions. From a baseline dataset of HRTFs, including more than 100 subjects, Guillon and Nicol (2008) proposed a pattern recognition algorithm for recovering individualized HRTFs from a small set of measurements.

6.4. HRTF Spatial Interpolation and Signal Mixing for Multi-channel Surround Sound

As discussed in Section 6.1.1, HRTFs are usually measured at discrete and finite directions. Under certain conditions, HRTFs at unmeasured directions can be reconstructed or estimated from measured HRTFs by various interpolation schemes. Similarly, in the multi-channel surround sound reproduction discussed in Sections 1.6.2 and 1.8.4, a finite number of loudspeakers (sound sources) are arranged in discrete spatial directions. A virtual source at both loudspeaker directions and intermediate directions can be created by changing the mixing or panning of loudspeaker signals on the basis of the principle of summing localization of multi-sound sources.

Traditionally, the binaural technique and multi-channel surround sound are classified as two different categories of spatial sound techniques. Accordingly, HRTF spatial interpolation and the signal mixing of multi-channel surround sound have been separately investigated. Both methods are relevant to the concepts and methods of spatial sampling and reconstruction; hence, they are closely related to each other (Poletti, 2000), which will be discussed in the succeeding section (Xie, 2006c).

6.4.1. Signal Mixing for Multi-channel Surround Sound

In an ideal, but impractical, multi-channel surround sound system, an infinite number of loudspeakers (real sources) are supposed to be continually arranged around the listener, and the virtual source at an arbitrary direction is created by the loudspeaker at the corresponding direction. In any practical system that includes the finite number of loudspeakers arranged in discrete spatial directions, the summing virtual source at various intermediate directions should be created by changing the mixing of loudspeaker signals.

In a multi-channel horizontal surround sound system, suppose that M loudspeakers are arranged in a horizontal circle around the listener; the azimuth of the ith loudspeaker is $\theta_i (i = 0,1, \ldots M - 1)$; the transfer functions (HRTFs) of the ith loudspeaker to two ears are $H_L(\theta_i, f)$ and $H_R(\theta_i, f)$, respectively [$H(\theta_i, f)$ for short]; and the signal amplitude for the ith loudspeaker is A_i. Then, the summing pressure at each ear is:

$$P' = E_0 \sum_{i=0}^{M-1} A_i H(\theta_i, f), \qquad (6.93)$$

where E_0 is a normalized constant associated with the total gain (or magnitude) of sound. A perceived virtual source at azimuth θ is created by appropriately choosing signal mixing or panning A_i

$= A_i(\theta)$, so that the summing pressure in Equation (6.93) equals the pressure generated by a point source at azimuth θ. Let $P' = P = E_0 H(\theta, f)$, yielding:

$$H(\theta,f) \approx \sum_{i=0}^{M-1} A_i H(\theta_i,f) = \sum_{i=0}^{M-1} A_i(\theta) H(\theta_i,f). \qquad (6.94)$$

A comparison of Equations (6.1) and (6.94) indicates that multi-channel surround sound can be regarded as a spatial sampling scheme, in which the azimuthally continuous source position is sampled at the M azimuthal directions of loudspeakers; the virtual source at intermediate directions is then created by methods that are similar to linear interpolation schemes. Some signal mixing methods are mathematically analogous to certain interpolation schemes. Conventionally, however, signal mixing methods for multi-channel surround sound are derived from virtual source localization theory or psychoacoustical experiments (Bernfled, 1975; Xie, 2001a). Alternatively, some signal mixing methods can be derived from Equation (6.94), based on HRTF spatial interpolation.

6.4.2. Pairwise Signal Mixing

Pairwise signal mixing is the most popular signal mixing method for multi-channel surround sound. A signal is fed into a single loudspeaker to create a virtual source at its direction, or fed to a pair of adjacent loudspeakers, by adjusting the signal amplitude ratio of adjacent loudspeakers, to create a virtual source at intermediate directions.

Suppose that the azimuths of a pair of adjacent loudspeakers, numbered i and $i + 1$, are θ_i and θ_{i+1}. The HRTFs from the two loudspeakers to the ear are $H(\theta_i, f)$ and $H(\theta_{i+1}, f)$, and the signal amplitudes fed into the loudspeakers are A_i and A_{i+1}. According to Equation (6.94), creating a virtual source at the target (intended) azimuth θ necessitates that the signal amplitudes A_i and A_{i+1} satisfy:

$$H(\theta,f) = A_{i+1} H(\theta_{i+1},f) + A_i H(\theta_i,f). \qquad (6.95)$$

Equation (6.95) is identical to Equation (6.5a). Therefore, pairwise signal mixing approximately creates the binaural pressures for target virtual source directions, by linear interpolation to the pressures created by the adjacent loudspeakers. In short, pairwise signal mixing is analogous to the adjacent linear interpolation of HRTFs.

Equation (6.95) indicates that the relative amplitudes A_i and A_{i+1} of loudspeaker signals can be derived by using a procedure similar to the derivation of Equation (6.5a). That is, $H(\theta, f)$ is expanded as a Taylor series of azimuth θ with the first-order term retained. The results are identical to that produced by Equation (6.5b). In multi-channel surround sound reproduction, however, the low-frequency interaural phase delay difference (ITD$_p$) is a dominant localization cue. The low-frequency ITD$_p$ is related to the virtual source azimuth θ in Equation (4.5). ITD$_p \approx 3a \sin\theta/c$, where a is the head radius and c is the sound speed. The spherical head HRTF provided by Equation (4.3), which is a rough approximation of actual HRTFs at low frequencies, is also a function of $\sin\theta$. Therefore, replacing the variable θ in Equation (6.95) with variable $x = \sin\theta$ is a more reasonable approach. Expanding Equation (6.95) as a Taylor series of $x = \sin\theta$ and retaining the first-order term yields:

$$H(x,f) \approx H(x_i,f) + \frac{H(x_{i+1},f) - H(x_i,f)}{x_{i+1} - x_i}(x - x_i)$$

where $x_i = \sin\theta_i$, $x_{i+1} = \sin\theta_{i+1}$. Comparing this equation with Equation (6.95) yields pairwise signal mixing:

$$A_{i+1} = A_{i+1}(\theta) = \frac{x - x_i}{x_{i+1} - x_i} = \frac{\sin\theta - \sin\theta_i}{\sin\theta_{i+1} - \sin\theta_i},$$

$$A_i = A_i(\theta) = 1 - \frac{x - x_i}{x_{i+1} - x_i} = 1 - \frac{\sin\theta - \sin\theta_i}{\sin\theta_{i+1} - \sin\theta_i}.$$

(6.96)

Equation (6.96) indicates the following:

1. If the target azimuth $\theta_i < \theta < \theta_{i+1}$ is close to front ($\theta = 0°$), then $\sin\theta \approx \theta$, and Equations (6.96) and (6.5b) lead to the same results; otherwise, different results occur. Therefore, replacing the weights A_i and A_{i+1} in the HRTF interpolation equation (6.5b) with the results provided by Equation (6.96) is a more appropriate strategy for retaining accurate low-frequency localization cues. That is, the conventional adjacent linear interpolation formula for HRTFs should be revised at low frequencies.

2. The smaller the adjacent loudspeaker's span angle and the smoother the $H(\theta, f)$ azimuthal variation, the more accurate the pressure at the ear for a virtual source at an intermediate azimuth. Accordingly, a loudspeaker pair with an excessive span is usually unfavorable for stereophonic and multi-channel sound reproduction in terms of virtual source quality.

3. When the target azimuth lies in the intermediate directions $\theta_i < \theta < \theta_{i+1}$ between two adjacent loudspeakers, Equation (6.96) yields $0 < A_i, A_{i+1} < 1$—that is, the signals of two adjacent loudspeakers are in phase.

4. Similar to predicting HRTFs at the target azimuth θ outside region $[\theta_i, \theta_{i+1}]$, the signal mixing in Equation (6.96) can be extended to creating a virtual source outside region $[\theta_i, \theta_{i+1}]$ of the loudspeaker pair, namely, the case of $\theta_i < \theta_{i+1} < \theta$ or $\theta < \theta_i < \theta_{i+1}$. In this case, the loudspeaker signals are out of phase with $A_{i+1} > 0$ and $A_i < 0$ (or $A_{i+1} < 0$ and $A_i > 0$), as stated in Section 1.6.1.

5. If a pair of adjacent loudspeakers are front-back symmetrically arranged on the side of the listener, $\theta_{i+1} = 90° + \theta'$, $\theta_i = 90° - \theta'$, the signal amplitudes A_i and A_{i+1} given by Equation (6.96) are infinite. Therefore, creating a summing virtual source between the directions of a pair of adjacent lateral loudspeakers by pairwise signal mixing is impossible. This conclusion is vital to designing the signal-mixing scheme for multi-channel surround sound. Theoretical analysis and localization experiments also confirm that the ITU (International Telecommunication Union) configuration of lateral loudspeakers (Figure 1.23) and the pairwise signal mixing used in 5.1 channel surround sound, result in a virtual source hole in lateral directions (Xie, 2001a). Similarly, precisely reconstructing lateral HRTFs by the adjacent linear interpolation scheme is impossible, a conclusion that has been neglected in some existing studies.

6. Conclusion (5) can also be drawn from the symmetry of HRTFs. As indicated in the analysis in Section 3.5.1, HRTFs are front-back symmetric below 1 kHz, making them approximately equal at front-back mirror directions, that is, $H(90° - \theta', f) \approx H(90° + \theta', f)$. Consequently, any weights A_i and A_{i+1} yield:

$$A_i H(90° - \theta', f) + A_{i+1} H(90° + \theta', f) \approx AH(90° - \theta', f) \approx AH(90° + \theta', f),$$

where A is a constant. Therefore, adjacent linear interpolation results in a scale of the original HRTFs, rather than of the HRTFs at the target lateral direction.

7. In Equation (6.96), an unnecessary assumption is that the azimuthal spans between all adjacent loudspeaker pairs are identical. This equation is therefore appropriate for

nonuniform loudspeaker configurations in the horizontal plane (for example, 5.1 channel configuration).

As a special case, the loudspeaker configuration for two-channel stereophonic sound is shown in Figure 1.16. The positions of two loudspeakers are $\theta_{i+1} = \theta_R = \theta_0$, $\theta_i = \theta_L = 360° - \theta_0$, and the signal amplitudes are $A_{i+1} = R$, $A_i = L$. From Equation (6.96), the summing virtual direction θ_I can be found by:

$$\sin\theta_I = -\frac{L-R}{L+R}\sin\theta_0,$$ (6.97)

Equation (6.97) is merely the stereophonic law of sines given by Equation (1.25). In contrast to Equation (1.25), Equation (6.97) is derived by linear interpolation to the ear pressures created by a pair of stereophonic loudspeakers, rather than by directly calculating ITD_p. Therefore, the analysis of Equations (6.96) and (6.97) is a more stringent derivation of the stereophonic law of sines.

This same analysis can be extended to three-dimensional spatial sound. For example, the spherical triangular interpolation (Freeland et al., 2004) discussed in Section 6.1.2 is analogous to vector base amplitude panning, a signal mixing technique for multi-channel spatial sound proposed by Pulkki (1997).

6.4.3. Sound Field Signal Mixing

Sound field signal mixing (Ambisonics) is another signal mixing method for multi-channel sound (Gerzon, 1985; Xie, 1982; Xie and Xie, 1996). This method is equivalent to the interpolation equation given by Equation (6.64). Suppose that M loudspeakers are uniformly arranged in a horizontal circle around the listener. The comparison of Equations (6.94) and (6.64) shows that, to create the virtual source at azimuth θ, the signal for the ith loudspeaker at azimuth $\theta_i(i = 0,1,2 \dots M - 1)$ should be:

$$A_i(\theta) = \frac{1}{M} \frac{\sin\left[\left(Q+\frac{1}{2}\right)(\theta-\theta_i)\right]}{\sin\left(\frac{\theta-\theta_i}{2}\right)} = \frac{1}{M}\left[1+2\sum_{q=1}^{Q}(\cos q\theta_i \cos q\theta + \sin q\theta_i \sin q\theta)\right],$$ (6.98)

where Q is the order of sound field signal mixing.

To obtain insight into the physical nature of Equation (6.98), the recording and reproduction of multi-channel surround sound signals are analyzed. The signal $A_i(\theta)$ in Equation (6.98) can be theoretically captured by a series of coincident microphones placed in the original sound field. Let θ denote the source azimuth in the original sound field. The second equality in Equation (6.98) indicates that the signal $A_i(\theta)$ for $Q \geq 1$-order sound field signal mixing consists of a linear combination of $(2Q + 1)$ independent signal components or azimuthal harmonics. Components 1, $\cos\theta$, and $\sin\theta$ are equivalent to the normalized outputs of an omnidirectional microphone and two bidirectional microphones, with their main axes pointing to the front and right in the original sound field, respectively; components $\cos q\theta$, $\sin q\theta$ ($q \geq 2$) are equivalent to the normalized outputs of higher-order multipole microphones. Figure 6.9 shows the polar patterns of the output response $A_i(\theta) = A(\theta - \theta_i)$ against $\Delta\theta = (\theta - \theta_i)$, with the azimuth off-axis of the source, for the preceding three orders $Q = 1,2,3$. In the figure, the maximum of $A_i(\theta)$ has been normalized to the unit. The main lobe is centered at $\Delta\theta = 0°$. Response $|A(\theta - \theta_i)|$ maximizes at the on-axis direction $\Delta\theta = 0°$, and then decreases with increasing $|\Delta\theta|$. As $\Delta\theta$ further increases, the responses exhibit the side

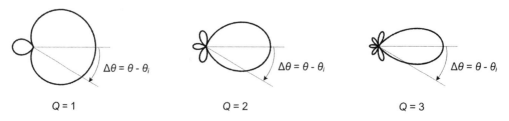

$Q = 1$ $\quad\quad\quad\quad\quad\quad$ $Q = 2$ $\quad\quad\quad\quad\quad\quad$ $Q = 3$

Figure 6.9 The polar pattern of output response $A_i(\theta) = A(\theta - \theta_i)$ against the $\Delta\theta = (\theta - \theta_i)$ for the order of $Q = 1,2,3$. The maximum of $A_i(\theta)$ has been normalized to unit

and rear lobes (in-phase or out-of-phase), as well as the null points, of the polar patterns at some azimuths. In addition, as order Q increases, both the width of the main lobe and the responses of the side and rear lobes decrease, so that an accurate polar pattern is derived. When order Q tends to infinity, Equation (6.98) tends to the asymptotic result:

$$\lim_{Q \to \infty} A_i(\theta) = A(\theta - \theta') = \frac{2\pi}{M}\delta(\theta - \theta') \quad with \quad \theta' = \theta_i, \tag{6.99}$$

where $\delta(\theta - \theta')$ is the Dirac delta function, and the discrete azimuthal parameter θ_i has been replaced by continuous θ'. Equation (6.98) represents a horizontal beamforming-based signal recording or mixing scheme, which enhances the output at the target azimuth $\theta = \theta_i$ while inhibiting the output at other azimuths. Azimuth θ_i is the point, or look, direction of beamforming. Beampattern accuracy improves as order Q increases. The asymptotic result given by Equation (6.99) represents an ideal beampattern. From the perspective of horizontal sound field sampling, Equation (6.99) is the ideal sampling function along a continuous azimuth. As shown in Equation (6.98), the actual sampling function is obtained by expanding Equation (6.99) into a Fourier series, truncating it up to a finite-order Q and letting $\theta' = \theta_i$. It is a function of continuous θ with the discrete parameter θ_i, and is centered at azimuth θ_i, but it spreads to both sides.

In multi-channel sound reproduction, $A_i(\theta) = A(\theta - \theta_i)$ in Equation (6.98) is the azimuthal distribution of the loudspeaker signal, with θ_i denoting the loudspeaker azimuth and θ representing the target virtual source azimuth. $A_i(\theta) = A(\theta - \theta_i)$ also represents the (relative) azimuthal distribution of M discrete incident waves with respect to the origin. Equation (6.64) indicates that the HRTF at the arbitrary target azimuth θ is approximated, or fitted, by a linear combination of measured HRTFs at M directions. Analogous to this condition, the binaural pressures caused by a real source at the target azimuth θ are approximated (or fit) by a combination of the binaural pressures caused by M loudspeakers in multi-channel sound reproduction (a similar fitting equation is held for the free-field pressure near the origin). $A_i(\theta)$ approaches the ideal Dirac delta distribution with increasing order Q, so that the upper frequency limit for accurate sound field reproduction increases. However, the number of independent signals $(2Q + 1)$ and the number of loudspeakers $M \geq (2Q + 1)$ also increase with order Q, a situation that complicates the system. When Q tends to infinity, $A_i(\theta)$ becomes the ideal beampattern given by Equation (6.99). By replacing the summation across the discrete azimuth θ_i in Equation (6.64) with the integral over the continuous azimuth θ' of source (loudspeaker) distribution, multiplying the free-field pressure P_0 at the head center and excluding denotation $\phi_0 = 0°$ yield the binaural pressures caused by ideal reproduction:

$$P(\theta, f) = H(\theta, f)P_0 = \int_0^{2\pi} P_0 H(\theta', f)\delta(\theta - \theta')d\theta'. \tag{6.100}$$

Equation (6.100) indicates that, in ideal reproduction, the virtual source at each azimuth $\theta' = \theta$ is created by a real loudspeaker at the corresponding azimuth alone. However, an ideal reproduction system requires an infinite number of independent signals (azimuthal harmonics) and loudspeakers. A practical system is characterized by a trade-off between complexity and perceived performance. In addition, the signals with higher-order polar characteristics can be synthesized by computer signal processing (Xie and Xie, 1996), and can be captured by the spherical microphone array discussed in Section 6.6. The investigation of the latter also provides insight into the nature of HRTF spatial interpolation and multi-channel sound reproduction from the perspective of sound field spatial sampling.

By extension, a similar analysis can be applied to multi-channel sound with nonuniform loudspeaker arrangement in the horizontal plane. This approach is analogous to HRTF interpolation at nonuniform azimuthal intervals. For brevity, however, this issue is not discussed.

The foregoing analysis can also be extended to HRTF interpolation at the three-dimensional directions (θ, ϕ) and multi-channel spatial surround sound. According to Equations (6.73) and (6.74), the $(L-1)$ order spatial beamforming signal that points to direction $\Omega_i = (\alpha_i, \beta_i)$ is given by:

$$A_i(\Omega) = \sum_{l=0}^{L-1} \sum_{m=-l}^{l} Y_{lm}^*(\Omega_i) Y_{lm}(\Omega). \tag{6.101}$$

Similar to the horizontal case, the signal $A_i(\Omega)$ in Equation (6.101) can be theoretically captured by a series of coincident microphones placed in the original sound field. Let Ω denote the source direction in the original sound field. $A_i(\Omega)$ consists of a linear combination of L^2 independent SH components $Y_{lm}(\Omega)$, which are equivalent to the normalized outputs of L^2 microphones with different three-dimensional directional characteristics and pointing directions. The accuracy of the beampattern improves as the $L-1$ order increases. When $(L-1)$ tends to infinity, Equation (6.101) tends to the ideal beampattern in three-dimensional directions:

$$A_i(\Omega) = A(\Omega - \Omega') \rightarrow \sum_{l=0}^{\infty} \sum_{m=-l}^{l} Y_{lm}^*(\Omega') Y_{lm}(\Omega) = \delta(\Omega - \Omega') \quad with \quad \Omega' = \Omega_i. \tag{6.102}$$

In multi-channel spatial sound reproduction, suppose that M loudspeakers are arranged on a spherical surface with a far-field radius according to some appropriate directional sampling schemes given in Appendix A (e.g., nearly uniform sampling or Gauss-Legendre sampling). The signal for the ith loudspeaker at direction $\Omega_i = (\alpha_i, \beta_i)$ is directly proportional to the beamforming signal that points in that direction; namely, $\lambda_i A_i(\Omega)$. The number of loudspeakers and the relative gain (weight) of each loudspeaker signal are selected according to the sampling scheme. In particular, the former should be selected to satisfy the requirement of the Shannon-Nyquist sampling theorem ($M \geq L^2$ for nearly uniform sampling). Under such a condition, the binaural pressures in multi-channel reproduction match those caused by a real source at direction $\Omega = (\alpha, \beta)$, with an accuracy of up to the $L-1$ order of SH decomposition, thereby rendering the virtual source at direction $\Omega = (\alpha, \beta)$.

$\lambda_i A_i(\Omega)$ are exactly the loudspeaker signals for $L-1$-order spatial Ambisonic sound reproduction. For example, the signals for first-order reproduction can be obtained by letting $(L-1) = 1$ in Equation (6.101) and using the expression of the complex-valued SHs given by Equation (A.5) in Appendix A:

$$\lambda_i A_i(\Omega) = \frac{\lambda_i}{4\pi} [1 + 3\cos\alpha_i \cos\alpha + 3\sin\alpha_i \cos\beta_i \sin\alpha\cos\beta + 3\sin\alpha_i \sin\beta_i \sin\alpha\sin\beta], \tag{6.103}$$

and by letting $\alpha = (\pi/2 - \phi)$, $\beta = \theta$, Equation (6.103) can be converted into the expression in the default coordinate (θ, ϕ) as:

$$\lambda_i A_i(\theta, \phi) = \frac{\lambda_i}{4\pi}[1 + 3\sin\phi_i \sin\phi + 3\cos\phi_i \cos\theta_i \cos\phi\cos\theta + 3\cos\phi_i \sin\theta_i \cos\phi\sin\theta], \quad (6.104)$$

where components 1, $\sin\phi$, $\cos\phi \cos\theta$, and $\cos\phi \sin\theta$ are four independent normalized signals for first-order spatial Ambisonics (B-format). These components are captured by an omnidirectional microphone and three three-dimensional bidirectional microphones, with their main axes pointing to the top, front, and right in the original sound field (Gerzon, 1985).

Multi-channel spatial Ambisonic reproduction can also be implemented using other loudspeaker configurations, in contrast to the sampling schemes provided in Appendix A. The corresponding loudspeaker signals can be derived by approximate calculations on the order of Equation (6.82) (Xie, 1988).

6.4.4. Further Discussion on Multi-channel Sound Reproduction

As indicated in Sections 6.4.2 and 6.4.3, some signal mixing methods in multi-channel sound reproduction are analogous to certain interpolation and reconstruction schemes for HRTFs from the perspective of directional sampling. In synthesizing virtual sources in VADs, the HRTFs at arbitrary target directions can be reconstructed from measured HRTFs using various linear interpolation schemes. In multi-channel sound reproduction, on the other hand, interpolation on binaural pressures is automatically carried out with reproduction. That is, the pressures caused by multiple loudspeakers are superposed, so that the summing pressures at the ears approximate those caused by a real source at the target direction. The analogy between multi-channel sound reproduction and HRTF interpolation enables interchanging some of the methods used by the two fields.

Some methods for evaluating HRTF interpolation performance can be used to analyze the multi-channel sound field. The sound field signal mixing for horizontal surround sound discussed in Section 6.4.3 provides an example. The azimuthal sampling theorem of HRTFs (Section 6.3.3) indicates that azimuthally continuous HRTFs can be reconstructed from $M \geq (2Q + 1)$ azimuthal measurements that are uniformly distributed in region $0 \leq \theta < 2\pi$, where Q is the highest harmonic in the azimuthal Fourier series representation of HRTFs. For KEMAR HRTFs, the azimuthal sampling theorem requires $M_{min} = (2Q + 1) = 65$ azimuthal measurements in the horizontal plane for HRTFs up to 20 kHz (Table 6.1, without arrival time correction and $\eta \geq 0.99$). If the azimuthal Fourier representation of HRTFs is regarded as a special case of the SH representation of HRTFs, Equation (6.76a) yields an identical result with $Q = (L - 1) = 32$ and $M_{min} = (2Q + 1) = 65$ for a head radius $a = 0.0875$ m; the highest frequency $f = 20$ kHz.

Analogously, for multi-channel sound reproduction with the sound field signal mixing given by Equation (6.98), if the number M of horizontal loudspeakers and the number of independent signal components $(2Q + 1)$ satisfy the requirement of azimuthal sampling theorem $M \geq (2Q + 1) = 65$, the binaural pressures in multi-channel sound reproduction precisely match those caused by a real source at a target azimuth θ up to 20 kHz, resulting in the accurate rendering of virtual sources. However, the resultant multi-channel sound system is too complicated to be implemented. This analysis focuses on mathematical accuracy or error, which may be useful for the physical analysis of HRTFs. For multi-channel sound reproduction, a more significant concern is perceivable error, which is helpful in reasonably simplifying a system. The results of a virtual source localization experiment indicate that, for a music stimulus, multi-channel sound reproduction using six or eight loudspeakers with $Q = 2$-order sound field signal mixing sufficiently recreates horizontal virtual sources around a centrally positioned listener with the full 360° azimuthal range (Xie and Xie, 1996). Contrastively, some methods for multi-channel sound can also be used to simplify HRTF-based virtual source synthesis, as will be discussed in the next section.

6.5. Simplification of Signal Processing for Binaural Virtual Source Synthesis

According to the discussion in Section 1.8.3, the binaural synthesis of a virtual source can be implemented by multiplying a monophonic (mono) stimulus with a pair of HRTFs at the target source direction (θ, ϕ) [Equation (1.36)] in the frequency domain, or, equivalently, by convoluting the mono stimulus with a pair of HRIRs in the time domain. Related signal processing is usually implemented with various practical HRTF-based filters.

In traditional binaural signal processing, synthesizing each virtual source necessitates two HRTF multiplication manipulations in the frequency domain, or, equivalently, two HRTF-based filters. Therefore, simultaneously synthesizing G independent virtual sources (such as the virtual auditory environment to be presented in Chapter 11) necessitates G pairs ($2G$) of HRTF-based filters. The outputs of each left and right filter are mixed to obtain corresponding binaural signals. Usually, computational cost rapidly increases with increasing virtual source number G. To simulate a moving virtual source, the HRTF-based filter pair should also be continually updated, which may cause some audible (commutation) artifacts (Chapter 12). If only the HRTFs at discrete directions are known, the pre-interpolation scheme described in the previous section is required. All these factors make designing hardware and software for signal processing difficult, especially in real-time processing. Thus, simplifying the signal-processing algorithm for simultaneously synthesizing multiple virtual sources is necessary.

6.5.1. Virtual Loudspeaker-based Algorithms

The analogous relationship between HRTF directional interpolation and the multi-channel sound reproduction discussed in Section 6.4 is applicable to simplifying the binaural synthesis of multiple virtual sources. Recall that, in multi-channel sound reproduction, a real loudspeaker at the target virtual source direction is not always required. Instead, the virtual source can be created by the summing localization of loudspeakers located at other directions. Analogous to the case of multi-channel sound, the virtual source in headphone-based binaural reproduction can be created by virtual loudspeakers. That is, the M virtual loudspeakers at appropriate directions are first created by the conventional HRTF-based algorithm for virtual source synthesis [Equation (1.36)]. The virtual sources at target directions are subsequently created by appropriate signal mixing or panning for M virtual loudspeakers, according to the principle of the summing localization of multiple sound sources (Jot et al., 1998; 1999). The virtual loudspeaker-based algorithm has two advantages over conventional implementation. One is its nearly constant computational cost for multiple virtual source synthesis; this constancy is attributed to the dependence of the required number of HRTF-based filters on only the number of virtual loudspeakers, rather than on the number of virtual sources. The other advantage is that the moving virtual source is created by changing signal mixing to virtual loudspeakers rather than by updating the HRTF-based filters. Audible artifacts are therefore easily avoided.

As an illustrative case (Xie et al., 2001b), Figure 6.10 shows a configuration of $M = 8$ virtual loudspeakers, including six in the horizontal plane with $\phi = 0°$ and $\theta_{RF} = 30°$, $\theta_R = 90°$, $\theta_{RB} = 150°$,

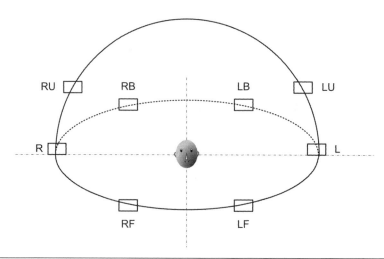

Figure 6.10 Configurations of eight virtual loudspeakers

$\theta_{LB} = 210°$, $\theta_L = 270°$, and $\theta_{LF} = 330°$, as well as two in the lateral plane with $\phi = 30°$ to $60°$ and $\theta_{LU} = 270°$, $\theta_{RU} = 90°$. The localization experiment by Theile and Plenge (1977) confirms that the horizontal $360°$ virtual sound source can be created by multi-channel sound reproduction using the six horizontal loudspeaker configurations in Figure 6.10 and pairwise signal mixing. This conclusion is further validated by using the summing localization law for multiple sound sources [Equation (1.27)]. With extension to three-dimensional directions, the eight virtual loudspeakers (Figure 6.10) and pairwise signal mixing create virtual sources in the horizontal and upper lateral planes.

Figure 6.11 shows the block diagram of the signal-processing algorithm. Given G independent mono input stimuli E_0, E_1 ... $E_{(G-1)}$, the goal is to synthesize corresponding virtual sources at G directions. As for a conventional audio mixing console, each mono stimulus E_g is first panned into two channel signals, with appropriate interchannel level difference, and then assigned to a pair of adjacent virtual loudspeakers. The virtual sources at different directions are created by appropriately selecting different pairs of adjacent virtual loudspeakers and changing the interchannel level difference. The input of each virtual loudspeaker may receive multiple signals coming from different mono stimuli E_g, depending on the distribution of G target virtual source directions. The mixed signal A_i ($i = 0,1, ... 7$) for each virtual loudspeaker is filtered by a pair of HRTFs at the virtual loudspeaker direction, in accordance with Equation (1.36). Finally, the resultant left and right signals from the eight virtual loudspeakers are mixed to form the overall binaural signals $E_{L\Sigma}$ and $E_{R\Sigma}$. The overall binaural signals are presented over a pair of headphones to simulate the binaural pressures caused by the eight virtual loudspeakers, yielding virtual sources at G directions.

The algorithm shown in Figure 6.11 requires $2M = 16$ HRTF-based filters to create $M = 8$ virtual loudspeakers. Taking advantage of the left-right symmetrical configuration of virtual loudspeakers and the assumption of roughly left-right symmetrical HRTFs, the algorithm is further simplified, ultimately leading to the shuffler implementation of signal processing (Cooper and Bauck, 1989; Bauck and Cooper, 1996). (According to the discussion in Section 3.5.2, although the assumption of a left-right symmetrical HRTF model does not always hold, it is a good approximation under certain conditions.)

The virtual loudspeaker pair at left-right mirror positions is considered together. The horizontal left and right virtual loudspeaker pair is taken as an example. Let H_{LL} and H_{RL} denote the HRTFs

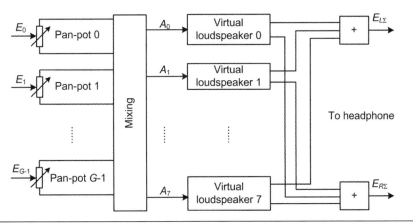

Figure 6.11 Block diagram of the eight virtual loudspeaker-based algorithm

from the left virtual loudspeaker to two ears; H_{LR} and H_{RR} represent the HRTFs from the right virtual loudspeaker to two ears; and $A_0 = A_L$ and $A_1 = A_R$ denote the signals for the left and right virtual loudspeakers, respectively. To synthesize the binaural pressures caused by the left and right virtual loudspeaker pair, the binaural signals fed to a pair of headphones should be:

$$\begin{bmatrix} E_{L\Sigma} \\ E_{R\Sigma} \end{bmatrix} = \begin{bmatrix} H_{LL} & H_{LR} \\ H_{RL} & H_{RR} \end{bmatrix} \begin{bmatrix} A_L \\ A_R \end{bmatrix}. \tag{6.105}$$

A direct implementation of the algorithm given by Equation (6.105) necessitates four HRTF-based filters. Taking advantage of left-right symmetry, the acoustic transfer matrix in Equation (6.105) is symmetric, with $H_{LL} = H_{RR} = H_\alpha$, $H_{LR} = H_{RL} = H_\beta$. Using the diagonalizing procedure for a symmetrical matrix confirms that Equation (6.105) is equivalent to:

$$\begin{bmatrix} E_{L\Sigma} \\ E_{R\Sigma} \end{bmatrix} = \begin{bmatrix} 0.707 & 0.707 \\ 0.707 & -0.707 \end{bmatrix} \begin{bmatrix} H_\alpha + H_\beta & 0 \\ 0 & H_\alpha - H_\beta \end{bmatrix} \begin{bmatrix} 0.707 & 0.707 \\ 0.707 & -0.707 \end{bmatrix} \begin{bmatrix} A_L \\ A_R \end{bmatrix}. \tag{6.106}$$

As shown in Equation (6.106), signals A_L and A_R are first mixed by a 2×2 MS (mid-side or sum-subtract) matrix (Rumsey, 2001). Then, the resultant signals $0.707\ (A_L + A_R)$ and $0.707\ (A_L - A_R)$ are filtered by two filters $\Sigma = (H_\alpha + H_\beta)$ and $\Delta = (H_\alpha - H_\beta)$, respectively, yielding two signals $0.707\ (A_L + A_R)\ (H_\alpha + H_\beta)$ and $0.707\ (A_L - A_R)\ (H_\alpha - H_\beta)$. Finally, additional MS mixing is applied to form the binaural signals. Two filters are required for the shuffler structure implementation in Equation (6.106). In this manner, synthesizing all four pairs of virtual loudspeakers in Figure 6.10 requires $2 \times 4 = 8$ filters, which is half the requirement for the conventional algorithm. Therefore, shuffler structure implementation is more efficient in simultaneously synthesizing four or more virtual sources. A virtual source localization experiment confirms the validity of the algorithm. Shuffler structure implementation in multiple virtual source synthesis is a general and efficient algorithm, which will be further discussed in Chapters 9 and 10.

In addition to pairwise signal mixing, the virtual loudspeaker-based algorithm is incorporated with other multi-channel signal mixing methods. When the sound field signal mixing method discussed in Section 6.4.3 is incorporated, it is called *virtual Ambisonics* or *binaural Ambisonics* (Travis, 1996; Jot et al., 1998; Leitner et al., 2000; Noisternig, 2003). Virtual spatial Ambisonics provides an example. Suppose that M virtual loudspeakers are arranged on a spherical surface, with a far-field radius according to some appropriate directional sampling schemes given in Appendix A.

The normalized signal for the ith virtual loudspeaker at direction $\Omega_i = (\alpha_i, \beta_i)$ is $\lambda_i A_i(\Omega)$ [Equation (6.101)]. Each virtual loudspeaker signal $\lambda_i A_i(\Omega)$ is filtered by a pair of HRTFs at the corresponding direction Ω_i and then mixed to form the binaural signal E:

$$E = \sum_{i=0}^{M-1} \lambda_i A_i(\Omega) H(\Omega_i, f) = \sum_{i=0}^{M-1} \lambda_i A_i(\theta, \phi) H(\theta_i, \phi_i, f). \tag{6.107}$$

The notation for the left or right ear is excluded from Equation (6.107) for simplicity. The virtual source at direction $\Omega = (\alpha, \beta)$ or (θ, ϕ) is created by reproducing the binaural signal in Equation (6.107) through a pair of headphones.

In Equation (6.107), the positions Ω_i of virtual loudspeakers and the associated HRTFs are invariable. The virtual sources at different directions are controlled by changing the directional parameter Ω in the signal mixing of the virtual loudspeakers given by Equation (6.101). The algorithm provided by Equation (6.107) requires M HRTF-based filters for each ear, or $2M$ for two ears. If left-right symmetry is considered, and signal processing is implemented in the shuffler structure similar to Equation (6.106), the number of HRTF-based filters required can be reduced to M for two ears. Nevertheless, the computational cost of single virtual source synthesis by virtual spatial Ambisonics is higher than that presented by the conventional algorithm.

In multiple virtual source synthesis, given G independent mono input stimuli with relative amplitudes $E_0, E_1 \ldots E_{(G-1)}$, the goal is to synthesize the corresponding virtual sources at G directions $\Omega_g = (\alpha_g, \beta_g)$ or (θ_g, ϕ_g) with $g = 0, 1 \ldots (G-1)$. According to Equation (6.101), the signal amplitude $\lambda_i A'_i$ for the ith virtual loudspeaker is the sum of those contributed by all the G mono stimuli:

$$\lambda_i A'_i = \lambda_i \sum_{g=0}^{G-1} E_g A_i(\Omega_g) = \lambda_i \sum_{g=0}^{G-1} \sum_{l=0}^{L-1} \sum_{m=-l}^{l} E_g Y^*_{lm}(\Omega_i) Y_{lm}(\Omega_g). \tag{6.108}$$

Substituting the $\lambda_i A_i(\Omega)$ in Equation (6.107) with the $\lambda_i A'_i$ in Equation (6.108) yields the algorithms for multiple virtual source synthesis. The algorithms are more efficient than the conventional one for the number of virtual sources $G > M$ (left-right symmetry has not been incorporated). This method will be applied to spherical microphone array recording and binaural reproduction in Section 6.6.

In addition to these cases, some other existing loudspeaker configurations for multi-channel sound may be applicable to the virtual loudspeaker-based algorithm. The subset selection method discussed in Section 6.2.6 is also applicable to selecting the privileged directions or configuration of virtual loudspeakers (Jot et al., 2006a). The performance of subset selection should be validated by psychoacoustical experiments.

The virtual loudspeaker-based algorithm can also be implemented by the multiple-input/multiple-output state-space system discussed in the last part of Section 5.3.3, in which the HRTFs at M virtual loudspeaker directions constitute the state space, and the signals for M virtual loudspeakers are system inputs. This algorithm more efficiently decreases computational burden (Adams and Wakefield, 2007).

The discussions in this section focus on applying multi-channel sound methods to simplifying binaural synthesis processing. Similar methods can be used to convert various existing multi-channel sound programs for binaural presentation. As stated in Section 1.8.4, additional signal conversion processing is needed prior to headphone presentation because multi-channel sound signals are originally intended for loudspeaker reproduction. This issue will be comprehensively discussed in Chapter 10.

6.5.2. Basis Function Decomposition-based Algorithms

The various basis function decompositions of HRTFs discussed in Sections 6.2 and 6.3 can be used to simplify virtual source synthesis (Larcher et al., 2000). According to Equation (6.14) or (6.57), each HRTF can be decomposed into a finite number of Q components, with each component including a directional function $w_q(\theta, \phi)$ cascaded with a filter $d_q(f)$. Accordingly, the virtual source synthesis processing in Equation (1.36) can be implemented by the algorithm shown in Figure 6.12 (for the left ear only; a similar block diagram is shown for the right ear). The mono input stimulus E_0 is weighted by Q directional functions $w_q(\theta, \phi)$ and then filtered by a parallel bank of Q filters $d_q(f)$ ($q = 1,2 \ldots Q$). The Q filters are invariable and the virtual source direction (θ, ϕ) is controlled by the Q directional functions or gains $w_q(\theta, \phi)$. Finally, the outputs of all Q filters are mixed to obtain the left ear signal.

To simultaneously synthesize multiple virtual sources at G directions (θ_g, ϕ_g) with $g = 0,1 \ldots G-1$, all G mono input stimuli share the same parallel bank of Q filters because of the source-independent characteristics of the filters. The only requirement is weighting each input stimulus with a different set of Q directional functions $w_q(\theta_g, \phi_g)$ ($q = 1,2 \ldots Q$). The resultant Q weighted stimuli for each input are then mixed with the stimuli associated with all other mono inputs and filtered by the Q filters. Similar to the virtual loudspeaker-based algorithms, basis function decomposition-based algorithms are beneficial to signal processing for synthesizing multiple virtual sources or rendering moving virtual sources.

In terms of efficiency, the conventional PCA decomposition discussed in Section 6.2 is especially suitable for implementing the basis function decomposition-based algorithm. The parallel bank of filters $d_q(f)$ can be designed on the basis of a small set of PC-vectors (spectral shape basis vectors or functions). Individualized signal processing can be achieved by substituting the individualized directional functions or PC-weights $w_q(\theta, \phi)$ in Figure 6.12, while keeping the parallel bank of filters unchanged. Chanda et al. (2005, 2006) represented the minimum-phase and diffuse-field equalized HRTFs of KEMAR by 7 PCs plus a mean spectral function $d_0(f) = H_{av}(f)$. They also used 13-order IIR filters or 15-order common-pole filters, designed by the BMT method (Section 5.3.3), to implement the algorithm. As mentioned in Section 6.2.4, Xie and Zhang (2012a) applied PCA to the near-field HRIRs of KEMAR at 493 source directions and 9 distances. On this basis, the multiple near-field virtual sources were synthesized by a parallel bank of $Q = 15$ filters that correspond to 15 PC-vectors, along with a filter that corresponds to the mean spectral function. A PCA-based algorithm for efficiently synthesizing multiple near-field virtual sources in dynamic VADs was also suggested. This algorithm enables multiple input stimuli to share the same bank of 16 parallel filters, thereby reducing the computational cost of multiple virtual source synthesis.

Alternatively, the SPCA introduced in Section 6.3.5 can be used to implement the algorithm. The

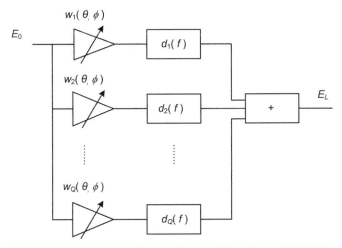

Figure 6.12   Block diagram of the basis function decomposition-based algorithm (for left ear only)

advantage of SPCA-based implementation is that the process and storage of directional information is nonindividualized. Directional information is encoded by a set of "universal" directional
functions $w_q(\theta, \phi)$, resulting in a set of individual-independent multi-channel signals. These multi-
channel signals can then be decoded into binaural signals by a set of individual filters $d_q(f)$ to
optimize the signals for each listener (Jot et al., 1998).

The azimuthal Fourier and SHF decomposition of HRTFs discussed in Section 6.3 are also applicable to virtual source synthesis processing (Jot et al., 1998; Menzies, 2002). SHF decomposition
provides an example. Given G mono input stimuli with relative amplitudes $E_0, E_1 \dots E_{(G-1)}$, the goal
is to synthesize corresponding virtual sources at the G directions $\Omega_g = (\alpha_g, \beta_g)$ or (θ_g, ϕ_g) with $g =$
$0, 1 \dots (G-1)$. According to Equation (1.36), conventional binaural synthesis processing is imple-
mented by:

$$E = \sum_{g=0}^{G-1} E_g H(\Omega_g, f),\tag{6.109}$$

where $H(\Omega_g, f)$ is the HRTF at direction Ω_g and can be decomposed by SHFs according to Equation
(6.72) thus:

$$H(\Omega_g, f) = \sum_{l=0}^{L-1} \sum_{m=-l}^{l} d_{lm}(f) Y_{lm}(\Omega_g).\tag{6.110}$$

Substituting Equation (6.110) into Equation (6.109) yields:

$$E = \sum_{l=0}^{L-1} \sum_{m=-l}^{l} [\sum_{g=0}^{G-1} E_g Y_{lm}(\Omega_g)] d_{lm}(f) = \sum_{l=0}^{L-1} \sum_{m=-l}^{l} W_{l,-m} d_{lm}(f),\tag{6.111}$$

where:

$$W_{l,-m} = \sum_{g=0}^{G-1} E_g Y_{lm}(\Omega_g) = \sum_{g=0}^{G-1} E_g Y^*_{l,-m}(\Omega_g) \quad l=0,1,2\dots, m=0,\pm1,\dots\pm l.\tag{6.112}$$

Equation (6.112) is the $(l, -m)$-order/degree SH coefficient of the amplitude distribution of all the
G virtual sources.

Equation (6.111), which consists of L^2 components, specifies the SH decomposition-based algorithm for multiple virtual source synthesis. Accordingly, signal processing can be implemented
by a parallel bank of L^2 filters designed from the corresponding SH coefficients $d_{lm}(f)$ derived
from HRTF decomposition. The L^2 filter inputs $W_{l,-m}$ are formed by G mono input stimuli, in accordance with Equation (6.112), and the outputs of the L^2 filters are then mixed to form signals
to the ear. The SH decomposition-based algorithm in Equation (6.111) is equivalent to the virtual
Ambisonic algorithm described in Equations (6.107) and (6.108), with $(L - 1)$ representing the
order of the virtual Ambisonics in Equation (6.111). Similar to SPCA-based implementation, the
directional information of the G virtual sources are encoded in the L^2 individual-independent
signals $W_{l,-m}$ in Equation (6.112). These universal signals are convenient for the processing and
storage of directional information. The universal signals can then be decoded into binaural signals
by a set of L^2 individual filters $d_{lm}(f)$ to optimize the signals for each listener.

According to the Shannon-Nyquist theorem (Section 6.3.4), the $(L - 1)$-order virtual Ambisonics described in Equations (6.107) and (6.108) requires $M \geq L^2$ virtual loudspeakers and $M \geq L^2$
filters for each ear. Because the HRTFs are represented by L^2 orthonormal spatial basis functions
in SH decomposition, the SH decomposition-based algorithm in Equation (6.111) requires L^2 filters for each ear; this requirement is the lower bound of the Shannon-Nyquist theorem. Similar to

the virtual Ambisonic algorithm in Equations (6.107) and (6.108), the SH decomposition-based algorithm is less efficient than the conventional HRTF-based method in single virtual source synthesis. Only when the number of virtual sources satisfies $G > L^2$ is the former superior to the latter (left-right symmetry has not been incorporated). Equations (6.76a) and (6.77a) indicate that the truncated order $(L - 1)$ in the SH representation of HRTFs depends on frequency. The first-order SH components exactly account for directional HRTF variations only up to 0.6 to 0.7 kHz. As frequency increases, more SH components are required. SH components up to order $(L - 1) = 33$ are required to precisely represent directional HRTF variations up to 20 kHz, which corresponds to SH components $L^2 = 1156$. Therefore, the PCA or SPCA decomposition-based algorithm is less complicated than the SH decomposition-based algorithm (or virtual Ambisonics). As stated in Section 6.3, SHFs are considerably less efficient than the optimal solution of spatial basis functions. However, virtual Ambisonics and the SH decomposition-based algorithm are necessary to the binaural reproduction of the recording signals of spherical microphone arrays. This issue will be discussed in the next section. In practice, the truncated order $(L - 1)$ is selected as a trade-off between the accuracy of high-frequency binaural signals and algorithm complexity. Such a trade-off is determined on the basis of certain psychoacoustical principles.

The relationship between virtual Ambisonics and the SH decomposition-based algorithm can be further explained from the standpoint of sound field analysis. Virtual Ambisonics uses a series of virtual loudspeakers to reproduce sound, and the signals for virtual loudspeakers are designed on the basis of sound field signal mixing. If all the virtual loudspeakers are located at a far-field distance, binaural pressures can be approximated by the summing pressures caused by a distribution of plane wave sources. Ideally, an infinite number of virtual loudspeakers should be uniformly distributed over a spherical surface, which is equivalent to decomposing the incident sound field into an infinite number of plane waves. In practice, only a finite number of virtual loudspeakers are available, which is equivalent to sampling the incident plane waves at discrete directions. The SH decomposition-based algorithm represents the incident sound field by SH or multipole components (Morse and Ingrad, 1968; Zotkin et al., 2007). Plane wave decomposition and SH decomposition are two representations of the same sound field, so that the weighting coefficients in two representations are related to each other. The relationship is reflected in the decoding stage of the SH decomposition-based algorithm. Sound field decomposition is also closely related to the beamforming model of HRTFs, to be discussed in the succeeding section.

Some researchers used the virtual Ambisonic algorithm and associated virtual far-field loudspeakers to synthesize near-field virtual sources (Daniel, 2003; Menzies and Marwan, 2007). This approach is equivalent to the range extrapolation of HRTFs (Section 6.3.4), in which SH decomposition is applied. The radiation from a point sound source at near-field distance r' relative to the head center (origin) can be decomposed into a weighted combination of incident plane waves from various directions. This radiation can then be captured by high-order Ambisonic signals and converted into binaural signals using far-field HRTF filters. The far-field HRTF filters model the scattering of plane waves by listeners. However, the plane wave decomposition of near-field point source radiation is valid only within region $r < r'$. For a source closer to the head center, part of the scattering body (torso) may lie outside the aforementioned region. Thus, binaural synthesis in which far-field HRTF filters are used may be inaccurate. In this case, a translation of the origin from head center to a new center is required to enlarge the valid region for plane wave decomposition. Accordingly, the complex-valued amplitude distribution of plane waves (virtual loudspeaker signals) should be corrected.

6.6. Beamforming Model for Synthesizing Binaural Signals and HRTFs

Microphone array beamforming is a general method for sound recording, in which some distinctive spatial microphone configurations are employed to capture sound field information. The signals from all the microphones are then weighted and summed to form the array output. The polar or directional pattern of array output is controlled by selecting different weights. The beamforming output of the array is characterized by polar patterns that enhance the output at the target direction while inhibiting the output at other directions, thereby improving the signal-to-noise/interference ratio of the output. Microphone array beamforming can also be incorporated into binaural reproduction; in other words, the spatial information of the sound field is recorded by an appropriate microphone array and then converted into binaural signals for headphone presentation or loudspeaker reproduction with crosstalk cancellation (Chapter 9). The theoretical foundation of this method is the spatial sampling, decomposition, and reconstruction of the sound field, which is closely related to the spatial sampling, decomposition, and reconstruction of HRTFs. Sound recording by microphone array beamforming is an important and promising field for sound recording and reproduction.

6.6.1. Spherical Microphone Array for Synthesizing Binaural Signals

As stated in Sections 6.4.3 and 6.5, the spatial variations of the sound field can be represented by SH components, and corresponding signals can, in principle, be recorded by a series of coincident microphones with various polar characteristics. The recorded signals can then be converted into binaural signals by virtual Ambisonics or the SH decomposition-based algorithm. However, microphones with higher-order polar characteristics are unavailable. Arranging a number of spatially-coincident microphones is also technically difficult. Alternatively, the higher-order SH components of the sound field can be recorded by spherical microphone array and then converted into binaural signals (Rafaely, 2005; Duraiswami et al., 2005; Li and Duraiswami, 2006; Song et al., 2008; Poletti and Svensson, 2008). Compared with other arrays, the spherical array presents the advantage of perfect symmetry.

With the notation similar to those applied in previous sections, let $\Omega = (\alpha, \beta)$ denote the incident direction of a plane wave with unit magnitude and $\Omega' = (\alpha', \beta')$ denote the direction of a field point on the spherical surface with radius a_0. Then, the pressure on the spherical surface that is caused by the incident plane wave can be represented by complex-valued (or real-valued) SHFs:

$$P(\Omega', \Omega, f) = \sum_{l=0}^{\infty} \sum_{m=-l}^{l} R_l(ka_0) Y_{lm}^*(\Omega) Y_{lm}(\Omega'), \qquad (6.113)$$

where $R_l(ka_0)$ is the l-order radial function. In addition, a plane wave with unit magnitude can also be expanded by:

$$\exp(-j\boldsymbol{k}\boldsymbol{r}') = \exp(jkr'\cos\Gamma) = \sum_{l=0}^{\infty} (2l+1) j^l j_l(kr') P_l(\cos\Gamma), \qquad (6.114)$$

where $\boldsymbol{k}$ is the wave vector that denotes the wave direction of propagation; r' is the position vector of the field point (rather than the source position) $|\boldsymbol{r}'| = r'$; Γ is the angle between the incident (source) direction and the ray from the origin to the field point. For incidents facing the field point, $\Gamma = 0°$, the direction of vector $\boldsymbol{k}$ is opposite to source direction and, thus, $\boldsymbol{k}\boldsymbol{r}' = -kr'$. Functions $j_l(kr')$

and $P_l(\cos\Gamma)$ are an l-order spherical Bessel function and l-degree Legendre polynomial, respectively. Using the addition equation of SHFs given by Equation (A.19) in Appendix A yields:

$$\frac{2l+1}{4\pi}P_l(\cos\Gamma) = \sum_{m=-l}^{l} Y_{lm}^*(\Omega)Y_{lm}(\Omega'),$$
(6.115)

and letting $r' = a_0$ yields the $R_l(ka_0)$ for an open sphere:

$$R_l(ka_0) = 4\pi j^l j_l(ka_0).$$
(6.116)

For a rigid sphere, substituting the head radius a in Equation (4.1) with the array radius a_0 and letting the magnitude of the incident plane wave $P_0 = 1$ enables the use of Equation (6.115) and:

$$\frac{dh_l(ka_0)}{d(ka_0)}j_l(ka_0) - \frac{dj_l(ka_0)}{d(ka_0)}h_l(ka_0) = \frac{-j}{(ka_0)^2},$$
(6.117)

to yield:

$$R_l(ka_0) = 4\pi j^l \left[j_l(ka_0) - \frac{dj_l(ka_0)/d(ka_0)}{dh_l(ka_0)/d(ka_0)}h_l(ka_0) \right],$$
(6.118)

where $h_l(ka_0)$ is the l-order spherical Hankel function of the second kind.

An ideal spherical array consists of an infinite number of omnidirectional microphones that are uniformly and continuously arranged on the spherical surface. The output signal of each microphone is directly proportional to the pressure at the microphone's location, as given by Equation (6.113). The array output is obtained by multiplying the pressure signal by a microphone location and frequency-dependent weight $\xi(\Omega', ka_0)$ and then integrating over the entire spherical surface as:

$$A(\Omega, ka_0) = \int \xi(\Omega', ka_0)P(\Omega', \Omega, f)d\Omega'.$$
(6.119)

Weights $\xi(\Omega', ka_0)$ can also be decomposed by SHFs:

$$\xi(\Omega', ka_0) = \sum_{l=0}^{\infty}\sum_{m=-l}^{l} \xi_{lm}(ka_0)Y_{lm}(\Omega'),$$
(6.120)

where $\xi_{lm}(ka_0)$ represents a set of SH coefficients for $\xi(\Omega', ka_0)$ and are related to $\xi(\Omega', ka_0)$ as:

$$\xi_{lm}(ka_0) = \int \xi(\Omega', ka_0)Y_{lm}^*(\Omega')d\Omega'.$$
(6.121)

Weights $\xi(\Omega', ka_0)$ are selected so that:

$$\xi_{lm}(ka_0) = \xi_{lm}(ka_0, \Omega_i) = \frac{Y_{lm}^*(\Omega_i)}{R_l(ka_0)}.$$
(6.122)

Substituting Equations (6.113), (6.120), and (6.122) into Equation (6.119) and applying the orthonormality of SHFs yield:

$$A(\Omega - \Omega_i) = \sum_{l=0}^{\infty}\sum_{m=-l}^{l} Y_{lm}^*(\Omega)Y_{lm}(\Omega_i) = \sum_{l=0}^{\infty}\sum_{m=-l}^{l} Y_{lm}^*(\Omega_i)Y_{lm}(\Omega).$$
(6.123)

In this case, the weighting integral of pressure signals over the spherical surface is independent of ka_0. Using Equation (A.18) converts Equation (6.123) into:

$$A_i(\Omega) = A(\Omega - \Omega_i) = \delta(\Omega - \Omega_i).$$
(6.124)

This equation is an ideal spherical beampattern or directional sampling function that points to the incident direction Ω_i for a plane wave. Equations (6.122) and (6.124) show that, by changing the parameter Ω_i in the SH coefficients of $\xi(\Omega', ka_0)$ given by Equation (6.122), spherical beamforming enables steering the beam toward any direction without changing the beampattern. This feature is an advantage of spherical beamforming.

A practical spherical array consists of a finite number of microphones, which is equivalent to sampling the pressure at a set of discrete positions on the spherical surface. Given the positions $\Omega'_{i'}$ with $i' = 0,1 \ldots (M'-1)$ of M' microphones, the weighted integral in Equation (6.119) is replaced by a weighted summation of directional samples of pressure signals over the spherical surface. If the positions of microphones are selected according to the directional sampling schemes given in Appendix A, the array output signal in Equation (6.119) becomes:

$$A(\Omega, ka_0) = \sum_{i'=0}^{M'-1} \lambda'_{i'} \xi(\Omega'_{i'}, ka_0) P(\Omega'_{i'}, \Omega, f). \tag{6.125}$$

Therefore, the array output is formed by filtering the pressure signal $P(\Omega'_{i'}, \Omega, f)$ of each microphone with $\xi(\Omega'_{i'}, ka_0)$, and then weighting and summing the filtered signals from all the microphones. Accordingly, with Equation (A.22), the beamforming output in Equations (6.123) and (6.124) is replaced by a summation of SH components up to a limited order $(L - 1)$:

$$A_i(\Omega) = A(\Omega - \Omega_i) = \sum_{l=0}^{L-1} \sum_{m=-l}^{l} Y^*_{lm}(\Omega_i) Y_{lm}(\Omega), \tag{6.126}$$

where the truncation order $(L - 1)$ is determined by the directional sampling scheme and the number M' of microphones. The Shannon-Nyquist theorem requires that, at the least, $L^2 \leq M'$. For the Gauss-Legendre sampling scheme, we have $2L^2 = M'$.

The foregoing discussion is intended for a single incident plane wave at a specified direction. For a general case, an arbitrary incident sound field can be decomposed into a finite or infinite number of plane waves that arrive at the spherical array from different directions. Let $W(\Omega)$ denote the complex-valued amplitude distribution function of plane waves against the incident direction Ω. The pressure at the spherical surface can be evaluated by weighing those caused by the incident plane wave with unit magnitude [Equation (6.113)] with $W(\Omega)$ and then integrating over all the incident directions Ω:

$$P_\Sigma(\Omega', f) = \int W(\Omega) P(\Omega', \Omega, f) d\Omega. \tag{6.127}$$

The ideal beamforming output in Equation (6.124) becomes:

$$A_{\Sigma,i} = A_\Sigma(\Omega_i) = \int W(\Omega) A(\Omega - \Omega_i) d\Omega = W(\Omega_i). \tag{6.128}$$

Equation (6.128) indicates that the output of ideal beamforming at direction Ω_i yields the complex-valued plane wave amplitude at that direction. Accordingly, the $(L - 1)$ order beamforming output in Equation (6.126) becomes:

$$A_{\Sigma,i} = A_\Sigma(\Omega_i) = \sum_{l=0}^{L-1} \sum_{m=-l}^{l} W_{lm} Y_{lm}(\Omega_i). \tag{6.129}$$

Equation (6.129) is only equal to the SH representation of $W(\Omega_i)$ up to the $(L - 1)$ order. W_{lm} is the SH coefficient of $W(\Omega)$ and is evaluated by Equation (6.130) using the orthonormality of SHFs:

$$W_{lm} = \int W(\Omega)Y_{lm}^*(\Omega)d\Omega. \tag{6.130}$$

In conclusion, for an arbitrary incident sound field, applying the beamforming process given by Equation (6.119) to the pressure signals of a practical spherical microphone array yields the $(L' - 1)$-order SH representation of the complex-valued amplitude of an incident plane wave at the point direction of beamforming.

To convert the signals captured by a spherical microphone array into binaural signals and render the signals over headphones, the directional sampling schemes discussed in Appendix A are employed again. The beamforming processing is carried out at M chosen directions Ω_i with $i = 0,1$ … $(M - 1)$, resulting in M signals $A_{\Sigma,i}$. Then, with the virtual Ambisonic algorithm presented in Section 6.5.1, the M signals $A_{\Sigma,i}$ are separately filtered by far-field HRTFs at M corresponding directions and then reproduced by M virtual loudspeakers. Substituting the $A_i(\Omega)$ in Equation (6.107) by the $A_{\Sigma,i}$ in Equation (6.129) yields the binaural signals for an arbitrary incident sound field:

$$E = \sum_{i=0}^{M-1} \lambda_i A_{\Sigma,i} H(\Omega_i, f) = \sum_{i=0}^{M-1} \lambda_i A_{\Sigma,i}(\theta, \phi) H(\theta_i, \phi_i, f). \tag{6.131}$$

This equation is the basic principle of synthesizing and rendering binaural signals from spherical microphone array recording.

Some complementary notes for spherical microphone array and binaural reproduction are:

1. As previously stated, two types of spherical arrays exist: the open-spherical array and the rigid-spherical array. Because the weights $\xi(\Omega', ka_0)$ in Equation (6.122) require the inversion of function $R_l(ka_0)$, the $R_l(ka_0)$ for an open sphere is given by Equation (6.116). Its inversion is divergent at some frequencies that are associated with the zeros of spherical Bessel functions $j_l(ka_0)$. This problem is avoided by using the rigid-spherical array, but this array interferes with the surrounding sound field, thereby creating errors in signal recording (Yu et al., 2012e). To overcome the disadvantage caused by the conventional open- or rigid-spherical omnidirectional microphone array, some researchers suggested using the open-spherical cardioid, rather than the omnidirectional, microphone array for sound field recording (Melchior et al., 2009).

2. Two directional sampling procedures are included in sound recording and reproduction. One is included at the stage during which the pressure at the spherical surface is sampled by M' microphones. The resultant M' pressure samples are then used for beamforming, as shown in Equation (6.125). The other procedure pertains to the stage of virtual Ambisonic reproduction described by Equation (6.131), in which binaural signals are created by the virtual loudspeakers at M sampled directions. The directional sampling schemes for recording and reproduction stages may be identical or different. In the former, $M = M'$ and $\lambda_i = \lambda'_i$, but this is not so in the latter.

3. Regardless of whether identical directional sampling schemes are applied, the number of directional samples in both recording and reproduction stages should at least satisfy the requirement of the Shannon-Nyquist theorem; that is, $M' \geq L^2$ and $M \geq L^2$. Because the number of HRTF measurements M is usually much higher than the number of microphones M' in a spherical array, the truncation order $(L - 1)$ in SH representation and corresponding upper frequency limit for the recording and reproduction of spatial information is determined by M'. For example, a spherical array with $M' = 64$ microphones can, at most, record the SH components of a sound field up to the $(L - 1) = 7$ order.

Accordingly, for a spherical radius of $a_0 = 0.1$ m, condition $ka_0 \leq (L - 1)$ yields an upper frequency limit of 3.8 kHz. High directional resolution and high signal-to-noise ratio at low frequencies require a large array radius, which, in turn, causes spatial aliasing at high frequencies. In practice, the array radius is selected as a trade-off between high- and low-frequency performance. In addition, some researchers suggested using a dual-radius open microphone array to overcome the conflicting requirements for high- and low-frequency performance (Balmages and Rafaely, 2007).

4. For a near-field sound source, the valid region of plane wave decomposition and its influence on conversion into binaural signals by far-field HRTF filters should be taken into consideration, as in the case of virtual Ambisonics discussed at the end of Section 6.5.2.

With the SH decomposition-based algorithm discussed in Section 6.5.2, spherical array recording and synthesizing binaural signals can be carried out in the SH domain. For an arbitrary incident sound field, a requirement is to identify the SH coefficients W_{lm} of complex-valued amplitude distribution function $W(\Omega)$ of plane wave decomposition. Substituting the SH representation of the $W(\Omega)$ in Equation (6.129) and the pressure $P(\Omega', \Omega, f)$ in Equation (6.113) into Equation (6.127) yields the spherical surface pressure by using the orthonormality of SHFs:

$$P_\Sigma(\Omega', f) = \sum_{l=0}^{\infty} \sum_{m=-l}^{l} R_l(ka_0) W_{lm} Y_{lm}(\Omega'). \tag{6.132}$$

Similarly, M' microphones are arranged on the spherical surface, resulting in M' pressure samples $P_\Sigma(\Omega'_{i'}, f)$ with $i' = 0,1 \dots (M'-1)$. With the discrete orthonormality of SHFs, we have:

$$W_{lm} = \sum_{i'=0}^{M'-1} \lambda'_{i'} \xi_{lm}(ka_0, \Omega'_{i'}) P_\Sigma(\Omega'_{i'}, f). \tag{6.133}$$

Substituting the resultant W_{lm} and known SH coefficients $d_{lm}(f)$ of HRTFs into Equation (6.111) yields the binaural signals:

$$E = \sum_{l=0}^{L-1} \sum_{m=-l}^{l} W_{l,-m} d_{lm}(f). \tag{6.134}$$

The truncation order $(L - 1)$ is determined by the directional sampling scheme for microphone configuration, and should, at the least, satisfy $L^2 \leq M'$. Equation (6.134) is the basic algorithm for synthesizing binaural signals from the beamforming output of spherical arrays in the SH domain.

Duraiswami et al. used a rigid-spherical array with 60 (close to 64) microphones to record the sound field and then converted the pressure signals on the spherical surface to binaural signals by beamforming and a KEMAR HRTF-based filter (Duraiswami et al., 2005; Li and Duraiswami, 2006). In addition, by converting its output into binaural signals, a rigid-spherical array can be used to measure and analyze binaural room acoustic parameters, such as the interaural cross-correlation coefficient (IACC) discussed in Section 1.7.2 (Rafaely and Avni, 2010).

A rigid-hemispherical microphone array was proposed for sound recording (Li and Duraiswami, 2006). The discussion thus far has focused on the recording and binaural reproduction of the spatial information of the three-dimensional spatial sound field. As a special case of the three-dimensional spatial sound field, the spatial information of the horizontal sound field can be recorded by a circular microphone array. The horizontal beamforming (Ambisonics) outputs of the array can be converted into binaural signals and then rendered over headphones (Poletti, 2000, 2005).

6.6.2. Other Array Beamforming Models for Synthesizing Binaural Signals and HRTFs

In addition to the spherical array, other microphone configurations can be employed to record the spatial information of the sound field. The recorded signals can also be converted into binaural signals and then rendered over headphones. Moreover, some array beamforming-based models can be designed to represent HRTFs. The theoretical foundation of these models is HRTF basis function decomposition. The first model was proposed by Chen et al. (1992); thereafter, various models have been suggested by other researchers (Bai and Ou, 2005a; Poletti and Svensson, 2008).

Suppose that Q microphones are arranged according to some array configurations. In the default coordinate of this book (but with the head absent), the position of the qth microphone is specified by vector $\mathbf{r}'_q$ or Cartesian coordinate (x'_q, y'_q, z'_q) with $q = 1,2 \ldots Q$, where x, y, and z are the axes of the coordinates directed to the right, front, and top, respectively. Suppose that the sinusoidal point source is far from the array, so that incidence can be approximated as a plane wave. Accordingly, the magnitudes of the pressure signals recorded by all microphones are identical and then normalized to unit for simplicity. By properly selecting the initial phase of the incident plane wave, the pressure signal recorded by the qth microphone is evaluated by:

$$P_q(\theta,\phi,f) = \exp(-j\mathbf{k}\mathbf{r}'_q) = \exp[-j2\pi f \tau_q(\theta,\phi)] \qquad q = 1,2\ldots Q, \tag{6.135}$$

where $\mathbf{k}$ is the wave vector; (θ, ϕ) is the incident direction; $\tau_q(\theta, \phi)$ is the delay for sound wave propagation to the qth microphone, and is evaluated by:

$$\tau_q(\theta,\phi) = -\frac{1}{c}[x'_q \sin\theta\cos\phi + y'_q \cos\theta\cos\phi + z'_q \sin\phi], \tag{6.136}$$

in which c is the speed of sound. Delay $\tau_q(\theta, \phi)$ depends on both incident direction and microphone position. The phase of the pressure signal $P_q(\theta, \phi, f)$ depends on incident direction, microphone position, and frequency. In the array model, the far-field HRTF at direction (θ, ϕ) is represented by a weighted combination or beamforming of the pressure signals of Q microphones caused by the plane wave incident from the same direction:

$$H(\theta,\phi,f) = \sum_{q=1}^{Q} d_q(f)P_q(\theta,\phi,f), \tag{6.137}$$

where $d_q(f)$, $q = 1,2 \ldots Q$, are frequency-dependent weights to be determined.

Given M incident plane wave directions (θ_i, ϕ_i) with $i = 0,1 \ldots (M\text{-}1)$, the pressure recorded by the qth microphone is denoted by $P_q(\theta_i, \phi_i, f)$, and the far-field HRTFs at corresponding M incident directions are denoted by $H(\theta_i, \phi_i, f)$. From Equation (6.137), we have:

$$H(\theta_i,\phi_i,f) = \sum_{q=1}^{Q} d_q(f)P_q(\theta_i,\phi_i,f) \qquad i = 0,1\ldots(M-1). \tag{6.138}$$

The unknown weights $d_q(f)$ can be solved from Equation (6.138). At each specified frequency f, the HRTFs at M directions can be represented by an $M \times 1$ matrix (column vector) as:

$$\mathbf{H} = [H(\theta_0,\phi_0,f), H(\theta_1,\phi_1,f)\ldots H(\theta_{M-1},\phi_{M-1},f)]^T. \tag{6.139}$$

The Q weights can also be denoted by a $Q \times 1$ matrix (column vector):

$$\mathbf{d}(f) = [d_1(f), d_2(f)\ldots d_Q(f)]^T. \tag{6.140}$$

Then, Equation (6.138) can be written as the matrix equation:

$$\boldsymbol{H} = [P]\boldsymbol{d}(f), \tag{6.141}$$

where $[P]$ is an $M \times Q$ matrix with element $P_{iq} = P_q(\theta_i, \phi_i, f)$.

At each specified frequency f, Equation (6.141) is a set of M linear algebra equations with Q weights $d_q(f)$ unknown. Equation (6.141) has a unique and exact solution only when it satisfies the following conditions:

1. $Q = M$, that is, the number of microphones are equal to the number of assumed incident directions (or HRTF directional measurements); and
2. the microphone configurations and assumed incident directions should be properly selected, so that matrix $[P]$ is of full rank and therefore invertible. In this case, the Q weights can be exactly evaluated by:

$$\boldsymbol{d}(f) = [P]^{-1}\boldsymbol{H}. \tag{6.142}$$

In most practical cases, however, these conditions are difficult to satisfy. First, identifying a microphone configuration and assumed incident directions that ensure the invertibility of matrix $[P]$ at all frequencies is difficult. Second, reducing HRTF dimensionality necessitates that the number of microphones selected be fewer than the number of assumed incident directions—that is, $Q < M$. Consequently, Equation (6.141) is an overdetermined case and has no exact solution. Alternatively, this equation can be solved by some approximate methods. Similar to Equation (6.82), Tikhonov regularization is used to identify the approximately least square error solution:

$$\boldsymbol{d}(f) = \{[P]^+[P] + \delta[I]\}^{-1}[P]^+\boldsymbol{H}, \tag{6.143}$$

where δ is the regularization parameter, and $[I]$ is the $Q \times Q$ identity matrix. Some other methods, such as the SVD (singular value decomposition) of the matrix, can also be applied to solve Equation (6.143).

This calculation is intended for a specified frequency, and Equation (6.143) should be solved at each frequency. Once the Q weights $d_q(f)$ at all frequencies are found, the array beamforming models of HRTFs can be obtained by substituting the resultant weights into Equation (6.137). This procedure is the basic principle of array beamforming models for synthesizing HRTFs.

Applying inverse Fourier transform to Equation (6.137) yields the array beamforming models of HRIRs:

$$h(\theta, \phi, t) = \sum_{q=1}^{Q} g_q(t) * p_q(\theta, \phi, t), \tag{6.144}$$

where "*" denotes convolution; $g_q(t)$ is the inverse Fourier transform of $d_q(f)$; and $p_q(\theta, \phi, t)$ is the inverse Fourier transform of $P_q(\theta, \phi, f)$, that is, the impulse response output of the qth microphone to the far-field or plane wave source at direction (θ, ϕ):

$$p_q(\theta, \phi, t) = \delta[t - \tau_q(\theta, \phi)] \tag{6.145}$$

Because $h(\theta, \phi, t)$ is a real-valued function, Equations (6.144) and (6.145) indicate that $g_q(t)$ is also a real-valued function.

The foregoing analysis and resultant Equation (6.137) can be further interpreted as being:

1. A joint spatial- and temporal-frequency filtering scheme. The far-field HRTF can be regarded as a spatial- and temporal-frequency response function. At each temporal-frequency f, $H(\theta, \phi, f)$ is described by a polar pattern against source direction (θ, ϕ).

At each source direction (θ, ϕ), $H(\theta, \phi, f)$ is described by a temporal-frequency filter. Equation (6.137) reveals that HRTFs can be synthesized from a sum of Q temporal-frequency filtered signals from the pressures spatially sampled by Q microphones. The responses of Q temporal-frequency filters are specified by weights $d_q(f)$ with $q = 1,2 \ldots$ Q. Therefore, Equation (6.137) specifies a joint spatial and temporal-frequency filtering scheme, whose characteristics are determined by microphone array configuration and the responses of Q temporal-frequency filters. The configuration and filters should be appropriately selected, so that the output of a joint spatial- and temporal-frequency filter matches the desired spatial- and temporal-frequency response of HRTFs.

2. An HRTF decomposition and interpolation scheme. Similar to the cases in Sections 6.2 and 6.3, the pressure signals $P_q(\theta, \phi, f)$ in Equation (6.137) are regarded as a set of Q basis functions that represent direction-dependent HRTFs. Then, the HRTF at the incident direction (θ, ϕ) can be represented by a weighted sum of Q basis functions $P_q(\theta, \phi, f)$. As indicated by Equation (6.135), the Q basis functions, which are determined by the array configuration, depend on frequency and incident direction. This condition contrasts with the decomposition based on direction-independent spectral shape basis functions or frequency-independent spatial basis functions that was presented in previous sections. Q weights $d_q(f)$ are determined by the array configuration and can be evaluated from the measurements at $M > Q$ assumed incident directions. The HRTF at the arbitrary direction (θ, ϕ) can be reconstructed by substituting the resultant $d_q(f)$ into Equation (6.137). Therefore, Equation (6.137) is also a directional interpolation, or fitting equation, for HRTFs. In addition, if a small set of basis functions is sufficient to account for the primary variations of HRTFs at all directions, then Equation (6.137) is a low-dimensional, or compacted, representation of HRTF data.

3. A method for sound field spatial information recording and binaural reproduction. Although Equation (6.137) is derived from a single-incidence or far-field source, it is applicable to multiple incidences (including direct and reflected sounds) because of the linear superposition principle of sound waves. Similar to the case of a spherical array, therefore, Equation (6.137) specifies a method for sound field spatial information recording and binaural reproduction. In this method, an artificial head or human subject is not required to obtain binaural signals. Provided that the individualized HRTFs are known, individual binaural signals can be synthesized from the array output by individually matching weights $d_q(f)$.

Some explanations should be supplemented to the interpretation of HRTF decomposition mentioned in item (2). According to Equations (6.136) and (6.137), HRTFs can be decomposed into a weighted sum of Q basis functions $P_q(\theta, \phi, f)$, which are spatial samples of the incident plane wave field. As previously stated, HRTFs can also be decomposed by other basis functions, such as SHFs. In terms of the acoustic principle of reciprocity, an HRTF is directly proportional to the pressure at a spatial field point that is caused by a source positioned at the ear canal. Therefore, the SH decomposition of HRTFs is equivalent to the multipole expansion of the spatial sound field that is caused by a source positioned at the ear canal. In addition, various basis function decomposition methods for HRTFs should, in principle, match one another. Let Equation (6.69) be equal to Equation (6.137). Then,

$$\sum_{l=1}^{\infty}\sum_{m=-l}^{l} d_{lm}(f) Y_{lm}(\Omega) = \sum_{q=1}^{Q} d_q(f)\exp(-j\boldsymbol{k}\boldsymbol{r}'_q) \tag{6.146}$$

Equation (6.146) relates the SH coefficients to the weights of plane wave sample decomposition, from which the former can be calculated from the latter, or vice versa. Accordingly, two sets of basis function representations of HRTFs can be converted into each other. A plane wave with unit magnitude can also be expanded by:

$$\exp(-j\boldsymbol{k}\boldsymbol{r}'_q) = \exp(jkr'_q \cos\Gamma) = \sum_{l=0}^{\infty}(2l+1)j^l j_l(kr'_q)P_l(\cos\Gamma), \tag{6.147}$$

where $k = |\boldsymbol{k}|$ is the wave number and $r'_q = |\boldsymbol{r}'_q|$; Γ is the angle between incident direction (opposite the direction of the wave vector $\boldsymbol{k}$) and the position vector $\boldsymbol{r}'_q$ for the qth microphone; and $j_l(kr'_q)$ and $P_l(\cos\Gamma)$ are the l-order spherical Bessel function and l-degree Legendre polynomial, respectively. Using the addition equation of SHFs similar to Equation (6.115), as well as the orthonormality of SHFs, yields the relationship between $d_{lm}(f)$ and $d_q(f)$:

$$d_{lm}(f) = 4\pi j^l \sum_{q=1}^{Q} d_q(f) j_l(kr'_q) Y_{lm}^*(\Omega'_q), \tag{6.148}$$

where $\Omega'_q = (\alpha'_q, \beta'_q) = (90° - \phi'_q, \theta'_q)$ is the qth microphone direction.

For practical array beamforming, the following factors should be taken into account:

1. Array configuration: Array configuration significantly influences the performance of HRTF reconstruction. Appropriately selecting array configurations is a complicated problem, and no well-developed procedure is available for this requirement. Accordingly, configurations are often selected by trial and by numerical simulations. However, the physical properties of HRTFs facilitate the selection of array configurations. To account for the front-back asymmetric nature of human HRTFs (Section 3.5) above 1–2 kHz, array configurations should be selected in such a way that they make the recorded pressure signals of some microphones asymmetric to the incidences from the front-back mirror directions; that is, $P_q(\theta, \phi, f) \neq P_q(180° - \theta, \phi, f)$. Otherwise, the synthesized HRTFs in Equation (6.137) are front-back symmetric. For example, the pressure signals recorded by a linear array along the left-right axis are always front-back symmetric.

2. Number of microphones and interval between adjacent microphones. The sound field is spatially sampled by a microphone array. The number of microphones should be selected so that the recorded signals or spatial samples represent the spatial variations of the sound field and HRTFs with the desired accuracy. This process is usually implemented by various numerical simulations, and the results depend on array configurations and the complexity of the spatial variations of HRTFs. Moreover, according to the Shannon-Nyquist theorem, the interval between adjacent microphones should not exceed half of the shortest wavelength, which is determined by the upper frequency limit of the model.

3. Number N of discrete temporal-frequency samples. A measured HRTF is usually represented by its value at N discrete temporal-frequencies, uniformly sampled from the lowest value of 0 Hz to the highest of Nyquist frequency, and corresponding to an N-point HRIR in the time domain. Accordingly, the $d_q(f)$ at N discrete temporal-frequencies should be calculated from Equation (6.141), corresponding to an N-point FIR filter implementation of the temporal-frequency filters. According to the illustration in Section 3.1.1, because the primary component of an HRIR persists for about 50 to 60 samples at a sampling frequency of $f_s = 44.1$ kHz, N selected should, at the least, be of such an order. In practice, the $d_q(f)$ at $(N/2 + 1)$ discrete frequencies require solving because of the conjugate symmetry of the $d_q(f)$ around the Nyquist frequency.

In an early work by Chen et al. (1992), an L-shaped array beamforming model was suggested to represent the horizontal HRTFs of a cat. A three-armed array model for simulating both the azimuthal and elevation dependencies of HRTFs was proposed by Bai and Ou (2005a). The array consisted of 16 microphones, with 5 microphones along each arm plus 1 at the origin. The space between the microphones was 8 mm. Each of $g_q(t)$ was modeled by a 50-tapped filter and the frequency response of the filter was uniformly sampled at 256 points over a bandwidth of 22 kHz. The model was applied to match the MIT-KEMAR HRTFs for elevation $-30° \leq \phi \leq 90°$ and ipsilateral azimuth $0° \leq \theta \leq 180°$. The least square error method was used to estimate the weights in Equation (6.141). The results demonstrate that compared with the original HRTFs, the mean relative energy error of the synthesized HRTFs over all directions [Equation (5.12)] is on the order of 12.5%. Therefore, the model represents original HRTFs with moderate accuracy. Some improvements over the array model, such as estimating the weights in Equation (6.141) by SVD with regularization and nonuniform temporal-frequency sampling, were also suggested. In addition, localization experiments were carried out to validate the model.

A similar scheme was applied to sound field simulation and auralization. When the sound field in a room is calculated by the finite-difference time-domain method (Section 11.1.4), the results are pressures at discrete spatial points on a regular three-dimensional grid. The pressure samples at discrete spatial points can be converted into binaural signals using the array beamforming method (Poletti and Svensson, 2008). The signals are then auralized over headphones (Section 14.2).

Finally, the Batteau model in Equation (1.13) can also be regarded as an array model, in which pinna response is synthesized by beamforming the pressure signals recorded by three microphones. Nevertheless, the Batteau model is too simple to describe detailed pinna responses.

6.7. Summary

A far-field HRTF is a continuous function of source direction. However, HRTFs are usually measured at discrete and finite directions, that is, sampled at directions around a spherical surface. Under certain conditions, the HRTFs at unmeasured directions can be reconstructed or estimated from measured ones by various interpolation schemes, including adjacent linear interpolation, bilinear interpolation, spherical triangular interpolation, and cubic spline interpolation. Directional interpolation can be applied to complex-valued HRTFs or HRTF magnitudes alone, as well as to head-related impulse responses (HRIRs). The arrival time correction of HRIRs prior to interpolation, as well as separate interpolation to HRTF magnitudes and unwrapped phases, improves interpolation performance.

Far-field HRTFs are complex-valued functions of multiple variables. They can be decomposed into a weighted sum of appropriate basis functions or vectors so that the dimensionality of the entire HRTF data is reduced. There are two primary types of HRTF linear decomposition: spectral shape basis function decomposition and spatial basis function decomposition.

Principal component analysis is a common and effective method for deriving spectral shape basis functions or vectors, through which HRTFs are decomposed into a weighted sum of spectral shape basis vectors (PC-vectors). The primary variations of HRTFs are then represented by a small set of principal components. Subset selection is another scheme for deriving spectral shape basis vectors, through which the HRTFs at arbitrary directions are decomposed into a weighted sum of HRTFs at some privileged directions.

Azimuthal Fourier series and spherical harmonics functions are two kinds of predetermined spatial basis functions, through which HRTFs can be decomposed into a weighted sum of azimuthal or spatial harmonics. From the azimuthal or spatial harmonic representation of HRTFs, the azimuthal or spatial sampling (Shannon-Nyquist) theorem is derived. Spatial principal component analysis is another algorithm for deriving spatial basis functions (vectors) and yields an efficient spatial basis representation of HRTFs. Spatial principal component analysis is applicable to recovering individualized HRTFs with high directional resolution from a considerably small set of measurements.

Both HRTF spatial interpolation and the signal mixing of multi-channel sound solve the problem of spatial sampling and reconstruction. They are closely related in terms of directional sampling. Some HRTF interpolation schemes are analogous to certain signal mixing methods for multi-channel sound. Consequently, some methods for multi-channel sound can be used to simplify HRTF-based synthesis of virtual source processing, resulting in some simplified algorithms, such as virtual loudspeaker-based and basis function decomposition-based algorithms.

Microphone array beamforming effectively records sound field spatial information. Beamforming array outputs can be converted into binaural signals. HRTF array beamforming models are closely related to the basis decomposition of HRTFs, in which different arrays correspond to different basis functions.

7

Customization of Individualized HRTFs

This chapter discusses the customization of individualized HRTFs. The anthropometric parameters and their correlation to auditory localization cues are analyzed. Anthropometry-based ITD models are presented. Various anthropometry-based and subjective selection-based HRTF customization methods are reviewed. The problems that arise in HRTF customization are addressed. The structural model of HRTF is also discussed.

As stated in Chapter 1, head-related transfer functions (HRTFs) vary across individuals. When a subject's own HRTFs are used in binaural synthesis, good localization performance is achieved; by contrast, when nonindividualized HRTFs are used, the performance of binaural synthesis depends upon the similarity between the nonindividualized HRTFs and the subject's HRTFs. To date, measurement has been a relatively accurate approach to obtaining individualized HRTFs. However, HRTF measurement necessitates sophisticated apparatuses and is time-consuming. Measuring the HRTFs of each individual is therefore impractical. This problem can be partially alleviated by the HRTF calculation presented in Chapter 4, but precise solutions of HRTFs can be obtained only for rare simplified head and/or torso models. The numerical computation of HRTFs is complex and computationally intensive. Thus, an essential requirement is to explore methods for customizing individualized HRTFs—that is, approximately estimating individualized HRTFs in a rapid and simple manner, while retaining the perceptually salient features contained in those HRTFs.

The methods for customizing individualized HRTFs are generally categorized into three types: anthropometry-based methods, subjective selection-based methods, and small measurement-based methods. The third was presented in Section 6.3.5, and this chapter focuses on the first and second methods. From the perspective of physics, HRTFs characterize the interaction between incident sound waves and human anatomical structures, such as the head, pinnae, and torso. That individualized HRTFs and individualized anatomical features are strongly correlated is, therefore, a reasonable hypothesis. Anthropometry-based methods approximately estimate individualized HRTFs from anthropometric measurements. Given that a set of HRTFs similar to a subject's HRTFs yield improved localization performance, we can also select a set of the best-matching HRTFs from a baseline HRTF database by subjective listening.

HRTFs are complex-valued functions that include magnitudes and phase spectra. Under minimum-phase approximation, a pair of HRTFs (one for each ear) can be represented by left and right minimum-phase HRTFs, along with a pure time delay (i.e., interaural time difference—ITD).

Left or right minimum-phase HRTFs are uniquely determined by left or right HRTF magnitudes, respectively. Correspondingly, customization is often separately implemented on HRTF magnitude and ITD. In addition, HRTF (or head-related impulse response—HRIR) magnitudes vary as a complex function of source position, frequency (or time), and individual. To reduce data dimensionality and thereby simplify analysis, some feature extraction or dimensional reduction algorithms (such as principle components analysis—PCA) are often applied prior to customization. Another method is to decompose HRTFs into several independent components that correspond to the contribution of different parts of human anatomical structures, such as the head, pinnae, and torso. Each component can be implemented by a filter that corresponds to a specific anatomical structure, after which customization can be separately applied to each component.

Section 7.1 discusses the anthropometric parameters and their correlation to localization cues. Individualized and anthropometry-based ITD models are presented in Section 7.2. Various anthropometry-based HRTF customization methods, including anthropometric parameter matching, frequency scaling, and multiple regression models, are described in Section 7.3. Section 7.4 briefly discusses subjective selection-based methods. The problems that occur in HRTF customization are discussed in Section 7.5. And finally, Section 7.6 discusses the structural model of HRTF.

7.1. Anthropometric Measurements and their Correlation with Localization Cues

7.1.1. Anthropometric Measurements

HRTFs describe the interaction between incident sound waves and human anatomical structures, and are therefore closely related to anthropometric parameters. Accordingly, customizing HRTFs by anthropometric measurement is a natural approach.

The anthropometric parameters of subjects can be measured by commonly-used rulers and optical or video cameras with relatively high accuracy. Ideally, anthropometric measurements should yield a set of complete and inter-independent anthropometric parameters that are indispensable to HRTFs. In practice, however, certain correlations exist among different anthropometric parameters, and a complete set of anthropometric parameters relevant to HRTFs has not been identified or verified. Different anthropometric parameters that are selected through semi-empirical methods are currently used in different studies. Moreover, the results of different studies cannot be compared because researchers use different definitions and measurement points of anthropometric parameters.

In some studies, anthropometric measurement points are labeled on simplified two-dimensional anatomic sketches, from which identifying the exact positions of anthropometric measurement points is difficult because human anatomy is three-dimensional and complicated. Moreover, in different studies, two anthropometric parameters with different measurement points or definitions may be described by the same term; alternatively, the same anthropometric parameter can be described by different terms. All these factors may cause confusion, and therefore require resolution.

As stated in Section 1.8.1, a KEMAR artificial head was designed with reference to the average anthropometric dimensions across adults mainly from Western countries and from the 1950s to the 1960s. The definitions of, and statistical results for, some important anthropometric dimensions are shown in Figure 7.1 and Table 7.1, respectively (Burkhard and Sachs, 1975). For example, the bitragion diameter is defined as the head diameter at a notch above the tragus, that is, at the anterior notch. External ear dimensions are also specified (Burkhard and Sachs, 1975).

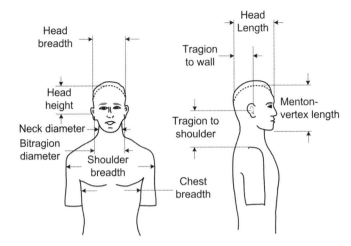

Figure 7.1 Some anthropometric parameters used in the design of KEMAR (from Burkhard and Sachs, 1975, by permission of J. Acoust. Soc. Am.)

Table 7.1 Anthropometric dimensions for average human adults and design of KEMAR (unit: cm)

Dimension	Median male	Median female	Average human	KEMAR
1. Head breadth	15.5	14.7	15.1	15.2
2. Head length	19.6	18.0	18.8	19.1
3. Head height	13.0	13.0	13.0	12.5
4. Bitragion diameter	14.2	13.5	13.85	14.3
5. Tragion to wall	10.2	9.4	9.8	9.65
6. Tragion to shoulder	18.8	16.3	17.55	17.5
7. Neck diameter	12.1	10.3	11.2	11.3
8. Shoulder breadth	45.5	39.9	42.7	44.0
9. Chest breadth	30.5	27.7	29.1	28.2
10. Menton vertex length	23.2	21.1	22.15	22.4

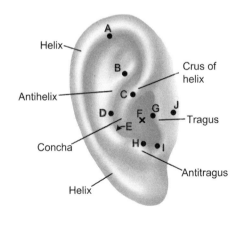

Figure 7.2 Measurement points of pinnae in Middlebrooks's study (from Middlebrook, 1999a, by permission of J. Acoust. Soc. Am.)

In Middlebrooks's (1999a) study on HRTF customization, eight anthropometric parameters, including six pinna-related parameters, were measured from 32 human subjects. The corresponding measurement points and results are shown in Figure 7.2 and Table 7.2, respectively. Head width is defined by the condyle of the mandible (J, in front of the tragus), which is close (but not identical) to the bitragion diameter defined by the tragus (G) in Figure 7.1.

The CIPIC HRTF database constructed by Algazi et al. (2001b) also includes 27 anthropometric parameters with 17 parameters for the head and torso and 10 for the pinnae (Figure 7.3; Algazi et al., 2001c). Table 7.3 lists the mean μ, standard deviation σ, and percentage variations ($2\sigma/\mu$) of the anthropometric parameters. (The anthropometric parameters for some subjects are unavailable in the public CIPIC database.)

Table 7.2 Results for eight anthropometric parameters in Middlebrooks' study (in mm)

Parameter	Definition	Minimum	Mean	Maximum	Standard deviation
1. Pinna cavity height	I to A	36.1	43.9	56.1	4.8
2. Notch to antihelix	I to B	21.6	26.6	32.5	2.8
3. Notch to crus	I to C	11.5	16.4	21.4	2.4
4. Concha depth	F to G	6.5	10.4	14.7	2.0
5. Tragus to antitragus	G to H	3.0	6.8	12.9	2.1
6. Pinna cavity breadth	G to E	8.2	11.9	16.7	1.9
7. Head width	Left-ear J to right-ear J	122	134	153	8
8. Body height		1520	1690	1930	102

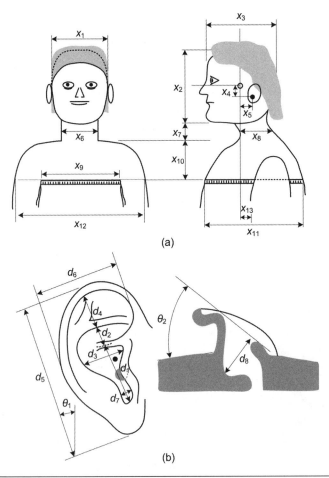

(a)

(b)

Figure 7.3 Anthropometric parameters in the CIPIC database (from Algazi et al., 2001b, by permission of IEEE): (a) Head and torso measurements; (b) Pinna measurement

Table 7.3 Results for the anthropometric parameters in the CIPIC database

No.	Parameter	Unit	Mean	Standard deviation	Percentage variation
x_1	Head width	cm	14.49	0.95	13
x_2	Head height	cm	21.46	1.24	12
x_3	Head depth	cm	19.96	1.29	13
x_4	Pinna offset down	cm	3.03	0.66	43
x_5	Pinna offset back	cm	0.46	0.59	254
x_6	Neck width	cm	11.68	1.11	19
x_7	Neck height	cm	6.26	1.69	54
x_8	Neck depth	cm	10.52	1.22	23
x_9	Torso top width	cm	31.50	3.19	20
x_{10}	Torso top height	cm	13.42	1.85	28
x_{11}	Torso top depth	cm	23.84	2.95	25
x_{12}	Shoulder width	cm	45.90	3.78	16
x_{13}	Head offset forward	cm	3.03	2.29	151
x_{14}	Height	cm	172.43	11.61	13
x_{15}	Seated height	cm	88.83	5.53	12
x_{16}	Head circumference	cm	57.33	2.47	9
x_{17}	Shoulder circumference	cm	109.43	10.30	19
d_1	Cavum concha height	cm	1.91	0.18	19
d_2	Cymba concha height	cm	0.68	0.12	35
d_3	Cavum concha width	cm	1.58	0.28	35
d_4	Fossa height	cm	1.51	0.33	44
d_5	Pinna height	cm	6.41	0.51	16
d_6	Pinna width	cm	2.92	0.27	18
d_7	Intertragal incisure width	cm	0.53	0.14	51
d_8	Cavum concha depth	cm	1.02	0.16	32
θ_1	Pinna rotation angle	degree	21.04	6.59	55
θ_2	Pinna flare angle	degree	28.53	6.70	47

The definition of the head width x_1 in Figure 7.3 appears to be similar to that of the head breadth in Figure 7.1 but different from that of the head width in Table 7.2. The mean value of the head width in Table 7.3 is 14.49 cm, which is smaller than the 15.1 cm mean value of the head breadth in Table 7.1. According to Dreyfuss (1993), the mean head width of American adult males is 15.5 cm and that of American adult females is 14.5 cm. Thus, the mean head width of American adults should be 15.0 cm, which is close to the mean head breadth given in Table 7.1.

Because human anatomy is a unified ensemble, the development of head and body dimensions is harmonious, closely relating some head parameters to body parameters (Liu et al., 1999). With the improvements in living conditions, the statistical results for some human anthropometric dimensions are likely to increase with time, or at least be invariant. The measurements for the KE-MAR design were carried out almost 40 years earlier than the CIPIC measurements, and the latter consists of more male subjects. Therefore, the measured results for the anthropometric dimensions in CIPIC are expected to be greater, or at least not less than, those for KEMAR. However, the head width in CIPIC measurements is less than the head breadth in KEMAR measurements, even though the two parameters appear identical in measurement schematic plots. This abnormality implies that consistency in defining measurement points is crucial to comparison.

Nishino et al. (2001, 2007) and Inoue et al. (2005) constructed a horizontal HRTF database for 96 Japanese subjects, including 9 anthropometric parameters for 86 subjects (71 males and 15 females). The statistical results for the anthropometric measurements are shown in Figure 7.4 and Table 7.4. The mean bitragion diameter in Table 7.4 is 148.0 mm, which is greater than the mean bitragion diameter of 13.85 cm in Table 7.1. This result seems abnormal because the bitragion diameter of Asians (including the Japanese) is expected to be less than that of Western subjects. There are four

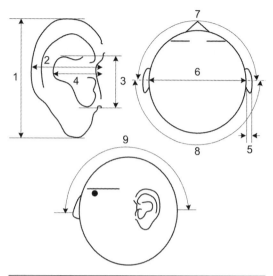

Figure 7.4 Measurements of anthropometric parameters in Nishino's study (adapted from Nishino et al., 2007)

possible reasons for this abnormality: (1) In Nishino's study, the sample was dominated by male subjects, which may lead to gender bias, and, therefore, to a greater mean value. (2) Although Nishino claims to follow the parameter definitions in the KEMAR design (i.e., Figure 7.1), deviation may still occur because measurement points are difficult to consistently select. (3) Some mean head-relevant anthropometric dimensions of populations are likely to increase with time. The statistical data used by Burkhard et al. in the KEMAR design were derived 40 years earlier than those used in Nishino's study. (4) Measurement errors and statistical fluctuations may also result in certain deviations.

The National Standard Agency of China has also published a set of standards for the terms and methods used for anthropometric measurement. These include the GB 3975-83, "human body

Table 7.4 Results for the anthropometric parameters in Nishino's study

Parameter	Mean	Maximum	Minimum	Standard Deviation
1. Ear length (mm)	65.6	82	55	4.9
2. Ear breadth (mm)	32.0	40	22	3.4
3. Concha length (mm)	18.4	30	10	3.5
4. Concha breath (mm)	18.4	24	13	2.2
5. Protrusion (mm)	22.1	30	15	3.4
6. Bitragion diameter (mm)	148.0	181	113	13.4
7. Radial distance between the bitragion and pronasale (mm)	307.3	371	270	19.1
8. Radial distance between the bitragion and opisthocranion (mm)	227.6	275	195	16.3
9. Radial distance among the pronasale, vertex, and opisthocranion (mm)	424.9	482	360	21.1

measurement term," GB 5703-85, "human body measurement method," and GB/T 2428-1998, "head-face dimensions of adults." Given that the parameters specified in these standards may not cover all HRTF-relevant parameters, we selected and measured 15 head- and pinna-related anthropometric parameters when constructing the Chinese-subject HRTF database with reference to GB/T 2428-1998 and the CIPIC database (Zhong, 2006b; Xie et al., 2007a). We also calculated two indirectly-measured parameters: pinna rotation angle and pinna flare angle. The 17 anthropometric parameters are listed in column 2 of Table 7.5. Total head height refers to the distance from the vertex to the gnathion; maximum head breadth refers to the distance between the left and right cranial side points; and bitragion breadth pertains to the distance between the left and right tragi (i.e., point G in Figure 7.2). We defined six of the 17 anthropometric parameters to precisely describe the HRTF-related anatomical features of pinnae. For example, measuring tragus helix distance and the corresponding pinna flaring distance entails determining pinna flaring angle; measuring the frontal arc length between two bitragions, back arc length between two pinna posteriors, and pinna and posterior to tragus distance involves ascertaining the front-back offsets of ears relative to the head center. Figure 7.5 shows a sketch of some anthropometric parameter measurements.

Table 7.5 lists the statistical results for the 17 anthropometric parameters across 52 human subjects (26 males and 26 females). Considering that gender may significantly influence statistical results, in the GB/T 2428-1998 standard and some other studies, the data on males and females are treated separately (Burkhard and Sachs, 1975; Henry Dreyfuss Assoc., 1993). The CIPIC database provides only the mean value across all subjects, with more male subjects. In our study, data on an equal number of males and females are used. The extent to which gender influences

Table 7.5 Results for the 17 anthropometric parameters in our study

No.	Parameter	Definition	Mean	Standard deviation	Significance of gender
1	Total head height x_1	GB/T	222(mm)	13.6(mm)	**
2	Maximum head breadth x_2	GB/T	157(mm)	6.8(mm)	**
3	Bitragion breadth x_3	GB/T	141(mm)	6.2(mm)	**
4	Head depth x_4	CIPIC	184(mm)	7.9(mm)	**
5	Pronasale opisthocranion distance x_5	GB/T	200(mm)	7.9(mm)	**
6	Tragion opisthocranion distance x_6	GB/T	92(mm)	11.6(mm)	×
7	Pinna width x_7	CIPIC	32(mm)	2.2(mm)	(*)
8	Pinna height x_8	CIPIC	61(mm)	5.0(mm)	*
9	Physiognomic pinna length x_9	GB/T	58(mm)	4.4(mm)	*
10	Pinna rotation angle x_{10}	CIPIC	19(°)	5.8(°)	×
11	Tragus helix distance x_{11}	self-defined	31(mm)	2.7(mm)	**
12	Pinna flaring distance x_{12}	self-defined	20(mm)	2.7(mm)	**
13	Pinna flaring angle x_{13}	CIPIC	42(°)	7.3(°)	(*)
14	Frontal-arc length between two Bitragions x_{14}	self-defined	274(mm)	11.0(mm)	**
15	Back-arc length between two pinna posteriors x_{15}	self-defined	210(mm)	13.1(mm)	**
16	Pinna posterior to tragus distance x_{16}	self-defined	24(mm)	2.7(mm)	*
17	Tragus shoulder distance x_{17}	self-defined	139(mm)	11.6(mm)	×

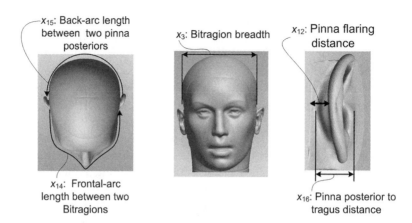

x_{15}: Back-arc length between two pinna posteriors

x_3: Bitragion breadth

x_{12}: Pinna flaring distance

x_{14}: Frontal-arc length between two Bitragions

x_{16}: Pinna posterior to tragus distance

Figure 7.5 Measurements of the anthropometric parameters in our study

anthropometric parameters is illustrated in column 6 of Table 7.5, where **, *, (*), and × denotes "highly significant," "significant," "less significant," and "no influence," respectively. Table 7.5 shows that gender significantly influences many anthropometric parameters, which further confirms the importance and necessity of employing data from an equal number of males and females.

Table 7.6 compares the mean values of some anthropometric parameters obtained in three different measurements for Chinese subjects; these measurements include the results from our work, a study on students of Han nationality in Shandong Province (Jin and Ding et al., 2000) and those from GB/T 2428-1998. As indicated by our results, the means of total head height, maximum head breadth, bitragion breadth, and head depth for males are larger than those for females. ANOVA results suggest that at a 0.05 level (e.g., $F = 14.95$, $p = 3.19 \times 10^{-4}$), the means of these four parameters

Table 7.6 Comparison of the mean values and standard deviations of some anthropometric parameters in three different measurements on Chinese subjects (in mm)

Data sources / Parameters	Our work		Students of Han nationality in Shandong Province		GB/T 2428-1998	
	Male	Female	Male	Female	Male	Female
Total head height	231 (9.1)	213 (11.0)	231.3 (10.8)	221.1 (10.4)	223 (10.68)	216 (9.77)
Maximum head breadth	160 (6.6)	153 (5.3)	161.3 (5.5)	156.6 (5.5)	154 (5.86)	149 (5.30)
Bitragion breadth	145 (5.1)	138 (5.0)	149.8 (5.3)	140.2 (6.0)	140 (3.99)	140 (3.75)
Head depth	188 (6.8)	180 (7.0)	183.1 (6.4)	176.0 (5.5)	–	–
Pronasale opisthocranion distance	204 (6.3)	196 (7.0)	–	–	219 (6.29)	201 (4.61)
Tragion opisthocranion distance	95 (13.0)	89 (9.6)	–	–	102 (4.92)	103 (3.44)
Physiognomic pinna length	59 (4.7)	56 (3.7)	–	–	56 (2.16)	53 (3.03)

for males and females significantly differ. These results are basically consistent with those derived from GB/T 2428-1998. For GB/T 2428-1998, the means of total head height and maximum head breadth for males are also larger than those for females, but the means of bitragion breadth for both genders are equal. The mean total head height for males and maximum head breadth for both genders in our study are larger than those derived from GB/T 2428-1998, although some statistical deviation may exist. This difference may be due to the fact that the GB/T 2428-1998 data were collected in 1988, 17 years before our data were collected (2005). Data validity is temporary. Our results are consistent with those measured from students of Han nationality in Shandong Province in 1999 (Jin and Ding et al., 2000).

7.1.2. Correlations among Anthropometric Parameters and HRTFs or Localization Cues

In Section 7.1.1, different researchers measured various sets of anthropometric parameters that reflect individual anatomical shapes and dimensions. Given that anthropometric parameters are closely related to HRTFs, individualized HRTFs can possibly be predicted or evaluated from an appropriate set of anthropometric parameters. However, HRTFs result from the overall scattering and diffraction effect of the head, pinnae, and torso, accordingly have a complicated correlation with anthropometric parameters. The manner by which an appropriate set of anthropometric parameters is identified for HRTF prediction or customization remains an unresolved issue.

Development is consistent across anatomical dimensions because of the unity of the human body. Some anthropometric parameters are consequently interdependent. However, these interdependent anthropometric parameters are not interchangeable, given the incomplete correlation among them (i.e., the normalized correlation coefficient is not a unity). Usually, a certain combination of anthropometric parameters is used for predicting or customizing individualized HRTFs. If HRTFs are linearly correlated with certain anthropometric parameters, then an algorithm for simple linear regression can be applied when predicting or customizing individualized HRTFs. The correlation between HRTFs and anthropometric parameters may not be completely linear, however. Linear regression and prediction are merely approximations.

Fels et al. (2009) investigated the influence of anthropometric parameters on HRTFs, with particular reference to children and their growth. Using a photogrammetric system, the authors measured more than 40 anthropometric parameters for 95 human subjects aged six months to 17.5 years. They also calculated the corresponding HRTFs within 0.1–8 kHz, using the boundary element method (BEM) described in Section 4.3. The analyses indicate that, among head and torso parameters, the distance between the ear and shoulder, the breadth of the head, and the back vertex considerably influence HRTFs; among pinnae parameters, the breadth and depth of the cavum conchae, as well as the rotation of the ears, substantially affect HRTFs.

HRTFs are correlated with anthropometric parameters in a complicated manner. Nevertheless, some HRTF-derived directional localization cues (such as ITD) are related to certain anthropometric parameters in a straightforward manner. Because ITD is a dominant directional localization cue at low frequencies, many researchers explored the relationship between ITD and anthropometric parameters, laying a foundation for the individualized ITD model to be presented in Section 7.2.

Middlebrooks (1999a) investigated the correlation between the maximal ITDs of 33 human subjects and anthropometric parameters. The ITDs were evaluated by cross-correlation. The maximal ITDs of the 33 subjects range from 657–792 μs (709 ± 32 μs). The statistical results show that maximal ITD is strongly correlated with the head width and body height given in Table 7.2, with correlation coefficients of 0.85 and 0.67, respectively. In view of its physical formation, ITD appears to have no direct correlation with body height. The strong correlation between maximal ITD and body height reflects an indirect correlation—namely, body height is correlated with head width, with a normalized correlation coefficient of 0.67. In other words, body height and head width are interdependent. The correlation between maximal ITD and body height results from that between body height and head width. Therefore, maximal ITD is essentially correlated with head width.

In establishing the CIPIC HRTF database, Algazi (2001b) also investigated the correlations among the different anthropometric parameters in Table 7.3. The results indicate that some anthropometric parameters exhibit a certain correlation. For example, the normalized correlation coefficient between d_1 (cavum concha height) and d_5 (pinna height) is 0.45, and that between x_3 (head depth) and d_3 (cavum concha width) is 0.33. By contrast, certain anthropometric parameters appear to be weakly correlated. For example, the normalized correlation coefficient between d_1 and d_3 is 0.25, and that between x_2 (head height) and d_5 is only 0.16. These results imply a relatively low correlation between head and pinna sizes. A subject with a large head may not have large pinnae.

The correlation between maximal ITD and anthropometric parameters was also calculated. The ITDs were evaluated by leading-edge detection with a threshold of 20% (see Section 3.2). The normalized correlation coefficient between maximal ITD and x_1 (head width) is 0.78. This result confirms the strong correlation between maximal ITD and head size. The methods for evaluating ITD and defining head width differ in the studies by Middlebrooks and Algazi, although a similar strong correlation between maximal ITD and head width is found in both.

The correlation between the pinna-notch frequency f_{pn} in Section 3.4.1 and anthropometric parameters also indicates that f_{pn} is related to pinna parameters, but the relationship is weak, with the normalized correlation coefficient between f_{pn} and d_1 (cavum concha height) being 0.33. This finding suggests that the scattering/diffraction of incident waves by pinnae is a complex process that is difficult to describe by the linear combination of a few pinna-related parameters.

The correlation between the maximal ITDs of all 52 subjects in the Chinese subject HRTF database and anthropometric parameters (Table 7.5) was also analyzed. The ITD was evaluated by the cross-correlation method discussed in Section 3.2.2. Table 7.7 lists the normalized correlation coefficients. A high correlation is found between maximal ITD (i.e., $|ITD|_{max}$) and maximum head breadth x_2, or bitragion breadth x_3, and a slight correlation is found between maximal ITD and total head height x_1; however, a low correlation exists between maximal ITD and head depth x_4. Nevertheless, the four parameters are not completely independent of one another. For example, x_1 and x_2, or x_2 and x_3 are correlated to some extent.

In conclusion, maximal ITD is related to maximum head breadth x_2 (or bitragion breadth x_3). As shown by Table 7.5 and the results in Section 3.2.2, a gender difference in mean anthropometric

Table 7.7 Correlation coefficients among $|ITD|_{max}$ and anthropometric parameters

	Total head height x_1	Maximum head breadth x_2	Bitragion breadth x_3	Head depth x_4
Maximum head breadth x_2	0.516			
Bitragion breadth x_3	0.373	0.544		
Head depth x_4	0.367	0.246	0.290	
$|ITD|_{max}$	0.580	0.741	0.748	0.435

dimensions causes a gender difference in mean maximal ITDs. The statistical difference in mean anthropometric dimensions is also responsible for the difference in mean maximal ITDs among different populations (a statistical difference in anthropometric dimensions exists among different populations). These results further confirm that establishing a Chinese subject HRTF database is necessary.

7.2. Individualized Interaural Time Difference Model and Customization

As stated in Sections 1.4 and 3.2, ITD is an important directional localization cue and closely related to the phases of a pair of HRTFs. Section 5.4 also showed that the linear-phase component of a pair of HRTFs is usually extracted and realized by a pure time delay ITD z^{-m}, thereby simplifying virtual source synthesis. Moreover, under the minimum-phase approximation of HRTFs, ITD and HRTF magnitude completely represent a pair of HRTFs.

ITD varies across individuals. Individualized ITDs can be evaluated from individualized HRTFs by using the definitions and methods presented in Section 3.2.1. When individualized HRTFs are unavailable, individualized ITDs can be approximated by customization. As discussed in Section 7.1.2, establishing an anthropometry-based ITD model to customize or predict individualized ITDs from anthropometric parameters is possible because of the high correlation between individualized ITDs and anthropometric parameters. As part of the discussion on HRTF customization, this section elucidates the methods of ITD customization with anthropometric parameters. For simplicity, a frequency-independent ITD that corresponds to the linear-phase component of HRTFs is analyzed here. This ITD also approximately corresponds to the result derived by leading-edge detection.

7.2.1. Extension of the Spherical Head ITD Model

The simplest ITD model is given by the spherical head-based Woodworth's formula [i.e., Equation (1.9)], from which the far-field and individualized ITDs in the horizontal plane can be predicted from the head radius a. Equation (1.9) can be conveniently extended to three-dimensional space by using the interaural polar coordinate system shown in Figure 1.3. Taking advantage of the fact that sound sources at the same Θ (i.e., on the cone of confusion in Figure 1.10) have identical ITDs in a spherical head model, we extend Equation (1.9) thus:

$$ITD(\Theta) = \frac{a}{c}(\sin\Theta + \Theta).\tag{7.1a}$$

In some studies, the sound source direction in Equation (7.1a) is described by the complementary angle of Θ, i.e., $\Gamma = \pi/2 - \Theta$. Equation (7.1a) then becomes:

$$ITD(\Gamma) = \frac{a}{c}\left(\cos\Gamma + \frac{\pi}{2} - \Gamma\right).\tag{7.1b}$$

After the default coordinate system in this book is converted, the ITD of a far-field sound source at (θ, ϕ) is derived as:

$$ITD(\theta,\phi) = \frac{a}{c}[\arcsin(\cos\phi\sin\theta) + \cos\phi\sin\theta)].\tag{7.2}$$

Saviojia et al. (1999) also derived a three-dimensional ITD formula by linearly fitting the excess phase component of measured HRTFs:

$$ITD(\theta,\phi) = \frac{a}{c}(\sin\theta + \theta)\cos\phi. \tag{7.3}$$

However, because the human head is not a perfect sphere, and the head radius is difficult to accurately determine, Woodworth's formula is only a rough approximation. When $a = 0.0875$ m, the result of Woodworth's formula is usually consistent with the measured mean for Western populations.

Using 20% leading-edge detection, Algazi et al. (2001c) calculated ITDs at $M = 1250$ directions for each of 25 human subjects, from which the optimal head radius a_{opt} of the spherical head model was computed by least squares, fitted between the measured ITD and the ITD derived by Woodworth's formula [Equation (7.1)]. For each subject, the ITD measurements at M directions can be expressed as an $M \times 1$ column vector $\boldsymbol{\tau} = [\tau_0, \tau_1, \ldots \tau_{M-1}]^T$. Let $X = (\sin\Theta + \Theta)$ and let $\boldsymbol{X} = [X_0, X_1, \ldots X_{M-1}]^T$ be the corresponding vector of the X values computed for all measurement directions. For M directions, all the ITD values calculated from Equation (7.1) consist of an $M \times 1$ column vector $\mathbf{ITD} = a\boldsymbol{X}/c$. Then, the optimal radius a_{opt} can be obtained by minimizing the square norm of the error vector $\varepsilon = \mathbf{ITD} - \boldsymbol{\tau} = a\boldsymbol{X}/c - \boldsymbol{\tau}$, yielding:

$$a_{opt} = c\frac{\boldsymbol{X}^T\boldsymbol{\tau}}{\boldsymbol{X}^T\boldsymbol{X}}. \tag{7.4}$$

The optimal head radius of the 25 subjects ranges from 0.079–0.095 m, with a mean value of 0.087 m. The mean result is close to the commonly used $a = 0.0875$ m, a value derived for Western subjects. The calculations indicate that the RMS error of the fit ranges from 22–47 μs, with a mean of 32 μs.

On this method, the optimal head radius a_{opt} is computed from the measured ITDs, which are usually unknown for each subject. In practice, estimating a head radius a_{opt} from anthropometry appears feasible. Algazi also derived the relationship between the optimal head radius a_{opt} and anthropometric parameters. The results indicate that the head radius can be estimated from a weighted sum of three head dimensions (the head width x_1, head height x_2, and head depth x_3 in Table 7.3) as:

$$a_{opt} = 0.51\frac{x_1}{2} + 0.019\frac{x_2}{2} + 0.18\frac{x_3}{2} + 32, \tag{7.5}$$

where mm is the unit used. As indicated by Equation (7.5), the dimension that provides the highest contribution is x_1, followed by x_3; x_2 is insignificant and can be excluded.

The ITD calculated from Woodworth's formula or its improved version is front-back and left-right symmetric, and cannot reflect the asymmetric nature of actual ITDs (Section 3.5.3). In particular, Equation (7.1) implies that the ITDs on the cone of confusion with a constant Θ are identical. However, the ITD measured from a human subject demonstrates a variation range of 120 μs (nearly 18% of the maximal ITD) on the cone of confusion with $\Theta = -55°$ (Duda et al., 1999). This result is attributed primarily to the features of the head and ears: the human head is not a perfect sphere with front-back symmetry, and the two ears are not exactly positioned at diametrically opposed directions on the head surface. As an improvement, Duda et al. proposed an ellipsoid head model and evaluated the corresponding ITD by geometric acoustics. Although the ellipsoid head model generates an improved result, the computational procedure is complex.

7.2.2. ITD Model Based on Azimuthal Fourier Analysis

As an improvement to Woodworth's formula, Zhong and Xie (2007a) proposed a novel ITD model. In contrast to aforementioned models, the proposed ITD model was derived without simplification of anatomical structures. On the basis of azimuthal Fourier analysis and experimental data fit, the authors analyzed measured ITDs and established a statistical relationship between anthropometric parameters and Fourier coefficients, generating a statistical ITD formula. The proposed model characterizes the left-right symmetry and front-back asymmetry of ITDs. This model can be used as a basis for evaluating and predicting individualized ITDs in the horizontal plane from three anthropometric parameters of the head and pinnae.

Similar to that implemented in the azimuthal Fourier analysis of HRTFs in Section 6.3.2, ITD is a periodic function of θ with period 2π at a given elevation plane. It can therefore be expanded as an azimuthal Fourier series as:

$$ITD\,(\theta) = a_0 + \sum_{q=1}^{\infty} \left[a_q \cos q\theta + b_q \sin q\theta \right] \quad 0 \le \theta < 2\pi , \tag{7.6}$$

and:

$$a_0 = \frac{1}{2\pi} \int_0^{2\pi} ITD(\theta)\,d\theta \qquad a_q = \frac{1}{\pi} \int_0^{2\pi} ITD(\theta)\cos q\theta\,d\theta$$

$$b_q = \frac{1}{\pi} \int_0^{2\pi} ITD(\theta)\sin q\theta\,d\theta . \tag{7.7}$$

The notation for elevation has been excluded in Equations (7.6) and (7.7). In Equation (7.6), ITD is decomposed as a weighted sum of infinite azimuthal harmonics. If coefficients $\{a_q, b_q\}$ equals 0 for $|q| > Q$, where Q is a positive integer, Equation (7.6) can be simplified as:

$$ITD\,(\theta) = a_0 + \sum_{q=1}^{Q} \left[a_q \cos q\theta + b_q \sin q\theta \right] \quad 0 \le \theta < 2\pi . \tag{7.8}$$

Then, ITD contains $(2Q + 1)$ azimuthal harmonics and is determined by the corresponding $(2Q + 1)$ coefficients $(a_0, a_q, b_q, q = 1, 2 ..., Q)$. Similar to the case in Section 6.3.2, each coefficient in Equation (7.8), and consequently, ITD, can be evaluated from the samples, if ITD is sampled from 0–2π at a uniform azimuthal interval of $2\pi/M$ with $M \ge (2Q + 1)$.

On the basis of the Chinese subject HRTF database presented in Section 2.3, we use 10% leading-edge detection to evaluate the ITDs of the HRIRs of each subject. The mean ITD across 52 subjects is shown in Figure 3.8(c). For each subject, the ITD in a given elevation plane consists of M uniform samples $ITD(\theta_m)$. Expanding the ITD as a Fourier series according to Equation (7.8) and evaluating the Fourier coefficients $(a_0, a_q, b_q, q = 1, 2 ... Q)$ from M samples determine the azimuthally continuous $ITD(\theta)$ of each subject. This is the individualized ITD formula for each subject. Furthermore, the relationship between Fourier coefficients and anthropometric parameters can be determined by linear multiple regression analysis, which results in a statistical ITD formula. Therefore, individualized ITDs can be evaluated by substituting corresponding anthropometric parameters into the statistical formula. This is the basic principle of using azimuthal Fourier analysis and experimental data fitting to obtain a statistical model for individualized ITDs.

ITD varies with maximal magnitude in the horizontal plane or nearby planes. Human localization resolution is also at its highest in the horizontal plane. As an example, therefore, we model the ITD of the horizontal plane. A similar modeling method can be applied to other arbitrary elevation planes.

Because the HRIRs in the horizontal plane are uniformly measured along 72 azimuthal directions, the corresponding sampling number M of ITD(θ) is 72. Primary evaluation and analysis show that $M = 72$ is likely redundant; that is, the highest order Q in Equation (7.8) is considerably lower than 35. We observe occasional abnormal high-order azimuthal harmonics for several subjects, a discrepancy that stems from abrupt subject movements during measurement sessions (Algazi et al., 2001c). To smooth out these irregularities, only the azimuthal harmonics from the 0th to 6th order in Equation (7.8) are retained. The evaluations show that such a truncation efficiently eliminates irregularities (especially at lateral directions) without visibly changing the characteristics of ITD(θ) (the average deviation over all azimuths is less than 6 μs). After smoothing, ITD(θ) is written as ITD′(θ):

$$ITD'(\theta) = a_0 + \sum_{q=1}^{6}\left[a_q\cos q\theta + b_q\sin q\theta\right] \quad 0 \leq \theta < 2\pi . \tag{7.9}$$

In what follows, ITD′(θ) is used as baseline, or reference, data to evaluate customization or prediction performance.

Equation (7.9) can be simplified by a consideration of human anatomical features. The anatomical features of the head are, by nature, approximately left-right symmetric but front-back asymmetric. As a result, ITD is also characterized by the same attributes. According to the left-right symmetry hypothesis, we have ITD(θ) = −ITD(360° − θ). Then, the a_q ($q = 0, 1\ldots6$) in Equation (7.9) should be 0. Moreover, according to the hypothesis of front-back asymmetry, we have ITD(θ) ≠ ITD(θ_1), where θ_1 = (540° − θ) for θ in the left half plane and θ_1 = (180° − θ) for θ in the right half plane. Therefore, one of the b_q ($q = 2, 4, 6$) should not be 0, at the least. Given that the auditory resolution of ITD in the horizontal plane is about 10 μs, we take |ITD′(θ) − ITD″(θ)| ≤ 10 μs as a criterion for simplifying Equation (7.9), and then obtain the ITD model:

$$ITD''(\theta) = \sum_{q=1}^{5} b_q\sin q\theta \quad 0 \leq \theta < 2\pi . \tag{7.10}$$

Equation (7.10) consists of five sine harmonics and is, therefore, left-right symmetric. Among the five azimuthal harmonics, those of {sinθ, sin(3θ), sin(5θ)} reflect the front-back symmetry of ITD, whereas the remaining harmonics {sin(2θ), sin(4θ)} reflect the front-back asymmetry of ITD. As a whole, the ITD model of Equation (7.10) is left-right symmetric but front-back asymmetric. If merely the front-back symmetric harmonic of $q = 1$ is retained in Equation (7.10) with $b_1 = 2a/c$, Equation (7.10) can be simplified into Equation (1.8).

To relate each coefficient in Equation (7.10) to anthropometric parameters, linear multiple regression analysis is applied to the coefficients of 52 subjects and their 17 anthropometric parameters (numbered x_1 to x_{17} in Table 7.5). The results show that only three anthropometric parameters are significant. The linear multiple regression formulas of each coefficient are listed in Table 7.8; these formulas are linearly significant at the level of 0.05. Substituting these coefficients into Equation (7.10) yields a novel ITD formula:

$$\begin{aligned} ITD'''(\theta) = &\, 128\sin\theta + 21\sin 2\theta - 2\sin 3\theta - 42\sin 4\theta - 21\sin 5\theta \\ &+ (3.02\sin\theta + 0.25\sin 5\theta)\times x_3 \\ &- 1.33\sin 2\theta \times x_{12} \\ &+ 0.15\times(-\sin 3\theta + \sin 4\theta)\times x_{15} \end{aligned} \tag{7.11}$$

where ITD‴ is in units of μs; x_3, x_{12}, and x_{15}, which denote the dimensions of the head, pinna flaring distance, and ear position, respectively, are in units of mm. Equation (7.11) indicates that individualized ITDs can be evaluated (predicted) from three anthropometric parameters.

Table 7.8 Linear multiple regression formulas of each coefficient in Eq. (7.11)

Regression formula	$b_1 = 128 + 3.02x_3$	$b_2 = 21 - 1.33x_{12}$	$b_3 = -2 - 0.15x_{15}$	$b_4 = -42 + 0.15x_{15}$	$b_5 = -21 + 0.25x_3$
F	74.4	8.7	9.9	17.8	28.0
P	1.8×10^{-11}	0.005	0.003	0.0001	2.7×10^{-6}

To validate Equation (7.11), we measured the HRTFs and anthropometric parameters of four "new" subjects (nos. 53–56) taken from outside the database, and calculated the corresponding ITD' in Equation (7.9). The ITD''' predicted from Equation (7.11) was compared to the reference ITD'. Figure 7.6 shows the results for subject number 53 (those for the other subjects are similar). The predicted ITD is almost identical to the reference ITD', except for small deviations at lateral directions.

To quantitatively evaluate error, we calculate the absolute difference between the reference ITD' and predicted ITD''' thus:

$$\Delta(\theta) = | ITD'(\theta) - ITD'''(\theta)|. \tag{7.12}$$

The calculations show that the means $\Delta(\theta)$ across 72 horizontal azimuths are 9, 8, 6, and 20 μs for each of the four subjects, respectively. Therefore Equation (7.11) guarantees a reasonable prediction performance, on the whole. The maximal $\Delta(\theta)$ for each of the four subjects occurs at (30 μs, 120°), (24 μs, 290°), (13 μs, 100°), and (56 μs, 300°). These results imply that maximal $\Delta(\theta)$ always appears at lateral directions (90° ± 30° and 270° ± 30°). Details on causation and on error analyses can be found in Zhong and Xie (2007a).

As a comparison, the maximal error of Woodworth's ITD formula [Equation (7.1)] and the optimal head radius evaluated according to Equation (7.5) is as large as 70 μs at lateral directions. Therefore, the improved Equation (7.1) with optimal head radius is inferior to Equation (7.11), in terms of prediction performance.

The prominent feature of Equation (7.11) is that it accounts for the front-back asymmetric nature of ITD. Taking the results of subject number 55 as an example, we determine that the reference ITDs at azimuth 35° and its front-back mirror azimuth 145° are 271 and 306 μs, whereas the

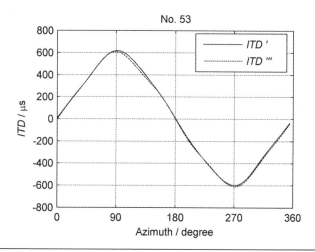

Figure 7.6 Predicted and referenced ITDs of subject no. 53

corresponding results predicted from Equation (7.11) are 272 and 303 μs. Those for the other three subjects are similar.

Equations (7.1) and (7.11) differ in the number of expansion components and the anthropometric parameters included. First, Equation (7.11) precedes five sine harmonics $\{\sin(q\theta), q = 1, 2, 3, 4, 5\}$ and discards higher-order harmonics with a trade-off in accuracy and efficiency. Consequently, Equation (7.11) is left-right symmetric and front-back asymmetric, which is consistent with the basic features of human anatomical structures. Equation (7.1) includes only the components of $\{\sin\theta\}$ and $\{\theta\}$, excluding the higher-order components that reflect the fine structures of ITD (e.g., front-back asymmetry). In general, a formula or model that consists of few and low-order components is always simple, but often inaccurate. This feature is the main reason that the accuracy of Equation (7.1) is inferior to that of Equation (7.11). Second, Equation (7.1) includes only the parameters of head dimensions, whereas Equation (7.11) also includes parameters that describe pinna protrusion and position, apart from head dimensions. Although the head plays a primary role in forming ITD, pinna position (or the position of the ear canal entrance) and protruding position also contribute to ITD. Therefore, Equation (7.11) is superior, from the perspectives of spatial asymmetry and the selection of relevant anatomical parameters. On the basis of spherical harmonics (SH) expansion, the ITD model from Equation (7.11) can be extended into three-dimensional space (Zhong and Xie, 2009b).

Similarly, an individualized interaural level difference (ILD) model was established (Watanabe et al., 2007). At a given elevation plane, ILD is also a periodic function of θ with period 2π, and can therefore be expanded as an azimuthal Fourier series. Once the relationship between expansion coefficients and anthropometric parameters is identified by multiple linear regression, individualized ILDs can be predicted through anthropometric parameters.

7.3. Anthropometry-based Customization of HRTFs

As stated in Section 7.1, HRTFs are closely related to certain anthropometric parameters and can therefore be customized from anthropometric measurements. This section addresses three anthropometry-based customization methods: anthropometry matching method, frequency scaling method, and linear regression method.

7.3.1. Anthropometry Matching Method

The primary hypothesis of the anthropometry matching method is that subjects with similar anatomical features have similar HRTFs. On the basis of matching some pinna-related anthropometric parameters with a baseline parameter/HRTF database, Zotkin et al. (2003; 2004) approximated individualized HRTFs by selecting the best match from the baseline database. This approach is a straightforward customization method.

Because pinnae encode the majority of HRTF individually, ear parameters are selected as indicators in matching. On the basis of a digital image of the ear taken by a video camera, Zotkin et al. measured seven ear-related parameters $d_1, d_2, \ldots d_7$ for a new subject according to Figure 7.3 and Table 7.3. Then, the ear parameter difference between the new subject and the subjects in the CIPIC database was evaluated by:

$$\varepsilon^k = \sum_{i=1}^{7} \frac{(d_i - d_i^k)^2}{\sigma_i^2}, \tag{7.13}$$

where d_i^k denotes the ith ear parameter of the kth subject in the CIPIC database, and σ_i^2 denotes the variations in the ith ear parameter across all the subjects in the database. Equation (7.13) represents the overall difference in ear parameters between the new and kth subjects in the CIPIC database. The HRTFs of the subject, whose parameters minimize Equation (7.13), were selected as an approximation of the individualized HRTFs of the new subject. The matching was implemented separately for the two ears, which sometimes resulted in different matching subjects for the left and right ears, given individualized anatomical asymmetry.

Equation (7.13) accounts only for ear matching, which is effective at high frequencies when ear size and sound wavelength are comparable. To obtain customized HRTFs in the entire audible frequency range, Zotkin et al. estimated HRTFs below 3 kHz by substituting the head and torso parameters of a new subject into the snowman model (Section 4.2). Thus, the combined HRTFs in the entire audible frequency range was:

$$A(f)=\begin{cases} A_l(f) & f<f_l \\ A_l(f)+\dfrac{A_h(f)-A_l(f)}{f_h-f_l}(f-f_l) & f_l<f<f_h, \\ A_h(f) & f>f_h \end{cases} \tag{7.14}$$

where f_l = 500 Hz, f_h = 3000 Hz; and $A(f)=\log_{10}|H(f)|$, $A_l(f)=\log_{10}|H_l(f)|$, and $A_h(f)=\log_{10}|H_h(f)|$ denote HRTF log-magnitudes for the new subject, snowman model, and the best match from the database, respectively. Thus, the individualized HRTFs of the new subject were customized from the HRTFs of the snowman model below 500 Hz, the HRTFs of the best match from the database above 3 kHz, and a progressive blending from 500 Hz–3 kHz. The results of a preliminary listening test indicate that this matching method improves both the localization accuracy and subjective perception of virtual auditory displays (VADs).

The matching customization method for individualized HRTFs is relatively simple for a few measurements of anthropometric parameters of a new subject. However, its performance is limited, unless an appropriate set of HRTF-relevant anthropometric parameters is pre-identified.

7.3.2. Frequency Scaling Method

According to the calculation formula of far-field and spherical-head HRTFs [Equation (4.1)], HRTF is a function of $ka = 2\pi fa/c$; that is, HRTF varies with the products of frequency f and head radius a (head parameter). Let $H(f, a_0)$ refer to the HRTF of a subject with the head radius a_0 at a given source direction (θ, ϕ). Then, the HRTF that corresponds to the head radius $a = a_0/\gamma$ can be evaluated from $H(f, a_0)$ as:

$$H(f,a)=H(f,a_0/\gamma)=H(f/\gamma,a_0). \tag{7.15}$$

In the linear frequency scale, therefore, the HRTF with the head radius $a < a_0$ ($\gamma > 1$) is obtained by expanding the entire $H(f, a_0)$ γ times to high frequency. Conversely, the HRTF with the head radius $a > a_0$ ($\gamma < 1$) is obtained by compressing the entire $H(f, a_0)$ γ times to low frequency. This way, frequency scaling connects head radius to HRTF features in the frequency domain and provides a method for customizing individualized HRTFs. Frequency scaling has long been introduced in acoustics, as in the scaled model for room acoustics (Kuttruff, 2000; Xiang and Blauert, 1991).

Middlebrooks (1999a; 1999b) introduced the frequency scaling method into individualized HRTF customization. The hypothesis was that the similar shapes but different anatomical dimensions of different subjects result in similar spectral features of HRTFs (such as peaks and notches) at different frequencies. Thus, reducing the intersubject spectral difference in HRTFs by frequency scaling is possible. This is the physical foundation of the frequency scaling customization method for individualized HRTFs.

Middlebrooks recruited 45 human subjects (23 males and 22 females), among which 34 were European, 9 were Asian, and 2 were African. For each subject, the individualized far-field HRTFs at 400 directions (elevation ϕ from $-70°$ to $+90°$) were measured in an anechoic chamber, with miniature microphones placed in both ears at approximately 5 mm inside the entrance of the ear canal. Middlebrooks used 512-point Golay codes at a sampling frequency of 50 kHz as stimuli. The measured HRTFs at $M = 393$ directions were used in analysis.

To isolate the direction-independent components (such as ear canal resonance and microphone response) in measured HRTFs, the HRTF was divided into two components: the common transfer function (CTF) and directional transfer function (DTF) as:

$$H(\theta,\phi,f) = CTF(f)\,DTF(\theta,\phi,f). \tag{7.16}$$

The magnitude of $CTF(f)$ was estimated by computing the RMS of $|H(\theta, \phi, f)|$ at each frequency and averaging across M source directions:

$$|CTF(f)| = \sqrt{\frac{1}{M}\sum_{i=0}^{M-1}|H(\theta_i,\phi_i,f)|^2}. \tag{7.17}$$

Then, the complex-valued $CTF(f)$ was formed according to minimum-phase reconstruction. Subsequently, $DTF(\theta, \phi, f)$ was calculated by dividing $H(\theta, \phi, f)$ by $CTF(f)$. The DTF defined here was equivalent to the diffuse field-equalized HRTF in Equation (2.47), and similar to the ideal of the CAPZ model of HRTFs discussed in Section 5.3.5. In Sections 5.3.4 and 6.2.2, the HRTF was divided into direction-dependent and direction-independent components. In contrast to that in Equation (7.17), however, the direction-independent component in Sections 5.3.4 and 6.2.2 was defined as the average magnitude, rather than as the root mean square (RMS) magnitude of HRTFs.

Applying frequency scaling to HRTFs or DTFs at logarithmic frequency scale is a convenient approach. First, a DTF is decomposed by a band-pass filter bank that consists of 85 triangular filters. The 3-dB bandwidth of the filters is 0.057 octave, the filter slopes are 105 dB per octave, and the central frequencies are spaced in an equal interval of 0.0286 octave from 3–16 kHz. An interval of 0.0286 octave is chosen to provide intervals of 2% at linear frequency [i.e., $\gamma = 1.02$ in Equation (7.15)]. Therefore, the filters belong to a constant ratio bandwidth filter and provide suitable resolution. Frequency zooming at linear frequency scale corresponds to frequency offsetting at logarithmic frequency scale. Upward scaling (or expansion) $\gamma = 1.02$, for instance, is accomplished by offsetting each component upward by 0.0286 octave (i.e., an interval of central frequencies of adjacent filters), and upward scaling $\gamma = 1.02^2$ corresponds to twice an upward interval of central frequencies of adjacent filters. Conversely, downward scaling $\gamma = 1/1.02$ is accomplished by offsetting each component downward an interval of central frequency of adjacent filters. Thus, the logarithm of a factor and the logarithm of its inverse are symmetric around the origin.

In Middlebrooks's study, the intersubject differences between DTFs are quantified by a metric computed across 64 filter bank components from 3.7 to 12.9 kHz because of the importance of this frequency range in hearing. The metric is defined as:

$$\Delta DTF(\theta,\phi,f) = 20\log_{10}|DTF_A(\theta,\phi,f)| - 20\log_{10}|DTF_B(\theta,\phi,f)| \quad (dB), \tag{7.18}$$

where f denotes the central frequency of each of the 64 band-pass filters, DTF is calculated from the RMS of magnitudes in each passband, and subscripts A and B denote different subjects. The differences in Equation (7.18) are of two types: the differences in overall gain that are constant across frequency, and the intersubject differences that vary with frequency. To eliminate the frequency-independent component, the mean variance of $\Delta DTF(\theta, \phi, f)$ across 64 frequency ranges is calculated as:

$$\sigma^2(\theta,\phi) = \frac{1}{64}\sum_f |\Delta DTF(\theta,\phi,f) - \overline{\Delta DTF(\theta,\phi,f)}|^2 \quad (dB^2)$$

$$\overline{\Delta DTF(\theta,\phi,f)} = \frac{1}{64}\sum_f \Delta DTF(\theta,\phi,f)$$

$$(7.19)$$

Finally, variance $\sigma^2(\theta, \phi)$ is averaged across $M = 393$ tested directions, yielding:

$$\overline{\sigma^2} = \frac{1}{M}\sum_{i=0}^{M-1}\sigma^2(\theta_i,\phi_i) \quad (dB^2).$$

$$(7.20)$$

Figure 7.7 shows the intersubject spectral difference between subjects S07 and S35 as a function of the relative frequency scale factor γ; that is, γ frequency scaling is applied to the DTFs of subject S07. The figure shows that the intersubject spectral difference decreases, from 17.5 dB^2 prior to scaling, to 5.4 dB^2 when the DTFs of subject S07 are scaled upward by 1.268 (called the optimal scale factor) relative to those of subject S35, indicating a 69.1% reduction in intersubject spectral difference. Therefore, appropriate frequency scaling can reduce the overall intersubject spectral difference in DTFs.

Analyzing the results for all the subjects, the following conclusions can be drawn:

1. Although the optimal scale factor γ_{opt} that minimizes the spectral difference between subjects A and B may differ slightly for each source direction, they are similar to the optimal scale factor for the overall M source directions as calculated above.

2. The analyses of the distribution of optimal scale factors across all 990 pairs of subjects indicate median and maximum optimal scale factors of 1.061 and 1.38, respectively. Scaling subject A to B with $\gamma_{opt} \geq 1$ is equivalent to scaling subject B to A with a complementary factor $1/\gamma_{opt} \leq 1$. Thus, the optimal scale factors $\gamma_{opt} \geq 1$ is chosen in the analyses.

3. Frequency scaling reduces spectral differences by 15.5% or more for half of the 990 subject pairs, and the spectral difference decreases by more than half for 9.5% of the subject pairs.

4. The optimal scale factors γ_{opt} for the subjects' left ears are strongly correlated with those of their right ears. For the 990 intersubject pairs, the correlation coefficient between the base-2 logarithms of γ_{opt} for right and left ears is 0.95. This observation suggests that ILD spectra can be scaled in frequency, similarly to the DTFs of each ear.

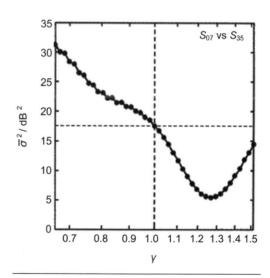

Figure 7.7 Intersubject spectral difference in sets of DTFs, from subjects S07 and S35, as a function of relative frequency scale factor γ (from Middlebrooks, 1999a, by permission of J. Acoust. Soc. Am.)

5. ITDs are produced by some of the same anatomical structures that produce the magnitude component of DTFs. A high correlation between γ_{opt} and ITD is therefore expected. The ITDs are calculated from raw HRTF data by the cross-correlation method presented in Section 3.2.1, and the maximal ITD ($|ITD|_{max}$) across M source directions are also obtained. For all 1056 pairwise combinations of 33 subjects, including both subject A/ subject B and subject B/subject A, the coefficient of correlation between $\log_2\gamma_{opt}$ and $\log_2[|ITD|_{max,A}/|ITD|_{max,B}]$ is 0.71.

Psychoacoustic experiments are also conducted to validate the perceptual influences caused by using individualized and nonindividualized DTFs. The results indicate that using nonindividualized DTFs causes localization errors that increase with the increasing difference between one subject's individualized DTF and another's DTF. This situation can be alleviated by employing frequency-scaled DTFs.

The foregoing analyses and frequency scaling were conducted with known DTFs for subjects A and B. In the customization of HRTFs or DTFs, the most important aspect is approximately estimating the unmeasured DTFs for a certain subject from the known DTFs of another. As stated in Section 7.1, Middlebrooks measured a set of anthropometric parameters for all subjects (Figure 7.2 and Table 7.2). Strong correlations are found between the logarithmic ratios of certain physical dimensions and those of optimal scale factors $\log_2\gamma_{opt}$ over all 1056 pairs of 33 subjects. The correlation coefficients between $\log_2\gamma_{opt}$ and either $\log_2(x_A/x_B)$ (x is referred to as head width in Table 7.2) or $\log_2(d_A/d_B)$ (d denotes pinna cavity height in Table 7.2) are 0.67 and 0.72, respectively. The correlation coefficients between $\log_2\gamma_{opt}$ and the other anthropometric parameters in Table 7.2 is less than 0.41, except for body height. A linear regression is performed on $\log_2(x_A/x_B)$ and $\log_2(d_A/d_B)$ with $\log_2\gamma_{opt}$, yielding:

$$\log_2\gamma_{opt} = 0.340\log_2\left(\frac{d_A}{d_B}\right) + 0.527\log_2\left(\frac{x_A}{x_B}\right). \tag{7.21}$$

If the $\log_2(x_A/x_B)$ and $\log_2(d_A/d_B)$ of subjects A and B, as well as the DTFs of subject A, are known, γ_{opt} can be estimated according to Equation (7.21). The DTFs of subject B can then be scaled from the DTFs of subject A. For the 1056 subject pairs, the scale factors estimated by this equation are highly correlated with acoustically measured optimal scale factors, with a correlation coefficient of 0.82 and an RMS difference of 0.0813 octave ($\gamma = 1.058$). In the spherical head HRTF model of Equation (7.15), the optimal scale factor γ_{opt} is evaluated from the ratio of the head radius. However, an actual head is not a perfect sphere, and pinnae also contribute to HRTFs. Therefore, optimal scale factors may be related to more than the head radius. In this sense, Equation (7.21) and corresponding frequency scaling are more effective approaches to customizing individualized HRTFs.

Frequency scaling assumes that different subjects have similar anatomical shapes with varying dimensions. This assumption is merely a rough approximation of actual conditions. As stated in Section 7.1.2, differences in anatomical shape (such as head shape) exist among different subjects because the correlation coefficients among head-relevant anthropometric parameters are not equal to unity. The differences in anatomical shape cause some specific differences in HRTF or DTF spectral features for different subjects. These differences cannot be eliminated by frequency scaling. As shown in Figure 7.7, the intersubject spectral difference σ^2 is still 5.4 dB^2 after frequency scaling. Further analyses show that a residue of spectral difference σ^2, averaging 6.2 dB^2 across all subject pairs, remains after scaling by an optimal factor. Possible components included in the residual spectral differences are the variability of experimental measurements and the specific spectral features generated by various anatomical shapes. This shortcoming is the main disadvantage of frequency scaling. Taking into account the intersubject differences in pinna orientation, Guillon et al. (2008a) reported that the customization performance of individualized HRTFs could be improved, by a combination of frequency scaling and rotation shifting.

7.3.3. Anthropometry-based Linear Regression Method

The anthropometry-based linear regression method estimates, or predicts, the individualized HRTFs of a new subject, using the linear regression relationship between HRTFs and anthropometric parameters. The linear regression relationship is derived from a baseline database that includes the HRTFs and anthropometric parameters of numerous subjects.

Given that an HRTF varies, as a complex function of source position, frequency, and individual, the statistical relationship between HRTFs and anthropometric parameters is highly complicated and cannot be described in a straightforward manner. To reduce the dimensionality of HRTFs, some basis function decomposition schemes are often applied prior to customization. As indicated in Sections 6.2 and 6.3, when a set of basis functions is specified, HRTFs are determined merely by the weights in decomposition. Accordingly, simplifying HRTF customization is possible, through the evaluation and customization of weights. In Section 7.2.2, ITD was first decomposed by azimuthal Fourier harmonics, and then the individualized weights were evaluated, or customized, from the multiple linear regression-derived relationship between weights and anthropometric parameters.

As confirmed in Section 6.2, PCA is an efficient method for HRTF decomposition, in which the HRTF at an arbitrary direction is represented by a weighted sum of a small set of spectral shape basis functions. Direction- and individual-dependent weights completely capture the characteristics of HRTFs. Wightman et al. (1993) applied the PCA decomposition in Section 6.2.3 to the log-magnitudes of the HRTFs of 15 subjects, and used the Euclidean distance between the PC-weights (the weights derived from principal component analysis) of different subjects to quantify the intersubject difference of HRTFs. Wightman et al. introduced multidimensional scaling analysis to evaluate the intersubject similarity of HRTFs.

Jin et al. (2000) first introduced PCA into the customization of individualized HRTFs. The results show that seven PCs account for 60% of the variation in individualized DTFs, yielding accurate subjective virtual source localization effects. The authors then analyzed the statistical relationship between PC-weights and anthropometric parameters using linear regression methods, yielding an anthropometry-based method for customizing individualized HRTFs. Beyond these details, however, no further explanations were offered in the study.

Similar studies were conducted by Nishino et al. (2001, 2007) and Inoue et al. (2005). In these works, the phases and magnitudes of HRTFs were separated. Under minimum-phase approximation, the excess phase component of HRTFs can be approximated as a linear-phase function and obtained by a direction-dependent, but frequency-independent, pure time delay. In practical use, the difference between left- and right-ear pure time delays is typically determined by the ITD model in Section 7.2.

However, the PCA performed by Nishino et al. slightly differs from those in Equations (6.44) to (6.51). Given that a baseline HRTF database includes the log-magnitudes of far-field HRTFs at M directions for both ears of each S subject, each HRTF magnitude is represented by its samples at N discrete frequencies. To emphasize the individuality of HRTFs, the $N \times N$ covariance matrix $[R]$ for S subjects is separately constituted for each direction and each ear, rather than for all directions and two ears. Accordingly, the mean log-magnitudes of the HRTF in Equation (6.51) are substituted by a mean across S subjects alone. The resultant PC-vectors (spectral shape basis vectors, derived from PCA) are denoted by $d_q(f, i)$, with $q = 1,2 \ldots Q$ representing the order of PC-vectors and $i = 0,1 \ldots (M-1)$ being M different directions. These PC-vectors are direction- and ear-dependent, but independent of each individual. Then, the log-magnitude of the HRTF of subject s, direction i, and a given ear are expressed as (the notation for ear has been excluded):

$$20\log_{10}|\hat{H}(\theta_i,\phi_i,f,s)| = \sum_{q=1}^{Q} w_{qi}(s)\,d_q(f,i) + d_0(f,i),\qquad(7.22)$$

where $d_0(f, i)$ represents the subject-independent but ear- and direction-dependent components in the log-magnitudes of HRTFs. For a specified direction and subject, the log-magnitude of HRTFs is determined by Q PC-weights $w_{qi}(s)$. For subjects included in the baseline database, PC-weights can be derived from known HRTFs.

Given that each S subject has L anthropometric parameters, the lth parameter of subject s is denoted by $x_l(s)$ with $l = 1,2\ldots L$ and $s = 1,2\ldots S$. Assuming that $w_{qi}(s)$ is linearly related to $x_l(s)$,

$$w_{qi}(s) = \beta_{qi,0} + \sum_{l=1}^{L} \beta_{qi,l}x_l(s).\qquad(7.23)$$

Then, coefficients $\beta_{qi,l}[q = 1,2 \ldots Q, i = 0,1 \ldots (M-1), l = 0,1\ldots L]$ are derived by applying linear regression analysis to $x_l(s)$ and known $w_{qi}(s)$.

For an arbitrary new subject taken from outside the baseline database, the individualized PC-weights $w_{qi}(s)$ and the log-magnitudes of HRTFs are predicted by substituting measured anthropometric parameters into Equation (7.23) and by using the PC-vectors derived from the baseline database. Complex-valued HRTFs can be recovered from HRTF magnitudes by minimum-phase reconstruction. This is the basic principle of the anthropometry-based linear regression method for HRTF customization.

Using the horizontal HRTFs at 72 azimuths (database no. 8 in Table 2.1) and 9 anthropometric parameters listed in Table 7.4, Nishino et al. analyzed the relationship between the log-magnitudes of HRTFs and anthropometric parameters. The HRTFs of 86 subjects were used in the analysis. Data on 82 subjects were used as baseline, and those on the remaining 4 subjects were used to test prediction performance.

The resultant coefficients $\beta_{qi,l}$ are unavailable, but some final results are given in Nishino's study. The findings indicate that if $Q = 5$ PCs are used in reconstruction, the mean spectral distortion SD_M [Equation (5.19)] of the HRTFs predicted over 72 horizontal azimuths is 4.0 dB within 0–8 kHz or 6.2 dB within 0–24 kHz.

Nishino et al. investigated only the prediction of HRTFs in the horizontal plane. However, the variances in HRTF high-frequency spectral characteristics with elevation (particularly in the median plane), which are important to individualized HRTFs, were excluded. According to Equations (7.22) and (7.23), the log-magnitude of the HRTFs of a specific subject s at azimuth θ_i is determined by certain terms, such as $x_l(s)d_q(f, i)$; that is, the product of anthropometric parameters and basis vectors. The nine relevant anthropometric parameters in Table 7.4 are in dimensions of distance. According to the discussion on the frequency scaling method in Section 7.3.2, HRTF magnitude (or log-magnitude) should be a function of the product of frequency (wave number) and anthropometric dimensions. Therefore, Equations (7.22) and (7.23) are inconsistent with the frequency scaling method's hypothesis. The aforementioned equations can be regarded as the first-order (linear) approximation in the Taylor expansion of HRTF magnitudes (or log-magnitudes), as a function of anthropometric dimensions. This approximation may cause some errors.

Similar methods were applied to the customization of the pinna-related transfer functions (PRTFs) defined in Section 4.3.3 (Rodríguez and Ramírez, 2005). Aside from pinna-related parameters, these methods include the products of some pinna-related parameters, with area and volume dimensions used in modeling.

In summary, although moderate success has been achieved with PCA (as well as other basis function linear decompositions) and anthropometry-based regression methods for HRTF or PRTF prediction, further study and improvement are still needed. In particular, because some linear decomposition methods for HRTFs (such as PCA) cannot fully represent the complex relationship between HRTFs and various variables (such as source position, frequency, and individual), some researchers introduced tensor framework representation for HRTF decomposition (Grindlay and Vasilescu, 2007).

7.4. Subjective Selection-based HRTF Customization

HRTFs can also be customized by auditory matching (Bernhard and Hugo, 2003). That is, binaural signals are synthesized using the HRTFs of different subjects in a baseline database and then reproduced via headphones. A new subject selects a set of HRTFs that generate optimal perceptual performance (in terms of localization accuracy and externalization, etc.) as the customized HRTFs.

To further improve the perceptual performance of customized HRTFs, some researchers proposed user-defined spectral manipulation (Tan and Gan, 1998). That is, the matched HRTFs perceptually selected from the baseline database are further subjected to band-pass filtering, with five parallel and parameterized band-pass filters. The new subject can tune filter parameters and change the magnitude spectra of signals to optimize perceived performance.

The optimal scale factor in the frequency scaling method presented in Section 7.3.2 can also be obtained by subjective experiments, aside from anthropometric measurements (Middlebrooks et al., 2000). That is, at some specific source directions, binaural signals are synthesized by HRTFs with various scale factors and then rendered to a new subject. The frequency scale factor that minimizes the statistical localization error for the intended source directions is selected as the optimal scale factor. The results indicate that the optimal scale factors from subjective experiments and measured DTFs are highly correlated, with a coefficient of 0.89.

Hwang et al. (2007, 2008) proposed an alternative method for HRTF customization—subjective tuning. Median plane HRIRs are first decomposed into a weighted sum of 12 PC-vectors in the time domain. The direction- and individual-independent PC-vectors are derived from the median-plane HRIRs of the CIPIC database, as stated in Section 6.2.4. The HRIRs of a new subject can then be customized, by tuning the PC-weights and by subjective listening. To simplify customization, only the preceding three PC-weights that correspond to the largest intersubject variations are tuned. The bound of tuning is set as the mean ±3 standard deviations of each PC-weight across all subjects in the CIPIC database. The remaining nine PC-weights are selected as the mean values across the subjects in the CIPIC database.

Considering the diversity of HRTFs, a baseline HRTF database should contain numerous subjects. However, applying the subjective matching method to a large database is tedious and time-consuming. Some improved schemes for subjective selection-based customization have been suggested. The basic idea behind these schemes is that a subset of representative data is identified, and then perceptually optimized HRTFs are selected from the subset by subjective evaluation. In a preliminary listening experiment, Katz and Parseihian (2012) identified seven sets of representative data from the HRTFs of 46 subjects. Their psychoacoustic experiment confirms that the

HRTFs perceptually selected from the seven sets of representative data somewhat improve localization performance.

Some similarities exist among the HRTFs of different individuals. Aside from spatial clustering tendencies (Section 6.2.6), some individual clustering tendencies have been observed in existing HRTF databases (So et al., 2010; Xie and Zhong, 2012d). These results are helpful for selecting representative HRTF subsets. Using spectral features that are important to simulating front and rear directional sounds, So et al. (2010) clustered 196 HRTFs of different individuals and ears (including 43 humans from CIPIC, a KEMAR artificial head, and 51 humans from the IRCAM lab) into 6 different orthogonal groups. The central HRTFs of each group were selected as the representative dataset. The listening results show that, unlike using the HRTFs of the KEMAR artificial head, employing representative HRTFs noticeably decreases back-front confusion rates and moderately reduces front-back confusion rates in headphone representation.

7.5. Notes on Individualized HRTF Customization

The discussion in Sections 7.3 and 7.4 focused on anthropometry-based and subjective selection-based HRTF customization methods. Although moderate success has been achieved, the performance of customized HRTFs remains inferior to that of measured or calculated HRTFs—in terms of accuracy. Some common problems exist in current research on HRTF customization.

In anthropometry-based customization, the results for measured anthropometric parameters are closely related to measurement points. Therefore, deviations may exist among various studies because of the differences in measurement points. Prior to customizing HRTFs by anthropometric parameters, formulating a clear definition of measurement points is essential. For a specific anthropometric parameter, if the measurement errors caused by the difference in measurement point selection are greater than, or equal to, actual intersubject differences, the customized HRTFs will be meaningless. Nonetheless, this possible error source has been only fractionally considered in some studies and publications. Although subjective selection-based HRTF customization is free of the errors caused by differences in the measurement points of anthropometric parameters, it is time-consuming and less accurate, with a significant degree of variance in perceptual evaluation (Schönstein and Katz, 2012).

As stated in the preamble of this chapter, individualized HRTFs can be customized from a small set of individualized HRTF measurements. In addition to the customization method discussed in Section 6.3.5, a small-measurement matching method was proposed by Fontana et al. (2006). In this method, optimally matched HRTFs for full spatial directions are selected from a baseline HRTF database by minimizing the global spectral difference, between candidate and measured individualized HRTFs, at a considerably small set of source directions. Let $H(\theta_i, \phi_i, f_k)$ with $i = 0$, $1 \ldots (M' - 1)$ denote the N-point discrete HRTF in the frequency domain at M' source directions ($M' = 6$ in Fontana's study), and $H_s(\theta_i, \phi_i, f_k)$ denote the corresponding HRTF of the sth subject in the baseline HRTF database. Then, the frequency-weighted RMS error is:

$$\varepsilon_s(\theta_i, \phi_i) = \frac{1}{N}\sqrt{\sum_{k=0}^{N-1} W(f_k)\,|\,H(\theta_i, \phi_i, f_k) - H_s(\theta_i, \phi_i, f_k)\,|^2}, \qquad (7.24)$$

where $W(f_k)$ is the frequency-dependent weight, and $W(f_k) = 1$ denotes a constant weight for all. The mean $\varepsilon_s(\theta_i, \phi_i)$ across M' source directions is used as a measure that represents the overall difference between the new subject and the sth subject in the baseline database. The calculation in

Equation (7.24) is separately applied to each subject in the baseline database. The HRTF of a specific subject in the baseline database that minimizes the mean $\varepsilon_s(\theta_i, \phi_i)$ is selected as the customized HRTF of the new subject. Small-measurement-based HRTF customization avoids the problems encountered in anthropometric measurement and subjective selection, but still requires HRTF measurement. Moreover, the measurement conditions for the new subject should be identical to those for the baseline database; otherwise, large errors may occur. The manner for selecting matching source directions requires further study.

Additionally, HRTF customization requires a baseline HRTF database that contains an adequate number of subjects. Even more important, the anthropometric parameters associated with HRTF data are required in anthropometry-based matching. Among the HRTF databases in the public domain, the CIPIC database is commonly used; it features a high directional resolution (1250 measured directions per subject), 43 subjects (27 males and 16 females), and 27 anatomical parameters (Table 2.1). Most existing studies related to HRTF customization were conducted using this database. The CIPIC database was primarily compiled from measurements on Western subjects. A statistically significant difference in certain anthropometric parameters exists between Western and Chinese subjects. Therefore, directly customizing individualized HRTFs of Chinese subjects with the CIPIC database, and vice versa, is logistically problematic. Therefore, the baseline database used in HRTF customization should represent the span of human beings; that is, it should satisfy diversity in constitution. Alternatively, different baselines for different populations are needed. One method for constructing a baseline database with large individualized samples is to combine the measured data from different public databases. However, different conditions, such as spatial coordinates, measurement methods, sampling frequencies, HRTF lengths, reference points, and equalization methods, were used to construct different databases. Thus, appropriate standardization procedures should be applied to all public databases to guarantee uniformity of combined sets (Andreopoulou and Roginska, 2011).

The manner by which customization performance is appropriately assessed is another important issue. Customization assessment generally includes physical assessment, with the various error criteria and analyses presented in Section 5.1, as well as subjective assessment by psychoacoustic experiment. As indicated by state-of-the-art HRTF measurement, the mean measurement error of HRTF magnitudes across directions and frequencies is on the order of 1–2 dB below 6 kHz; this error increases to 3–5 dB in the range 6–10 kHz (Section 2.2.9). For even higher frequencies, the error becomes more prominent. However, some studies tend to underestimate measurement errors and regard a reduction of 1–2 dB or less in calculated mean customization error as an obvious improvement, even without statistical testing. A reduction of mean customization errors of less than 1 dB is unlikely to cause substantial improvement in auditory perception. Moreover, some studies evaluated customization errors in the range 0 Hz to the Nyquist frequency (e.g., 22.05 kHz at a sampling frequency of 44.1 kHz). Limited by the response of practical HRTF measurement systems (such as loudspeakers), measured HRTFs are less accurate at very low (below 0.1–0.2 kHz) and very high frequencies (above 20 kHz). Therefore, the frequency ranges of inaccurate measurements should be excluded from customization error evaluation. Considering the limited frequency resolution of human hearing, we suggest introducing some preprocessing procedures, such as auditory filter bandwidth smoothing, prior to HRTF customization. Moreover, customization errors should be calculated in the frequency range that is of perceptual importance and include few measurement errors from the original HRTFs (i.e., from hundreds of Hz to 12–16 kHz).

For subjective assessment, corresponding experimental procedures and statistical analyses should be standardized. Specific attention should be paid to headphone equalization. These issues will be comprehensively discussed in Chapters 8 and 13, respectively.

Finally, these discussions are applicable to static VAD applications, without the dynamic localization cues introduced by head movement. Thus, individualized HRTFs with accurate high-frequency spectral cues are a compelling requirement. In the dynamic VADs to be presented in Chapter 12, some dynamic cues captured by head tracking can partially alleviate the dependence of high-frequency spectral cues on individualized HRTFs. Thus, a relatively large customization error of individualized HRTFs is tolerable in dynamic VADs.

7.6. Structural Model of HRTFs

7.6.1. Basic Idea and Components of the Structural Model

The structural model is an effective representation and simplification of HRTFs. HRTFs are acoustical transfer functions of a linear-time-invariant (LTI) system and reflect the overall filtering effect of anatomical structures, including the head, pinnae, and torso. HRTFs can be modeled by various filter structures (Chapter 5). If the effect of each anatomical component can be isolated, the contribution of each component is individually described by a parameterized substructure or model, and the composition of all substructures represents the entire HRTF or HRIR. In addition, if each substructure can be implemented by simple delays and filters, the entire HRTF is realized by a simplified model. Individualized HRTFs are realized by adjusting the parameters of each substructure to accommodate variations among the corresponding anatomy of different individuals. This is the basic idea of the structural model of HRTFs.

Genuit (1984) proposed an early structural model of HRTFs. Thereafter, Algazi, Duda, Brown and other researchers collaborated to improve and complete the model. Figure 7.8 shows the structural model of HRTFs described by Brown and Duda (1998). The effects of anatomical structures are classified into three: effects of the head model, torso model, and pinna model.

The head model accounts for the transmission delay around the head and the diffraction/shadow effect caused by the head. For a spherical head with radius a and an incident plane wave, the propagation delay at a receiving point on the head surface, relative to that at the head center with the head absent, is evaluated similarly to the evaluation in Woodworth's formula in Sections 1.4.1 and 7.2.1:

$$\Delta T_H(\Gamma) = \begin{cases} -\dfrac{a}{c}\cos\Gamma & 0 \le |\Gamma| < \dfrac{\pi}{2} \\ \dfrac{a}{c}\left(|\Gamma| - \dfrac{\pi}{2}\right) & \dfrac{\pi}{2} \le |\Gamma| < \pi \end{cases}, \qquad (7.25)$$

where Γ is the angle between incident direction (the ray from sphere center to source) and the ray from the sphere center to the receiving point on the sphere surface (Figure 4.1). The propagation delay provided by Equation (7.25) corresponds to the linear-phase component of HRTFs and roughly corresponds to the onset delays of an HRIR evaluated by leading-edge detection. To ensure causality in practical implementation, a linear delay of a/c should be supplemented to Equation (7.25).

The logarithmic HRTF magnitudes of a rigid-spherical head model for an incident plane wave are shown in Figure 4.2 (Section 4.1). The logarithmic magnitudes approach 0 dB at low frequencies and all directions. At ipsilateral incident directions, HRTF magnitude increases with rising frequency. At incident direction that exactly faces the receiving point, high-frequency magnitude increases to 6 dB over low-frequency magnitude. Meanwhile, the head shadow effect causes the high-frequency magnitude to roughly roll off, similar to a low-pass filter, with maximal attenuation

Input

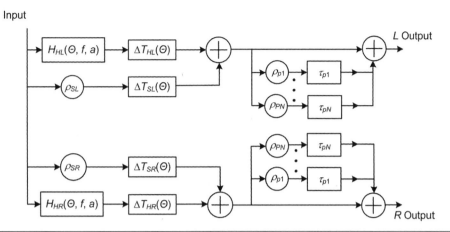

Figure 7.8 Structural model of HRTFs (adapted from Brown and Duda, 1998)

occurring around $\Gamma = 150°$. In the structural model, these features of spherical head HRTFs are approximated by a single-zero and single-pole filter as

$$H_H(\Gamma, f, a) = \frac{2\omega'_0 + j\alpha(\Gamma)\omega'}{2\omega'_0 + j\omega'} \qquad 0 \le \alpha(\Gamma) \le 2, \tag{7.26}$$

where the transfer function of the filter is expressed in the analog domain for convenience. In practice, the filter is realized in the digital domain. $\omega' = 2\pi f$ is the analog angular frequency and $\omega'_0 = c/a$, with c as the speed of sound. Given the head radius a, the pole location is specified. Parameter $\alpha(\Gamma)$ is related to source direction. It controls the zero location and determines the transfer response of the filter. $\alpha(\Gamma) = 2$ corresponds to a 6-dB increase at high frequencies, and $\alpha(\Gamma) < 1$ corresponds to high-frequency attenuation. Therefore, the directional variation in HRTF magnitude is realized by changing the zero location while keeping the pole intact. This characteristic is similar to the CAPZ model of HRTFs described in Section 5.3.5 and is beneficial to simulating moving virtual sources. Parameter $\alpha(\Gamma)$ is selected as:

$$\alpha(\Gamma) = \left(1 + \frac{\alpha_{min}}{2}\right) + \left(1 - \frac{\alpha_{min}}{2}\right)\cos\left(\frac{\Gamma}{\Gamma_{min}}180°\right). \tag{7.27}$$

When we select $\alpha_{min} = 0.1$ and $\Gamma_{min} = 150°$, the magnitude response in Equation (7.26) relatively matches those in Figure 4.2.

The filter in Equation (7.26) also provides the appropriate phase responses of HRTFs at low frequencies. Under minimum-phase approximation, the phase of an HRTF consists of two components: linear and minimum phases. The propagation delay introduced by Equation (7.25) accounts for the contribution of the linear-phase component. The minimum-phase component increases the interaural phase delay difference (ITD_p) at low frequencies (Section 4.1.2). The phase characteristics of Equation (7.26) only account for the contribution of a minimum-phase component to an HRTF.

In Equations (7.25) and (7.27), Γ denotes the source direction, with respect to the receiving point (concerned ear). In the interaural polar coordinate shown in Figure 1.3, source direction is specified by (Θ, Φ). For two ears, diametrically opposed on a spherical head, angle Γ is given by:

$$\Gamma_L = 90° + \Theta \qquad \Gamma_R = 90° - \Theta. \tag{7.28}$$

For offset ear locations, given the source direction (Θ, Φ) and ear positions (Θ_L, Φ_L), (Θ_R, Φ_R), Γ_L and Γ_R can be identified. Subsequently, the ΔT_H in Equation (7.25) and H_H in Equation (7.26) can be calculated for the left and right ears, respectively, and the results are denoted by $\Delta T_{HL}(\Theta)$, $\Delta T_{HR}(\Theta)$, $H_{HL}(\Theta, f, a)$, and $H_{HR}(\Theta, f, a)$. For simplicity, source or ear position is denoted by a single notation, Θ. Generally, it should be denoted by (Θ, Φ). The interaural polar coordinate can be converted into the default coordinate (θ, ϕ) of this book. For convenience, however, the interaural polar coordinate is still used in the remainder of this section. In addition, in the spherical head model, individualized HRTFs can be obtained by changing four sets of parameters, namely, a, α_{min}, and Γ_{min}, as well as (Θ_L, Φ_L) and (Θ_R, Φ_R).

In Figure 7.8, the contribution of the torso is described by the shoulder reflection (echo) model. Shoulder reflections are characterized by the reflection coefficients ρ_{SL} and ρ_{SR}, as well as by the delays $\Delta T_{SL}(\Theta)$ and $\Delta T_{SR}(\Theta)$ for the left and right sides, respectively. These physical parameters are relevant to source direction. Strictly speaking, the diffracted/reflected sounds generated by the head and shoulder should pass though two different pinna models because they arrive at the ear canal from different directions. This approach, however, results in a complex model. In the preliminary work of Brown et al., shoulder reflection is discarded because of its nondominant contribution to localization.

The effects of pinnae include reflection, scattering, and resonance; such effects are, therefore, complex. A series of six reflections are used to simulate the multiple reflections within pinnae. The nth reflection is characterized by the corresponding reflection coefficient ρ_{pn} and delay τ_{pn}. Informal listening tests indicate that the values of the reflection coefficients are not critical, and τ_{pn} depends on source direction. For sources at the frontal-hemispherical space, the empirical formula for τ_{pn} is:

$$\tau_{pn}(\Theta, \Phi) = A_n \cos\left(\frac{\Theta}{2}\right)\sin[D_n(90° - \Phi)] + B_n$$

$$-90° \leq \Theta \leq 90° \qquad -90° \leq \Phi \leq 90°, \tag{7.29}$$

where A_n and B_n are the constant amplitude and offset across subjects, respectively; and D_n is a scaling factor that should be adapted to different individuals. The study of Brown et al. lists the values of all the parameters in Equation (7.29). It also discusses the implementation of the structural model and provides the preliminary validation of psychoacoustic experiments.

7.6.2. Discussion and Improvements of the Structural Model

The contribution of each anatomical component to HRTFs is isolated in the structural model, and each component in the model is represented by a simple parameterized submodel. Individualized HRTFs are modeled by changing the parameters in the submodels. From the perspective of application, this model efficiently implements HRTF signal processing. However, it employs considerable simplification of sound transmission and scattering, thereby reducing model accuracy.

The effect of each anatomical component is considered separately in the structural model. Strictly speaking, however, the effect of each anatomical component on incident sound waves cannot be isolated because of the interactions among the waves scattered/reflected/diffracted from all anatomical components. The structural model can only be valid when the higher-order interactions among the contributions of all anatomical components are negligible. Fortunately, as stated in Appendix B, retaining the first-order, while discarding the higher-order, reflections/scattering between the head and torso yields an appropriate approximation of far-field HRTFs.

Algazi et al. (2001d) measured the responses of an isolated pinna model, generating the pinna-related transfer functions (PRTFs) described in Section 4.3.3. The principle of PRTF measurement is similar to that of HRTFs. The pinna is placed on a circular plate surrounded by a rectangular table, to approximate a pinna mounted on an infinite acoustical baffle. A microphone is placed in the ear canal of the pinna. The PRTFs at various source directions are measured, by changing the relative position between source (small loudspeaker) and pinna. The results indicate that the responses of the HRTFs (cascading the responses in the frequency domain) composed from KEMAR without pinnae and measured PRTFs are close to those of the measured HRTFs of KEMAR with pinnae. This experiment therefore confirms that the simple composition of a head/torso and pinna model yields reasonable results.

Similar to the pinna model suggested by Batteau (Section 1.4.4), the structural model used by Faller et al. (2005, 2010) simulates HRTFs and decomposes pinna responses into a summation of damped and delayed sinusoids. Individualized HRTFs can be customized by appropriately selecting the initial duration and length of the time window, as well as the amplitude, frequency, phase, and damping factor of the sinusoids. Satarzadeh et al. (2007) used some second-order band-pass filters, to model the resonance of the external ear, and used reflections with delay [similar to Equation (1.13)] to model high-frequency pinna notches. In most cases, model parameters and PRTFs can be evaluated by anthropometric measurements on pinnae.

As stated in Section 7.6.1, shoulder reflection was discarded for Brown's original model. Subsequent psychoacoustic experiments show that introducing shoulder reflections into the structural model is helpful to vertical localization below 3 kHz, for sources outside the median plane (Avendano, 1999). In a later version of the structural model (Algazi et al., 2002b), the effects of the head and torso were simulated by the snowman model with the spherical head radius a_A and spherical torso radius a_B (Section 4.2). Depending on source direction, the effects of the torso are classified into two groups: torso reflection and torso shadow (Figure 4.10). Accordingly, the torso model in Figure 4.9 alternates, between torso reflection submodels and torso shadow submodels, in accordance with the source that lies outside or inside the region of the torso shadow (only one ear is illustrated in the model).

Figure 7.9(b) shows the torso reflection submodel. The submodel involves two components that are superimposed at the ear, the direct component and the torso reflection component. The direct component represents the sound waves directly arriving at the head, and is described by the propagation delay $\Delta T_H(\Theta)$ as well as a single-zero and single-pole head filter $H_H(\Theta, f, a_A)$. The delay and filter are specified by Equations (7.25) to (7.27). In contrast to the elements in Figure 7.8, aside from being characterized by the torso reflection coefficient ρ and torso propagation delay $\Delta T_T(\Theta)$, torso reflection components are also subjected to the head filter $H_H(\Theta', f, a_A)$ and propagation delay $\Delta T_H(\Theta')$ that is generated by the head. The directional parameters in the head filter and delay $\Delta T_H(\Theta')$ for torso reflection differ from those for the direct component because torso reflections and the direct component arrive at the head from different directions.

Figure 7.9(c) shows the torso shadow submodel, which consists of the torso (shadow) filter $H_T(\Theta, f, a_B)$, head filter $H_H(\Theta_1, f, a_A)$, and overall propagation delay $\Delta T_H(\Theta_2)$. The torso filter and head filter are similar to those in Equation (7.26), but with different incident directions. For details, refer to Algazi et al. (2002b).

The contributions of only the head and torso are included in the model shown in Figure 7.9. Similar to that in Figure 7.8, the pinna response can be cascaded into the head and torso models in Figure 7.9. In practice, similar to the case of Equation (7.14), the structural model in Figure 7.9 is used to simulate low-frequency HRTFs, which crossfade to the HRTFs measured at high

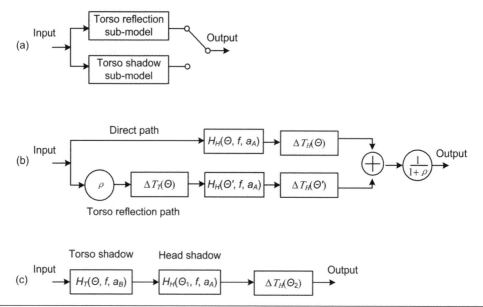

Figure 7.9 Structural model of the head and torso derived from the snowman model (adapted from Algazi et al., 2002b)

frequencies, thereby retaining individualized spectral features. At the same time, the structural model resolves the deficiency of the HRTFs measured at low frequencies.

Compared with the model in Figure 7.8, that in Figure 7.9 is closer to the actual course of scattering/diffraction by anatomical structures. The latter is, therefore, more accurate, but also more complex. The requirements for accuracy and simplicity are often incompatible with this model.

7.7. Summary

HRTFs are physical variables with prominent individuality. Developing customization methods, beyond measurement and computation, for approximately obtaining individualized HRTFs is necessary because of the complicated measurement apparatuses and procedures or potentially huge computational costs involved in obtaining individualized HRTFs. Given the complexity of HRTFs, some feature extraction or dimensional reduction algorithms (such as principal component analysis, minimum-phase approximation, and magnitude smoothing) are often applied, prior to customization, to reduce data and, thereby, simplify analysis.

HRTFs are closely related to anatomical structures and dimensions, and can therefore be estimated or customized from anthropometric parameters. Ideally, the involved anthropometric parameters should be a set of complete and inter-independent parameters indispensable to HRTFs; such parameters, however, have not been completely identified. The anthropometric parameters measured and analyzed in existing studies were primarily chosen according to experience and vary from study to study. Among different studies, some anthropometric parameters with identical or similar terms may denote different anthropometric dimensions; or, conversely, some anthropometric parameters with different terms may refer to the same anthropometric dimensions. Thus, the results of different studies cannot be compared, potentially giving rise to confusion.

Measurements in international and Chinese studies have also confirmed the significant gender differences among many head- and face-related parameters.

Statistical analyses confirm that a certain intercorrelation exists among the anthropometric parameters used in existing HRTF studies, and that high-degree correlations are observed between some anthropometric parameters and the directional localization cues contained in HRTFs. For example, many studies reported a strong correlation between interaural time difference (ITD) and head width; the gender difference in mean maximal values of ITD is mainly attributed to the gender difference in mean head dimensions. These results also facilitate the establishment of individualized ITD models that include anthropometric parameters.

Woodworth's ITD formula, based on the spherical head model, is the simplest and most commonly used interaural time difference formula. However, it is merely a rough approximation because the human head is not a perfect sphere. As an improvement, the head radius in Woodworth's formula was replaced by the weighted mean of head width, head height, and head depth. An ellipsoid head model was also employed to calculate ITD. The ITD model recently proposed, on the basis of azimuthal Fourier analysis, can completely reflect the spatial left-right symmetry and front-back asymmetry of ITD, through which individualized ITD are conveniently estimated from three head- and pinna-relevant parameters.

The methods for customizing individualized HRTFs can be divided into three categories: anthropometry-based methods, subjective selection-based methods, and small measurement-based methods. Anthropometry-based HRTF customization includes anthropometric parameter matching, frequency scaling, and linear regression. In anthropometric parameter matching, individualized HRTFs are approximated by selecting the best match from baseline anthropometric parameters and HRTF databases.

The frequency scaling method hypothesizes that different subjects have similar anatomical shapes with varying dimensions, leading to similar spectral features of HRTFs at different frequencies. Therefore, the unknown HRTF of a new subject can be estimated from the known HRTF of another subject by frequency scaling. The optimal scale factor can be derived from the ratios of anthropometric parameters. However, differences in anatomical shape generate specific differences in HRTF and spectral features of directional transfer function for different subjects. Such differences cannot be eliminated by frequency scaling.

In the linear regression method, the log-magnitude of HRTFs is first decomposed by principal component analysis to simplify data, and then linear regression analysis is applied to PC-weights and anthropometric parameters. Finally, individualized HRTFs are estimated by substituting the measured anthropometric parameters into the derived linear regression equation. In essence, this method decomposes the log-magnitude of HRTFs into a Taylor series, with respect to anthropometric parameters. Subsequently, it takes the first-order expansion component as an approximation. This low-order approximation inevitably causes certain errors.

Anthropometry-based HRTF customization is relatively simple, but suffers from measurement errors of anthropometric parameters. By contrast, subjective selection-based HRTF customization avoids anthropometric parameter measurement, but is time-consuming and less accurate. All these problems are avoided by small-measurement-based HRTF customization. However, a specific apparatus is needed for this approach.

Generally, moderate success has been achieved in HRTF customization, but performance accuracy remains inferior to HRTF measurement and computation. Common problems, such as the diversity and representativeness of the baseline database, appropriate assessment of customization performance, and standardization of assessment paradigms in psychoacoustic experiments, exist in current research.

The structural model of HRTFs isolates the effect of different anatomical components on HRTFs. The contribution of each anatomical component is represented by a simple and parameterized substructure or model. The composition of all substructures represents an entire HRTF. The parameter of each submodel can be adjusted according to a subject's anthropometric parameters. Therefore, the structural model simplifies HRTF signal processing and provides an effective way to customize individualized HRTFs. However, it implements considerable simplification of sound transmission and scattering, which may reduce model accuracy. Some improvements to model accuracy can be implemented, making the model complex. The requirements for accuracy and simplicity are often incompatible with this model.

8

Binaural Reproduction through Headphones

> *This chapter deals with issues concerning binaural reproduction through headphones. The principle and method of headphone-to-ear-canal transmission equalization are presented. Individuality and repeatability of headphone-to-ear-canal transfer functions are explored. Some problems with headphone reproduction, including perceived directional errors, in-head localization, and control of the perceived virtual source distance, are also addressed.*

Binaural signals, which are suitable for headphone presentation, are signals obtained by binaural recording or synthesis. As stated in Sections 1.8.2 and 1.8.3, equalizing (or compensating for) headphone-to-ear canal transfer functions (HpTFs) is necessary in headphone reproduction. Some problems, such as errors in virtual source directions and in-head localization, may also arise in headphone reproduction. This chapter addresses these issues in detail. Section 8.1 presents the underlying principle of headphone equalization. The measurement, individuality, and repeatability of HpTFs are discussed in Section 8.2. Section 8.3 investigates the possible origin and solution of the perceived directional errors associated with headphone-based binaural reproduction, especially front-back confusion and error in elevation directions. The problems regarding in-head localization and control of the perceived distance of virtual sound sources are investigated in Section 8.4.

8.1. Equalization of the Characteristics of Headphone-to-Ear Canal Transmission

8.1.1. Principle of Headphone Equalization

Møller (1992) comprehensively analyzed the characteristics of headphone-to-ear canal transmission and relevant equalization. In the current work, the subscripts that denote the left and right ears are excluded for simplification.

The acoustical transmission processes, from a headphone to a listener's eardrum, are characterized by the headphone response, as well as by the acoustical coupling between the headphone and the external ear. The corresponding analogue model is shown in Figure 8.1. According to Thevenin's theorem, the acoustical characteristics in headphone reproduction are specified by the

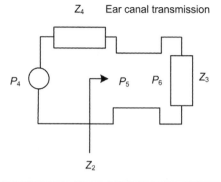

Figure 8.1 Analogue model of headphone-external ear coupling, as well as sound transmission in the ear canal (adapted from Møller, 1992)

radiation impedance Z_4 and the "open circuit" pressure P_4 (Figure 1.15). Z_4 is the impedance observed, outward at the entrance of the ear canal, and it includes a possible influence of volume enclosed by a circumaural headphone, as well as by the contribution of the electrical, mechanical, and acoustical transfer characteristics of the headphone (all have been transferred to the acoustical side); P_4 is the "open" pressure at the entrance to the ear canal with zero volume velocity ("current"), which does not exist in actual cases but can be determined if the ear canal is physically blocked (Møller, 1992). The open-circuit pressure P_4 is related to the actual pressure P_5 at the entrance to the ear canal as:

$$\frac{P_5}{P_4} = \frac{Z_2}{Z_2 + Z_4} \tag{8.1}$$

where Z_2 is the impedance observed inward (into the ear canal). The relationship of P_6 (the pressure at the eardrum) with P_4 and P_5 is expressed by:

$$\frac{P_6}{P_5} = \frac{Z_3}{Z_2} \qquad \frac{P_6}{P_4} = \frac{Z_3}{Z_2 + Z_4} \tag{8.2}$$

where Z_3 is the impedance of the eardrum.

The binaural signal E_x can be obtained by recording at a specific reference point x along the ear canal entrance to the eardrum or by filtering with the head-related transfer functions (HRTFs) obtained at the same reference point (Section 1.5). It is proportional to the pressure P_x at the reference point generated by a sound source:

$$E_x = M_1 P_x, \tag{8.3}$$

where M_1 is the response of the recording microphone. P_x refers to P_1, P_2, and P_3 (Figure 1.15) when the entrance of the blocked ear canal, the entrance of the open ear canal, and the eardrum, respectively, are chosen as the measurement reference points.

The P_6 at the listener's eardrum is generated by the binaural signal E_x with the use of headphones. In desired sound reproduction, this pressure should be identical to P_3 (which is caused by a real sound source) at the eardrum in Figure 1.15. That is,

$$P_6 = P_3. \tag{8.4}$$

However, even when the anatomical difference between the listener and the head, employed in binaural recording or synthesis is neglected, directly reproducing E_x using headphones cannot guarantee Equation (8.4) because of the nonideal acoustical transmission characteristics from headphone to eardrum, as well as the nonideal response of the recording microphone. As stated in Sections 1.8.2 and 1.8.3, E_x should be equalized before being sent to a headphone. Let $F = F(f)$ denote the transfer function of the equalization filter. The actual signal sent to the headphone is:

$$E = FE_x. \tag{8.5}$$

According to Equations (8.3), (8.4), and (8.5), the pressure P_6 at the listener's eardrum can be expressed as:

$$P_6 = \frac{P_6}{E}E = \frac{P_6}{E}FM_1P_x = \frac{P_6}{P'_x}\frac{P'_x}{E}FM_1P_x,$$

where P'_x refers to the pressure at the reference point x in headphone reproduction. Ideally, P_6 in this equation should satisfy Equation (8.4):

$$\frac{P_6}{P'_x}\frac{P'_x}{E}FM_1P_x = P_3 = \frac{P_3}{P_x}P_x. \tag{8.6}$$

When a given point from the open ear canal to the eardrum is selected as the reference point x, we obtain $P_6/P'_x = P_3/P_x$ because of the one-dimensional sound transmission in the ear canal (Figures 1.15 and 8.1). Then,

$$F = F(f) = \frac{1}{M_1(P'_x/E)}. \tag{8.7}$$

The situation becomes complex when the entrance of the blocked ear canal is selected as the reference point. Taking $P_x = P_1$ in Figure 1.15 and $P'_x = P_4$ in Figure 8.1, substituting Equations (1.17) and (8.2) into Equation (8.6) yields:

$$F = F(f) = \frac{Z_2 + Z_4}{Z_1 + Z_2}\frac{1}{M_1(P'_x/E)}. \tag{8.8}$$

Equation (8.8) involves the radiation impedance Z_1 observed outward, from the entrance of the ear canal to the free air, the impedance Z_2 of the ear canal, and the radiation impedance Z_4 observed outward, at the entrance of the ear canal, in the case of headphone reproduction. Measuring the aforementioned impedances is slightly complicated and difficult. Fortunately, if the impedances satisfy either one of these conditions—(1) $Z_1 \ll Z_2$, $Z_4 \ll Z_2$, (2) $Z_1 \approx Z_4$—then, Equation (8.8) can be simplified. However, actual measurements reveal that condition (1) is invalid at some frequencies, whereas Z_1 is small below 1 kHz, satisfying $Z_1 \ll Z_2$. If the following condition is satisfied,

$$Z_1 \approx Z_4 \quad and \quad Z_4 \ll Z_2 \quad (f < 1\,kHz), \tag{8.9}$$

then Equation (8.8) is simplified as:

$$F = F(f) = \frac{1}{M_1(P'_x/E)} = \frac{1}{M_1(P_4/E)}. \tag{8.10}$$

The headphone that satisfies Equation (8.9) is called an open headphone because the headphone positioned at some distance from the ear does not disturb the radiation impedance observed at the entrance of the ear canal. This definition differs from that of the commercial open headphone, which enables sounds from outside to be heard. To avoid ambiguity, the headphone that satisfies Equation (8.9) is called the headphone with free-air equivalent coupling to the ear, or FEC-headphone.

Møller et al. (1995a) measured the pressure division ratio (PDR) P_5/P_4 (Figure 8.1) in headphone reproduction, as well as the pressure division ratio P_2/P_1 (Figure 1.15) of 14 headphones and 40 human subjects. The authors then calculated the impedance ratio in Equation (8.8) thus:

$$PDR = \frac{Z_2 + Z_4}{Z_1 + Z_2}.$$

(8.11)

The results indicate that the *PDR* of the 14 headphones is close to unity (i.e., 0 dB) at frequencies of up to 2 kHz. Above 2 kHz, the *PDR*s of all the headphones, except one, deviate from 0 dB on the order of 2–4 dB. The measurements above 7 kHz are unreliable. In practical uses, whether a headphone can be considered an FEC-headphone depends upon acceptable error.

As indicated in Equations (8.7), (8.8), and (8.10), the characteristics of the headphone equalization filter are related to P'_x/E, which refers to the characteristics of transmission from the electric input signal of the headphone to the pressure at the reference point x in the ear canal. Hence, P'_x/E is called the *headphone-to-ear-canal transfer function* (*HpTF*) and denoted by $Hp(f)$:

$$Hp(f) = \frac{P'_x}{E}.$$

(8.12)

In particular, when the position of the eardrum is selected as the reference point x, $Hp(f)$ is called the headphone-to-eardrum transfer function (HETF).

Given that $Hp(f)$ and the response M_1 of a microphone in binaural recording (or a microphone in HRTF measurements) are known, headphone equalization can be implemented according to Equations (8.7), (8.8), and (8.10). If the microphone has an ideal transmission response, then M_1 in the aforementioned equations can be disregarded. In the practical measurement of $Hp(f)$, the same microphone as that employed for binaural recording or HRTF measurements can be used. In this case, substituting the microphone output $E'_x = M_1 P'_x$ into Equation (8.7) or Equation (8.10) yields:

$$F = F(f) = \frac{1}{E'_x/E}.$$

(8.13)

Here, the final result is irrelevant to the transmission response of the microphone. Therefore, the influence of M_1 can be cancelled, using either a microphone with an ideal transmission response or the same microphone used for binaural recording (or HRTF measurements). Then, Equations (8.7) and (8.10) become:

$$F = F(f) = \frac{1}{Hp(f)}.$$

(8.14)

Some notes are worth mentioning:

1. The disturbance imposed by the microphone on ear canal pressure is disregarded in these analyses. This condition is allowable in blocked ear canal measurements, in measurements with probe microphones, or in measurements with microphones positioned at the end of an ear canal simulator. Generally, however, the influence of such a disturbance should be calibrated.
2. If the transmission characteristics of the left and right ears, or left and right microphones differ, then headphone equalization should be separately applied to the left and right ears.

3. As indicated in Equation (8.14), $F(f)$ is the inverse function of $Hp(f)$. When $Hp(f)$ is a minimum-phase function, it is invertible and the resultant $F(f)$ is causal. Otherwise, some time delays should be incorporated into $F(f)$, or the inverse of the minimum-phase version of $Hp(f)$ should be adopted to ensure the causality of the resultant equalization filter. Some researchers also introduced the Wiener filter when deriving the inverse of $Hp(f)$ (Kim and Choi, 2005).

8.1.2. Free-field and Diffuse-field Equalization

As presented in Section 8.1.1, the binaural signals for headphone presentation are equalized by the inverse transfer function, from the headphone to a reference point in the ear canal. In binaural playback and virtual auditory displays (VADs), equalization processing is also often implemented, using the inverse transfer function from a specific sound field to a reference point in the ear canal (Møller, 1992; Larcher et al., 1998). The commonly used reference sound fields are the free field and the diffuse field, which correspond to free-field equalization and diffuse-field equalization, respectively.

Free-field equalization is implemented for a free-field transfer function $H(\theta_0, \phi_0, f) = P_x/P_0$ that is measured from a specific far- and free-field sound source to a reference point x (identical to that in binaural recording or HRTF measurements) in the ear canal:

$$F_{free,1}(f) = \frac{1}{M_1(P_x/P_0)} = \frac{1}{M_1 H(\theta_0, \phi_0, f)}, \tag{8.15}$$

where P_0 denotes the free-field sound pressure at the head center with the head absent. Thus, $H(\theta_0, \phi_0, f)$ denotes the HRTF related to the reference point x at source direction (θ_0, ϕ_0), and the direct front $(\theta_0 = 0°, \phi_0 = 0°)$ is usually selected as the reference direction. M_1 denotes the microphone response.

For binaural recording with an artificial head, the head whose output is equalized by Equation (8.15) is called the free field-equalized artificial head. The HRTF involved in a free field-equalized artificial head is just the free field-equalized HRTF in Equation (2.46).

Free-field equalization has the following characteristics:

1. The eardrum is usually selected as the reference point. Given that the reference point in binaural recording (or HRTF measurements) is identical to that in free-field equalization, the result of free-field equalization is independent of the reference point because transmission characteristics, from the ear canal entrance to the eardrum, are one-dimensional. Free-field equalization can eliminate the influence of the transmission response M_1 of the microphone used in binaural recording or HRTF measurements.
2. Free-field equalization enables the binaural signals obtained by binaural recording or HRTF-based synthesis to have a flat frequency response at a certain direction (θ_0, ϕ_0).

Diffuse-field equalization is implemented with respect to the root mean square (RMS) of the HRTF magnitudes across all source directions:

$$F_{diff,1}(f) = \frac{1}{M_1 \sqrt{\dfrac{1}{M} \displaystyle\sum_{i=0}^{M-1} |H(\theta_i, \phi_i, f)|^2}}, \tag{8.16}$$

Given the uniform directional distribution of energy density and the random distribution of phase in the diffused field, the RMS value in the denominator of Equation (8.16) is proportional to the diffuse-field sound pressure at the reference point x in the ear canal. The reference point x selected is identical to that in binaural recording or HRTF measurements.

For binaural recording with an artificial head, the head whose output is equalized using Equation (8.16) is called the diffuse field-equalized artificial head. The HRTF involved in a diffuse field-equalized artificial head is just the diffuse field-equalized HRTF in Equation (2.47). Similar to free-field equalization, diffuse-field equalization has the following characteristics:

1. The eardrum is usually selected as the reference point. Given that the reference point in binaural recording (or HRTF measurements) is identical to that in diffuse-field equalization, the result of diffuse-field equalization is independent of the reference point. Diffuse-field equalization can eliminate the influence of the transmission response magnitude $|M_1|$ of the microphone used in binaural recording (or HRTF measurements).
2. Diffuse-field equalization enables the binaural signals to have a flat frequency response in the diffuse field.

The original purpose of free- or diffuse-field equalization is to guarantee compatibility between binaural signals and loudspeaker reproduction. Because HRTFs are frequency dependent, using a loudspeaker to directly render the binaural signals obtained by binaural recording or HRTF-based binaural synthesis causes timbre coloration. This situation can be improved by free-field equalization at a desired direction because the equalization yields a flat frequency response of output at the desired direction. In particular, selecting $(\theta_0 = 0°, \phi_0 = 0°)$ generates a flat output response for directly frontal sources. In practice, although sound sources may not always be located at directly frontal directions, free-field equalization can reduce timbre coloration in loudspeaker reproduction when sources are centered on the front direction.

In room environments, the reflection field dominates when the head for binaural recording is located far from the sound source. In this case, timbre coloration in loudspeaker reproduction can be alleviated by diffuse-field equalization.

Binaural signals are originally suitable for headphone presentation. Researchers have provided inconsistent opinions concerning whether free- or diffuse-field equalization should be used in compatible loudspeaker reproduction. Some studies suggest that the selection should depend on practical context, especially the sound field being recorded or simulated. As mentioned in Section 1.8.4, rendering binaural signals directly through loudspeakers causes both timbre coloration and impairs the spatial information contained in the signals. The former can be reduced by the equalization method discussed above, and the latter can be eliminated by further signal processing. The complete scheme for converting binaural signals into loudspeaker reproduction will be discussed in Chapter 9.

If the free- or diffuse-field equalized binaural signals are rendered through headphones, an alternative equalization method, rather than those from Equations (8.7) to (8.10), should be adopted. Ideally, the headphone equalization function $F_{free,2}(f)$ of free-field equalized binaural signals should satisfy:

$$F(f) = F_{free,1}(f) F_{free,2}(f), \tag{8.17}$$

where $F(f)$ and $F_{free,1}(f)$ are given in Equation (8.7) or Equations (8.8) and (8.15). Equation (8.17) consists of two components: the free-field equalization of binaural signals using $F_{free,1}(f)$ and headphone equalization using $F_{free,2}(f)$, which illustrates that the left- and right-ear equalization methods are identical (Figure 8.2). However, the left and right ears should be separately equalized under the situation of left-right asymmetry. Equations (8.7), (8.15), and (8.17) yield:

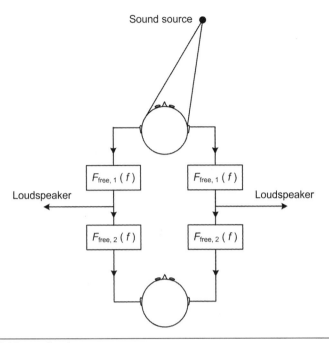

Figure 8.2 Free-field equalization

$$F_{free,2}(f) = \frac{F(f)}{F_{free,1}(f)} = \frac{H(\theta_0,\phi_0,f)}{P'_x / E}. \tag{8.18}$$

Similarly, the headphone equalization function of diffuse-field equalized binaural signals satisfies:

$$F(f) = F_{diff,1}(f) F_{diff,2}(f), \tag{8.19}$$

where $F_{diff,1}(f)$ is given in Equation (8.16), and $F_{diff,2}(f)$ is the corresponding headphone equalization function. Similar to the conditions of free-field equalization, Equation (8.19) consists of two components: the diffuse-field equalization of binaural signals and headphone equalization using $F_{free,2}(f)$. Equations (8.7), (8.16), and (8.19) yield:

$$F_{diff,2}(f) = \frac{F(f)}{F_{diff,1}(f)} = \frac{\sqrt{\dfrac{1}{M}\sum_{i=0}^{M-1}|H(\theta_i,\phi_i,f)|^2}}{P'_x / E}. \tag{8.20}$$

The result of Equation (8.18) or Equation (8.20) is independent of the reference point. In some literature, the headphone processed with Equation (8.18) or (8.20) is called the free- or diffuse-field equalized headphone. Numerous commercial models of free- or diffuse-field equalized headphones are currently available, but product properties and quality considerably vary.

No additional equalization is needed for reproducing free- or diffuse-field equalized binaural signals with free- or diffuse-field equalized headphones. This result is convenient for reproducing the binaural signals obtained by free- or diffuse-field equalized artificial head recording, or by equally free- or diffuse-field equalized HRTF-based binaural synthesis. In practice, for headphones with responses similar to those of free- or diffuse-field equalized headphones, applying free- or diffuse-field equalization to binaural signals can also alleviate timbre coloration in reproduction.

To reduce timbre coloration in headphone presentation, Merimaa (2009, 2010) proposed a method based on reducing the variations in the RMS spectral sum of a pair of HRTFs while preserving interaural time difference (ITD) and interaural level difference (ILD). For nonindividualized HRTFs, this method minimally influences virtual source localization, whereas for individualized HRTFs, it produces a slight subject-dependent influence on localization. This concept is similar to the constant-power equalization method of loudspeaker reproduction that will be presented in Section 9.5.2.

8.2. Repeatability and Individuality of Headphone-to-ear-canal Transfer Functions

8.2.1. Repeatability of HpTF Measurement

As discussed in Section 8.1.1, the transfer function of the equalization filter in headphone reproduction is exactly the inverse of an HpTF. Therefore, designing the equalization filter necessitates a known HpTF, which can be obtained by measurements similar to the HRTF measurements described in Chapter 2. First, excitation signals (such as maximal length sequence) are transmitted to a headphone placed on an artificial head or a human subject. Second, the binaural signals recorded by a pair of microphones positioned at the two ears are sent to a computer after being passed through an amplifier and an analog-to-digital converter. Finally, HpTFs are calculated from the recorded signals, after the transfer responses of the microphones are eliminated. In HpTF measurement, the issues that require attention, such as reference point selection, are similar to those in HRTF measurements (see Section 2.2.3).

Numerous researchers have measured HpTFs. In particular, Møller et al. (1995a) measured the HpTFs of 40 human subjects with 14 different types of headphones, as well as with miniature and probe microphones positioned at the ear canal. The authors collected a set of relatively complete data.

Because the equalization filter for headphone presentation is completely determined by HpTFs, the repeatability of HpTF measurement substantially influences successful headphone equalization. Even though a number of studies have been conducted on such repeatability, no general consensus has been achieved.

Pralong and Carlile (1996) studied the repeatability of HpTF measurement on a customized human head and ear model, as well as on 10 human subjects. The authors used a probe microphone positioned near the eardrum of the artificial model or 6 mm from the eardrums of the human subjects. The repeatability of a circumaural headphone (Sennheiser Linear HD 250) was examined. For the head and ear model, the standard deviation of six different placements of the headphones are at their worst levels (around 2 dB) at 8 kHz and are below 1 dB for the other regions of the spectrum below 12 kHz. This result demonstrates that the placement of this type of headphones on the model is highly reproducible. Conversely, the standard deviation of a human subject for six different placements of the headphones is at its worst level (about 4 dB) around 14 kHz, about 2 dB at frequencies between 8 and 12 kHz, and below 1 dB at frequencies below 7 kHz. Thus, the repeatability of this type of headphone on human subjects is also satisfactory. By contrast, the repeatability of a type of supra-aural headphone (Realistic Nova 17) on the head and ear model is less satisfactory, with a standard deviation of 8–9 dB around 8–12 kHz for six repetitions of HpTF measurements performed on the head and ear model.

Kulkarni and Colburn (2000) surveyed the repeatability of the HpTF measurement associated with the popular Sennheiser HD-520 headphone. The measurements were performed on a KEMAR artificial head equipped with large pinnae and DB-100 Zwislocki occluded-ear simulators (refer to Section 1.8.1), fitted with 12.7 mm (1/2 in) Etymotic ER-11 microphones. The HETF was then obtained, with the eardrum selected as the reference point. The results of 20 repetitions of headphone placements vary considerably, with the HETF magnitudes having a standard deviation of up to 9 dB at some frequencies between 8 and 14 kHz. The authors claim that the inverse filter based on a single measurement ineffectively equalizes headphone characteristics and sometimes generates results that are worse than when no equalization is performed. Although a mean inverse filter computed by averaging the responses of several headphone placements exhibits good performance on average, it still inadequately compensates for the extreme variations in HETFs.

Rao and Xie (2006) investigated the repeatability of HpTF measurement for different types of headphones. The measurement was carried out on KEMAR, with a method and condition similar to those in Kulkarni's work, except that small pinnae were used. Three types of headphones were employed:

1. Sennheiser Linear HD 250 II, a circumaural headphone with a frequency range of 10 Hz to 25 kHz and a circumaural cavity volume (diameter × depth, 8.5 cm × 2.5 cm);
2. Sony MRD 7506, a circumaural headphone with a frequency range of 10 Hz to 20 kHz and a circumaural cavity volume (length × width × depth, 6 cm × 4.5 cm × 1.2 cm); and
3. Sennheiser MX 500, an in-ear headphone with a frequency range of 18 Hz to 22 kHz.

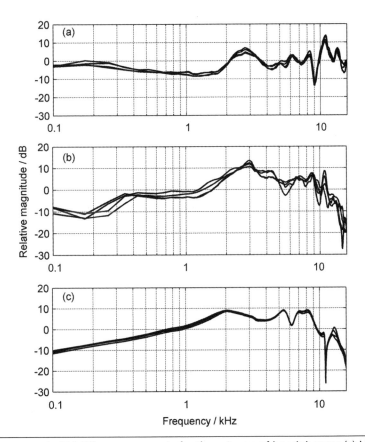

Figure 8.3 Four repeated HpTF measurements for three types of headphones: (a) HD 250 II; (b) MRD 7506; (c) MX 500.

The ends of the occluded-ear simulators were selected as reference points. Figure 8.3 shows the magnitudes of the HpTFs of four headphone placements and three types of headphones.

As shown in Figure 8.3(a), the spectral characteristics of the HD 250 II headphones and HRTF (Figure 3.3) are comparable. The common peak around 2–3 kHz and the common notch around 8–10 kHz are generally attributed to ear canal resonance and the pinna effect. The headphone mounted on the outer ear forms an enclosed space, so that the HpTF obtained at the eardrum exhibits the filtering properties of the ear canal and pinna. This result is consistent with Pralong's and Kulkarni's findings. Given that HpTFs and HRTFs share similar spectral characteristics, careful headphone equalization is required to avoid impairing the important localization cues encoded in HRTF spectra. The four measured curves also show high consistency overall, with a standard deviation of less than 2 dB at 0.1 to 16 kHz (not shown in the figure). Thus, the repeatability of this headphone is desirable and agrees with the results of Pralong et al.

Figure 8.3(b) shows the HpTF magnitudes associated with the MRD 7506 headphone. In addition to some features similar to HRTFs, a drop in magnitude at low frequencies below 0.2–0.3 kHz was also observed. This decrease is possibly due to low-frequency sound leakage. The deviation among the four curves in Figure 8.3(b) demonstrates the relatively poor repeatability of HpTF measurement for this headphone. In most cases, the standard deviation above 6 kHz exceeds 2 dB with a maximum value of 8 dB, which is similar to the result for the supra-aural headphone in the works of Pralong and Kulkarni et al.

Figure 8.3(c) indicates that the in-ear headphone of MX 500 shows extremely high repeatability with a standard deviation of less than 1 dB below 10 kHz.

Some underlying principles can be determined by comparing headphone structures. Because the large volume of the HD 250 II cavity causes less compressive deformation of pinnae, the HpTFs among different headphone placements slightly differ. This type of headphone also exhibits reasonable repeatability in individualized HpTF measurement, when subjects are allowed to place the headphone at a comfortable fit. The standard deviation of equivalent rectangular bandwidth-smoothed HpTF magnitudes across 10 repeated measurements is within 2 dB at a frequency of up to 20 kHz (Zhong et al., 2010). The volume of the MRD 7506 cavity is relatively smaller, making the pinnae prone to compression deformation. This deformation varies with each measurement. Under such a situation, the measured HpTF varies with headphone placement, especially at high Q-value spectral peaks and notches. An even worse condition occurs with supra-aural headphones. However, the HpTF of the in-ear MX 500 headphone is immune to different headphone placements because the headphone is directly inserted into the ear canal, rather than placed around the outer areas of the ear. The repeatability of HpTF measurement is therefore closely related to the extent of pinna deformation. Generally, repeatability ranks in descending order: in-ear headphone, large-cavity circumaural headphone, small-cavity circumaural headphone, supra-aural headphone. This sequence should be considered in headphone selection.

McAnally and Martin (2002) evaluated the repeatability of HpTF measurement with a cochlear model. The HpTFs of three human subjects were measured by a blocked ear canal technique and a supra-aural Sennheiser HD 520 headphone. The HpTFs were also measured using microphones on an artificial head (Head Acoustics HMS II.3). The standard deviations of the HpTF magnitudes associated with 20 headphone placements on each of the six human ears are generally smaller than 2.5 dB for frequencies up to 10 kHz, with one exception (a peak of 6.3 dB at 8.5 kHz for one ear). Above 10 kHz, the standard deviations are larger and range up to about 9 dB. Similar results were generated for the artificial head. The measured HpTFs were also convolved with Gaussian noise and passed through a cochlear filter model used to stimulate the frequency smoothing effect of the inner ear.

The standard deviations of the filtered HpTF magnitudes associated with 20 headphone placements on each of the six human ears are generally smaller than those of unfiltered HpTFs. For the three human subjects, the standard deviations are smaller than 1.4 dB for frequencies up to 10 kHz, and up to 3.6 dB for frequencies above 10 kHz. Therefore, the authors concluded that the variability of HpTFs across headphone placements is unlikely to have an adverse effect on the fidelity of VADs. However, more listening tests should be conducted to give this conclusion greater validity.

8.2.2. Individuality of HpTFs

The analyses in Section 8.2.1 do not consider individuality in HpTFs. In nature, the HpTFs of circumaural headphones are strongly related to the size and shape of the external ear, and therefore differ from individual to individual, as for HRTFs. As previously mentioned, Møller et al. (1995a) determined the HpTFs of 40 human subjects with 14 different headphones. Considerable variations between the subjects with the same headphone were observed.

Pralong and Carlile (1996) also obtained the HpTFs of the left and right ears of 10 subjects with a Sennheiser linear HD 250 circumaural headphone. The corresponding magnitude responses are

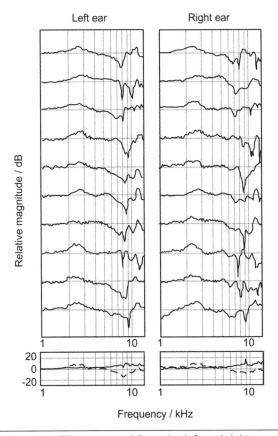

Figure 8.4 Top: Magnitudes of HpTFs measured from the left and right ears of 10 human subjects. The transfer functions are displaced by 40 dB on the *y*-axis to facilitate comparison. Bottom: The mean (solid line) and standard deviation (thick dotted line) of the HpTF magnitudes for each ear (Pralong and Carlile, 1996, by permission of J. Acoust. Soc. Am.)

illustrated in Figure 8.4, with the standard deviation at the bottom of the figure. As discussed in Section 8.2.1, this type of headphone demonstrates high repeatability. As indicated by Figure 8.4, considerable variability in responses occurs at frequencies above 6 kHz, with an interindividual standard deviation peaking up to 17 dB at frequencies around 9 kHz for the right ear. In particular, the frequency and depth of the first spectral notch vary between 7.5 and 11 kHz and −15 and −40 dB, respectively. The interindividual difference in HpTFs at high frequencies is very similar to that in the HRTFs discussed in Section 3.4.1. With circumaural headphones, the ear is entirely surrounded by cushion. Therefore, that the response of this type of headphone captures many of the external ear filtering effects similar to those in HRTFs is not surprising.

Because the high-frequency spectral characteristics of HRTFs are important localization cues and are obviously individual dependent, individualized HRTFs should, ideally, be employed in binaural synthesis. However, the high-frequency spectral characteristics of HpTFs are similar to those of HRTFs; thus, individualized HpTFs should also be employed in headphone equalization to ensure accurate replication of the binaural signals at the eardrum. Otherwise, nonindividualized HpTF equalization is likely to impair the localization cues contained in individualized HRTFs.

8.3. Directional Error in Headphone Reproduction

In headphone presentation, an accurate virtual source can be rendered if the sound pressures of a real sound source are precisely replicated at the eardrums. However, numerous experimental results indicate that subject-dependent directional errors or distortions generally exist. Examples of such errors include:

1. *Reversal error (front-back* or *back-front confusion).* A virtual source intended for the front hemisphere is perceived at a mirror position in the rear hemisphere, or, less frequently, the reverse. Sometimes, confusion arises regarding up and down source positions; such confusion is termed *up-down* or *down-up confusion.*
2. *Elevation error.* For example, the angle of a virtual source in the front median plane is usually elevated to higher positions.

As stated in Section 1.4, interaural cues, such as ITD and ILD, determine only a cone of confusion, rather than a well-defined spatial position of sound sources. The high-frequency spectral cues introduced by pinnae, along with the dynamic cues introduced by head movement, are important to the disambiguation of front-back confusion and vertical localization. However, the binaural signals obtained by conventional static binaural recording or binaural syntheses lack the dynamic cues introduced by head movement. Under such a condition, pinna-relevant high-frequency spectral cues are particularly critical. Unfortunately, high-frequency spectral cues are highly individual dependent, indicating that they are prone to impairment—either by the use of a recording head that is incompatible with a listener's head or by the use of nonindividualized HRTFs in binaural synthesis. High-frequency spectral cues will also be destroyed if the unwanted frequency response of the playback chain cannot be completely and correctly equalized, such as in the case wherein nonindividualized HpTFs are used in headphone equalization or under circumstances in which no headphone equalization is performed. Additionally, high-frequency errors in HRTF and HpTF measurements, and in every step of signal processing, may generate incorrect high-frequency spectral cues. All these factors are possible sources of directional error in headphone reproduction.

Directional error in headphone reproduction has been investigated by many researchers. Even though Wightman and Kistler (1989b) used both individualized HRTFs and HpTFs, the authors observed a substantial increase in front-back confusion in headphone reproduction, with a confusion rate of 11%, in contrast to 6% for real sources, in free-field localization. Similar qualitative results were found by Bronkhorst (1995), but with significant differences in quantification. Using individualized HRTFs and HpTFs, Martin et al. (2001) studied the localization performance of headphone reproduction in terms of front-back confusion rate and mean localization error. A localization performance equivalent to that of free-field localization was achieved. Martin et al. attributes this result to accurately measured HRTFs and HpTFs, made possible with the use of the blocked ear canal technique in his work. By contrast, in the studies of Wightman and Bronkhorst, probe microphones were used in HRTFs and HpTFs measurement, so that the results were sensitive to various errors. However, only three subjects participated in Martin's experiment; thus, the results require further verification.

Wenzel et al. (1993a) studied the influence of using nonindividualized HRTFs on virtual source localization for headphone reproduction. The experimental results show that, compared with the rates of front-back and up-down confusion in free-field real source localization, those in headphone reproduction, in which representative (nonindividual) HRTFs are used, increase from 19% and 6% to 31% and 18%, respectively. Further statistical testing confirms the increase in confusion rate with headphone reproduction. Similar results were observed in studies using speech as a stimulus (Begault and Wenzel, 1993). Pralong and Carlile (1996) argue that using nonindividual headphone equalization can partly account for the increased cases of the mislocalizations reported by Wenzel et al. (1993a).

Møller et al. (1996b) compared the performance of individual and nonindividual binaural recordings in single sound source localization. The experimental results show that the localization performance of individualized binaural recording, together with individual headphone equalization, is better than that of nonindividualized binaural recording. Nonindividualized binaural recording increases the errors for sound sources in the median plane, as well as increases front-back confusion.

Kulkarni and Colburn (1998) reported that a preferable reproduction performance can be achieved with open-canal tube phones inserted in the ear canal. Headphone equalization also appears to be unnecessary when tube phones are used. A similar method was proposed by Riederer and Niska (2002).

As stated in Section 7.4, So et al. (2010) revealed that using representative HRTFs selected by the clustering method reduces back-front confusion, but minimally influences front-back confusion.

Eliminating or reducing the perceived virtual source direction errors in headphone reproduction therefore necessitates careful reduction of the errors introduced in binaural signal recording/synthesis and reproduction stages to as few as possible. As will be detailed in Chapter 12, incorporating dynamic localization cues is also essential for reducing front-back or up-down confusion.

Considering the significant individual differences in the performance of free-field real source localization, some researchers suggest that subjects with prominent individual spectral features of HRTFs exhibit better localization. The same scholars then proposed adopting these "better" subjects in binaural recording or using their HRTFs in binaural synthesis. Some experiments confirm that this method improves localization accuracy in headphone reproduction, especially in vertical and front-back localization (Butler and Belendiuk, 1977; Wenzel et al., 1988). However, this result is not supported by Møller et al. (1996b).

Some researchers attempted to modify nonindividualized HRTFs to improve localization performance. A method proposed by Zhang et al. (1998) involves enlarging the spectral difference of HRTFs at front-back mirror directions, in which the modified HRTF $H'(\theta, f)$ is written as:

$$H'(\theta, f) = W(\theta, f) H(\theta, f),$$ (8.21)

where $H(\theta, f)$ is the original HRTF, and $W(\theta, f)$ is a direction- and frequency-dependent weighting function with the form:

$$W(\theta, f) = \left| \frac{H(\theta, f)}{\max_f |H(\theta, f)|} \right|^m.$$ (8.22)

In Equation (8.22), m is a positive constant, and $0 \leq W(\theta, f) \leq 1$. Multiplying $H(\theta, f)$ by $W(\theta, f)$ alters HRTF magnitudes while preserving phase. This manipulation also deepens the notches in HRTF magnitudes but basically retains peaks. This way, the front-back difference in HRTFs is enlarged. The corresponding experiments, in which modified HRTFs were derived from MIT-KEMAR HRTFs with $m = 0.618$, indicate that this enlargement method reduces front-back confusion in the horizontal plane to some extent.

A similar algorithm with a different weighting function was proposed by Park et al. (2005). To reduce front-back confusion, Lee et al. (2003) put forward a method for increasing frontal-hemisphere HRTFs while attenuating back-hemisphere HRTFs with a weighting function. Brungart and Romigh (2009) divided HRTF magnitudes within the cone of confusion into the products of a lateral localization-relevant component and a vertical localization-relevant component, and then enhanced the latter. The listening experiments indicate that this enhancement method yields improved localization performance in vertical directions, regardless of whether individual or nonindividualized HRTFs are used. Gupta et al. (2002) assumed that the preferable localization performance within the cone of confusion of some subjects may be due to protruding pinnae, which enhance the spectral difference between front-back mirror directions. The authors verified this assumption by measuring a solid sphere (bowling ball), with attached cardboard flaps representing the ears. Accordingly, HRTF modification was implemented to yield enhanced front-back differentiation in headphone reproduction.

However, the spectral cues for localization are highly individual dependent. Human hearing determines source directions by comparing received localization cues with stored patterns that are formed from previous auditory experiences. Therefore, whether a listener can adapt to modified HRTFs remains a problem. The enhancement method that involves deepening the notches in HRTF magnitudes seems insufficient, particularly because the HRTFs of different subjects differ not only in depth of notches but also in frequency of notch occurrence, among other factors. Comparing received localization cues with previous auditory experience is a process of self-adaptation; thus, a listener may gradually adapt to modified HRTFs or to the HRTFs of other individuals. The experiments conducted by Shinn-Cunningham (1998a, 1998b) show that listeners can reduce the bias between the source direction determined by HRTFs (acoustic cues) and the physical source direction derived from visual feedback by training. Hofman et al. (1998) confirmed that listeners adapt to distorted spectral cues after several weeks of training. Zahorik et al. (2001) also revealed that the localization error caused by the use of nonindividualized HRTFs can be reduced through short-term training (two 45-min training sessions). Parseihian and Katz (2012) demonstrated rapid HRTF adaptation using a virtual auditory environment, in which participants using nonindividualized HRTFs reduced localization errors in elevation by 10° in three 12-min sessions. Other adaptation methods were also proposed, such as the technique put forward by Susnik et al. (2008). The psychoacoustic experiments conducted by Mendonça et al. (2012) indicate that mere exposure to virtual sources synthesized with generic HRTFs do not improve subjects' performances in sound source localization, but short training periods that involve active learning and feedback yield significantly better results. Mendonça et al. recommend that the use of nonindividualized HRTFs for binaural synthesis be preceded by a learning period. The adaptation degree, training method, and time consumed are, therefore, interesting issues requiring further investigation.

8.4. Externalization and Control of Perceived Virtual Source Distance in Headphone Reproduction

8.4.1. In-head Localization and Externalization

Conventional stereophonic reproduction over headphones often causes an auditory event inside the head, leading to an unnatural hearing experience. A similar problem often occurs when the binaural signals obtained by binaural recording or synthesis are reproduced through headphones, even in the case of a far field with an intended virtual source distance r greater than 1.0–1.2 m. In many cases, however, the virtual source appears on the head surface (especially for front source directions), rather than completely inside the head. In this phenomenon, virtual sources cannot be externalized in headphone reproduction; to avoid confusing this occurrence with localization in three-dimensional space, it is generally defined as *in-head localization, intracranial lateralization*, or *lateralization*, as mentioned in Section 1.3. The elimination of in-head localization is called *externalization*.

Many researchers agree that lateralization is caused by incorrect spatial information on both ears in sound reproduction (Plenge, 1974). Human hearing determines the direction and distance of a sound source by comparing the localization cues encoded in binaural pressures with previous auditory experiences. When binaural pressure signals fail to provide correct localization cues (as for those of real sound sources), the hearing system likely creates an error, or illusion, that the sound source is inside the head. Because stereophonic signals are originally suitable for loudspeaker reproduction, errors in binaural pressures occur when they are reproduced through headphones, an issue to be comprehensively discussed in Section 10.1. However, although the binaural signals obtained by binaural recording or synthesis are appropriate for headphone reproduction, they also suffer from in-head localization.

Binaural pressure errors originate from numerous sources, thereby causing in-head localization. Some studies indicate that nonindividualized HRTFs and inappropriate headphone equalization accounts for this problem (Durlach et al., 1992). Using random noise as a sound stimulus, Kim and Choi (2005) conducted psychoacoustic experiments under the following conditions:

1. Nonindividual HRTFs (of a KEMAR artificial head; the same for the succeeding conditions) were used in binaural synthesis without headphone equalization.
2. Nonindividualized HRTFs were used in binaural synthesis with headphone equalization.
3. Individualized HRTFs were used in binaural synthesis without headphone equalization.
4. Individualized HRTFs were used in binaural synthesis with headphone equalization.

The results indicate that it is important to use individualized HRTFs and headphone equalization for externalization, as well as for controlling the perceived virtual source distance (to be discussed in Section 8.4.2). Through headphone reproduction associated with individualized HRTFs and HpTFs, Wightman and Kistler (1989b) successfully replicated an auditory experience that is equivalent to the condition of free-field real sources. The correct binaural cues (Hartmann and

Wittenberg, 1996) and dynamic cues caused by head movement (Loomis et al., 1990; Durlach et al., 1992; Wenzel, 1996) are also considered essential for externalization.

Many studies point out that environmental reflection is also highly important to externalization. In addition to direct sound, reflection is crucial to spatial hearing in reality. As mentioned in Sections 1.4.6 and 1.7.2, environmental reflection is key to distance perception. Free-field HRTFs (or head-related impulse responses—HRIRs) were used in the preceding binaural synthesis discussion; therefore, the resultant binaural signals only contained direct sound without environmental reflection. To eliminate the effect of in-head localization, HRIRs can be replaced by binaural room impulse responses (BRIRs) in binaural synthesis. BRIRs can be obtained, not only from measurements in which an artificial head or a human subject is used, but also from binaural room acoustic modeling or artificial reverberation algorithms (Chapter 11). Binaural signals with environmental reflections can be directly recorded using an artificial head or a human subject.

The psychoacoustic experiment conducted by Begault (1992) verifies that incorporating the reflection generated by room acoustic modeling into nonindividual binaural synthesis can reduce the likelihood of in-head localization in headphone reproduction, but at the cost of increased incorrect direction localization. Using speech as a reproduced signal, Begault et al. (2001) compared the effect of head tracking (introducing dynamic cues), reflective properties, and individualized HRTFs on externalization. The results show that only reflection is paramount, contrary to the reports of previous studies. In Begault's study, the conclusion applies only to speech stimuli, whereas, in other studies, other kinds of stimuli (such as noise) were used. The experiment also shows that simulating several preceding-order early reflections in VADs causes externalization. This result presents significant practical implications. The duration of the BRIR that has the same order of room reverberation time (about hundreds of milliseconds in a normal room, even longer for a hall) is usually longer than the duration of free-field HRIRs (no more than 10 ms). Therefore, convolution with a pair of complete BRIRs in binaural synthesis incurs a large computational cost; simulating only several preceding-order early reflections effectively simplifies this signal processing approach.

Some studies show that incorporating the dynamic cues caused by head movements is important to the externalization of wideband stimuli (Durlach et al., 1992; Wenzel, 1996). Stimulating some random movements of sound sources is also helpful to externalization (Wersenyi, 2009), but further experimental validation is needed.

Another subjective experiment, in which the binaural signals recorded from an artificial head were used, shows that asymmetric pinnae remarkably enhance the possibility of external localization (Brookes and Treble, 2005). The underlying physical reasons are unclear, however, prompting the need for stronger evidence and further research.

The foregoing discussions indicate that in headphone reproduction, many cues are relevant to externalization; conclusions vary, or sometimes contradict, depending upon the study. However, a definite deduction is that environmental reflection plays an important role in externalization. Details concerning the simulation of environmental reflections in VADs will be presented in Chapter 11.

8.4.2. Control of Perceived Virtual Source Distance in Headphone Reproduction

The previous discussions focused primarily on binaural synthesis with far-field HRTFs, in which directional cues were considered without regard for distance cues. Thus far, researchers have successfully synthesized virtual sources at various directions. However, the sound source position in three-dimensional space is determined by both direction and distance; an ideal VAD should be able to render virtual sources at various directions and distances. Studies on rendering virtual sources at different distances are fewer than those on rendering virtual sources at different directions because auditory distance perception depends upon the combination of multiple cues; the mechanisms behind auditory distance perception are less well understood than those underlying direction localization. Section 8.4.1 showed that in-head localization often occurs in headphone reproduction, and externalization is the basis for controlling perceived virtual source distances. Moreover, as indicated in Equation (1.14), perceived distance is inconsistent with physical distance, even for real sound sources. In headphone reproduction by signal processing, all the aforementioned factors result in relatively difficult simulations of virtual sources at different distances.

The control of perceived virtual source distances has been investigated by some researchers. Sections 1.4.6 and 1.7.2 showed that many cues contribute to distance perception; such cues include subjective loudness, the high-frequency attenuation caused by air absorption, the distance dependence of near-field ILD and the spectral cues caused by head and pinna diffraction, and environmental reflection. Among these cues, subjective loudness is the easiest to simulate under free-field conditions because of the simple inversely proportional relationship between source distance and pressure magnitude. Simulating high-frequency air attenuation with the use of filters is also easy, when absorption as a function of frequency and distance is known (Morse and Ingrad, 1968). However, these two cues are dependent upon the subject's degree of familiarity with sound sources, and they contribute only to relative distance perception.

In relevant signal processing, controlling environmental reflection is an effective way to manipulate the perceived distance of virtual sources, especially under a direct-to-reverberant energy ratio. This control is realized by artificial reflection, and reverberation algorithms are often used in practical cases that allow for certain errors.

The discussion in Section 3.6 indicated that for a lateral sound source with a distance r of less than 1.0 m, the relationship between r and near-field ILD is a cue for absolute distance perception. Thus, some studies used near-field HRTFs to synthesize virtual sound sources at various distances (Brungart, 1998). This method is valid only within 1 m, and its accuracy diminishes when the virtual source approaches the median plane with a small ILD value.

As indicated in Section 2.4, measuring a full set of near-field HRTFs is difficult and tedious; therefore, near-field HRTF databases are rare. Algorithms for approximately deriving near-field HRTFs from far-field HRTFs have been developed. These algorithms can be used to control distance perception. On the basis of the spherical harmonic decomposition of HRTFs (Section 6.3.4), Kan et al. (2006a, 2009) proposed a simplified method for deriving near-field HRTFs, in which a distance variation function (DVF) is defined:

$$DVF = \frac{P(r,\theta,\phi,f)}{P(r_1,\theta,\phi,f)} = \frac{r_1}{r}\frac{H(r,\theta,\phi,f)}{H(r_1,\theta,\phi,f)},\tag{8.23}$$

where $P(r, \theta, \phi, f)$ and $P(r_1, \theta, \phi, f)$ refer to the left or right ear sound pressure caused by the harmonic point sources located at (r, θ, ϕ) and (r_1, θ, ϕ), respectively. When the DVF is known, the HRTF at distance r can be calculated using the HRTF measured at distance r_1 thus:

$$H(r,\theta,\phi,f) = \frac{r}{r_1}DVF\, H(r_1,\theta,\phi,f).\tag{8.24}$$

In practice, $H(r, \theta, \phi, f)$ can be realized by a filter of $H(r_1, \theta, \phi, f)$ and then by a DVF filter with distance gain. According to Equation (8.24), Kan et al. obtained near-field HRTFs by using the DVF calculated from a rigid-spherical head model (Section 4.1.4), along with measured individualized HRTFs. The localization experiments show that, in headphone reproduction, both the methods in Equation (8.24) and in Section 6.3.4 can be used to accurately generate virtual sources at a distance $r \leq 0.5$ m (0.4–0.6 m). As mentioned in Section 6.5, virtual Ambisonics or virtual loudspeaker-based algorithms, as well as principal components analysis based algorithms, can also be used to synthesize near-field virtual sources (Daniel, 2003; Menzies et al., 2007; Xie and Zhang, 2012a).

Given that multiple cues contribute to auditory distance perception, an ideal approach is to simultaneously simulate all relevant cues. BRIRs can be used to synthesize binaural signals because they contain most relevant factors for distance perception. Zahorik (2002a) used the BRIRs measured from an artificial head to stimulate distance perception in headphone reproduction, and achieved subjective performance equivalent to that of a real sound source. Alternatively, BRIRs can be obtained by computer modeling, as will be presented in Chapter 11.

8.5. Summary

Headphone equalization is implemented by filtering binaural signals with the inverse of headphone-to-ear canal transfer functions (HpTFs). In headphone-based binaural reproduction, this equalization is required to eliminate the influence of headphone-to-ear canal transmission. The measurement positions (reference points) of HpTFs should be identical to those in HRTF measurement. In practical applications, binaural signals can be free- or diffuse-field equalized and then reproduced using corresponding free- or diffuse-field equalized headphones. The repeatability of HpTF measurement is related to the extent of the compressive deformation of pinnae during measurement, an issue that should be considered in selecting headphones suitable for binaural reproduction.

HpTFs vary among individuals. Because the spectral features and interindividual differences in high-frequency HpTFs are comparable to those of HRTFs, individualized HpTFs should, in principle, be incorporated into headphone equalization. Otherwise, the localization information encoded in individualized HRTFs may be impaired.

Direction errors, especially those at front-back mirror and elevation directions, often occur in static binaural reproduction through headphones. These errors arise from the lack of dynamic localization cues, along with spectral distortion at high frequencies. Nonindividualized HRTFs and HpTFs employed in binaural synthesis and equalization, and the high-frequency errors introduced in measurement and signal-processing stages may cause the perceived direction errors of virtual sources.

Another problem that frequently occurs in headphone reproduction is in-head localization, which is probably caused by binaural pressure errors. Incorporating environmental reflections is crucial to externalization.

Some studies have been devoted to controlling the perceived distance of virtual sound sources; such control can be partially realized by processing with near-field HRTFs, manipulating near-field interaural level difference, and providing environmental reflections. Compared with research on controlling virtual source directions, however, that on controlling virtual source distances is relatively rare and requires more attention.

9

Binaural Reproduction through Loudspeakers

This chapter addresses issues concerning binaural signal conversion for loudspeaker reproduction. The concepts of crosstalk cancellation and transaural processing for loudspeaker reproduction are introduced. Some problems in loudspeaker reproduction, including stability against head movement, the influence of mismatched HRTFs and loudspeaker pairs, and coloration and timbre equalization, are analyzed. The design and implementation of crosstalk cancellation and transaural processing are also discussed.

Heretofore, a constant assumption has that binaural signals are presented through headphones. In practice, however, some listeners prefer loudspeakers to headphones. Because binaural signals are originally intended for headphone presentation (Section 1.8), additional signal conversion/ processing is needed prior to loudspeaker reproduction. Loudspeaker reproduction cannot stably deliver all spatial information, which will be comprehensively discussed in this chapter. Practical loudspeaker reproduction also frequently suffers from coloration, which was preliminarily discussed in Section 8.1.2 and will be examined further in this chapter. Section 9.1 introduces crosstalk cancellation and transaural processing for reproduction with two frontal loudspeakers, and then presents the extension of this approach to general cases of multiple-loudspeaker reproduction. Stability and its improvement in loudspeaker reproduction are addressed in Sections 9.2 to 9.4. Section 9.2 analyzes the stability of crosstalk cancellation against head rotation. Section 9.3 examines the stability of crosstalk cancellation against head translation and then introduces the stereo dipole concept. Section 9.4 evaluates the influence of mismatched head-related transfer functions (HRTFs) and loudspeaker pairs in reproduction. Coloration and timbre equalization in loudspeaker reproduction are discussed in Section 9.5. Sections 9.6 and 9.7 discuss the design and implementation of crosstalk cancellation and transaural processing. Specifically, Section 9.6 discusses implementation methods, and Section 9.7 presents approximate methods for solving the crosstalk cancellation matrix.

9.1. Basic Principle of Binaural Reproduction through Loudspeakers

9.1.1. Binaural Reproduction through a Pair of Frontal Loudspeakers

Binaural signals from either binaural recording or synthesis are originally intended for headphone presentation. When binaural signals are reproduced through a pair of left and right loudspeakers arranged in front of a listener, unwanted *crosstalk*, from each loudspeaker to the ear opposite, occurs. Crosstalk impairs the directional information encoded in binaural signals. Schroeder and Atal (1963) introduced *crosstalk cancellation* for binaural reproduction through loudspeakers. Prior to loudspeaker reproduction, binaural signals are precorrected or filtered to cancel transmission from each loudspeaker to its opposite ear.

Let $E_L(f)$ and $E_R(f)$, or simply E_L and E_R, denote the binaural signals in the frequency domain. As illustrated in Figure 9.1, binaural signals are prefiltered by a 2×2 crosstalk cancellation matrix (crosstalk canceller) and then reproduced through loudspeakers. The loudspeaker signals are given by:

$$\begin{bmatrix} L' \\ R' \end{bmatrix} = \begin{bmatrix} A_{11} & A_{12} \\ A_{21} & A_{22} \end{bmatrix} \begin{bmatrix} E_L \\ E_R \end{bmatrix}, \tag{9.1}$$

where A_{11}, A_{12}, A_{21}, and A_{22} are the four transfer functions or filters that form the crosstalk cancellation matrix.

Let H_{LL}, H_{RL}, H_{LR}, and H_{RR} denote the four acoustic transfer functions from two loudspeakers to two ears. These four transfer functions are determined by loudspeaker configuration and listener location. According to Equation (9.1), the reproduced pressures at the two ears are expressed as:

$$\begin{bmatrix} P_L' \\ P_R' \end{bmatrix} = \begin{bmatrix} H_{LL} & H_{LR} \\ H_{RL} & H_{RR} \end{bmatrix} \begin{bmatrix} L' \\ R' \end{bmatrix} = \begin{bmatrix} H_{LL} & H_{LR} \\ H_{RL} & H_{RR} \end{bmatrix} \begin{bmatrix} A_{11} & A_{12} \\ A_{21} & A_{22} \end{bmatrix} \begin{bmatrix} E_L \\ E_R \end{bmatrix}. \tag{9.2}$$

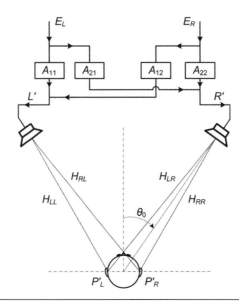

Figure 9.1 Binaural reproduction through a pair of frontal loudspeakers with crosstalk cancellation

The transfer characteristics of the crosstalk cancellation matrix are appropriately selected, so that the product of two 2×2 matrices in Equation (9.2) equals an identity matrix:

$$\begin{bmatrix} H_{LL} & H_{LR} \\ H_{RL} & H_{RR} \end{bmatrix} \begin{bmatrix} A_{11} & A_{12} \\ A_{21} & A_{22} \end{bmatrix} = \begin{bmatrix} 1 & 0 \\ 0 & 1 \end{bmatrix}, \tag{9.3}$$

or:

$$[H][A] = [I]. \tag{9.4}$$

Then, we have:

$$P_L' = E_L \qquad P_R' = E_R. \tag{9.5}$$

In this case, crosstalk is completely cancelled out, and the desired binaural signals are precisely transmitted to a listener's two ears.

If the acoustic transfer matrix $[H]$ in Equation (9.4) is non-singular, and therefore invertible, the crosstalk cancellation matrix is calculated by the equation:

$$\begin{bmatrix} A_{11} & A_{12} \\ A_{21} & A_{22} \end{bmatrix} = \begin{bmatrix} H_{LL} & H_{LR} \\ H_{RL} & H_{RR} \end{bmatrix}^{-1} = \frac{1}{H_{LL}H_{RR} - H_{LR}H_{RL}} \begin{bmatrix} H_{RR} & -H_{LR} \\ -H_{RL} & H_{LL} \end{bmatrix}. \tag{9.6}$$

Equation (9.6) depicts the basic principle of crosstalk cancellation for loudspeaker reproduction.

Suppose that binaural signals are obtained by synthesis according to Equation (1.36). Thus,

$$E_L(f) = H_L(\theta, f)E_0(f) \qquad E_R(f) = H_R(\theta, f)E_0(f). \tag{9.7}$$

The source distance dependence r is excluded and the target virtual source direction is simply denoted by azimuth θ in Equation (9.7), which will be explained in detail in Section 9.2. Substituting Equations (9.6) and (9.7) into Equation (9.1) yields:

$$\begin{bmatrix} L' \\ R' \end{bmatrix} = \frac{1}{H_{LL}H_{RR} - H_{LR}H_{RL}} \begin{bmatrix} H_{RR} & -H_{LR} \\ -H_{RL} & H_{LL} \end{bmatrix} \begin{bmatrix} H_L(\theta, f) \\ H_R(\theta, f) \end{bmatrix} E_0(f). \tag{9.8}$$

The signal processing in Equation (9.8) involves two stages. The monophonic (mono) input stimulus $E_0(f)$ is first filtered by a pair of HRTFs at the target source direction according to Equation (1.36); then, the resultant binaural signals E_L and E_R are prefiltered by a 2×2 crosstalk cancellation matrix to obtain signals L' and R', which are reproduced through a pair of loudspeakers to create virtual sources at target direction θ.

In most practical cases, loudspeaker configuration is left-right symmetric with respect to listener. If the HRTFs are assumed to be left-right symmetric, the transfer functions from each loudspeaker to the ipsilateral and contralateral ears satisfy $H_{LL} = H_{RR} = H_\alpha$ and $H_{LR} = H_{RL} = H_\beta$, respectively. Then, the general solution in Equation (9.6) for the crosstalk cancellation matrix is simplified as the symmetric solution:

$$A_{11} = A_{22} = \frac{H_\alpha}{H_\alpha^2 - H_\beta^2} \qquad A_{12} = A_{21} = \frac{-H_\beta}{H_\alpha^2 - H_\beta^2}, \tag{9.9}$$

and Equation (9.8) becomes:

$$\begin{bmatrix} L' \\ R' \end{bmatrix} = \frac{1}{H_\alpha^2 - H_\beta^2} \begin{bmatrix} H_\alpha & -H_\beta \\ -H_\beta & H_\alpha \end{bmatrix} \begin{bmatrix} H_L(\theta, f) \\ H_R(\theta, f) \end{bmatrix} E_0(f). \tag{9.10}$$

The azimuthal span of loudspeakers in Equation (9.10) is arbitrary. The discussions here are therefore not limited to conventional stereophonic loudspeaker configuration, where two loudspeakers are located at $\theta_L = 330°$ and $\theta_R = 30°$ with an azimuthal span of $60°$ with respect to a listener.

If the signal processing is initially designed to create appropriate loudspeaker signals, the two stages of binaural synthesis and crosstalk cancellation can be merged. The matrix product in Equation (9.8) yields:

$$L' = G_L(\theta, f)E_0(f) \qquad R' = G_R(\theta, f)E_0(f), \tag{9.11}$$

where:

$$G_L(\theta, f) = \frac{H_{RR}H_L(\theta, f) - H_{LR}H_R(\theta, f)}{H_{LL}H_{RR} - H_{LR}H_{RL}}$$

$$G_R(\theta, f) = \frac{-H_{RL}H_L(\theta, f) + H_{LL}H_R(\theta, f)}{H_{LL}H_{RR} - H_{LR}H_{RL}}. \tag{9.12a}$$

In the symmetric case, we have:

$$G_L(\theta, f) = \frac{H_\alpha H_L(\theta, f) - H_\beta H_R(\theta, f)}{H_\alpha^2 - H_\beta^2} \qquad G_R(\theta, f) = \frac{-H_\beta H_L(\theta, f) + H_\alpha H_R(\theta, f)}{H_\alpha^2 - H_\beta^2}. \tag{9.12b}$$

Equation (9.11) demonstrates that the loudspeaker signals L' and R' for the target virtual sources at direction θ can be directly synthesized by filtering a mono stimulus $E_0(f)$ with a pair of filters $G_L(\theta, f)$ and $G_R(\theta, f)$. This condition is the basic principle of binaural virtual source synthesis for reproducing sound through a pair of loudspeakers.

Equation (9.8) can also be derived from an alternate method. As illustrated in Figure 9.2 and Equation (1.20), the frequency-domain binaural pressures caused by a source at direction θ are given as:

$$P_L = H_L(\theta, f)P_0(f) \qquad P_R = H_R(\theta, f)P_0(f). \tag{9.13}$$

Let L' and R' denote the loudspeaker signals. Then, the binaural pressures in reproduction are:

$$\begin{bmatrix} P_L' \\ P_R' \end{bmatrix} = \begin{bmatrix} H_{LL} & H_{LR} \\ H_{RL} & H_{RR} \end{bmatrix} \begin{bmatrix} L' \\ R' \end{bmatrix}. \tag{9.14}$$

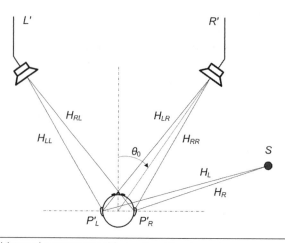

Figure 9.2 Control of binaural pressures with a pair of loudspeakers

If signals L' and R' are appropriately chosen to guarantee that the binaural pressures in reproduction are identical or directly proportional to those caused by a real source at direction θ, a perceived virtual source at direction θ is then created. Accordingly, Equations (9.11) and (9.12a) can be directly derived by letting Equation (9.14) be equal to Equation (9.13)—$P_L = P_L'$, $P_R = P_R'$.

The two methods for deriving Equations (9.11) and (9.12a) are mathematically equivalent, but are physically different. Through the control of loudspeaker signals, the second method can be interpreted as a technique for controlling binaural pressures, and, therefore, synthesized virtual sources (Sakamoto et al., 1981, 1982); this method is called *transaural technique* or *transaural stereo* (Cooper and Bauck, 1989; Bauck and Cooper, 1996). Correspondingly, the scheme for loudspeaker signal synthesis is called *transaural synthesis*, and the filters specified by G_L and G_R in Equation (9.12a) are called *transaural synthesis filters*.

9.1.2. General Theory for Binaural Reproduction through Loudspeakers

Bauck and Cooper (1996) generalized the theory of crosstalk cancellation and transaural synthesis to cases with multiple loudspeakers and listeners. Ultimately, their work formulated the general theory for binaural reproduction through loudspeakers with various configurations.

As illustrated in Figure 9.3, K input signals are denoted by E_1, E_2 ... E_K, which are prefiltered by an $M \times K$ crosstalk cancellation matrix $[A]$ and then reproduced through M loudspeakers. The signals for M loudspeakers are:

$$
\begin{bmatrix} S_1 \\ S_2 \\ \vdots \\ S_M \end{bmatrix} = \begin{bmatrix} A_{11} & A_{12} & .. & A_{1K} \\ A_{21} & A_{22} & .. & A_{2K} \\ \vdots & \vdots & \vdots & \vdots \\ A_{M1} & A_{M2} & .. & A_{MK} \end{bmatrix} \begin{bmatrix} E_1 \\ E_2 \\ \vdots \\ E_K \end{bmatrix},
\tag{9.15a}
$$

or:

$$
S = [A]E,
\tag{9.15b}
$$

where S and E are the $M \times 1$ and $K \times 1$ matrices or column vectors of the loudspeaker signals and input signals, respectively.

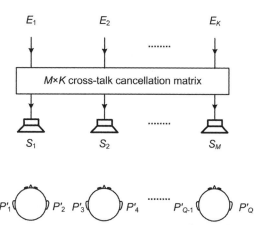

Figure 9.3 Crosstalk cancellation for multiple loudspeakers and listeners

Suppose that Q receiving points (ears, or $Q/2$ listeners where Q is an even number) exist. The acoustic transmission from M loudspeakers to Q receiving points is described by a $Q \times M$ transfer matrix $[H]$ with element $H_{qm}(q = 1,2 \dots Q, m = 1,2 \dots M)$, which represents the transfer function from the mth loudspeaker to the qth receiving point. If the multiple scattering among two or more subjects is negligible, these transfer functions are approximated by corresponding HRTFs. Then, the pressures at Q receiving points are:

$$
\begin{bmatrix} P_1' \\ P_2' \\ \vdots \\ P_Q' \end{bmatrix} = \begin{bmatrix} H_{11} & H_{12} & .. & H_{1M} \\ H_{21} & H_{22} & .. & H_{2M} \\ \vdots & \vdots & \vdots & \vdots \\ H_{Q1} & H_{Q2} & .. & H_{QM} \end{bmatrix} \begin{bmatrix} S_1 \\ S_2 \\ \vdots \\ S_M \end{bmatrix},
\tag{9.16a}
$$

or:

$$
\mathbf{P}' = [H]\mathbf{S},
\tag{9.16b}
$$

where $\mathbf{P}'$ is a $Q \times 1$ matrix or column vector of pressures at receiving points. Combining Equations (9.15b) and (9.16b) yields:

$$
\mathbf{P}' = [H][A]\mathbf{E}.
\tag{9.17}
$$

In addition, we assume that the desired pressures at Q receiving points are related to the K input signals by a known $Q \times K$ matrix $[Z]$ as:

$$
\begin{bmatrix} P_1 \\ P_2 \\ \vdots \\ P_Q \end{bmatrix} = \begin{bmatrix} Z_{11} & Z_{12} & .. & Z_{1K} \\ Z_{21} & Z_{22} & .. & Z_{2K} \\ \vdots & \vdots & \vdots & \vdots \\ Z_{Q1} & Z_{Q2} & .. & Z_{QK} \end{bmatrix} \begin{bmatrix} E_1 \\ E_2 \\ \vdots \\ E_K \end{bmatrix},
\tag{9.18a}
$$

or:

$$
\mathbf{P} = [Z]\mathbf{E},
\tag{9.18b}
$$

where $\mathbf{P}$ is a $Q \times 1$ matrix or column vector of desired pressures, and matrix $[Z]$ is composed of $Q \times K$ desired transfer functions.

Let the pressures in Equation (9.17) be equal to those in Equation (9.18b), yielding the matrix equation:

$$
[H][A] = [Z],
\tag{9.19}
$$

which is a matrix equation for determining the $M \times K$ crosstalk cancellation matrix $[A]$. Given that $[Z]$ is a $Q \times K$ known matrix, Equation (9.19) is equivalent to a set of $Q \times K$ linear algebraic equations, with the $M \times K$ elements of matrix $[A]$ as unknowns. The solution to Equation (9.19) is classified into the following cases:

1. $Q = M$—that is, the number of receiving points equals that of loudspeakers, leading to an equal number of $M \times K$ equations and unknowns in Equation (9.19). Provided that matrix $[H]$ is full rank, with rank $[H] = M$ and therefore invertible, Equation (9.19) has a unique and exact solution:

$$
[A] = [H]^{-1}[Z].
\tag{9.20}
$$

2. $Q < M$—that is, fewer receiving points than loudspeakers exist, leading to an insufficient number of equations relative to the number of unknowns. In this case, Equation (9.19) is

underdetermined. Assuming rank $[H] = Q$, an infinite set of exact solutions to Equation (9.19) exists, among which the pseudoinverse solution is:

$$[A] = [H]^+ \{[H][H]^+\}^{-1}[Z], \qquad (9.21)$$

where superscript "+" denotes the transpose associated with the conjugation manipulation of the matrix. Equation (9.21) can be proven to be a least square solution that minimizes the total power of the loudspeaker signals provided by Equation (9.15a):

$$W = W_{\min} = \min\left[\sum_{m=1}^{M} |S_m|^2\right]. \qquad (9.22)$$

This characteristic is a significant feature of the solution given by Equation (9.21). In the underdetermined case, some constraints can also serve as a supplement to yield a solution with the desired properties.

3. $Q > M$—that is, more receiving points than loudspeakers exist, leading to an excess number of equations relative to the number of unknowns. In this case, provided that rank $[H]= M$, Equation (9.19) is overdetermined, without exact solutions. An approximate solution is given by the pseudoinverse:

$$[\hat{A}] = \{[H]^+[H]\}^{-1}[H]^+[Z], \qquad (9.23)$$

which is a solution that minimizes the square norm of the error matrix:

$$\min\|[\varepsilon]\|^2 = \min\|[H][\hat{A}] - [Z]\|^2 = \min\sum_{q=1}^{Q}\sum_{i=1}^{K}|\{[H][\hat{A}]\}_{qi} - [Z]_{qi}|^2. \qquad (9.24)$$

The approximate solution given by Equation (9.23) may be reasonable in a mathematical sense, but not in a perceptual sense.

4. If rank $[H] < \min(M, Q)$, the situation is complicated. Some algorithms, such as the singular value decomposition of a matrix, are required to solve the problem. For brevity, the details of that decomposition are excluded here.

The first example of the foregoing analysis is reproducing binaural signals $E_1 = E_L$ and $E_2 = E_R$ through a pair of frontal loudspeakers, as shown in Figure 9.1. In this case, $K = M = Q = 2$—that is, two input signals, two loudspeakers, and two ears (one listener) exist. Theoretically, the matrix $[Z]$ in Equation (9.18a) should be a 2×2 identity matrix, and the 2×2 crosstalk cancellation matrix $[A]$ can be precisely derived from Equation (9.20), as shown in Equation (9.6).

The second example is reproducing binaural signals $E_1 = E_L$ and $E_2 = E_R$ through M loudspeakers for $Q/2$ listeners. In this case, $K = 2$ and $M = Q \geq 4$ is an even number. Ideally, to deliver desired binaural signals to each listener, matrix $[Z]$ should take the form of the $Q \times 2$ matrix:

$$[Z] = \begin{bmatrix} 1 & 0 \\ 0 & 1 \\ 1 & 0 \\ 0 & 1 \\ \vdots & \vdots \\ \vdots & \vdots \\ 1 & 0 \\ 0 & 1 \end{bmatrix}. \qquad (9.25)$$

If the loudspeaker configuration and listener locations are chosen so that rank $[H] = M$, the crosstalk cancellation matrix $[A]$ can be precisely obtained from Equation (9.20). M loudspeakers can, therefore, simultaneously create the desired binaural signals for $M/2$ listeners, which is an important conclusion.

The third example is reproducing binaural signals $E_1 = E_L$ and $E_2 = E_R$ through three frontal loudspeakers for one listener (two ears). In this case, $K = 2$, $M = 3$, and $Q = 2$—fewer receiving points than loudspeakers exist. For simplicity, the left-right symmetric case is considered. Let $H_{LL} = H_{RR} = H_\alpha$, $H_{RL} = H_{LR} = H_\beta$, and $H_{LC} = H_{RC} = H_\gamma$ denote the transfer functions from each of the three loudspeakers to two ears (Figure 9.4). Ideally, the matrix $[Z]$ in Equation (9.18a) should be a 2×2 identity matrix. Then, the minimum least square solution of the 3×2 crosstalk cancellation matrix $[A]$ is given by Equation (9.21) as:

$$A = \begin{bmatrix} A_{11} & A_{12} \\ A_{21} & A_{22} \\ A_{31} & A_{32} \end{bmatrix}, \tag{9.26}$$

and:

$$A_{11} = A_{32} = X_1 = \frac{M_1 H_\alpha^* - M_2 H_\beta^*}{M_1^2 - M_2^2} \qquad A_{12} = A_{31} = X_2 = \frac{M_1 H_\beta^* - M_2 H_\alpha^*}{M_1^2 - M_2^2},$$

$$A_{21} = A_{22} = X_3 = \frac{M_3 H_\gamma^*}{M_1^2 - M_2^2}, \tag{9.27}$$

$$M_1 = |H_\alpha|^2 + |H_\beta|^2 + |H_\gamma|^2 \qquad M_2 = |H_\gamma|^2 + H_\alpha H_\beta^* + H_\beta H_\alpha^*,$$

$$M_3 = |H_\alpha|^2 + |H_\beta|^2 - H_\alpha H_\beta^* - H_\beta H_\alpha^*.$$

where superscript "*" denotes complex conjugation. The loudspeaker signals are given by:

$$\begin{bmatrix} L' \\ C' \\ R' \end{bmatrix} = \begin{bmatrix} A_{11} & A_{12} \\ A_{21} & A_{22} \\ A_{31} & A_{32} \end{bmatrix} \begin{bmatrix} E_L \\ E_R \end{bmatrix}. \tag{9.28}$$

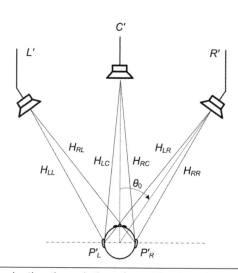

Figure 9.4 Transaural reproduction through three frontal loudspeakers

If the binaural synthesis in Equation (1.36) is merged with crosstalk cancellation, then the transaural synthesis in Equation (9.28) becomes:

$$L' = (X_1 H_L + X_2 H_R)E_0 \quad C' = X_3(H_L + H_R)E_0 \quad R' = (X_2 H_L + X_1 H_R)E_0. \quad (9.29)$$

Therefore, loudspeaker signals L', C', and R' for the virtual sources at direction θ can be synthesized by directly filtering a mono stimulus $E_0(f)$ with the three transaural filters specified in Equation (9.29). Researchers have demonstrated other cases of binaural or transaural reproduction through multiple loudspeakers. For details, refer to Bauck and Cooper (1996).

9.2. Head Rotation and Loudspeaker Reproduction

9.2.1. Virtual Source Distribution in Two-Front Loudspeaker Reproduction

Binaural pressures or signals contain primary localization information. Provided that binaural signals are precisely replicated, headphone-based binaural reproduction can recreate perceived virtual sources at arbitrary horizontal, or even three-dimensional, directions. As discussed in Section 9.1, loudspeaker-based binaural reproduction, in which crosstalk cancellation is incorporated, can also generate the same binaural pressures as those produced by a real source. Theoretically, therefore, this reproduction technique can also recreate perceived virtual sources at arbitrary horizontal or three-dimensional directions. In practical situations, however, such recreation is not implemented. Some authors reported that, under a series of critical conditions (e.g., individualized HRTF processing, restriction of head movement, reproduction in anechoic rooms, etc.), a pair of frontal loudspeakers can recreate perceived virtual sources at all horizontal or three-dimensional directions for listeners, to some extent (Takeuchi et al., 1998). The experimental results indicate that, in reproduction performed in a reflective room or with nonindividualized HRTF processing, perceived virtual source positions are restricted in the region of frontal-horizontal quadrants (i.e., $\phi = 0°$, $0° \leq \theta \leq 90°$ or $270° \leq \theta \leq 360°$) (Nelson et al., 1996; Gardner, 1997). The virtual sources intended for rear-horizontal quadrants are often perceived at the mirror position in frontal-horizontal quadrants. For example, the horizontal virtual source intended for $\theta = 135°$ is often perceived at $\theta = 45°$. The virtual sources intended for outside the horizontal plane ($\phi \neq 0°$) are often perceived in frontal-horizontal quadrants in the same cone of confusion (Section 1.4.3).

As stated in Sections 1.4 and 3.4.1, the spectral cues introduced by pinna reflection and diffraction are essential to front-back and vertical localization, but these cues are highly individual dependent and sensitive to various errors. Moreover, the pinna effect is effective only for high-frequency sound waves with wavelengths comparable to the pinna dimension. HRTFs are approximately front-back symmetric below 1 kHz. The dynamic cue caused by head turning is another important cue for front-back and vertical localization, but has been disregarded in conventional static binaural reproduction through loudspeakers. Even if pinnae-induced spectral cues are carefully manipulated in loudspeaker reproduction, such cues work within a very limited listening region because high frequencies have short wavelengths. A distance deviation of 1/4 to 1/2 a wavelength from the optimal (default) listening position, or "sweet point," likely causes a radical change in binaural pressure. Therefore, the high-frequency spectral cues caused by pinnae cannot be stably replicated in loudspeaker reproduction. Conventional reproduction with two (or three) frontal loudspeakers cannot recreate stable perceived virtual sources in rear-horizontal quadrants or full three-dimensional directions. By contrast, if slight head rotation is allowed in reproduction with a pair of frontal loudspeakers, the incorrect but dominant dynamic localization cue caused by head rotation makes the virtual source intended for rear-horizontal quadrants to be perceived at the mirror azimuth in frontal-horizontal quadrants.

9.2.2. Transaural Synthesis for Four-Loudspeaker Reproduction

To recreate perceived virtual sources in full 360° (all 360° in horizontal directions), researchers suggested some methods for binaural or transaural reproduction through multiple loudspeakers. The most straightforward method is to reproduce transaural signals through four loudspeakers (Gierlich, 1992). That is, the transaural signals for creating virtual sources in frontal-horizontal quadrants are synthesized according to Equations (9.11) and (9.12b), and then reproduced by a pair of frontal loudspeakers. The transaural signals for creating virtual sources in rear-horizontal quadrants are synthesized similar to Equations (9.11) and (9.12b) (with H_α and H_β replaced by the transfer functions from each rear loudspeaker to the ipsilateral and contralateral ears, respectively), and then reproduced by a pair of rear loudspeakers. This approach implies that the front and rear loudspeaker pairs are responsible for the virtual sources in frontal- and rear-horizontal quadrants, respectively.

To incorporate correct dynamic localization cues, Hill et al. (2000) analyzed the ITD (interaural time difference) variations associated with head rotation, and proposed an alternative algorithm for transaural synthesis and reproduction through four loudspeakers. The HRTFs of a rigid-spherical head model with offset ears located at 100°/260° (Section 4.1.3) were used in the analysis.

As shown in Figure 9.5, four loudspeakers, labeled 1, 2, 3, and 4, are symmetrically arranged at the front and rear of the horizontal plane. A mono input stimulus $E_0(f)$ is filtered with four transaural filters: $G_1(\theta,f)$, $G_2(\theta,f)$, $G_3(\theta,f)$, and $G_4(\theta,f)$. The resultant four signals—S_1, S_2, S_3, and S_4—are reproduced by four loudspeakers:

$$\begin{bmatrix} S_1 \\ S_2 \\ S_3 \\ S_4 \end{bmatrix} = \begin{bmatrix} G_1(\theta,f) \\ G_2(\theta,f) \\ G_3(\theta,f) \\ G_4(\theta,f) \end{bmatrix} E_0(f) \tag{9.30}$$

or simply:

$$S = GE_0, \tag{9.31}$$

where S and G are a 4 × 1 signal vector (column matrix) and filter response vector, respectively.

Four receiving (field) points with slightly forward or backward deviations from real ear locations are of interest: point 1 at azimuth 265°, point 2 at azimuth 95°, point 3 at azimuth 255°, and point 4 at azimuth 105°. Let H_{qm} ($q = 1, 2, 3, 4$, $m = 1, 2, 3, 4$) denote the transfer functions from the mth loudspeaker to the qth receiving point. Then, the pressures at the four receiving points are given by:

$$\begin{bmatrix} P_1' \\ P_2' \\ P_3' \\ P_4' \end{bmatrix} = \begin{bmatrix} H_{11} & H_{12} & H_{13} & H_{14} \\ H_{21} & H_{22} & H_{23} & H_{24} \\ H_{31} & H_{32} & H_{33} & H_{34} \\ H_{41} & H_{42} & H_{43} & H_{44} \end{bmatrix} \begin{bmatrix} S_1 \\ S_2 \\ S_3 \\ S_4 \end{bmatrix} = \begin{bmatrix} H_{11} & H_{12} & H_{13} & H_{14} \\ H_{21} & H_{22} & H_{23} & H_{24} \\ H_{31} & H_{32} & H_{33} & H_{34} \\ H_{41} & H_{42} & H_{43} & H_{44} \end{bmatrix} \begin{bmatrix} G_1(\theta,f) \\ G_2(\theta,f) \\ G_3(\theta,f) \\ G_4(\theta,f) \end{bmatrix} E_0(f) \tag{9.32a}$$

or simply:

$$\tilde{P}' = [\tilde{H}]S = [\tilde{H}]GE_0, \tag{9.32b}$$

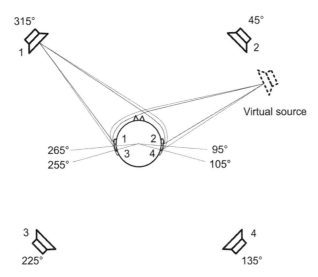

Figure 9.5 Method for transaural reproduction through four loudspeakers (adapted from Hill et al., 2000)

where $\tilde{P}'$ is a 4 × 1 pressure vector (column matrix) and $[\tilde{H}]$ is a 4 × 4 transfer matrix. The notation "~" denotes the analysis results for the four receiving points, to distinguish these from the results for the two receiving points (ears) in subsequent discussions.

Let $H_1(\theta, f)$, $H_2(\theta, f)$, $H_3(\theta, f)$, and $H_4(\theta, f)$ denote the four acoustic transfer functions from a real source at direction θ to the four receiving points. The real source-induced pressures at the four receiving points are calculated by:

$$\begin{bmatrix} P_1 \\ P_2 \\ P_3 \\ P_4 \end{bmatrix} = \begin{bmatrix} H_1(\theta, f) \\ H_2(\theta, f) \\ H_3(\theta, f) \\ H_4(\theta, f) \end{bmatrix} E_0(f) \tag{9.33a}$$

or simply:

$$\tilde{P} = H_\theta E_0. \tag{9.33b}$$

Let the summing pressures at the four receiving points in loudspeaker reproduction match those caused by a real source, $\tilde{P}' = \tilde{P}$. (This condition also results in binaural pressures of loudspeaker reproduction that are approximate to those of a real source in terms of adjacent linear interpolation.) We have:

$$[\tilde{H}]G = H_\theta. \tag{9.34}$$

If matrix $[\tilde{H}]$ is non-singular and therefore invertible, the 4 × 1 vector for the responses of four transaural filters are given by:

$$G = [\tilde{H}]^{-1} H_\theta. \tag{9.35}$$

Accordingly, the resultant 4 × 4 crosstalk cancellation matrix is:

$$[A] = [\tilde{H}]^{-1}. \tag{9.36}$$

However, matrix $[\tilde{H}]$ is inherently frequency dependent and may be non-invertible at some frequencies. Therefore, the method for calculating vector G in Equation (9.35) or matrix $[A]$ in Equation (9.36) is not always feasible. Hill et al. (2000) minimized the cost function to obtain the solution of Equation (9.34), as will be discussed in Section 9.7.1.

Lee et al. (2004) suggested using four or five microphones, installed on the surface of a rigid sphere, to record sound pressure signals and then reproduce these signals through four or five loudspeakers after crosstalk cancellation. The basic principle of this suggestion is similar to that of Hill et al. (2000).

9.2.3. Analysis of Dynamic Localization Cues in Loudspeaker Reproduction

The variations in binaural pressures with head rotation are examined to analyze the dynamic localization cues in loudspeaker reproduction. For transaural reproduction through a pair of frontal loudspeakers, the binaural pressures of a head oriented, by default, toward the front are calculated according to Equations (9.2) to (9.10):

$$
\begin{bmatrix} P_L{}' \\ P_R{}' \end{bmatrix} = \begin{bmatrix} H_{LL} & H_{LR} \\ H_{RL} & H_{RR} \end{bmatrix} \begin{bmatrix} L' \\ R' \end{bmatrix} = \begin{bmatrix} H_{LL} & H_{LR} \\ H_{RL} & H_{RR} \end{bmatrix} \begin{bmatrix} A_{11} & A_{12} \\ A_{21} & A_{22} \end{bmatrix} \begin{bmatrix} E_L(\theta,f) \\ E_R(\theta,f) \end{bmatrix}, \tag{9.37a}
$$

or simply:

$$
P' = [H]S = [H][A]E, \tag{9.37b}
$$

where $P' = [P_L', P_R']^T$, $S = [L', R']^T$, $E = [E_L, E_R]^T$ are the three 2×1 vectors (column matrices) that represent binaural pressures, loudspeaker signals, and synthesized binaural signals, respectively. $[H]$ is a 2×2 transfer matrix that consists of the transfer functions from two loudspeakers to two ears, and $[A]$ is a 2×2 crosstalk cancellation matrix. Under the ideal condition $[H][A] = [I]$ (2×2 identity matrix), crosstalk is completely cancelled out. Matrix $[A]$ is designed according to this ideal condition. In this case, the binaural pressures in reproduction are consistent with those of real sources.

Head rotation or other factors (such as head translation or mismatched HRTFs, to be discussed in Sections 9.3 and 9.4) alter the transfer functions from loudspeakers to two ears, as well as binaural pressures. Thus, the condition for complete crosstalk cancellation does not hold. The 2×2 transfer matrix $[H]$ in Equation (9.37b) is then replaced by a new transfer matrix $[H']$, and the binaural pressures become:

$$
\begin{bmatrix} P_L{}' \\ P_R{}' \end{bmatrix} = \begin{bmatrix} H'_{LL} & H'_{LR} \\ H'_{RL} & H'_{RR} \end{bmatrix} \begin{bmatrix} L' \\ R' \end{bmatrix} = \begin{bmatrix} H'_{LL} & H'_{LR} \\ H'_{RL} & H'_{RR} \end{bmatrix} \begin{bmatrix} A_{11} & A_{12} \\ A_{21} & A_{22} \end{bmatrix} \begin{bmatrix} E_L(\theta,f) \\ E_R(\theta,f) \end{bmatrix}, \tag{9.38a}
$$

or simply:

$$
[P'] = [H']S = [H'][A]E. \tag{9.38b}
$$

To evaluate the variations in binaural pressures with head rotation, the case of a real source is analyzed first. For a fixed head oriented toward the front, the binaural pressures of a real source at the horizontal azimuth θ are calculated according to Equation (1.20) thus:

$$
P_L = H_L(\theta,f)P_0(f) \qquad P_R = H_R(\theta,f)P_0(f). \tag{9.39}
$$

When the head rotates around the vertical axes with azimuth $\Delta\theta$ (a positive value of $\Delta\theta$ represents rotation to the right and a negative value denotes rotation to the left), the binaural pressures become:

$$P_L = H_L(\theta - \Delta\theta, f)P_0(f) \qquad P_R = H_R(\theta - \Delta\theta, f)P_0(f). \qquad (9.40)$$

In reproduction with two frontal loudspeakers, the binaural pressures or HRTFs associated with each loudspeaker vary with head rotation in the same manner as that observed in Equation (9.40). For example, for a right loudspeaker at azimuth 30°, head rotation changes the original $H_{LR} = H_L(30°, f)$ into $H_L(30° - \Delta\theta, f)$. The resultant binaural pressures are calculated by substituting the HRTFs that correspond to the rotated head orientation into Equation (9.38a).

The preceding analysis can be generalized to reproduction with more than two loudspeakers. Let $E = [E_L, E_R]^T$ denote synthesized binaural signals. The M loudspeaker signals $S = [S_1, S_2, \ldots S_M]^T$ are derived from an $M \times 2$ crosstalk matrix $[A]$ and E thus:

$$S = [A]E. \qquad (9.41)$$

Let $[H]$ denote the $2 \times M$ transfer matrix from M loudspeakers to two ears. Similar to the binaural pressures calculated in Equation (9.38b), those of a fixed head oriented toward the front are calculated by:

$$P' = [H]S = [H][A]E. \qquad (9.42)$$

Matrix $[A]$ is usually designed accurately or approximately according to the ideal condition $[H][A] = [I]$ (2×2 identity matrix), so that crosstalk is completely cancelled out, resulting in binaural pressures that are consistent with those of a real source. When the head rotates around the vertical axes with azimuth $\Delta\theta$, the transfer functions from loudspeakers to two ears change, and binaural pressures can be calculated similarly to the computations in Equation (9.38b):

$$P' = [H']S = [H'][A]E. \qquad (9.43)$$

To analyze the dynamic localization cues introduced by head rotation, the normalized interaural cross-correlation function $\Phi_{LR}(\tau)$ and corresponding ITD are calculated and the results of a real source and transaural reproduction are compared. Let $p_L(t)$ and $p_R(t)$ denote the binaural pressures in the time domain. $\Phi_{LR}(\tau)$ can be calculated similarly to the computation in Equation (3.16):

$$\Phi_{LR}(\tau) = \frac{\displaystyle\int_{-\infty}^{+\infty} p_L(t+\tau)p_R(t)dt}{\left\{\left[\displaystyle\int_{-\infty}^{+\infty} p_L^2(t)dt\right]\left[\displaystyle\int_{-\infty}^{+\infty} p_R^2(t)dt\right]\right\}^{1/2}}, \qquad (9.44)$$

which can be converted into the frequency domain by Fourier transform:

$$\Phi_{LR}(\tau) = \frac{\displaystyle\int_{-\infty}^{+\infty} P_L(f)P_R^*(f)\exp(j2\pi f\tau)df}{\left\{\left[\displaystyle\int_{-\infty}^{+\infty} |P_L(f)|^2\,df\right]\left[\displaystyle\int_{-\infty}^{+\infty} |P_R(f)|^2\,df\right]\right\}^{1/2}}. \qquad (9.45)$$

Then, the interaural cross-correlation coefficient—IACC, the maximal value of $\Phi_{LR}(\tau)$ within the range $|\tau| \le 1$ ms, is evaluated, and the corresponding time value τ_{max} is defined as ITD.

$\Phi_{LR}(\tau)$ is calculated here in terms of binaural pressures, and the results (as well as the ITD) are related to the power spectrum $|P_0(f)|^2$ of the sound source [Equation (9.39)]. $\Phi_{LR}(\tau)$ is calculated in terms of HRIRs (head-related impulse responses) or HRTFs in Equations (3.16) and (3.18), a computation that is equivalent to assuming that $|P_0(f)|^2$ is a constant.

Hill et al. (2000) compared the ITDs and their variations with head rotation in three cases: a real source, transaural reproduction through two frontal loudspeakers, and transaural reproduction through four loudspeakers (Figure 9.5). In the reproduction with two frontal loudspeakers, the loudspeakers were arranged at horizontal azimuths of 330° and 30°, as in conventional stereophonic loudspeaker configuration. In the four loudspeaker reproduction, the loudspeakers were arranged at horizontal azimuths of 315°, 45°, 135°, and 225°.

In all three cases, the $\Phi_{LR}(\tau)$ and ITD of a head oriented by default toward the front and the rotated head orientation with an azimuth of $\Delta\theta = \pm 5°$ were analyzed. The input stimulus $E_0(f)$ in Equation (9.11) or Equation (9.30) was random noise within a frequency range of 20 Hz to 20 kHz. $\Phi_{LR}(\tau)$ and ITD were evaluated at target horizontal virtual source azimuths, varying from 0° to 355° at increments of 5°. The HRTFs used in the analysis were calculated from a rigid-spherical head with radius $a = 0.0875$ m, backward ears at $\theta_R = 100°$ and $\theta_L = 260°$, and an incident plane wave.

The calculation results indicate that, for a default head orientation, two- and four-loudspeaker reproductions yield ITDs that are consistent with that of a real source, resulting in correct ITD cues for localization. However, ITD alone does not provide information sufficient for resolving front-back ambiguity in horizontal localization (see Figure 4.6).

For four-loudspeaker reproduction, the variations in ITD with head rotation exhibit a pattern (positive or negative change) similar to that of a real source. The difference in ITD variation patterns between target virtual sources in front and rear quadrants provides correct dynamic cues for resolving front-back ambiguity in horizontal localization. Two ears are located at $\theta_R = 100°$ and $\theta_L = 260°$ on a fixed head surface. Hill et al. (2000) designed transaural processing for four-loudspeaker reproduction, so that the pressures at the four receiving points (100° ± 5°, 260° ± 5°) adjacent to two ears match those caused by a real source (Section 9.2.2). When the head rotates $\Delta\theta = +5°$ or −5°, the locations of the two ears changes to new locations, which are coincident with two of four receiving points of interest. Consequently, the resultant binaural pressures are equal to those caused by the real source.

In reproduction with two frontal loudspeakers, however, the variations in ITD with head rotation exhibit a similar pattern as that of the real source in front quadrants, despite the target virtual source being located at the front or rear quadrants. Therefore, head rotation provides correct dynamic localization cues only for target virtual sources in front quadrants; it provides inconsistent or conflicting dynamic cues for target virtual sources at rear quadrants. Accordingly, target virtual sources in front quadrants are perceived in the correct quadrant, and target virtual sources at rear quadrants are perceived in the mirror-reflected direction in front quadrants, as stated in Section 9.2.1. By contrast, transaural reproduction through four loudspeakers can resolve front-back confusion, thereby recreating virtual sources in a full 360° horizontal plane. Hill et al. (2000) conducted a localization experiment to validate the aforementioned findings.

9.2.4. Stability of the Perceived Virtual Source Azimuth against Head Rotation

As stated in previous sections, dynamic cues, especially the variations in ITD with head rotation, are crucial to resolving front-back confusion in loudspeaker reproduction. For reproduction with two frontal loudspeakers, dynamic cues enable the virtual source intended for front quadrants to be perceived in the correct quadrant. However, because ITD variations exhibit merely similar patterns, rather than quantitatively matching those of a real source, the perceived virtual source azimuth relative to a fixed spatial coordinate may vary with head rotation, causing instability in perceived virtual source directions. Therefore, analyzing the variation patterns of ITDs alone is insufficient, and the effect of head rotation on the perceived virtual source azimuth should be quantitatively examined.

Takeuchi et al. (2001a) analyzed the variations in ITD and spectral cues with head rotation for reproduction with two frontal loudspeakers and different span angles. The procedure for analyzing ITD was similar to those discussed in previous sections, except that MIT-KEMAR HRTFs were used. The synthesized binaural signals E_L and E_R in Equation (9.37a) make up a pair of simultaneous Dirac delta functions $\delta(t)$. The results indicate that ITD is 0 for a default head orientation. When the head rotates around the vertical axes, ITD deviates from its initial 0 value. For conventional loudspeaker configuration with a 60° span angle, the variation rate of ITD with head rotation is about 1.2 μs/°, whereas for loudspeaker configuration with a narrow 10° span angle (stereo dipole, see Section 9.3.2), the variation rate of ITD is about 0.4 μs/°. Therefore, loudspeaker configuration with narrow spans reduces the variations in ITD with head rotation. However, head rotation alters the source direction relative to a listener and, therefore, changes the ITD—even for a real source at fixed spatial positions. It is the variations in ITD with head rotation that generate dynamic localization cues. However, head rotation does not alter the absolute perceived direction of a real source relative to a fixed spatial coordinate. Therefore, a low ITD variation rate in two-loudspeaker reproduction does not translate to correct dynamic localization cues or perceived virtual source positions that are stable against head rotation.

Xie (2005b) analyzed the variations in the perceived virtual source azimuth with head rotation in reproduction with two and three frontal loudspeakers. As stated in Section 9.2.3, for two-loudspeaker reproduction and a default head orientation, crosstalk is cancelled out, and the binaural pressures in transaural reproduction are identical to those of a real source. In this case, the virtual sources intended for frontal-horizontal quadrants are perceived at the correct direction.

When the head rotates around the vertical axes with an azimuth $\Delta\theta$, the transfer functions from loudspeakers to two ears change, and binaural pressures are recalculated using Equation (9.38a). The interaural cross-correlation function $\Phi_{LR}(\tau)$ and corresponding ITD are reevaluated using the procedures similar to those in Section 9.2.3. Suppose that ITD is a dominant azimuthal localization cue. The perceived virtual source azimuth is then evaluated by comparing the ITD of loudspeaker reproduction and that of a real source (the perceived azimuth evaluated in this manner can be regarded as a rough approximation of the "mean" or "central" position of a virtual source with a certain spectral bandwidth). The perceived azimuth θ' obtained here is relative to the new head orientation (after rotation). To compare θ' with the initial or target virtual source azimuth θ_S (before rotation), the former should be converted back into the initial spatial coordinate without head rotation—$\theta_I = \theta' + \Delta\theta$, with $\Delta\theta$ denoting the azimuth of head rotation.

Using similar procedures, Xie (2005b) also evaluated the variations in perceived virtual source azimuth with head rotation in three-loudspeaker reproduction (Figure 9.4). The author analyzed four kinds of loudspeaker configurations, including reproduction with three kinds of two-front loudspeakers and span angles of 60°, 30°, and 10° (stereo dipole), as well as reproduction with three frontal loudspeakers with a 60° span angle between left and right loudspeakers. The input stimulus was pink noise, which was subjected to low-pass filtering and had a cutoff frequency of 3.0 kHz. The HRTFs calculated from a rigid-spherical head model (Section 4.1.1) were used. The calculations indicate that the four kinds of loudspeaker configurations exhibit similar variation patterns, summarized as follows:

1. For virtual sources intended for the frontal region of $0° \leq \theta_S \leq +45°$ or $315° \leq \theta_S < 360°$, head rotation with a moderate azimuth causes only a slight change in the perceived virtual source azimuth. For example, at a counterclockwise (to the left) head rotation with an azimuth $\Delta\theta = -20°$, the variation in perceived azimuth $|\theta_I - \theta_S|$ is less than $4.0°$ for all target azimuths. In this case, the perceived virtual source is stable against head rotation.

2. For virtual sources intended for lateral directions, head rotation toward the intended virtual source causes the perceived virtual source to move toward the front. For example, for a two front loudspeaker configuration with a $60°$ span angle, a counterclockwise (to the left) head rotation with an azimuth $\Delta\theta = -20°$ causes the intended virtual source at azimuth $\theta_S = 270°$ to be perceived at azimuth $\theta_I = 288°$.

3. For virtual sources intended for lateral directions (e.g., $\theta_S = 75°$ or $90°$) and head rotation opposite the target virtual source, the resultant absolute value of ITD exceeds the possible maximal ITD with a real source at $\theta = 90°$ or $270°$. Therefore, the perceived azimuth cannot be predicted by calculation. This case corresponds to two practical possibilities. One is that no definite position exists for auditory events or perceived virtual sources. The other is that an unnatural auditory event or virtual source is perceived at certain directions.

4. The virtual sound localization experiment validates these results.

In conclusion, head rotation imposes less influence on the perceived virtual source azimuth intended for the front region, but exerts a considerable effect on the virtual sources intended for lateral directions, an issue to be discussed further in Section 9.4. Four kinds of loudspeaker configurations exhibit similar results, suggesting that no loudspeaker configuration presents an advantage over another, with regard to the stability of the perceived virtual source azimuth against head rotation. Therefore, optimal loudspeaker configurations should be selected in accordance with criteria other than the variations in ITD with head rotation. This conclusion contrasts with the results of Takeuchi et al., insofar as Takeuchi analyzed the variations in ITD rather than the actual perceived virtual source azimuth generated by head rotation. Using an artificial head, other researchers measured the influence of head rotation (and translation) on crosstalk cancellation (Prodi and Velecka, 2003).

9.3. Head Translation and Stability of Virtual Sources in Loudspeaker Reproduction

As discussed in Sections 9.1 and 9.2, the key to binaural or transaural reproduction through loudspeakers is crosstalk cancellation, or, equally, control over binaural pressures. However, crosstalk cancellation exhibits position-dependent performance. For a given loudspeaker configuration, crosstalk cancellation is effective in a very limited region (the sweet point). Similar to the case of head rotation, head deviation from the default (optimal) position negatively affects crosstalk cancellation and thereby alters binaural pressures. The limited size of the listening region is a common defect in loudspeaker reproduction. This section analyzes the stability of loudspeaker reproduction against head translation and discusses an important loudspeaker configuration, the stereo dipole.

9.3.1. Preliminary Analysis of Head Translation and Stability

Although deriving the quantitative relationship between head translation and the stability of loudspeaker reproduction is a complex task, a qualitative analysis of this issue is relatively simple, from which we can obtain insight into the problem of loudspeaker configuration and the size of the listening region (Xie et al., 2005c).

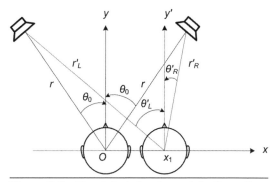

Figure 9.6   Head deviation from default position

A pair of loudspeakers with a span angle of $2\theta_0$ are arranged at azimuths $\theta_L = 360° − \theta_0$ and $\theta_R = \theta_0$ (Figure 9.6). If a listener is located at the default position (central line), the distances of two loudspeakers relative to the head center (origin of coordinates) are identical. When the head center laterally deviates from the default position to the right and is located at coordinate $(x_1, 0)$, the distances of the two loudspeakers relative to the head center become unequal with:

$$r_L' = \left[(r\sin\theta_0 + x_1)^2 + (r\cos\theta_0)^2 \right]^{1/2} \qquad r_R' = \left[(r\sin\theta_0 - x_1)^2 + (r\cos\theta_0)^2 \right]^{1/2}.$$

Correspondingly, the difference in path length is calculated by $\Delta r = r_L' - r_R'$ or the difference in arrival time is:

$$\Delta t = \frac{\Delta r}{c} = \frac{r_L' - r_R'}{c}$$

where c is the speed of sound. This arrival time difference is equivalent to an additional phase difference between the loudspeaker signals:

$$\Delta\psi = 2\pi f \Delta t \approx \frac{4\pi f x_1 \sin\theta_0}{c}, \tag{9.46}$$

where $x_1 \ll r$ is assumed in the second approximate equality. When head translation exceeds a certain distance so that $|\Delta\psi| \geq \pi/2$, that is, when the difference in path length exceeds a quarter of a wavelength, the localization information encoded in the relative phase between the left and right loudspeaker signals is destroyed. Accordingly, the size of the listening area can by estimated by:

$$|x_1| \leq x_{max} = \frac{c}{8 f \sin\theta_0}, \tag{9.47}$$

where x_{max}, which is inversely proportional to frequency, is regarded as the limited distance for tolerable lateral head translation. As a result, the listening region is small at high frequencies. x_{max} is also inversely proportional to $\sin\theta_0$. Thus, a narrow span angle between left and right loudspeaker pairs expands the listening region. This conclusion is important, as it suggests that the x_{max} for loudspeaker configuration with a 30° span angle is about twice that of conventional loudspeaker configuration with a 60° span angle. For example, at a frequency of 1.5 kHz, $x_{max} = 0.11$ m for a 30° span angle and $x_{max} = 0.06$ m for a 60° span angle.

Alternatively, some authors suggested using a pair of elevated loudspeakers, symmetrically arranged left and right in the lateral plane, to improve the stability of transaural reproduction (Takeuchi et al., 2002a). However, elevated loudspeaker configuration neither introduces conflicting dynamic localization cues with head rotation nor provides useful dynamic cues for resolving front-back confusion. Therefore, this method is effective only under restricted head positions. Parodi and Rubak (2010) evaluated the size of the listening region for elevated loudspeakers arranged at different elevation angles.

9.3.2. Stereo Dipole

Early in 1998, Kirkeby et al. (1998a, 1998b) suggested that frontal loudspeaker configuration with narrow span angles enhances the stability of virtual sources against lateral head translation. The authors drew this conclusion by analyzing the reproduced wave front in free fields, or by evaluating binaural pressures with a rigid-spherical head model. In contrast to this approach, however, the analysis given in Section 9.3.1 is straightforward.

To expand the listening region, Kirkeby et al. proposed a pair of frontal loudspeakers at $\theta_L = 355°$ and $\theta_R = 5°$ with span $2\theta_0 = 10°$ for binaural or transaural reproduction. This concept is called the *stereo dipole* (Figure 9.7).

Crosstalk cancellation is also required for a stereo dipole, as for other loudspeaker configurations. The cancellation for a stereo dipole is shown in Equations (9.1) and (9.6). Transaural synthesis processing for a stereo dipole is provided by Equations (9.11) and (9.12a). Figure 9.8 shows the magnitude responses of the transaural synthesis filters $G_L(90°, f)$ and $G_R(90°, f)$ of the stereo dipole for a target horizontal virtual source azimuth $\theta = 90°$. For comparison, Figure 9.8 illustrates the corresponding magnitude responses of a conventional 60° loudspeaker span. The HRTFs are calculated from a rigid-spherical head with radius $a = 0.0875$ m and two ears located diametrically opposite across a spherical head (Section 4.1.1).

The magnitude responses of the stereo dipole visibly increase at low frequencies. That is, recreating a lateral virtual source at azimuth 90° necessitates that the low-frequency component of the input stimulus $E_0(f)$ in Equation (9.11) be considerably boosted, a task that makes for difficult signal processing. Here, an illustrative case for target virtual sources at azimuth 90° is discussed. The closer the target virtual source to the lateral direction, the larger the boost required for the low-frequency component of input stimulus. Furthermore, for target virtual sources at azimuth 90°, two loudspeaker signals are out of phase at low frequencies. This phenomenon is similar to the case in Section 1.6.1, in which the summing localization of virtual sources outside the span of a conventional stereophonic loudspeaker pair is recreated by out-of-phase loudspeaker signals at low frequencies. In addition, a closely spaced loudspeaker pair with out-of-phase signals serves as an acoustical dipole, which is the origin of the term "stereo dipole." The close space between two loudspeakers results in a slight difference in path lengths, from the two loudspeakers to either ear. As a result, the out-of-phase loudspeaker signals, small difference in path length, and large wavelength generate small summing low-frequency pressures at either ear. Generating reasonable ear pressures at low frequencies necessitates a considerable increase in the low-frequency component of input stimulus. From the perspective of mathematics, the differences among the HRTFs from two close loudspeakers to two ears are small at low frequencies, so that the transfer matrix $[H]$ in Equation (9.6) is ill-conditioned and therefore nearly non-invertible. These features substantially increase the magnitude responses of a crosstalk canceller and transaural synthesis filters at low frequencies. Therefore, a stereo dipole improves the stability

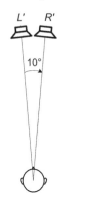

Figure 9.7 The stereo dipole

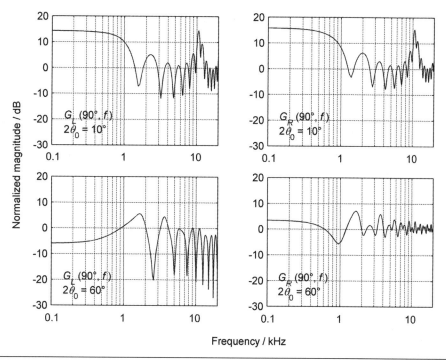

Figure 9.8   Magnitude responses of the transaural synthesis filters for the stereo dipole with $2\theta_0 = 10°$ and conventional loudspeaker configuration with $2\theta_0 = 60°$; the target virtual source azimuth is $\theta = 90°$

of virtual sources against lateral head translation and expands the listening region, but at the cost of difficult signal processing. In practice, all these aspects should be comprehensively considered in choosing the span of loudspeakers.

9.3.3. Quantitative Analysis of Stability against Head Translation

When the head laterally deviates from the default position (central line) (Figure 9.6), the transfer functions from two loudspeakers to two ears become H'_{LL}, H'_{RL}, H'_{LR}, and H'_{RR}. Similar to those in Equation (9.38b), the binaural pressures in loudspeaker reproduction are:

$$P' = [H']S = [H'][A]E, \tag{9.48}$$

where $[H']$ represents the new transfer matrix from two loudspeakers to two ears after head translation, instead of the head rotation in Equation (9.38b).

Various studies analyzed the effect of lateral head translation on the stability of virtual sources in loudspeaker reproduction. Takeuchi et al. (2001a) examined variations in ITD and spectral cues with lateral head translation. The ITDs were evaluated using MIT-KEMAR HRTFs, and the procedures were similar to those discussed in Section 9.2.3. The synthesized binaural signals E_L and E_R in Equation (9.37a) make up a pair of simultaneous Dirac delta functions $\delta(t)$. The ITD is 0 for a default head position, but deviates from this value when the head laterally deviates from the default position. For conventional loudspeaker configuration with a 60° span angle, the variation rate of ITD with head translation is about 2.7 μs/mm, whereas that with a stereo dipole is about 0.2 μs/mm, assuming that the noticeable difference in ITD is 10 μs. A 4-mm lateral head translation in conventional loudspeaker configuration is sufficient to cause a noticeable variation in ITD.

By contrast, a 50-mm lateral head translation is required for the stereo dipole. The ITD variation rate for the stereo dipole is considerably less than that for conventional loudspeaker configuration. Therefore, a loudspeaker configuration with a narrow span angle reduces ITD variations with lateral head translation. This conclusion holds for virtual sources at other intended horizontal azimuths.

Similar to head rotation, head translation alters the source direction relative to a listener and therefore changes the ITD—even for a real source at a fixed spatial location. However, when head translation is considerably less than source distance, the variations in real source direction relative to a listener are small. Therefore, a low variation rate of ITD with head translation partially accounts for the stability of the perceived virtual source azimuth.

Some other studies focused on the influence of head translation on crosstalk cancellation. Ideal crosstalk cancellation is achieved in a default head position. Moderate crosstalk occurs when the head deviates from the default position, a phenomenon that can be quantitatively described by channel separation or channel crosstalk in resultant binaural signals. Under the ideal condition, the matrix $[R] = [H'][A]$ in Equation (9.48) is a 2×2 identity matrix. Let:

$$[R] = \begin{bmatrix} R_{11} & R_{12} \\ R_{21} & R_{22} \end{bmatrix}$$

The non-zero value of any non-diagonal element of matrix $[R]$ indicates crosstalk. Then, channel separation is defined as:

$$J_L = 20\log_{10}\left|\frac{R_{11}}{R_{12}}\right| \quad (dB) \qquad J_R = 20\log_{10}\left|\frac{R_{22}}{R_{21}}\right| \quad (dB). \tag{9.49}$$

Equation (9.49) describes the ratio (in units of dB) between the desired ear pressures and the ear pressure error caused by crosstalk in reproduction. Left-right symmetry yields $J_L = J_R$. However, this condition does not hold when the head laterally deviates from the default position. Similarly, channel crosstalk is defined as:

$$D_L = -J_L = 20\log_{10}\left|\frac{R_{12}}{R_{11}}\right| \quad (dB) \qquad D_R = -J_R = 20\log_{10}\left|\frac{R_{21}}{R_{22}}\right| \quad (dB). \tag{9.50}$$

Parodi and Rubak (2012) evaluated the minimum audible channel crosstalk in loudspeaker reproduction to determine the acceptable bound of crosstalk. They simulated channel crosstalk in reproduction with two frontal loudspeakers by deliberately introducing crosstalk in headphone-based binaural reproduction. They also simulated 12° and 60° loudspeaker span angles, as well as symmetrical and asymmetrical listener locations, with respect to the loudspeaker. The results of subjective discrimination experiments indicate that avoiding lateralization effects necessitates channel crosstalk levels that are below −15 dB for most (band-pass) stimuli and around −20 dB for broadband noise.

Lopez et al. (2001) evaluated the performance of crosstalk cancellation against lateral head translation in two-front loudspeaker reproduction. Channel separation was measured per octave, from 63 Hz to 4 kHz, using a pair of microphones mounted on an artificial head. The results for loudspeaker pairs with span angles $2\theta_0 = 60°$, 30°, and 10° (stereo dipole) show that the 30°-span loudspeaker pair exhibits an appropriate overall performance. Although the stereo dipole improves stability at mid- and high-frequency ranges, its low-frequency performance is unsatisfactory.

Bai and Lee (2006b) calculated and measured channel separation in two-front loudspeaker reproduction with span angles of 120°, 60°, and 10° at default and deviated head positions. The results also indicate that channel separation with the stereo dipole is unsatisfactory at low frequencies. By contrast, a wide span configuration enables large and stable channel separation within wide frequency ranges because of the natural high-frequency crosstalk attenuation caused by the head shadow effect on contralateral loudspeakers. However, the performance of transaural reproduction is not uniquely determined by channel separation. Lateral head translation also causes a difference in the path between the left and right loudspeakers to the head center, especially for wide-span angle configurations. This difference, in turn, introduces a relative arrival time difference between loudspeaker signals, thereby damaging localization information.

In the loudspeaker-based dynamic binaural reproduction to be discussed in Section 12.2.5, the head may be located at an off-central line (asymmetric) position, and crosstalk cancellation automatically adopts the temporary head position. Rose et al. (2002) analyzed the stability of crosstalk cancellation (or size of the listening region) in a stereo dipole designed for asymmetric head positions. The results show that, in terms of both channel separation and ITD variation, crosstalk cancellation is stable and effective against lateral head translation from the designed position.

9.3.4. Linear System Theory for the Stability of Crosstalk Cancellation

Crosstalk cancellation and loudspeaker reproduction can be regarded as a linear system with multiple inputs and outputs. Crosstalk cancellation is designed in accordance with ideal or default conditions. Changes in system conditions, such as head rotation and translation, as well as mismatched HRTFs (Section 9.4), inevitably influence the performance of crosstalk cancellation. The stability of crosstalk cancellation can be regarded as the stability or robustness of a system to perturbation. It can therefore be analyzed by linear system theory.

According to Equation (9.42), the binaural pressures of a listener in a general case of M loudspeaker reproduction are provided by:

$$[\boldsymbol{P}'] = [H]\boldsymbol{S} = [H][A]\boldsymbol{E}. \tag{9.51}$$

The $M \times 2$ crosstalk cancellation matrix $[A]$ is usually designed according to the ideal or default condition of:

$$[H][A] = [I]. \tag{9.52}$$

Here, crosstalk is completely cancelled.

When some perturbations, such as head rotation or translation, alter the transfer functions from M loudspeakers to two ears, the $2 \times M$ transfer matrix becomes $[H']$ and the binaural pressures in reproduction become:

$$\boldsymbol{P}' = [H']\boldsymbol{S} = [H'][A]\boldsymbol{E}. \tag{9.53}$$

The transfer chain from the inputs of a crosstalk canceller to a listener's two ears can be modeled by a two-input and two-output linear system, with the overall transfer matrix $[H][A]$ (Ward and Elko, 1999; Nelson and Rose, 2005). Ideally, $[H][A]$ should satisfy Equation (9.52). Perturbations convert the transfer matrix $[H]$ into $[H']$. According to linear system theory, system stability against perturbations is evaluated by the condition number of matrix $[H]$ as follows:

$$cond[H] = \frac{\gamma_{max}\{[H]\}}{\gamma_{min}\{[H]\}}, \tag{9.54}$$

where $\gamma_{max}\{[H]\}$ and $\gamma_{min}\{[H]\}$ are the largest and smallest singular values of the $2 \times M$ transfer matrix $[H]$, respectively (see Section 6.2.2 for the singular value of a matrix). A small cond$[H]$ represents a system that is robust to perturbations, that is, stable crosstalk cancellation. Therefore, cond$[H]$ can be used as a criterion for determining loudspeaker configurations with stable crosstalk cancellation.

Given loudspeaker configuration and head position, matrix $[H]$ is determined, and cond$[H]$ can then be evaluated. To simplify the analysis, Ward et al. (1999) excluded head diffraction and scattering, and approximated two ears as two points separated by $2a$ in free space. That is, a shadowless head model was used to calculate the HRTFs and transfer matrix. The results show that, for a two-front loudspeaker configuration, the highest robustness is achieved at a half-span angle θ_0 that satisfies:

$$\sin\theta_0 = \frac{qc}{8fa} \quad q = 1,3,5....., \tag{9.55}$$

where a is the head radius, f denotes the frequency, and c represents the speed of sound. The worst robustness occurs at a half-span angle θ_0 that satisfies:

$$\sin\theta_0 = \frac{qc}{8fa} \quad q = 2,4,6...... \tag{9.56}$$

Therefore, the robustness of crosstalk cancellation is related to both span angle and frequency. For a given span angle, robust and non-robust crosstalk cancellation alternately occurs with increasing frequency, a behavior that corresponds to the variations in integer q alternately being odd and even. For example, for a conventional $2\theta_0 = 60°$ span, $a = 0.0875$ m, and $c = 340$ m/s, crosstalk cancellation is at its most robust at frequencies of 971, 2914, and 4857 Hz, whereas it is at its weakest at frequencies of 1942, 3885, and 5829 Hz. Therefore, the optimal loudspeaker span depends on frequency, or a given loudspeaker span is optimal only at some specific frequencies or frequency bands. Ideally, therefore, loudspeaker span should vary with frequency. Otherwise, matrix $[H]$ is ill-conditioned at, or near, some frequencies that correspond to the most non-robust performance. As a result, the magnitudes of some elements in the crosstalk matrix remarkably expand at certain frequencies, leading to excessive magnitudes (gains) of loudspeaker signals. To avoid overloaded loudspeaker signals, the magnitudes of the elements of the crosstalk cancellation matrix are often hard-limited when such magnitudes exceed a certain threshold. Therefore, non-robust crosstalk cancellation influences reproduction performance in terms of two aspects. One is that crosstalk cancellation is highly sensitive to slight perturbations; the other is that the dynamic range in reproduction is compressed.

The conclusion drawn here slightly differs from that derived in previous sections. As indicated in the analysis in previous sections, a narrow loudspeaker span enables stable reproduction against lateral head translation and expands the listening region. The lowest frequency at which non-robustness in crosstalk cancellation occurs can be evaluated by letting $q = 2$ in Equation (9.56):

$$f_l = \frac{c}{4a\sin\theta_0}, \tag{9.57}$$

which corresponds to the case wherein the path difference $2a \sin \theta_0$ between a loudspeaker and two ears is equal to half of a wavelength. The lowest frequency f_l is inversely proportional to $\sin \theta_0$, with $f_l \approx 2$ kHz for $2\theta_0 = 60°$ and $f_l \approx 11.1$ kHz for $2\theta_0 = 10°$ (stereo dipole). From this perspective, a narrow loudspeaker span increases the lowest frequency at which non-robust crosstalk cancellation occurs. This increase improves stability, but, as stated in Section 9.3.2, a stereo dipole improves stability at mid and high frequencies, but at the cost of low-frequency performance.

Head diffraction and scattering are excluded in the foregoing analysis. Orduna et al. (2000) and Bustamante et al. (2001) used the HRTFs of a spherical and an artificial head to analyze stability in loudspeaker reproduction, and obtained similar results. Yang and Gan (2000) used the aforementioned method to analyze the three-loudspeaker configuration depicted in Figure 9.4. The results indicate that a three-loudspeaker configuration more strongly improves stability in reproduction than does conventional two-loudspeaker configuration. This result is of practical importance, as discussed in Section 10.4.3. Kim et al. (2006) extended the analysis to reproduction with multiple loudspeakers and listeners. By analyzing the condition number of the transfer matrix from loudspeakers to two ears, Rose et al. (2002) confirmed that crosstalk cancellation with a stereo dipole designed for asymmetric head positions is stable (refer to the last paragraph in Section 9.3.3). Ward and Elko (1998) used the cost functions of the left and right ear pressures to analyze stability in two-front loudspeaker reproduction and generated similar results.

Given that the optimal span between two frontal loudspeakers is frequency dependent, wideband signals should, ideally, be reproduced by a series of loudspeaker pairs with continuous variations in span angle. In practice, several loudspeaker pairs with different spans can be used, with each loudspeaker pair reproducing signals at specific frequency ranges. This compromise enables stable crosstalk cancellation within a wide frequency range (or, at least, avoids the most non-robust performance). This method for improving stability is the basic principle that underlies the *optimal source distribution* reproduction proposed by Takeuchi et al. (2001b; 2002b; 2008), in which three pairs of loudspeakers with span angles $2\theta_0 = 180°$, $30°$ (or $32°$), and $6°$ (or $6.2°$), as well as corresponding crosstalk cancellers, are adopted to reproduce signals below 500 Hz, 500–3500 Hz, and above 3500 Hz, respectively. Optimal source distribution reproduction avoids the nearly ill-conditioned transfer matrix $[H]$ at certain frequencies, so that loudspeaker signals do not exceed the dynamic range of electroacoustic reproduction systems.

9.4. Effects of Mismatched HRTFs and Loudspeaker Pairs

9.4.1. Effect of Mismatched HRTFs

As stated in Section 8.3, binaural synthesis with nonindividualized HRTFs (i.e., mismatched with a listener's HRTFs) increases front-back confusion and elevation error in headphone reproduction. The discussion in Section 9.2.1 indicates that pinna-related high-frequency spectral cues cannot be stably replicated in loudspeaker reproduction. Consequently, transaural reproduction through two frontal loudspeakers can recreate only stable virtual sources in frontal-horizontal quadrants. The virtual source azimuth is determined by interaural cues, especially ITD. Therefore, transaural synthesis with mismatched HRTFs also yields mismatched ITDs and errors in the perceived azimuth of virtual sources intended for frontal horizontal quadrants.

For two-front loudspeaker reproduction, Xie (2002c) analyzed the effect of a mismatched head radius and the resultant ITD on the perceived azimuth of virtual sources intended for frontal-horizontal quadrants. As indicated by Equations (9.1) to (9.10), when the HRTFs used in transaural synthesis match a listener's HRTFs, ideal crosstalk cancellation generates binaural pressures and resultant ITDs that are identical to those of a real source.

Suppose that the HRTFs used in transaural synthesis do not match those of a listener. The HRTFs used in transaural synthesis are denoted by $H_L, H_R, H_{LL} = H_{RR} = H_\alpha$, and $H_{LR} = H_{RL} = H_\beta$. The listener's HRTFs are denoted by $H_L', H_R', H'_{LL} = H'_{RR} = H_\alpha', H'_{LR} = H'_{RL} = H'_\beta$. Similar to the binaural pressures in Equation (9.38b), those in two-loudspeaker reproduction are given by:

$$P' = [H']S = [H'][A]E, \tag{9.58}$$

where $[H']$ is the actual transfer matrix from loudspeakers to two ears:

$$[H'] = \begin{bmatrix} H'_{LL} & H'_{LR} \\ H'_{RL} & H'_{RR} \end{bmatrix}. \tag{9.59}$$

Similar to the case in Section 9.2.3, the normalized interaural cross-correlation function $\Phi_{LR}(\tau)$ and ITD are evaluated using binaural pressures P_L' and P_R'. Then, the virtual source azimuth can be estimated by comparing the ITD in two-loudspeaker reproduction with that of a real source.

The far-field HRTFs calculated from a rigid-spherical head model are used in the analysis (Section 4.1.1). The angle span between a pair of frontal loudspeakers is $2\theta_0 = 60°$. Let a_0 and a denote the head radius of the model used in transaural synthesis and the listener's actual head radius, respectively; θs and θ_I denote the target and perceived virtual source azimuths in frontal-horizontal quadrants, respectively. Similar results are obtained for four kinds of stimuli, including sinusoidal (or narrow-band) signals of 0.4, 1.25, and 2.5 kHz, as well as pink noise (subjected to band-pass filtering) within 0.1–3.0 kHz. The results are summarized as follows:

1. A matched head radius with $a = a_0$ yields $\theta_I = \theta s$—that is, the perceived azimuth is consistent with the target azimuth.

2. For $a < a_0$, that is, the real head radius is smaller than that of the model, and the perceived azimuth error is minimal for the target azimuth in the front region $0° < \theta s \leq 30°$. The perceived azimuth error is clearly observable for the target azimuth in the lateral region. Even the resultant ITD exceeds the maximal possible ITD with a real source at $\theta = 90°$. Therefore, the perceived azimuth cannot be predicted by calculation, a phenomenon that may give rise to the following results: no definite location of auditory events exists, or an unnatural perception of virtual sources arises.

3. For $a > a_0$, that is, the real head radius is larger than that of the model, and the perceived azimuth error is minimal for the target azimuth in the front region $0° < \theta s \leq 30°$. The perceived azimuth error is clearly observable for the target azimuth in the lateral region, especially when θs is close to $90°$. The virtual source intended for the lateral direction moves toward the front, and the perceived azimuth error increases with the difference between a and a_0. For the sinusoidal (or narrow-band) signal at 1.25 kHz and $a_0 = 0.070$ m, the perceived azimuths θ_I are $15°$ and $15.6°$ for $a = 0.075$ m and $a = 0.080$ m, respectively, at a target azimuth $\theta s = 15°$. At a target azimuth $\theta s = 90°$, the perceived azimuths are $67°$ and $59°$ for $a = 0.075$ m and $a = 0.080$ m, respectively.

Therefore, a mismatched head radius only minimally affects the perceived azimuth of virtual sources intended for frontal regions, but considerably influences that of virtual sources intended for lateral regions. This result stems from a small ITD error that corresponds to a large variation in the perceived azimuth at lateral directions (Section 3.2.2). This conclusion is supported by the results of virtual source localization experiments. The findings of other experiments also indicate large perceived azimuth errors for lateral virtual sources in reproduction with two frontal loudspeakers (Nelson et al., 1996a). Thus, a mismatched head radius or HRTFs are factors that give rise to large errors in the perceived lateral azimuth in reproduction with two frontal loudspeakers.

Akeroyd et al. (2007) analyzed and measured the effect of mismatched HRTFs on the performance of an optimal source distribution system (Section 9.3.4). The results indicate that mismatched HRTFs diminish the performance of crosstalk cancellation, resulting in decreased wideband average cancellation, from about 25 dB for matched HRTFs to only 13 dB for mismatched HRTFs. Moreover, mismatched HRTFs generate resultant binaural localization cues (including ITD and ILD—interaural time difference and interaural level difference) that differ from intended values.

Therefore, an optimal source distribution system can deliver accurate binaural cues only when matched HRTFs are used.

Moore et al. (2010) conducted a psychoacoustic experiment to determine whether a horizontal virtual source at $\theta = 0°$ created with individualized crosstalk cancellation can be distinguished from a real sound source. For noise and click train stimuli, weak—almost completely unreliable—discrimination cues exist. Therefore, loudspeaker-based binaural reproduction with individualized crosstalk cancellation is suitable for binaural perceptual experiments.

9.4.2. Effect of Mismatched Loudspeaker Pairs

In the preceding discussion, the responses of loudspeaker pairs are assumed ideal and matched. In this case, the effect of loudspeaker responses can be disregarded and transaural processing can be implemented according to Equation (9.11). Let $Y_L(f)$ and $Y_R(f)$ denote the complex-valued responses of two loudspeakers (from input signal to free-field pressure). Provided that the responses are perfectly matched, that is, $Y_L(f) = Y_R(f)$, a pair of loudspeakers with nonideal responses (non-flat magnitude response and nonlinear phase response) continues to deliver correct interaural cues, such as ITD and ILD. Such a pair of loudspeakers imposes an imperceptible influence on the perceived virtual source azimuth, but with possible timbre change.

However, a mismatched loudspeaker pair with $Y_L(f) \neq Y_R(f)$ influences the perceived virtual source azimuth in reproduction. The ratio between the responses of two loudspeakers is:

$$D(f) = \frac{Y_R(f)}{Y_L(f)}. \tag{9.60}$$

Then,

$$d(f) = 20\log_{10}|D(f)| = 20\log_{10}\left|\frac{Y_R(f)}{Y_L(f)}\right| \quad (dB), \tag{9.61}$$

where $d(f)$ represents the degree of difference between the magnitude responses of two loudspeakers, and $d(f) > 0$ indicates that the magnitude response of the right loudspeaker is larger than that of the left loudspeaker, and vice versa. Furthermore,

$$\Delta\varphi(f) = \arg D(f) = \arg Y_R(f) - \arg Y_L(f), \tag{9.62}$$

where $\Delta\varphi(f)$ represents the difference between the phase responses of two loudspeakers. The right loudspeaker leads in phase for $\Delta\varphi(f) > 0$, and vice versa. For a matched loudspeaker pair, $d(f) = 0$ dB and $\Delta\varphi(f) = 0$.

Chi et al. (2009) analyzed the effect of the magnitude responses of mismatched loudspeakers on the perceived virtual source azimuth. Similar to the case in Equation (9.58), the binaural pressures generated by a mismatched loudspeaker pair can be calculated by:

$$P' = [H][Y]S, \tag{9.63}$$

where $[H]$ is the transfer matrix from loudspeakers to two ears, and:

$$[Y] = \begin{bmatrix} Y_L(f) & 0 \\ 0 & Y_R(f) \end{bmatrix}. \tag{9.64}$$

As indicated in the analysis in Section 9.4.1, the normalized interaural cross-correlation function $\Phi_{LR}(\tau)$ and ITD are evaluated using binaural pressures P_L' and P_R'. Then, the virtual source azimuth can be estimated by comparing the ITD in two-loudspeaker reproduction with that of a real source. (The constant-power equalization algorithm discussed in Section 9.5.1 is incorporated here.)

The calculation results indicate that a moderate mismatched magnitude response of a loud-speaker pair slightly influences the horizontal virtual source intended for the frontal region. A small mismatched response (on the order of 1 dB) may cause a large perceived azimuthal error for the horizontal virtual source intended for the lateral region. Therefore, a mismatched magnitude response of loudspeakers is another factor that causes large perceived lateral azimuth errors in two-loudspeaker reproduction. A well-matched loudspeaker pair is required for practical use. Another experiment confirms that correcting or equalizing loudspeaker magnitude responses reduces lateral azimuth localization errors to a certain extent (Chi et al., 2008).

The summary of the results in this section, as well as those in Sections 9.4.1 and 9.2.4, shows three factors that mutually determine the usual localization errors of lateral virtual sources in two-loudspeaker reproduction:

1. mismatched head radius or HRTFs;
2. head rotation around the vertical axes as in typical cases; and
3. mismatched loudspeaker pairs.

A straightforward approach to reducing lateral localization errors is to use individualized HRTFs and matched loudspeaker pairs (or equalized loudspeaker pairs). In practice, reducing the average lateral localization error benefits from the use of HRTFs that correspond to appropriate average anatomical dimensions. Taking ITD variations with head rotation into account, using four-loudspeaker reproduction (Section 9.2) also facilitates the reduction of lateral localization errors.

In the preceding discussion, loudspeakers are approximated as omnidirectional point sources. Qiu et al. (2009) took the directivity and scattering characteristics of loudspeakers into account. The analysis results indicate that both the directivity and scattering characteristics of loudspeakers influence the performance of crosstalk cancellation to a certain extent, but that channel separation is most sensitive to misaligned span angles of loudspeakers.

9.5. Coloration and Timbre Equalization in Loudspeaker Reproduction

9.5.1. Coloration and Timbre Equalization Algorithms

As stated in Section 8.1.2, coloration often occurs when binaural signals are directly reproduced through a pair of stereophonic loudspeakers. Consequently, the free- or diffuse-field equalization of binaural signals was suggested for loudspeaker compatibility in timbre. In addition to coloration, however, crosstalk is a major problem in binaural reproduction through loudspeakers. Crosstalk damages the spatial information included in binaural pressures. Free- or diffuse-field equalization partially compensates for coloration in loudspeaker reproduction, but not for crosstalk.

Ideally, complete crosstalk cancellation yields the same binaural pressures as those of a real source. Nevertheless, the analysis in previous sections indicates that completely cancelling out crosstalk is difficult within a full audible frequency range. In practice, factors—such as slight head translation or rotation, unmatched HRTFs, and room reflection—inevitably lead to incomplete crosstalk cancellation. Thus, the binaural pressures in reproduction deviate from those of a real source. The magnitude responses of the transaural synthesis filters specified by Equation (9.12a) are visibly frequency dependent. This dependence modifies the overall power spectra of input signal $E_0(f)$, thereby altering the balance among the spectral components of the resultant loudspeaker signals provided by Equation (9.11). Consequently, the incomplete crosstalk cancellation caused by perturbations leads to virtual source localization errors and perceived coloration, especially at high frequencies and for off-center listeners. These defects commonly occur in binaural reproduction through loudspeakers, therefore necessitating additional timbre equalization.

The principle of timbre equalization in two-loudspeaker reproduction is explained as follows. Given the difficulty in robustly rendering the fine high-frequency spectral cues to listeners' ears in loudspeaker reproduction, the perceived virtual source direction is dominated by interaural cues (especially ITD) and is limited to frontal-horizontal quadrants. The interaural cues are controlled by the relative, rather than the absolute, magnitude and phase of left and right loudspeaker signals. Scaling both loudspeaker signals with identical frequency-dependent coefficients does not alter their relative magnitude and phase, or the perceived virtual source direction. However, this manipulation alters the overall power spectra of loudspeaker signals and therefore equalizes timbre. All the succeeding equalization algorithms are based on this principle.

Hawksford (2002) proposed four equalization algorithms denoted as equalization algorithms 1, 2, 3, and 4:

1. *Equalization algorithm* **1**: In this algorithm, equalization is performed by the modulus of the sum of left and right HRTFs. That is, in Equations (9.11) and (9.12b), original $H_L(\theta, f)$ and $H_R(\theta, f)$ are replaced by the equalized HRTFs:

$$H'_L(\theta, f) = \frac{H_L}{|H_L + H_R|} W \qquad H'_R(\theta, f) = \frac{H_R}{|H_L + H_R|} W, \qquad (9.65)$$

where W is a normalized gain. The H_α and H_β in Equations (9.11) and (9.12b) are equalized in a similar manner (this approach also holds for the next three algorithms).

2. *Equalization algorithm* **2**: Equalization is performed by the sum of left and right HRTFs. The procedure is similar to that in equalization algorithm 1, but the equalized HRTFs are given by:

$$H'_L(\theta, f) = \frac{H_L}{H_L + H_R} W \qquad H'_R(\theta, f) = \frac{H_R}{H_L + H_R} W. \qquad (9.66)$$

3. *Equalization algorithm* **3**: Equalization is carried out by the minimum-phase spectrum of the sum of left and right HRTFs. The procedure is similar to that in equalization algorithm 1, but the equalized HRTFs are provided by:

$$H'_L(\theta, f) = \frac{H_L}{\exp\{conj[hilbert\,(\ln|H_L + H_R|)]\}} W$$
$$H'_R(\theta, f) = \frac{H_R}{\exp\{conj[hilbert(\ln|H_L + H_R|)]\}} W \qquad (9.67)$$

where "conj" denotes complex conjugation and "hilbert" denotes Hilbert transform (Oppenheim et al., 1999).

4. *Equalization algorithm* **4**: Equalization is conducted by the minimum-phase spectrum of the sum of the modulus of left and right HRTFs. The procedure is also similar to that in equalization algorithm 1, but the equalized HRTFs are given by:

$$H'_L(\theta, f) = \frac{H_L}{\exp\{conj[hilbert(\ln(|H_L| + |H_R|))]\}} W$$
$$H'_R(\theta, f) = \frac{H_R}{\exp\{conj[hilbert(\ln(|H_L| + |H_R|))]\}} W \qquad (9.68)$$

Xie et al. (2005c) proposed an equalization algorithm called *equalization algorithm* **5** or *constant-power equalization algorithm*, in which the responses of the transaural synthesis filters $G_L(\theta, f)$ and $G_R(\theta, f)$ in Equations (9.11) and (9.12b) are equalized by their root mean square:

$$L' = G_L{}'(\theta, f) E_0(f) \qquad R' = G_R{}'(\theta, f) E_0(f), \tag{9.69}$$

$$
\begin{aligned}
G_L{}'(\theta, f) &= \frac{G_L(\theta, f)}{\sqrt{|G_L(\theta, f)|^2 + |G_R(\theta, f)|^2}} \\
&= \frac{H_\alpha H_L - H_\beta H_R}{\sqrt{|H_\alpha H_L - H_\beta H_R|^2 + |-H_\beta H_L + H_\alpha H_R|^2}} \frac{|H_\alpha{}^2 - H_\beta{}^2|}{H_\alpha{}^2 - H_\beta{}^2}, \\
G_R{}'(\theta, f) &= \frac{G_R(\theta, f)}{\sqrt{|G_L(\theta, f)|^2 + |G_R(\theta, f)|^2}} \\
&= \frac{-H_\beta H_L + H_\alpha H_R}{\sqrt{|H_\alpha H_L - H_\beta H_R|^2 + |-H_\beta H_L + H_\alpha H_R|^2}} \frac{|H_\alpha{}^2 - H_\beta{}^2|}{H_\alpha{}^2 - H_\beta{}^2}.
\end{aligned}
\tag{9.70}
$$

The loudspeaker signals given by Equation (9.69) satisfy the constant-power spectral relationship,

$$|L'|^2 + |R'|^2 = E_0^2. \tag{9.71}$$

Therefore, the overall power spectra of loudspeaker signals are equal to those of the input stimulus, thereby reducing reproduction coloration.

9.5.2. Analysis of Timbre Equalization Algorithms

He et al. (2006) carried out subjective and objective analyses of two-loudspeaker transaural synthesis with algorithms 1–5 and without timbre equalization. The authors used a method for rating subjective attribution (Section 13.4), in which the subjective timbre was graded by a group of listeners, and the statistical results were analyzed. An experiment was conducted in a listening room with a reverberation time of 0.15 s. The results indicate significant differences in timbre among the five equalization algorithms. For loudspeaker configuration with a conventional 60° span angle or a narrow 30° span angle, as well as for speech and music stimuli, equalization algorithm 5 receives the highest score in subjective timbre with the least coloration, followed by equalization algorithm 4. Equalization algorithm 1 exhibits the lowest score, sometimes registering a performance that is inferior to the case without timbre equalization.

The results of He's psychoacoustic experiment can be further explained from the perspective of signal processing. According to Equations (9.11) and (9.12b), the unequalized loudspeaker signals L' and R' are obtained by filtering the mono stimulus $E_0(f)$ with $G_L(\theta, f)$ and $G_R(\theta, f)$, respectively. For convenience, $G_L(\theta, f)$ and $G_R(\theta, f)$ are transformed into the functions of complex variable z in the Z-domain by letting $z = \exp(j2\pi f/f_s)$, where f_s is the sampling frequency. Similar to the infinite impulse response (IIR) model of HRTF in Equation (5.23), $G_L(\theta, f)$ or $G_R(\theta, f)$ can be modeled by an IIR filter as:

$$G_L(\theta, z) = \frac{H_\alpha H_L - H_\beta H_R}{H_\alpha^2 - H_\beta^2} = K \frac{\displaystyle\prod_{q=1}^{Q}(z - z_q)}{\displaystyle\prod_{p=1}^{P}(z - z_p)}, \tag{9.72}$$

where K is a constant, z_p and z_q denote the poles and zeros in the Z-plane, and P and Q are the number of poles and zeros, respectively. The poles and zeros at $z = 0$ are excluded from Equation (9.72).

The causality and stability of transaural synthesis in Equation (9.11) require that all the poles of filters $G_L(\theta, z)$ and $G_R(\theta, z)$ lie within the unit circle in the Z-plane. The magnitude response of $G_L(\theta, f)$ or $G_R(\theta, f)$ is closely related to the locations of the zeros and poles. Some zeros and poles close to the unit circle cause deep notches and prominent peaks in magnitude response, respectively, thereby causing obvious coloration in reproduction. The resonances introduced by the poles close to the unit circle also result in difficult signal processing, an issue to be addressed in Section 9.6.

The goals of timbre equalization are to cancel the poles and zeros close to the unit circle and smooth or eliminate the peaks and notches in the magnitude responses of $G_L(\theta, z)$ and $G_R(\theta, z)$. Appropriate timbre equalization also enables the easy implementation of signal processing. Therefore, the performance of timbre equalization depends upon the effect of pole and zero cancellation.

The numerator and denominator of the transaural filter responses $G_L(\theta, z)$ and $G_R(\theta, z)$ in Equation (9.12a) or Equation (9.12b) comprise HRTFs. If each associated head-related impulse response (HRIR) is approximated by an N-point impulse response, the corresponding HRTF can be modeled by a finite impulse response (FIR) filter in the Z-domain by:

$$H(z) = \sum_{n=0}^{N-1} h(n)z^{-n} = \sum_{n=0}^{N-1} b_n z^{-n}, \tag{9.73}$$

where $b_n = h(n)$ are FIR filter coefficients that are related to source direction θ. Except for the poles at $z = 0$, no other poles are introduced by the HRTFs in the FIR model. The zeros and poles of $G_L(\theta, z)$ or $G_R(\theta, z)$ are determined by the zeros of its numerator and denominator, respectively. For example, when no timbre equalization is implemented, the zeros of $G_L(\theta, z)$ are determined by the numerator $(H_\alpha H_L - H_\beta H_R)$ in the first formula of Equation (9.12b), and the poles of $G_L(\theta, z)$ are determined by the zeros of denominator $(H_\alpha^2 - H_\beta^2)$.

We analyze the performance of timbre equalization algorithms by evaluating the effect of pole and zero cancellation. MIT-KEMAR HRTFs (Table 2.1) are used in the succeeding analysis.

1. For conventional transaural synthesis without timbre equalization, the denominator of Equation (9.12b) is given by $(H_\alpha^2 - H_\beta^2) = (H_\alpha - H_\beta)(H_\alpha + H_\beta)$. The zeros of either component $(H_\alpha - H_\beta)$ or $(H_\alpha + H_\beta)$ generate the poles of filters $G_L(\theta, z)$ and $G_R(\theta, z)$. Some of these poles are close to the unit circle and may not always be cancelled by the zeros in the numerators of $G_L(\theta, z)$ and $G_R(\theta, z)$. This attribute boosts the signal components close to the frequency of poles, thereby causing coloration in reproduction and complexity in signal processing. For example, Figure 9.9(a) shows the magnitude response of $G_L(\theta, z)$ for a

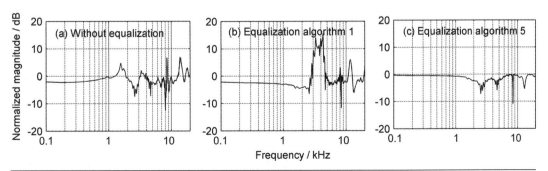

Figure 9.9 Magnitude responses of $G_L(\theta, f)$ with and without timbre equalization (for $\theta = 345°$, $2\theta_0 = 60°$)

conventional $2\theta_0 = 60°$ loudspeaker configuration with a target horizontal source azimuth $\theta = 345°$. The abscissa is the analog frequency f in kHz. As expected, peaks (and notches) are observed in the response.

2. In equalization algorithm 1, where the H_L and H_R in Equation (9.12b) are replaced by the H'_L and H'_R in Equation (9.65), H_α and H_β are replaced by $H_\alpha' = H_\alpha/|H_\alpha + H_\beta|$ and $H_\beta' = H_\beta/|H_\alpha + H_\beta|$, respectively. The loudspeaker signals become:

$$L' = G_L'(\theta,z)E_0(z) \qquad R' = G_R'(\theta,z)E_0(z), \tag{9.74}$$

$$G_L'(\theta,z) = \frac{(H_\alpha H_L - H_\beta H_R)}{(H_\alpha - H_\beta)|H_L + H_R|}\frac{|H_\alpha + H_\beta|}{(H_\alpha + H_\beta)}$$

$$G_R'(\theta,z) = \frac{(-H_\beta H_L + H_\alpha H_R)}{(H_\alpha - H_\beta)|H_L + H_R|}\frac{|H_\alpha + H_\beta|}{(H_\alpha + H_\beta)}. \tag{9.75}$$

In Equation (9.75), although the poles caused by $(H_\alpha + H_\beta)$ in the denominator are cancelled by the zeros of $|H_\alpha + H_\beta|$ in the numerator, the poles caused by $(H_\alpha - H_\beta)$ in the denominator remain intact. Moreover, the zeros of $|H_L + H_R|$ in the denominator may introduce additional poles in $G_L'(\theta, z)$ and $G_R'(\theta, z)$. For example, Figure 9.9(b) shows the magnitude response of $G'_L(\theta, z)$ for a conventional $2\theta_0 = 60°$ loudspeaker configuration with a target horizontal source azimuth $\theta = 345°$. The figure shows a 15 dB peak at 3–5 kHz, which is more prominent than that observed when no timbre equalization is implemented. This result is attributed to the additional poles introduced by the factor $|H_L + H_R|$ in the denominator. The prominent peak of $G_L(\theta, z)$ causes coloration in reproduction and complexity in signal processing. These problems account for the low score in subjective timbre of equalization algorithm 1.

3. For equalization algorithm 5, given by Equation (9.70), the poles of $G_L'(\theta, z)$ and $G_R'(\theta, z)$ caused by $(H_\alpha^2 - H_\beta^2)$ in the denominator are cancelled by the zeros of $|H_\alpha^2 - H_\beta^2|$ in the numerator. The two other factors within the root sign of the denominator satisfy $|H_\alpha H_L - H_\beta H_R|^2 \geq 0$ and $|-H_\beta H_L + H_\alpha H_R|^2 \geq 0$. Except for these two factors being simultaneously equal to 0, the root term in the denominator does not introduce additional poles. When these two factors are simultaneously equal to 0, the numerators of $G_L'(\theta, z)$ and $G_R'(\theta, z)$ are also equal to 0. Therefore, equalization algorithm 5 cancels the poles in the conventional transaural filters $G_L(\theta, z)$ and $G_R(\theta, z)$, and does not introduce additional poles, thereby reducing coloration in reproduction. As an example, Figure 9.9(c) shows the magnitude response of $G'_L(\theta, z)$ for a conventional $2\theta_0 = 60°$ loudspeaker configuration with a target horizontal source azimuth $\theta = 345°$. No obvious peaks are observed in the figure. Several shallow notches will not cause obvious coloration or difficult signal processing. The notch near 8 kHz is deep, but narrow; thus, it will not cause obvious coloration because of the finite frequency resolution of human hearing at high frequencies.

4. For brevity, similar analyses of other equalization algorithms are excluded.

These analyses of the equalization algorithms show that timbre equalization is required to reduce the perceived coloration in practical loudspeaker reproduction. Significant differences among the performance of the five equalization algorithms are found, with algorithm 5 (constant-power equalization) registering the best performance. Inappropriate equalization may worsen timbre. From a signal-processing standpoint, the timbre in loudspeaker reproduction strongly depends on the locations of the poles (and zeros) of transaural synthesis filters. An effective equalization algorithm should eliminate or cancel the close-unit circle poles (and zeros) of filters to smooth or eliminate response peaks (and notches).

The pinna effect and ear canal resonance are included in the HRTFs used in Figure 9.9; these factors manifest as notches and peaks in the HRTF magnitude responses. As shown in Figure 3.3 (Section 3.1.2), a peak at 2–3 kHz and a notch near 8 kHz are observed for the horizontal right ear HRTF magnitudes at azimuths 0°, 30°, and 60°. This peak and notch correspond to the pole and zero of the HRTF IIR model. The peak at 2–3 kHz is caused by ear canal resonance and source direction independence, and the notch near 8 kHz is caused by pinna and source direction dependence. In principle, the constant-power equalization algorithm can efficiently cancel the close-unit circle poles and zeros of the transaural filter $G_L'(\theta, z)$, thereby smoothing the peaks and notches in resultant loudspeaker signals. Because of the finite length of HRTFs (512 points at a sampling frequency of 44.1 kHz in the case of Figure 9.9), limited frequency resolution may cause errors in the cancellation of poles and zeros, resulting in remnant zeros in $G_L'(\theta, z)$. Such an error causes the notch in the magnitude response of $G_L'(\theta, z)$ near 8 kHz.

Ear canal resonance is independent of source direction, and robustly rendering fine high-frequency spectral cues to listeners' ears is difficult. Thus, He et al. (2007) used blocked ear canal HRTFs without pinnae for transaural synthesis with the constant-power equalization algorithm. The HRTFs were measured from KEMAR without pinnae (Zhong and Xie, 2006a). The theoretical analysis demonstrates that using the HRTFs without pinnae effectively eliminates the notch in the magnitude responses of the transaural filter near 8 kHz, consequently further reducing coloration in reproduction. The psychoacoustic experiment also indicates that the HRTFs with and without pinnae yield almost the same localization performance in frontal-horizontal quadrants, but the latter exhibits less coloration, making it suitable for practical use.

For a very narrow loudspeaker span, such as a stereo dipole, however, the constant-power equalization algorithm may cause perceptible low-frequency attenuation because of the out-of-phase cancellation of two loudspeaker signals. In practical use, such as the virtual reproduction of multichannel sound to be discussed in Section 10.4, transaural synthesis and constant-power equalization can be applied above a certain frequency (e.g., above 200 Hz), and the low-frequency component is separately reproduced through an optional subwoofer.

9.6. Some Issues on Signal Processing in Loudspeaker Reproduction

Since Schroeder and Atal (1963) introduced crosstalk cancellation, numerous researchers have realized binaural reproduction through loudspeakers (Damaske, 1971; Iwahara and Mori, 1978; Sakamoto et al., 1981, 1982; Hamada et al., 1985; Cooper and Bauck, 1989; Møller, 1989; Griesinger, 1989; Koring and Schmitz, 1993; Gardner, 1995c; Jot, 1995; Mouchtaris et al., 2000). In the early stages of research, signal processing for crosstalk cancellation or transaural synthesis was implemented by various analog circuits. During the past two decades, analog signal processing has been replaced by digital approaches. Various algorithms may be similar in basic principle, but differ in design and implementation. This section focuses on the design and implementation of signal processing for loudspeaker reproduction.

9.6.1. Causality and Stability of a Crosstalk Canceller

As shown in Equations (9.6) and (9.12a), the 2×2 crosstalk cancellation matrix or transaural synthesis filters for two-loudspeaker reproduction includes a common denominator, or the inverse of function $\varXi = (H_{LL}H_{RR} - H_{LR}H_{RL})$ to all terms. This inclusion does not influence crosstalk cancellation but influences overall equalization. Above all, function $\varXi$ determines the causality and stability of signal processing. This problem is analyzed in the Z-domain. According to signal processing theory (Oppenheim et al., 1999), a causal and stable crosstalk canceller requires that all its poles lie within the unit circle in the Z-plane, implying that all the zeros of the common denominator $\varXi$ must lie within the unit circle. This condition can be guaranteed if $\varXi$ is a minimum-phase function, so that all its zeros and poles lie within the unit circle. Nevertheless, a practical $\varXi$ is usually a non-minimum phase function. First, the outside-unit circle zeros (common to all HRTFs) generate the corresponding zeros of function $\varXi$, but if all the HRTFs can be approximated by their minimum-phase functions (as indicated in Section 3.1.3), this problem can be avoided. Second, the manipulation of subtraction within function $\varXi = (H_{LL}H_{RR} - H_{LR}H_{RL})$ may introduce some zeros on or outside the unit circle, even if all the HRTFs are approximated as minimum-phase functions. Therefore, the zeros of function $\varXi$ may cause non-causality and instability in the crosstalk canceller.

If function $\varXi$ comprises some zeros that are outside but far from the unit circle, the crosstalk canceller is stable but non-causal. Approximately obtaining a causal, and therefore, realizable crosstalk canceller usually necessitates an appropriate linear delay. Accordingly, the condition for ideal crosstalk cancellation given by Equation (9.4) becomes:

$$[H][A] = z^{-d}[I]. \tag{9.76}$$

or:

$$[A] = z^{-d}[H]^{-1}. \tag{9.77}$$

In Equation 9.77, d is the number of delay samples. In practice, implementing an approximately causal FIR filter in a crosstalk canceller is achieved by applying an appropriate circle delay and time window to the inverse of the acoustical transfer matrix $[H]$. Similarly, for multiple-loudspeaker reproduction, an appropriate linear delay is cascaded with the crosstalk canceller obtained by the pseudoinverse in Equation (9.21). For simplicity, the linear delay is excluded in the succeeding discussion.

If function $\varXi$ contains some zeros on the unit circle, the acoustical transfer matrix $[H]$ in Equation (9.6) is singular, and therefore non-invertible at certain frequencies. If some zeros of function $\varXi$ lie close to the unit circle, the crosstalk canceller becomes unstable, leading to excessive signal gain at certain frequencies. The narrow peaks in Figure 9.9(a), for example, are caused by this phenomenon. The simplest solution to this problem is hard-limiting the magnitude of inverse function $\varXi^{-1}$, that is, the magnitude of $\varXi^{-1}$ is attached to a constant when it exceeds a certain threshold. However, processing by hard-limiting magnitude inevitably degrades the performance of crosstalk cancellation. The singularity of the acoustical transfer matrix $[H]$ can also be manipulated by other methods, as will be discussed in Section 9.7.

The problem of stability in crosstalk cancellation or transaural synthesis filters is closely related to the timbre equalization discussed in Section 9.5. The constant-power equalization algorithm reduces coloration in loudspeaker reproduction by cancelling the close-unit circle poles of transaural synthesis filters. The cancellation also improves the stability of the system, another advantage of the fifth algorithm.

In addition, because all the items in the numerator and denominator of Equation (9.12a) contain a product of two HRTFs, any factor common to the HRTFs will be cancelled out in the final responses $G_L(\theta, f)$ and $G_R(\theta, f)$ of the transaural synthesis filter. In principle, therefore, all measured HRTFs can be used for transaural synthesis processing with the same results, regardless of the reference point used in measurement (from the blocked or open entrance of the ear canal to the eardrum) and equalization methods (measurement, free- or diffuse-field equalization, or no equalization). The only constraint is that all HRTFs used in processing are measured and equalized under the same condition, a feature that facilitates practical applications. Beyond the effective frequency range of the electroacoustic system for HRTF measurement, all measured HRTFs are invalid and some other equalization algorithms should be used as a supplement.

9.6.2. Basic Implementation Methods for Signal Processing in Loudspeaker Reproduction

As indicated in Equation (9.1), the crosstalk canceller for two-loudspeaker reproduction is characterized by the transfer functions A_{11}, A_{12}, A_{21}, and A_{22}. Given these transfer functions, the crosstalk cancellation can be directly realized by the lattice structure illustrated in Figure 9.1. The associated four transfer functions in the figure can be implemented by four FIR or IIR filters. Similarly, the transaural synthesis specified by Equations (9.11) and (9.12) can also be directly realized by a pair of FIR or IIR filters $G_L(\theta, f)$ and $G_R(\theta, f)$, as shown in Figure 9.10. The methods described in Chapter 5 (i.e., windowing or the frequency sampling method for FIR filters and Prony's method for IIR filters) are applicable to designing filters with appropriate orders. For example, MacCabe and Furlong (1991) designed a pair of (20, 20)-order IIR filters to model $G_L(\theta, f)$ and $G_R(\theta, f)$.

As stated in Section 9.6.1, the zeros in the denominator $\Xi = (H_{LL}H_{RR} - H_{LR}H_{RL})$ of Equation (9.6) or Equation (9.8) result in poles in the crosstalk cancellation filters. The resonances introduced by the poles, especially by those close to the unit circle, prolong the impulse response of filters. The duration of the impulse responses of the crosstalk canceller typically ranges from 5–10 ms (Jot, 1995), roughly corresponding to 256–512 points at a sampling frequency of 44.1 or 48 kHz, which is longer than the duration of an HRIR (Section 5.2.3). From this viewpoint, recursive IIR filters are more suitable for transaural synthesis, including binaural synthesis and crosstalk cancellation. However, the poles close to the unit circle should be carefully manipulated. Otherwise, instability in IIR filters may occur. Alternatively, the approximate implementation of transaural synthesis by FIR filters avoids instability. Nevertheless, the long duration of the impulse response caused by the poles close to the unit circle yields a high-order FIR filter, and a low-order FIR filter causes large errors.

For transaural synthesis with the constant-power equalization introduced in Section 9.5, the poles of filters are effectively cancelled, so that the impulse response decreases to 1.3–2.7 ms. Therefore, signal processing can be simply implemented by a pair of FIR filters. Psychoacoustic experiments indicate that 128- or 64-point FIR filters yield satisfactory subjective performance (Xie et al., 2006b).

Four filters are required for the lattice structure implementation of the crosstalk canceller in Figure 9.1. In the left-right symmetric case, the implementation can be simplified into a shuffler

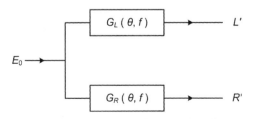

Figure 9.10 Direct implementation of transaural synthesis for two-front loudspeaker reproduction

structure (Cooper and Bauck, 1989; Bauck and Cooper, 1996). As indicated by Equation (9.9), left-right symmetry yields a symmetric crosstalk cancellation matrix $[A]$ with $A_{11} = A_{22}$ and $A_{12} = A_{21}$. Accordingly, similar to that calculated in Equation (6.106), matrix $[A]$ can be diagonalized as:

$$[A] = \begin{bmatrix} A_{11} & A_{12} \\ A_{21} & A_{22} \end{bmatrix} = \begin{bmatrix} 0.707 & 0.707 \\ 0.707 & -0.707 \end{bmatrix} \begin{bmatrix} A_{11} + A_{12} & 0 \\ 0 & A_{11} - A_{12} \end{bmatrix} \begin{bmatrix} 0.707 & 0.707 \\ 0.707 & -0.707 \end{bmatrix}. \tag{9.78}$$

Substituting Equation (9.9) into Equation (9.78) yields:

$$[A] = \begin{bmatrix} 0.707 & 0.707 \\ 0.707 & -0.707 \end{bmatrix} \begin{bmatrix} \dfrac{1}{H_\alpha + H_\beta} & 0 \\ 0 & \dfrac{1}{H_\alpha - H_\beta} \end{bmatrix} \begin{bmatrix} 0.707 & 0.707 \\ 0.707 & -0.707 \end{bmatrix}. \tag{9.79}$$

Equation (9.79) denotes the shuffler structure implementation of crosstalk cancellation with two filters.

The conventional design of a crosstalk canceller is based on Z or Fourier transform, in which a uniform frequency scale is used. As stated in Section 9.3.2, a significant low-frequency increase is required in the crosstalk canceller for loudspeaker configuration with a narrow span angle (such as a stereo dipole). Obtaining adequate low-frequency resolution necessitates a sufficiently long filter; otherwise, low-frequency performance is degraded. Therefore, the conventional filters for crosstalk cancellation are complex, and the uniform frequency scale does not accord with the non-uniform frequency resolution of human hearing. To improve low-frequency performance, some authors implemented crosstalk cancellation by frequency-warped FIR or IIR filters, similar to the cases discussed in Section 5.5 (Kirkeby et al., 1999a; Jeong et al., 2005). Another benefit of frequency-warped filters is that the narrow peaks and notches at high frequencies are smoothed, so that the filters are simplified and stability improves. The basic methods for two-loudspeaker signal processing discussed here can be extended to multiple loudspeakers.

9.6.3. Other Implementation Methods for Signal Processing in Loudspeaker Reproduction

Aside from directly implementing a lattice or shuffler structure, other structures can also be used to execute crosstalk cancellation or transaural synthesis. Gardner (1997) provided an extensive overview of this issue.

Using Equation (9.8) as basis, Schroeder and Atal (1963) drew a block diagram of the structural decomposition of transaural synthesis (Figure 9.11). The procedures for signal processing are as follows.

1. The mono input stimulus E_0 is filtered with H_L and H_R, yielding the binaural signals E_L and E_R, respectively.
2. The resultant E_L and E_R are passed through a two-input and two-output network that comprises filters H_{LL}, H_{LR}, H_{RL}, and H_{RR}.
3. The two outputs of the network are separately equalized by a pair of identical inverse filters $\Xi^{-1} = (H_{LL}H_{RR} - H_{LR}H_{RL})^{-1}$, yielding the loudspeaker signals L' and R'.

Each of these steps was implemented by analog filters in some early studies. For digital signal processing, the filters in the first and second steps can be directly designed from corresponding HRTFs and realized in FIR or IIR form. The inverse filter in the third step, which equalizes the

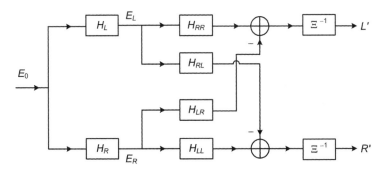

Figure 9.11 Structural decomposition implementation of transaural synthesis

overall power spectra of the loudspeaker signals, may be realized in recursive IIR form. Because the inverse filter is identical for left and right channels, equalization can be implemented by a single filter prior to binaural synthesis.

Dividing the numerator and the denominator of Equation (9.6) by $H_{LL}H_{RR}$, the crosstalk cancellation matrix becomes (Møller, 1992):

$$[A] = \begin{bmatrix} 1/H_{LL} & 0 \\ 0 & 1/H_{RR} \end{bmatrix} \begin{bmatrix} 1 & -ITF_R \\ -ITF_L & 1 \end{bmatrix} \frac{1}{1 - ITF_L ITF_R}, \tag{9.80}$$

where:

$$ITF_L = \frac{H_{RL}}{H_{LL}} \qquad ITF_R = \frac{H_{LR}}{H_{RR}}, \tag{9.81a}$$

or simply,

$$ITF = \frac{H_\beta}{H_\alpha}. \tag{9.81b}$$

The *ITF* is the interaural transfer function defined by Equation (5.46), and it represents the ratio of the contralateral HRTF to the ipsilateral HRTF. Because the components common to all HRTFs are cancelled in the ratio of the two HRTFs, the *ITF*s are independent of reference point and equalization method in HRTF measurement.

We can obtain insight into the physical perspective of crosstalk cancellation from Equation (9.80). The essence of crosstalk cancellation is exhibited by the $-ITF$ as the off-diagonal elements in the second matrix in Equation (9.80), which introduces the out-of-phase signal of each channel to the opposite channel. For example, the product of the right input signal with ITF_R in the frequency domain predicts only the crosstalk from the right loudspeaker to the left ear; this crosstalk can be cancelled by introducing a corresponding out-of-phase signal to the left channel. However, each crosstalk cancellation signal is inevitably transmitted to the opposite ear. For example, the crosstalk cancellation signal $-ITF_R E_R$ introduced to the left channel is also transmitted to the right ear, leading to a higher-order crosstalk. Consequently, more out-of-phase signals are required to cancel the higher order crosstalk. This cancellation is performed by the common term $(1 - ITF_L ITF_R)^{-1}$ on the right side of Equation (9.80). According to the series expansion:

$$\frac{1}{1 - F(z)} = 1 + F(z) + F^2(z) + \dots\dots$$

$ITF_R/(1 - ITF_L ITF_R)$ is expanded as:

$$ITF_R \frac{1}{1 - ITF_L ITF_R} = ITF_R + ITF_R ITF_L ITF_R + ITF_R (ITF_L ITF_R)^2 + \dots \quad (9.82)$$

Each term on the right side of Equation (9.82) represents the first, second, and third-order cross-talk to the right ear, respectively. The out-of-phase signals of this ear are introduced to the left channel to compensate for each order crosstalk. Because the *ITF* represents the ratio of the contra-lateral HRTF to the ipsilateral HRTF, the head shadow effect causes an ITF magnitude of less than a unit. The *ITF* magnitude is close to a unit at low frequencies and attenuates at high frequencies, so that it can be approximated by a low-pass filter. The effect of common item $(1 - ITF_L ITF_R)^{-1}$ is a low-frequency boost. The series in Equation (9.82) rapidly converges at high frequencies, so that it can be well approximated by a few preceding-order terms. That is, high-order crosstalk can be disregarded at high frequencies. The effect of the first diagonal matrix on the right side of Equation (9.80) is to equalize channel (loudspeaker) signals with the inverse of the ipsilateral HRTF. For example, the left channel is equalized by $1/H_{LL}$. According to Equations (9.80), (9.11), and (9.12a), if binaural synthesis and crosstalk cancellation are merged, the transaural synthesis filters are given by (Møller, 1992):

$$G_L(\theta, f) = \left[\left(\frac{H_L}{H_{LL}} \right) - \left(\frac{H_R}{H_{LL}} \right) ITF_R \right] \frac{1}{1 - ITF_L ITF_R},$$

$$G_R(\theta, f) = \left[\left(\frac{H_R}{H_{RR}} \right) - \left(\frac{H_L}{H_{RR}} \right) ITF_L \right] \frac{1}{1 - ITF_L ITF_R}. \quad (9.83)$$

In addition to the structure decomposition implementation illustrated in Figure 9.11, two recursive structure implementations of crosstalk cancellation are depicted in Figure 9.12 (Iwahara and Mori, 1978). Both implementations yield the same transfer characteristics as those given by Equation (9.6).

In the left-right symmetric case, the HRTF from the loudspeaker to the ipsilateral ear is $H_{LL} = H_{RR} = H_\alpha$, and that to the contralateral ear is $H_{LR} = H_{RL} = H_\beta$. Then,

$$ITF_L = ITF_R = ITF = \frac{H_\beta}{H_\alpha}. \quad (9.84)$$

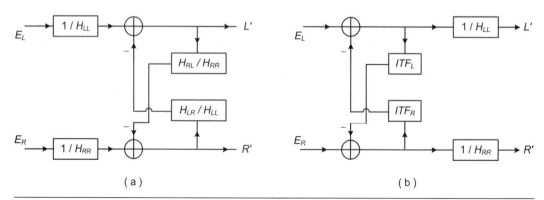

Figure 9.12 Two recursive structural implementations of crosstalk cancellation: (a) recursive structure 1; (b) recursive structure 2

The crosstalk cancellation in Equation (9.80) becomes:

$$[A] = \begin{bmatrix} 1 & -ITF \\ -ITF & 1 \end{bmatrix} \frac{1}{1-ITF^2} \frac{1}{H_\alpha}. \tag{9.85}$$

Figure 9.13 shows a block diagram of a symmetric and recursive structure implementation of Equation (9.85).

Similar to the case in Equation (6.106), the matrix in Equation (9.85) can be diagonalized, and the corresponding crosstalk canceller can be implemented by a shuffler structure, as shown in Equation (9.86) (Cooper and Bauck, 1989; Bauck and Cooper, 1996). The shuffler structure implementation requires only two filters, along with the ipsilateral equalized filters $1/H_\alpha$.

$$[A] = \begin{bmatrix} 0.707 & 0.707 \\ 0.707 & -0.707 \end{bmatrix} \begin{bmatrix} \dfrac{1}{1+ITF} & 0 \\ 0 & \dfrac{1}{1-ITF} \end{bmatrix} \begin{bmatrix} 0.707 & 0.707 \\ 0.707 & -0.707 \end{bmatrix} \frac{1}{H_\alpha}. \tag{9.86}$$

The *ITF* included in foregoing equations can be approximately modeled by various low-order FIR or IIR low-pass filters, which makes signal processing simple. In particular, under the minimum-phase approximation of HRTFs in Equation (3.6), the *ITF* defined by Equation (9.81b) becomes:

$$ITF = \frac{H_{\beta,min}}{H_{\alpha,min}} \exp(-j2\pi f\, |ITD|), \tag{9.87}$$

where $H_{\beta,min}$ and $H_{\alpha,min}$ are the minimum-phase contralateral and ipsilateral HRTFs, respectively; and ITD is the interaural time difference associated with the linear-phase component of HRTFs, which are evaluated according to ITD definition 5, or, sometimes, according to definition 4 or 7 in Section 3.2.1. Given that the ratio $H_{\beta,min}/H_{\alpha,min}$ is also a minimum-phase function, the *ITF* in Equation (9.87) can be implemented by a minimum-phase filter cascaded with a linear-phase or pure-delay |ITD|. In practice, the minimum-phase filter is usually realized in low-order IIR form according to Prony's method in Section 5.3.2.

9.6.4. Bandlimited Implementation of Crosstalk Cancellation

The crosstalk cancellation discussed in the preceding sections is implemented in the full audio frequency range. However, crosstalk cancellation in bandlimited range is also preferred to simplify signal processing. The following factors prompt such preference: First, the pinna effect causes HRTFs to rapidly vary at high frequencies with narrow notches and peaks that correspond to the

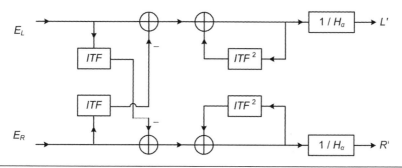

Figure 9.13   Symmetric and recursive structural implementation of crosstalk cancellation

zeros and poles close to the unit circle in the Z-plane. This feature makes for difficult crosstalk cancellation. Second, as stated in Section 9.2.1, robustly delivering fine high-frequency spectral cues to listeners' ears in loudspeaker reproduction is difficult, and the perceived azimuthal direction is determined by interaural cues, especially ITD. Third, for conventional two-front loudspeaker configuration, the head shadow serves as a low-pass filter for the incident wave, so that the high-frequency pressure at the ear contralateral to the loudspeaker is visibly attenuated. In other words, the natural head shadow inhibits crosstalk at high frequencies.

A straightforward method is to implement crosstalk cancellation below a certain frequency f_0, enabling the transmission of intact binaural signals above f_0 to loudspeakers. In this case, the crosstalk cancellation matrix of Equation (9.6) becomes (Gardner, 1997):

$$[A] = H_{LP} \begin{bmatrix} H_{LL} & H_{LR} \\ H_{RL} & H_{RR} \end{bmatrix}^{-1} + H_{HP} \begin{bmatrix} 1 & 0 \\ 0 & 1 \end{bmatrix}, \tag{9.88}$$

where $H_{LP} = H_{LP}(f, f_0)$ and $H_{HP} = H_{HP}(f, f_0)$ are the transfer functions of low-pass and high-pass filters, respectively. The selected crossover frequency is $f_0 = 6$ kHz.

Cooper and Bauck (1990) proposed another algorithm for the bandlimited implementation of crosstalk cancellation. The crosstalk cancellation matrix is:

$$[A] = \begin{bmatrix} H_{LL} & H_{LP}H_{LR} \\ H_{LP}H_{RL} & H_{RR} \end{bmatrix}^{-1}. \tag{9.89}$$

Below the frequency f_0, Equation (9.89) is converted into the results of Equation (9.6). Above f_0, Equation (9.89) becomes:

$$[A] = \begin{bmatrix} 1/H_{LL} & 0 \\ 0 & 1/H_{RR} \end{bmatrix}. \tag{9.90}$$

Let $H_L = H_L(\theta, f)$ and $H_R = H_R(\theta, f)$ be a pair of HRTFs for the target virtual source at direction θ. According to Equations (9.1), (9.7), and (9.90), above frequency f_0, the loudspeaker signals are:

$$\begin{bmatrix} L' \\ R' \end{bmatrix} = \begin{bmatrix} H_L/H_{LL} \\ H_R/H_{RR} \end{bmatrix} E_0(f). \tag{9.91}$$

Equation (9.91) is equivalent to filtering the mono input stimulus with a pair of free-field equalized HRTFs with respect to the ipsilateral loudspeaker direction. In the left-right symmetric case, the loudspeaker signals are similar to the free-field equalized binaural signals discussed in Section 8.1.2. When the virtual source is located at the direction of either loudspeaker, such as the direction of the left loudspeaker, we have $\theta = \theta_L$, $H_L(\theta, f) = H_{LL}$, and $H_R(\theta, f) = H_{RL}$. Then, Equation (9.91) becomes:

$$L' = E_0(f) \qquad R' = \frac{H_{RL}}{H_{RR}} E_0(f). \tag{9.92}$$

In this case, the response of the left loudspeaker signal with respect to input $E_0(f)$ is flat, as expected. However, the uneven response of the right loudspeaker signal with respect to input $E_0(f)$ causes the loudspeaker signals to deviate from power panning properties, that is, the overall power (spectrum) of two loudspeaker signals is not equal to $|E_0(f)|^2$. Moreover, the power of resultant binaural signals differs from that of a real source at direction θ, which is undesirable.

To solve this problem, Gardner (1997) proposed a high-frequency power transfer model for the bandlimited implementation of crosstalk cancellation. Above frequency f_0, the scaling gains g_L and g_R are introduced into the left and right loudspeaker signals, respectively. Equation (9.91) is then rewritten as:

$$\begin{bmatrix} L' \\ R' \end{bmatrix} = \begin{bmatrix} g_L H_L / H_{LL} \\ g_R H_R / H_{RR} \end{bmatrix} E_0(f). \tag{9.93}$$

As derived with Equation (9.2), the binaural pressures in loudspeaker reproduction are:

$$\begin{bmatrix} P_L' \\ P_R' \end{bmatrix} = \begin{bmatrix} H_{LL} & H_{LR} \\ H_{RL} & H_{RR} \end{bmatrix} \begin{bmatrix} g_L H_L / H_{LL} \\ g_R H_R / H_{RR} \end{bmatrix} E_0(f). \tag{9.94}$$

To obtain g_L and g_R, Equation (9.94) should be replaced by a corresponding power transfer equation. The input stimulus is assumed to be stationary white noise with power spectra (or variances) σ_0^2 and the transfer functions of two ears are uncorrelated (an assumption that is invalid at low frequencies, and partially valid at high frequencies). Corresponding to Equation (9.94), the power spectra $\sigma_L'^2$ and $\sigma_R'^2$ of binaural pressures in loudspeaker reproduction can be calculated by:

$$\begin{bmatrix} \sigma_L'^2 \\ \sigma_R'^2 \end{bmatrix} = \begin{bmatrix} W_{LL} & W_{LR} \\ W_{RL} & W_{RR} \end{bmatrix} \begin{bmatrix} g_L^2 W_L / W_{LL} \\ g_R^2 W_R / W_{RR} \end{bmatrix} \sigma_0^2, \tag{9.95}$$

where W_{LL}, W_{LR}, W_{RL}, W_{RR}, W_L, and W_R are power transfer functions, from the loudspeakers or virtual source to the two ears. These power transfer functions are related to N-point HRIRs or HRTFs in the form:

$$W = \sum_{n=0}^{N-1} h^2(n) = \frac{1}{N} \sum_{k=0}^{N-1} |H(k)|^2. \tag{9.96}$$

Equation (9.95) can also be written as:

$$\begin{bmatrix} \sigma_L'^2 / \sigma_0^2 \\ \sigma_R'^2 / \sigma_0^2 \end{bmatrix} = \begin{bmatrix} W_{LL} & W_{LR} \\ W_{RL} & W_{RR} \end{bmatrix} \begin{bmatrix} g_L^2 W_L / W_{LL} \\ g_R^2 W_R / W_{RR} \end{bmatrix}. \tag{9.97}$$

The scaling gains g_L and g_R can be determined by substituting the $\sigma_L'^2 / \sigma_0^2$ and $\sigma_R'^2 / \sigma_0^2$ in Equation (9.97) with the desired power transfers W_L and W_R to the ears thus:

$$\begin{bmatrix} g_L^2 \\ g_R^2 \end{bmatrix} = \begin{bmatrix} W_{LL} / W_L & 0 \\ 0 & W_{RR} / W_R \end{bmatrix} \begin{bmatrix} W_{LL} & W_{LR} \\ W_{RL} & W_{RR} \end{bmatrix}^{-1} \begin{bmatrix} W_L \\ W_R \end{bmatrix}. \tag{9.98}$$

If either row on the right side of Equation (9.98) is negative, the resultant scaling gain is an imaginary number. In this case, the gain of the negative-valued row should be set to 0, and another scaling gain can be selected according to Equation (9.97) to ensure that the overall binaural power spectrum is equal to the desired value—in other words, another scaling gain is solved according to the equation:

$$W_L + W_R = g_L^2 \frac{W_L}{W_{LL}} (W_{LL} + W_{RL}) + g_R^2 \frac{W_R}{W_{RR}} (W_{LR} + W_{RR}). \tag{9.99}$$

Additionally, Bai and Lee (2006a) proposed a 5.5-kHz bandlimited implementation of crosstalk cancellation for multiple loudspeakers (an array) by subband filters.

9.7. Some Approximate Methods for Solving the Crosstalk Cancellation Matrix

9.7.1. Cost Function Method for Solving the Crosstalk Cancellation Matrix

As stated in Section 9.6.1, the transfer matrix $[H]$ in two-front loudspeaker reproduction is assumed non-singular and therefore invertible. However, this assumption does not always hold. The practical matrix $[H]$ is singular, or nearly singular, at some frequencies (such as the matrix $[H]$ for a stereo dipole at low frequencies), which causes large magnitudes of some elements in the crosstalk cancellation matrix and an excessive gain for loudspeaker signals at corresponding frequencies. To address this problem, Kirkeby et al. (1999b, 1999c) proposed an alternative scheme for solving the crosstalk cancellation matrix in the frequency domain.

Binaural pressures with a real source at direction θ are given by Equation (9.13). Let $\boldsymbol{P} = [P_L, P_R]^{\mathrm{T}}$ and $\boldsymbol{H}_\theta = [H_L, H_R]^{\mathrm{T}}$ be two 2×1 vectors (matrices), and let $P_0(f) = E_0(f)$. Then, Equation (9.13) can be written as:

$$\boldsymbol{P} = \boldsymbol{H}_\theta E_0. \tag{9.100}$$

For reproduction with two frontal loudspeakers, binaural pressures are derived according to Equations (9.2) and (9.7) thus:

$$\boldsymbol{P}' = [H][A]\boldsymbol{H}_\theta E_0. \tag{9.101}$$

Here, $P' = [P_L', P_R']^{\mathrm{T}}$ is a 2×1 binaural pressure vector, and $[H]$ is a 2×2 transfer matrix from two loudspeakers to two ears (which should not be confused with $\boldsymbol{H}_\theta$).

The crosstalk cancellation matrix $[A]$ is determined by minimizing the cost function:

$$J = \varepsilon + \lambda V(f), \tag{9.102}$$

where ε is the sum of square errors for the left and right ear pressures:

$$\varepsilon = (\boldsymbol{P}' - \boldsymbol{P})^+ (\boldsymbol{P}' - \boldsymbol{P}) = |P_L' - P_L|^2 + |P_R' - P_R|^2. \tag{9.103}$$

$\lambda \geq 0$ is a regularization parameter, with $\lambda = 0$ corresponding to the common scheme in Section 9.1.1. Function $V(f)$, which represents a frequency-dependent effort penalty, introduces frequency-dependent regularization. The magnitudes of the elements in the crosstalk matrix at some frequencies are suppressed by a large $V(f)$ value at the same frequencies. In practice, function $V(f)$ is derived from the loudspeaker signals in Equation (9.1) because the magnitudes of the elements in the crosstalk cancellation matrix should be suppressed at the frequencies that correspond to large loudspeaker signal magnitudes. Generally, we let $S = [L', R']^{\mathrm{T}}$ denote loudspeaker signals. Using Equations (9.1) and (9.7), we define function $V(f)$ as:

$$V(f) = \boldsymbol{S}_b^+ \boldsymbol{S}_b, \qquad \boldsymbol{S}_b = [B]\boldsymbol{S} = [B][A]\boldsymbol{H}_\theta E_0, \tag{9.104}$$

where $[B]$ is a 2×2 matrix.

Minimizing the cost function J in Equation (9.102) leads to an approximate solution of the crosstalk cancellation matrix $[A]$ as:

$$[A] = \{[H]^+[H] + \lambda[B]^+[B]\}^{-1}[H]^+. \tag{9.105}$$

Especially if $[B]$ is selected as a 2×2 identity matrix, then $[B] = [I]$, $S_b = S$ and:

$$V(f) = |L'|^2 + |R'|^2 . \tag{9.106}$$

In this case, function $V(f)$ is the overall power spectrum of two loudspeaker signals, and the crosstalk cancellation matrix in Equation (9.105) becomes:

$$[A] = \{[H]^+ [H] + \lambda[I]\}^{-1} [H]^+ . \tag{9.107}$$

Papadopoulos and Nelson (2010) analyzed the influence of inverse filter parameters on crosstalk cancellation performance. The preceding scheme can be extended to multiple-loudspeaker reproduction. Hill et al. used this scheme to obtain the crosstalk cancellation matrix for four-loudspeaker reproduction (Section 9.2.2), and Bai et al. (2005b) used it, together with a genetic algorithm, to design a robust crosstalk canceller.

9.7.2. Adaptive Inverse Filter Scheme for Crosstalk Cancellation

Nelson et al. (1992) proposed an adaptive inverse filter scheme for crosstalk cancellation. Here, binaural reproduction through a pair of frontal loudspeakers provides an example. The binaural pressures in reproduction are given by Equation (9.2) as:

$$\begin{bmatrix} P_L' \\ P_R' \end{bmatrix} = \begin{bmatrix} H_{LL} & H_{LR} \\ H_{RL} & H_{RR} \end{bmatrix} \begin{bmatrix} A_{11} & A_{12} \\ A_{21} & A_{22} \end{bmatrix} \begin{bmatrix} E_L \\ E_R \end{bmatrix} . \tag{9.108}$$

In Equation (9.108), H_{LL}, H_{RL}, H_{LR}, and H_{RR} are the transfer functions from two loudspeakers to the ears. Let $P_1 = P_L'$, $P_2 = P_R'$, $E_1 = E_L$, $E_2 = E_R$, $H_{11} = H_{LL}$, $H_{12} = H_{LR}$, $H_{21} = H_{RL}$, and $H_{22} = H_{RR}$. The aforementioned equation can be rewritten in the form:

$$\begin{bmatrix} P_1 \\ P_2 \end{bmatrix} = \begin{bmatrix} E_1 H_{11} & E_1 H_{12} & E_2 H_{11} & E_2 H_{12} \\ E_1 H_{21} & E_1 H_{22} & E_2 H_{21} & E_2 H_{22} \end{bmatrix} \begin{bmatrix} A_{11} \\ A_{21} \\ A_{12} \\ A_{22} \end{bmatrix}, \tag{9.109a}$$

or:

$$\begin{bmatrix} P_1 \\ P_2 \end{bmatrix} = \begin{bmatrix} R_{111} & R_{121} & R_{112} & R_{122} \\ R_{211} & R_{221} & R_{212} & R_{222} \end{bmatrix} \begin{bmatrix} A_{11} \\ A_{21} \\ A_{12} \\ A_{22} \end{bmatrix}, \tag{9.109b}$$

where:

$$R_{lmq} = H_{lm} E_q \qquad l, m, q = 1, 2. \tag{9.110}$$

When Equation (9.109) is converted from the frequency domain or Z-domain to the time domain, the frequency-domain multiplications are replaced by time-domain convolution. Thus, the discrete time representation of Equation (9.109) is given by:

$$\begin{bmatrix} p_1(n) \\ p_2(n) \end{bmatrix} = \begin{bmatrix} r_{111}(n) & r_{121}(n) & r_{112}(n) & r_{122}(n) \\ r_{211}(n) & r_{221}(n) & r_{212}(n) & r_{222}(n) \end{bmatrix} * \begin{bmatrix} a_{11}(n) \\ a_{21}(n) \\ a_{12}(n) \\ a_{22}(n) \end{bmatrix} . \tag{9.111}$$

Each function in Equation (9.111) is related to the corresponding function in Equation (9.109) by inverse Fourier transform, and notation "*" denotes time-domain convolution. Suppose that each of the four impulse responses $a_{11}(n)$, $a_{21}(n)$, $a_{12}(n)$, and $a_{22}(n)$ can be described by an N-point FIR filter. Then, the convolution in Equation (9.111) can be written as:

$$r_{lmq}(n) * a_{mq}(n) = \sum_{s=0}^{N-1} a_{mq}(s) r_{lmq}(n-s) \qquad l, m, q = 1, 2. \tag{9.112}$$

We define a $4N \times 1$ vector that denotes the impulse responses of crosstalk cancellation filters:

$$a = [a_{11}(0)....a_{11}(N-1), a_{21}(0)....a_{21}(N-1), a_{12}(0)....a_{12}(N-1), a_{22}(0)....a_{22}(N-1)]^T. \tag{9.113}$$

Then, Equation (9.111) can be written as:

$$p(n) = [R(n)]a, \tag{9.114}$$

where:

$$p(n) = \begin{bmatrix} p_1(n) \\ p_2(n) \end{bmatrix}, \tag{9.115}$$

is a 2×1 binaural pressure vector, and $[R(n)]$ is a $2 \times 4N$ matrix, whose elements can be derived by solving Equation (9.112).

Ideally, vector a should be selected so that binaural pressures $p(n)$ in reproduction match the intended or desired binaural pressures (signals) in the time domain:

$$e(n) = \begin{bmatrix} e_1(n) \\ e_2(n) \end{bmatrix}, \tag{9.116}$$

where $e_1(n) = e_L(n)$ and $e_2(n) = e_R(n)$ are the inverse Fourier transform of E_1 and E_2, respectively.

In practice, vector a can be obtained by minimizing the square error between vector $p(n)$ and $e(n)$ as:

$$\begin{aligned} \min\{|\varepsilon(n)|^2\} &= \min E\{[e(n) - p(n)]^+ [e(n) - p(n)]\} \\ &= \min E\{[e(n) - [R(n)]a]^+ [e(n) - [R(n)]a]\}, \end{aligned} \tag{9.117}$$

where "E" denotes the expectation operation, and superscript "+" denotes the transpose and conjugation of the matrix. Vector a can be solved by the pseudoinverse method, but this approach is complex because of the high dimension of the matrix involved.

Nelson et al. (1992) presented an adaptive signal processing scheme for solving vector a. From Equation (9.117), the gradient of square error is given by:

$$\frac{\partial |\varepsilon|^2}{\partial a} = -2E\{[R(n)]^+ \varepsilon(n)\}, \tag{9.118}$$

$$\varepsilon(n) = e(n) - [R(n)]a. \tag{9.119}$$

Let a_i denote the vector a obtained in the ith iteration, and a_{i+1} denote the vector a obtained in the $(i + 1)$th iteration. Then, the adaptive signal processing yields the iterative equation:

$$a_{i+1} = a_i + \lambda [R_i(n)]^+ \varepsilon_i(n), \tag{9.120}$$

where λ is the coefficient that controls step length and convergence; $\varepsilon_i(n)$ represents the error vector after the ith iteration; and $[R_i(n)]$ denotes the matrix $[R(n)]$ related to the ith iteration. By appropriately choosing $\varepsilon(n)$ in Equation (9.116) as the training input stimulus and $\boldsymbol{a}_0$ as the initial $\boldsymbol{a}$ for crosstalk cancellation processing, the binaural pressures in Equation (9.115) for loudspeaker reproduction can be derived by calculation or measurement (using microphones mounted on an artificial head). Then, the error is evaluated by:

$$\varepsilon_0(n) = e(n) - \boldsymbol{p}_0(n). \tag{9.121}$$

Vector $\boldsymbol{a}_1$ is obtained by substituting $\varepsilon_0(n)$ into Equation (9.120). Finally, a convergent vector $\boldsymbol{a}$ is obtained after iterative operations. This method can be extended to binaural reproduction with multiple loudspeakers and listeners.

9.8. Summary

Binaural signals from either recording or synthesis are originally intended for headphone presentation. When they are reproduced through loudspeakers, unwanted crosstalk occurs, damaging the directional information contained in the binaural signals. Prior to loudspeaker reproduction, therefore, binaural signals should be precorrected or filtered in order to cancel out crosstalk. The transaural technique, which merges the two stages of binaural synthesis and crosstalk cancellation, is equivalent to controlling binaural pressures in reproduction and synthesizing the virtual source by controlling loudspeaker signals.

The crosstalk cancellation matrix for conventional reproduction with two frontal loudspeakers can be obtained by inverting the acoustical transfer matrix from two loudspeakers to two ears (assuming that the inversion exists). Binaural pressure control and crosstalk cancellation can be extended to multiple loudspeakers and listeners, resulting in the general theory of transaural reproduction.

Pinna-related high-frequency spectral cues cannot be stably replicated in loudspeaker reproduction. Incorrect dynamic cues often cause back-front confusion in static binaural reproduction through a pair of frontal loudspeakers. In contrast to headphone reproduction, two-front loudspeaker reproduction can recreate stably perceived virtual sources only in frontal-horizontal quadrants, rather than in full three-dimensional directions. Transaural reproduction through four horizontal loudspeakers can resolve front-back confusion and improve horizontal localization performance.

The performance of crosstalk cancellation is sensitive to listening position relative to loudspeakers. For a given loudspeaker configuration and signal processing, crosstalk cancellation is effective only at a default listening position. Head rotation and translation influence the performance of crosstalk cancellation and alter binaural pressures in reproduction. Therefore, a common defect in binaural or transaural reproduction through loudspeakers is the limited size of the listening region. Researchers evaluated the stability of crosstalk cancellation and the perceived virtual source direction in loudspeaker reproduction. The results indicate that reproduction stability depends upon loudspeaker configuration and frequency. A narrow span of two frontal loudspeakers improves stability against lateral head translation. This improvement is the principle behind the stereo dipole. The performance of a stereo dipole is desirable at mid and high frequencies, but unsatisfactory at low frequencies.

Mismatched HRTFs influence the perceived horizontal virtual source azimuth at lateral directions. Mismatched loudspeaker pairs and head rotation exacerbate this situation.

Under ideal conditions, transaural reproduction yields the same binaural pressures as those of a real source. Given various errors, however, completely satisfying ideal conditions in all audible frequency ranges is difficult to accomplish. Transaural filters modify the overall power spectrum of input signals and alter the balance among the spectral components of resultant loudspeaker signals. As a result, nonideal conditions (e.g., lateral head translation and rotation, room reflection, etc.) cause perceived coloration and localization error. Perceived coloration is also a common defect in transaural reproduction. Additional timbre equalization is therefore required. Significant differences among the performances of different equalization algorithms are found. Inappropriate equalization may worsen timbre. An effective equalization algorithm should eliminate or cancel the close-unit circle poles (and zeros) in transaural filters to smooth or eliminate response peaks (and notches). This approach can also simplify signal processing. The constant-power equalization algorithm satisfies the aforementioned requirements.

Crosstalk cancellation or transaural synthesis can be implemented by various schemes on the basis of digital signal processing. Such schemes are similar in basic principle but different in terms of implementation and design. The causality and stability of a crosstalk canceller should be carefully manipulated. To solve this problem, researchers developed approximation methods for crosstalk cancellation.

10

Virtual Reproduction of Stereophonic and Multi-channel Surround Sound

This chapter presents a catalog of specific applications of virtual auditory display, that is, the signal processing and schemes for virtual reproduction of stereophonic and multi-channel surround sound. Schemes for headphone-rendered stereophonic and multi-channel surround sound are addressed. The problems of correction of stereophonic reproduction through nonstandard loudspeaker configurations, as well as virtual reproduction of multi-channel surround sound through a few loudspeakers, are discussed.

Conventional stereophonic (as well as multi-channel surround) sound and virtual auditory displays (VADs) are two different categories of spatial sound systems. Two-channel stereophonic sound is the most popular, and 5.1-channel surround sound is recommended by the ITU (International Telecommunication Union) as the standard for multi-channel sound systems, with and without accompanying pictures. As stated in Section 6.4, VADs and multi-channel surround sound are closely related to each other; thus, some methods for these technologies are interchangeable. In Section 6.5, some principles of multi-channel sound were applied to simplifying binaural virtual source synthesis. Meanwhile, this chapter applies the principles of binaural and transaural synthesis processing to reproducing stereophonic and multi-channel surround sound, so as to accommodate headphone presentation or various loudspeaker configurations.

The principles and methods discussed in this chapter are similar to those in Section 6.5, given that both resolve the problems encountered for virtual loudspeakers. However, this chapter focuses on a category of special application—the virtual reproduction of stereophonic or multi-channel surround sound. Two problems are addressed. One is the algorithm for converting stereophonic or multi-channel surround sound signals; the conversion is intended to accommodate headphone reproduction. The other is the algorithm for converting stereophonic or multi-channel surround sound signals; this conversion is designed to accommodate loudspeaker reproduction with various numbers or configurations of loudspeakers in practical applications. Section 10.1 discusses headphone presentation of stereophonic and multi-channel surround sound signals. Section 10.2 presents correction of stereophonic reproduction through nonstandard loudspeaker configurations. Section 10.3 outlines some stereophonic enhancement systems. The virtual reproduction of multi-channel surround sound through loudspeakers is addressed in Section 10.4.

10.1 Binaural Reproduction of Stereophonic and Multi-channel Surround Sound through Headphones

VADs and conventional stereophonic (or multi-channel surround) sound are based on different principles. The binaural signals in a VAD are originally intended for headphone presentation, and crosstalk cancellation is needed to convert binaural signals into signals for loudspeaker reproduction (Chapter 9). By contrast, stereophonic and multi-channel surround sound signals are originally intended for loudspeaker reproduction. This section addresses the issue of binaural synthesis processing for converting stereophonic and multi-channel surround sound signals for headphone presentation.

10.1.1 Binaural Reproduction of Stereophonic Sound through Headphones

Most existing audio programs are recorded in conventional stereophonic format. They are originally intended for reproduction through a pair of frontal loudspeakers with a standard span angle $2\theta_0 = 60°$. In some practical applications, such as MP3 (Moving Pictures Experts Group, Audio Layer III) and other portable players, however, conventional stereophonic signals are often reproduced through headphones, causing in-head localization or lateralization. This unnatural spatial auditory perception is caused by the following factors:

1. As stated in Section 1.6.1, when two channel stereophonic signals are reproduced through a pair of loudspeakers, these signals are received by both ears after being scattered/diffracted by the head and pinnae. The pressure signal at each ear is the combination of those from two loudspeakers (Figure 1.16). The interchannel level difference between the two loudspeaker signals is then converted into interaural time difference (ITD) in binaural pressure signals, resulting in the summing localization of virtual sources. By contrast, when reproduction is conducted through headphones, the stereophonic signals are directly fed into two ears, generating interaural level difference (ILD), rather than ITD. This phenomenon differs from the case of a real source, in which ITD is generated at low frequencies, and both ITD and ILD are produced at high frequencies. Unnatural binaural pressure signals generate in-head localization.
2. In contrast to loudspeaker reproduction, headphone reproduction does not cause listening room reflections. The presence of such reflections is important for externalization.
3. Static headphone reproduction lacks the dynamic information caused by head movement.

Binaural processing is applicable to improving conventional stereophonic sound reproduction through headphones. HRTF-based filtering (that is, filtering based on head-related transfer function) is used to simulate acoustic transmission from loudspeakers to two ears, or reproduce stereophonic signals through a pair of virtual loudspeakers (sources). Such processing presents correct stereophonic information to two ears. Bauer (1961) introduced the basic idea of binaural processing for reproducing stereophonic signals through headphones. In early works, signal processing was implemented by various analog circuits, but with low accuracy (Thomas, 1977; Sakamoto et al., 1976). During the past two decades, analog signal processing has been replaced by digital techniques. The basic principles of various algorithms are similar, but implementation may differ (Zhang et al., 2000; Kirkeby, 2002).

Let H_{LL}, H_{RL}, H_{LR}, and H_{RR} denote the transfer functions (HRTFs) from two frontal loudspeakers to two ears in stereophonic reproduction; let L and R be the stereophonic or loudspeaker signals (Figure 10.1). The binaural pressures in reproduction are given by:

$$P_L = H_{LL}L + H_{LR}R \qquad P_R = H_{RL}L + H_{RR}R. \tag{10.1}$$

In headphone reproduction, if signals L and R are preprocessed according to Equation (10.2a), the resultant binaural pressures are equal, or directly proportional, to those in loudspeaker reproduction:

$$E_L = H_{LL}L + H_{LR}R \qquad E_R = H_{RL}L + H_{RR}R. \tag{10.2a}$$

Equation (10.2a) can also be written in a matrix form:

$$\begin{bmatrix} E_L \\ E_R \end{bmatrix} = \begin{bmatrix} H_{LL} & H_{LR} \\ H_{RL} & H_{RR} \end{bmatrix} \begin{bmatrix} L \\ R \end{bmatrix}, \tag{10.2b}$$

in which acoustic transmission from loudspeakers to two ears is simulated in a similar manner to that in actual loudspeaker reproduction. Thus, correct stereophonic information is presented to two ears.

Subjective experiments indicate that the processing provided by Equation (10.2) can recreate summing virtual sources in frontal regions, thereby relatively improving the spatial auditory experience presented by stereophonic headphone reproduction (Zhang et al., 2000). However, the virtual sources generated by static and free-field virtual loudspeaker processing are usually located on the head surface, rather than completely externalized. Incorporating reflections and reverberations to binaural signals improves externalization to a certain extent. In processing, binaural room impulse responses (BRIRs) can be used to replace free-field HRTFs (or head-related impulse responses) to simulate transmission from stereophonic loudspeakers to two ears in a reflective listening room (Mickiewicz and Sawicki, 2004). Nonetheless, the discussion in Section 10.1.2 indicates that reflections should be carefully introduced to processing. Otherwise, unnatural auditory perception may occur.

Signal processing with Equation (10.2) is equivalent to simultaneously synthesizing two virtual sources (loudspeakers) in headphone reproduction. All the problems encountered in headphone-based binaural reproduction (Chapter 8), such as reversal and elevation errors in localization, may occur as stereophonic signals are binaurally reproduced through headphones. The various methods discussed in Chapter 8 improve the performance of binaural stereophonic reproduction through headphones. In particular, the equalization of headphone-to-ear canal transmission is preferable. According to the discussion in Section 8.2.2, however, ideal equalization necessitates individualized HpTF (headphone-to-ear canal transfer function) processing, which is difficult to accomplish in most practical applications. Thus, the equalization of headphone-to-ear canal transmission is excluded in most applications of binaural stereophonic reproduction. This exclusion inevitably degrades subjective localization performance.

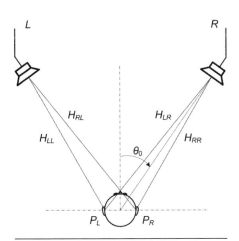

Figure 10.1 Transmission from two front loudspeakers to two ears

Using stereophonic music programs, Lorho et al. (2002) conducted an experiment to evaluate the subjective preference of nine algorithms for binaural stereophonic reproduction and unprocessed headphone reproduction. The results indicate that in most cases listeners prefer unprocessed headphone reproduction to binaural stereophonic reproduction. An experiment on subjective attribute rating shows that timbre attributes enable the discrimination of binaural stereophonic reproduction from unprocessed headphone reproduction (Lorho et al., 2005). Although no definitive conclusion has been reached, these experimental results imply that binaural stereophonic reproduction tends to cause changes in timbre. A similar problem occurs in the binaural reproduction of 5.1-channel surround sound through headphones (Lorho and Zacharov, 2004), an issue to be discussed in Section 10.1.2.

Similar to the case in Section 8.1.2, the changes in timbre during binaural stereophonic reproduction can be reduced using free- or diffuse-field equalized binaural signals. The signals can then be presented through free- or diffuse-field equalized headphones. In some conventional applications, unprocessed stereophonic signals are directly reproduced through free- or diffuse-field equalized headphones. In this case, the resultant binaural pressures are equivalent to those caused by a pair of real stereophonic loudspeakers in the free or diffuse field, but transmission from each loudspeaker to the opposite ear is disregarded. This method reduces changes in timbre during headphone reproduction, but cannot recreate correct spatial information on stereophonic sound.

10.1.2 Basic Algorithm for Headphone-based Binaural Reproduction of 5.1-Channel Surround Sound

The scheme in Section 10.1.1 can be extended to the binaural reproduction of multi-channel surround sound (e.g., 5.1-channel) through headphones. Some related techniques and patents have been introduced, including the Dolby headphones developed by Dolby Laboratories (http://www .dolby.com). Although implementation details vary among different techniques and patents, the basic principles are similar.

We take the binaural reproduction of 5.1-channel surround sound through headphones as an example. As stated in Section 1.8.4, the 5.1-channel system has five independent channels with full audio bandwidths—left (L), right (R), center (C), left surround (LS), and right surround (RS) channels—plus an optional low-frequency effect (LFE) channel. The ITU-recommended 5.1-channel loudspeaker configuration is illustrated in Figure 1.23, and the horizontal azimuths of loudspeakers are:

$$\theta_L = 330° \quad \theta_R = 30° \quad \theta_C = 0° \quad \theta_{LS} = 250° \pm 10° \quad \theta_{RS} = 110° \pm 10°. \tag{10.3}$$

When directly reproduced through headphones, 5.1-channel signals are downmixed into two channel headphone signals E_L and E_R as:

$$E_L = L + 0.707C + LS \qquad E_R = R + 0.707C + RS \tag{10.4}$$

where signals L and LS are mixed with −3 dB-attenuated signals C and then fed into the left headphone; and signals R and RS are mixed with −3 dB-attenuated signals C and then fed into the right headphone. The −3 dB attenuation ensures a constant overall power for signal C. Signal LFE is processed in a similar manner to signal C, and is therefore excluded in the succeeding discussion. Similar to directly reproducing stereophonic signals through headphones, directly reproducing 5.1-channel signals through headphones degrades spatial information and results in in-head localization.

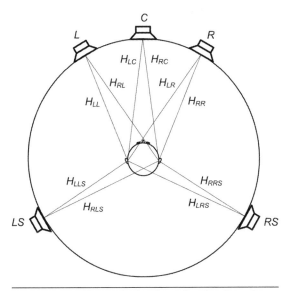

Figure 10.2 Transmission from five loudspeakers in 5.1-channel surround sound to two ears

Binaural synthesis can also be used to improve the effect of 5.1-channel surround sound reproduction through headphones. That is, virtual loudspeaker-based algorithms that are similar to those in Section 6.5.1 are incorporated, so that 5.1-channel signals are reproduced through corresponding virtual loudspeakers or sources. This reproduction generates correct spatial information.

In actual 5.1-channel loudspeaker reproduction, let H_{LL}, H_{RL}, H_{LR}, and H_{RR} denote the HRTFs from the left and right loudspeakers to two ears; let H_{LC} and H_{RC} denote the HRTFs from the central loudspeaker to two ears; and let H_{LLS}, H_{RLS}, H_{LRS}, and H_{RRS} denote the HRTFs from the left and right surround loudspeakers to two ears (Figure 10.2). The binaural pressures P_L and P_R are the sum of these caused by each loudspeaker, expressed as:

$$P_L = H_{LL}L + H_{LC}C + H_{LR}R + H_{LLS}LS + H_{LRS}RS$$
$$P_R = H_{RL}L + H_{RC}C + H_{RR}R + H_{RLS}LS + H_{RRS}RS$$

(10.5)

Similar to the case in Equation (10.2), the 5.1-channel signals are filtered with five pairs of HRTFs that correspond to the locations of five loudspeakers. The signals are then mixed and reproduced through headphones thus:

$$E_L = H_{LL}L + H_{LC}C + H_{LR}R + H_{LLS}LS + H_{LRS}RS$$
$$E_R = H_{RL}L + H_{RC}C + H_{RR}R + H_{RLS}LS + H_{RRS}RS$$

(10.6)

Through the simulation of acoustic transmission from five loudspeakers to two ears, the binaural pressures generated in headphone reproduction are equal, or directly proportional, to those in actual loudspeaker reproduction, resulting in correct spatial information. The algorithm given by Equation (10.6) is complex because it requires 10 HRTF-based filters. Taking advantage of left-right symmetry, we can simplify the algorithm by shuffler structure implementation (Cooper and Bauck, 1989; Bauck and Cooper, 1996; Xie et al., 2005d).

In real 5.1-channel surround sound reproduction, the central signal C can be reproduced by the phantom center channel method. That is, signal C is attenuated by −3 dB and mixed with left and right signals. The mixed signals are then reproduced by left and right loudspeakers. When all the 5.1-channel signals are zeros except C, both left and right loudspeaker signals are 0.707 C. In this case, the summing virtual source is located in the front. Accordingly, in headphone reproduction, signal C is attenuated by −3 dB and mixed with left and right signals. The mixed signals are then processed by HRTF-based filters. Equation (10.6) can therefore be simplified as:

$$E_L = H_{LL}(L + 0.707C) + H_{LR}(R + 0.707C) + H_{LLS}LS + H_{LRS}RS$$
$$E_R = H_{RL}(L + 0.707C) + H_{RR}(R + 0.707C) + H_{RLS}LS + H_{RRS}RS$$

(10.7)

Left-right symmetry yields $H_{LL} = H_{RR} = H_{\alpha1}$, $H_{LR} = H_{RL} = H_{\beta1}$, $H_{LLS} = H_{RRS} = H_{\alpha2}$, $H_{LRS} = H_{RLS} = H_{\beta2}$. Then, Equation (10.7) can be written in the matrix form:

$$\begin{bmatrix} E_L \\ E_R \end{bmatrix} = \begin{bmatrix} H_{\alpha1} & H_{\beta1} \\ H_{\beta1} & H_{\alpha1} \end{bmatrix} \begin{bmatrix} L+0.707C \\ R+0.707C \end{bmatrix} + \begin{bmatrix} H_{\alpha2} & H_{\beta2} \\ H_{\beta2} & H_{\alpha2} \end{bmatrix} \begin{bmatrix} LS \\ RS \end{bmatrix}$$

and, with the symmetry of the two 2×2 matrices on the right side, this equation can be diagonalized, as in Equation (6.106). Thus, this equation is equivalent to:

$$\begin{bmatrix} E_L \\ E_R \end{bmatrix} = \begin{bmatrix} 0.5 & 0.5 \\ 0.5 & -0.5 \end{bmatrix} \left\{ \begin{bmatrix} H_{\Sigma1} & 0 \\ 0 & H_{\Delta1} \end{bmatrix} \begin{bmatrix} 1 & 1 \\ 1 & -1 \end{bmatrix} \begin{bmatrix} L+0.707C \\ R+0.707C \end{bmatrix} + \begin{bmatrix} H_{\Sigma2} & 0 \\ 0 & H_{\Delta2} \end{bmatrix} \begin{bmatrix} 1 & 1 \\ 1 & -1 \end{bmatrix} \begin{bmatrix} LS \\ RS \end{bmatrix} \right\}, \quad (10.8)$$

where:

$$H_{\Sigma1} = H_{\alpha1} + H_{\beta1} \qquad H_{\Delta1} = H_{\alpha1} - H_{\beta1} \qquad H_{\Sigma2} = H_{\alpha2} + H_{\beta2} \quad H_{\Delta2} = H_{\alpha2} - H_{\beta2}. \quad (10.9)$$

Only four HRTF-based filters are required for the shuffler structure implementation in Equation (10.8). Therefore, the algorithm is simplified and contributes to real-time implementation.

Equation (10.8) can be expressed in the time domain as:

$$\begin{bmatrix} e_L \\ e_R \end{bmatrix} = \begin{bmatrix} 0.5 & 0.5 \\ 0.5 & -0.5 \end{bmatrix} \left\{ \begin{bmatrix} h_{\sigma1} & 0 \\ 0 & h_{\delta1} \end{bmatrix} * \begin{bmatrix} 1 & 1 \\ 1 & -1 \end{bmatrix} \begin{bmatrix} l+0.707c \\ r+0.707c \end{bmatrix} + \begin{bmatrix} h_{\sigma2} & 0 \\ 0 & h_{\delta2} \end{bmatrix} * \begin{bmatrix} 1 & 1 \\ 1 & -1 \end{bmatrix} \begin{bmatrix} ls \\ rs \end{bmatrix} \right\}, \quad (10.10)$$

where the lowercase notations denote the time-domain responses or signals, which are related to those in Equation (10.8) in the frequency domain by inverse Fourier transform. Notation "*" denotes convolution manipulation. Figure 10.3 shows a block diagram of the signal processing that is designed according to Equation (10.8).

Direct (free-field) transmission from loudspeakers to two ears is simulated using the signal-processing algorithm in Equation (10.6) or Equation (10.10), and the reflections in a listening room are disregarded. Similar to binaural stereophonic reproduction, simulating listening room reflections is often incorporated into the binaural reproduction of 5.1-channel surround sound (such as that implemented with Dolby headphones). Introducing this simulation is critical to the externalization of auditory events. However, introducing reflections to processing causes two problems:

1. In loudspeaker reproduction, the strong reflections from a listening room degrade the spatial information in the original recording of a surround sound program. According to the recommendation of the ITU (ITU-R Rec. BS 775-1, 1994), controlling the reflections from a listening room necessitates the appropriate treatment of sound absorption. Appropriately controlling the simulated reflections in the binaural reproduction of

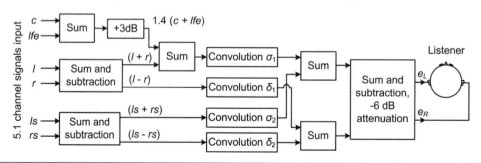

Figure 10.3 Block diagram of signal processing for the binaural reproduction of 5.1-channel surround sound through headphones

5.1-channel surround sound through headphones is difficult to accomplish. The contribution of weakly simulated reflections is insufficient for the externalization of virtual sound sources. Strongly simulated reflections, on the other hand, degrade the spatial information in original program recording, or cause unnatural perceived effects. In particular, some reflection information is often mixed with original 5.1-channel signals. In practical programs, the central signal C often records speech (such as dialogue in film). Adding superfluous reflections to speech signals may diminish intelligibility.

2. The length of room impulse responses is in the order of reverberation time. Accurately simulating the reflections in a listening room necessitates the convolution of input signals with BRIRs. For a reverberation time of 0.3 s and a sampling frequency of 48 kHz, for example, the length of BRIRs is in the order of $48000 \times 0.3 = 14400$ points. The real-time convolution of such a long impulse response is difficult. In practice, only a few early reflections are precisely simulated, which may degrade the perceived performance in reproduction.

10.1.3 Improved Algorithm for Binaural Reproduction of 5.1-Channel Surround Sound through Headphones

In many 5.1-channel surround sound programs, especially 5.1-channel music programs, some reflections are recorded in two surround channel signals. If this reflection information is enhanced in binaural reproduction through headphones, externalization can also be achieved, without incorporating reflections from a listening room. For most 5.1-channel programs with pictures (such as films and digital television—DTV—programs), the central channel signal usually records speech (dialogue), which is unsuitable for incorporating superfluous reflections for externalization. In this case, reintroducing room reflections for externalization of the central channel is not always necessary because the picture captures a listener's attention.

In domestic use, two surround signals in 5.1-channel surround sound are reproduced through a pair of surround loudspeakers, in accordance with the ITU configuration (Figure 1.23). By contrast, cinema applications feature two surround signals reproduced through a series of loudspeakers arranged at the side and rear (Figure 1.24). Decorrelated processing can be applied to the surround signals, thereby enhancing subjective envelopment.

On the basis of these considerations, Xie et al. (2005d; 2005e) suggested an improved algorithm for binaurally reproducing 5.1-channel surround sound through headphones, in which decorrelated surround signals are reproduced by a series of virtual loudspeakers. The main points are as follows:

1. Signals L and R are mixed with a −3 dB-attenuated central signal C and then filtered with free-field HRTFs at corresponding virtual loudspeakers. The reflections from a listening room are not simulated.
2. The surround signals LS and RS are decorrelated and then reproduced by multiple virtual surround loudspeakers.

Figure 10.4 shows an illustrative case with six virtual surround loudspeakers. The figure also shows the horizontal azimuth of virtual surround loudspeakers. The decorrelated left surround signals LS_1, LS_2, and LS_3 derived from the original left surround signal LS are reproduced through three virtual left surround loudspeakers. The decorrelated right surround signals RS_1, RS_2, and RS_3 derived from the original right surround signal RS are reproduced through three virtual right surround loudspeakers. The configuration of virtual surround loudspeakers is not confined to the case of Figure 10.4. For example, the six virtual surround loudspeakers can be arranged at horizontal azimuths of 90°, 120°, 150°, 210°, 240°, and 270°. More virtual surround loudspeakers can also be simulated. In practical applications, decorrelation and multiple virtual surround loudspeakers are optional functions; a listener can select these functions or two conventional virtual surround loudspeakers in headphone reproduction.

Some schemes for surround signal decorrelation are available. The simplest involves applying appropriate delay to the original signal in order to reduce correlation to a certain extent. The delay period should be selected in accordance with psychoacoustic rules. This scheme was previously adopted in our algorithm, but it tends to create timbre coloration in reproduction. To reduce coloration, decorrelation can be implemented by filtering signals with special all-pass filters, including all-pass filters with random phases

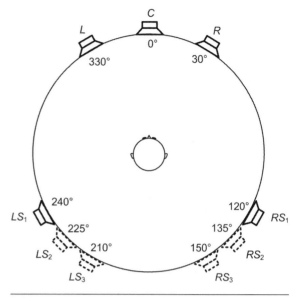

Figure 10.4 Illustrative case of multiple virtual surround loudspeakers

(Kendall, 1995), all-pass filters with random delay imposed on each critical band (Bouéri and Kyirakakis, 2004), and reciprocal maximal-length sequence filters (Xie et al., 2012c).

In practice, the surround signals LS and RS often contain reflections of the original sound field. Low correlation among the six signals (LS_1, LS_2, LS_3, RS_1, RS_2, and RS_3) improves subjective envelopment. However, two surround signals may sometimes record certain localization information, such as the effect of a helicopter flying over/around a listener. In these cases, the original surround signals LS and RS are correlated. To take this problem into account, a left-right symmetric decorrelation process is suggested. That is, the all-pass filters for creating the left surround signals LS_1, LS_2, and LS_3 are identical to those for creating the right surround signals RS_1, RS_2, and RS_3. When original LS and RS are a pair of low-correlation signals, decorrelation processing yields six signals with low correlation among one another. When original LS and RS are a pair of correlated signals, decorrelation processing retains the correlation between each pair of signals (i.e., LS_1 and RS_1, LS_2 and RS_2, LS_3 and RS_3), but eliminates the correlation among different pairs of signals. Therefore, left-right symmetric decorrelation preserves some localization information encoded in the original surround signals LS and RS.

The six uncorrelated surround signals are reproduced through corresponding virtual surround loudspeakers. Similar to the case in Equation (10.6), a straightforward method is to create each virtual surround loudspeaker by HRTF-based filters. Synthesizing the three virtual frontal loudspeakers and six virtual surround loudspeakers in Figure 10.4 necessitates 18 HRTF-based filters. Even if the phantom center channel method is used and signal processing is implemented by the shuffler structure similar to Equation (10.8), eight HRTF-based filters are required, which is still a complicated process.

The virtual loudspeaker-based algorithm discussed in Section 6.5.1 is applicable to simplifying signal processing. In conventional intensity stereo (or multi-channel surround sound), the virtual source at the direction without loudspeakers can be created by summing localization. The perceived virtual source direction is controlled by loudspeaker signal mixing (panning). For an actual loudspeaker configuration (as in Figure 10.4), when identical signals (except for a difference in amplitude) are reproduced by two surround loudspeakers at $\theta_{LS} = 240°$ and $\theta_{RS} = 120°$, the azimuth of the low-frequency summing virtual source can be evaluated according to Equation (1.27) thus:

$$\sin\theta_I = \frac{RS - LS}{RS + LS}\sin 120° = \frac{RS/LS - 1}{RS/LS + 1}\sin 120°. \tag{10.11}$$

Therefore, the virtual source azimuth within $[120°, 240°]$ can be continuously controlled by adjusting RS/LS.

This method is applicable to binaurally reproducing multiple decorrelated surround signals through headphones. A pair of virtual surround loudspeakers at $\theta_{LS} = 240°$ and $\theta_{RS} = 120°$ are separately simulated by HRTF-based filtering. The input signals LS' and RS' for the virtual surround loudspeaker pair are the weighted mix of decorrelated surround signals:

$$LS' = LS_1 + g_1 LS_2 + g_2 LS_3 + g_3 RS_2 + g_4 RS_3 \tag{10.12}$$
$$RS' = RS_1 + k_1 LS_2 + k_2 LS_3 + k_3 RS_2 + k_4 RS_3$$

Here g_i and k_i with $i = 1, 2, 3, 4$ are the weights or gains for signal mixing. For example, when $LS_1 = LS_3 = RS_1 = RS_2 = RS_3 = 0$, signal LS_2 is mixed into LS' and RS', according to:

$$LS' = g_1 LS_2 \qquad RS' = k_1 LS_2$$

Gains g_1 and k_1 are identified by replacing LS and RS with the LS' and RS' in Equation (10.11), as well as letting $\theta_I = 225°$ and combining with a restriction of constant signal power $LS'^2 + RS'^2 = LS_1^2$. Similar procedures are applied to signal LS_3, RS_2, and RS_3, yielding:

$$g_1 = k_3 = 0.999 \qquad k_1 = g_3 = 0.101 \qquad g_2 = k_4 = 0.966 \qquad k_2 = g_4 = 0.259. \tag{10.13}$$

To summarize, instead of being directly reproduced through a pair of virtual surround loudspeakers according to Equation (10.7) or Equation (10.8), the original surround signals LS and RS are decorrelated, after which the resultant mixing signals LS' and RS' are reproduced through a pair of virtual surround loudspeakers. That is, the signals LS and RS in Equation (10.7) or Equation (10.8) are replaced by the mixing signals LS' and RS' in Equation (10.12). Figure 10.5 shows a block diagram for surround signal decorrelation and multiple virtual loudspeaker reproduction. Actual signal processing is implemented by replacing the signals ls and rs in Figure 10.5 with ls' and rs' (the time domain counterparts of signals LS' and RS').

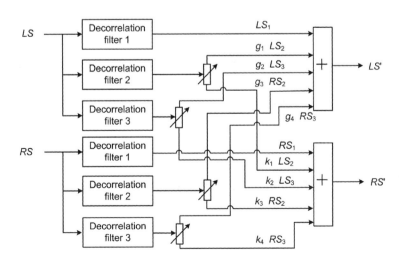

Figure 10.5 Block diagram of surround signal decorrelation and multiple virtual loudspeaker reproduction

10.1.4 Notes on Binaural Reproduction of Multi-channel Surround Sound

Previous sections focused on binaurally reproducing stereophonic and 5.1-channel surround sound through headphones. Similar algorithms have been designed for the binaural reproduction of other types of multi-channel surround sound, such as 7.1-channel surround sound, Dolby Pro-Logic surround sound, and Ambisonics. The principle that underlies the binaural reproduction of Ambisonics was discussed in Section 6.5. Various algorithms for the binaural reproduction of stereophonic or multi-channel surround sound are applicable to audio players with headphones, such as MP3 players and multimedia computers.

In contrast to reproduction through a pair of frontal loudspeakers (Section 10.4), binaural reproduction through headphones can recreate virtual sources in the entire horizontal plane or in three-dimensional space. As stated in Chapter 8, however, using nonindividualized HRTFs in binaural synthesis often causes localization errors in perceived virtual source (virtual loudspeaker) directions; examples of such errors are reversal and elevation errors. Implementing individualized HRTF processing relatively improves localization performance.

Some researchers suggested binaurally reproducing 5.1-channel surround sound through quadraphonic headphones (Shiu et al., 2012). The central channel signal C is attenuated by −3 dB and mixed with left and right channel signals; signal LFE is attenuated −3 dB and mixed with left and right surround channel signals. The four channel downmixing signals are subjected to binaural synthesis and then presented through quadraphonic headphones.

Incorporating the dynamic information generated by head movement into binaural processing also visibly improves localization performance (Chapter 12). Nevertheless, these methods cause the system and algorithm to be complex, preventing them from being widely used in consumer electronics. The future application of these methods depends upon technique development and production cost.

Similar to the case of binaural stereophonic reproduction in Section 10.1.1, the equalization of headphone-to-ear canal transmission is a strict requirement for binaurally reproducing multi-channel surround sound. Incorporating appropriate headphone-to-ear canal transmission equalization improves the perceived performance. According to the discussion in Section 8.2.2, however, ideal equalization necessitates individualized HpTF processing. Nevertheless, this processing is difficult to accomplish in practical situations, and is therefore excluded in most applications.

Signal processing for the binaural reproduction of stereophonic and multi-channel surround sound can be implemented by various filters in the frequency domain (Chapter 5). Most multi-channel surround sound signals are currently recorded and transmitted using subband coding techniques, such as MPEG (Moving Pictures Experts Group) and Dolby digital. Signal processing for the binaural reproduction of multi-channel surround sound is more efficient when implemented in the subband domain. Some works suggested the realization of binaural processing for MPEG 5.1-channel surround sound signals in the quadrature mirror filter domain (Virette et al., 2007; Yu et al., 2007). Other researchers proposed the implementation of binaural processing for Dolby digital 5.1-channel surround sound signals in the modified discrete cosine transform domain (Lee et al., 2009).

10.2 Algorithms for Correcting Nonstandard Stereophonic Loudspeaker Configurations

As shown in Figure 1.16, the stereophonic loudspeakers of a standard configuration are symmetrically arranged with a 60° span angle relative to the listener. Stereophonic programs are made for the standard loudspeaker configuration. In some applications, however, the stereophonic loudspeakers may not be arranged in compliance with the standard configuration because of the limitations imposed by practical conditions. For example, the stereophonic loudspeakers may be arranged with a narrow span or in an asymmetric manner. Using a nonstandard loudspeaker configuration in stereophonic reproduction degrades perceived performance. The transaural synthesis discussed in Section 9.1.1 solves this problem. That is, a pair of actual loudspeakers with a nonstandard configuration is used to simulate a pair of virtual loudspeakers with a standard configuration, through which stereophonic signals are reproduced. This approach is the basic principle that underlies the correction of nonstandard stereophonic loudspeaker configurations (Bauck and Cooper, 1996).

Let L and R be the loudspeaker signals originally intended for a stereophonic loudspeaker configuration with a standard $60°$ span angle. The transfer functions from two loudspeakers in the standard configuration to two ears are denoted by H_{LL}^1, H_{RL}^1, H_{LR}^1, and H_{RR}^1. Then, the reproduced binaural pressures are identified by:

$$\begin{bmatrix} P_L \\ P_R \end{bmatrix} = \begin{bmatrix} H_{LL}^1 & H_{LR}^1 \\ H_{RL}^1 & H_{RR}^1 \end{bmatrix} \begin{bmatrix} L \\ R \end{bmatrix}. \tag{10.14}$$

For a nonstandard stereophonic loudspeaker configuration, let H_{LL}, H_{RL}, H_{LR}, and H_{RR} denote the transfer functions from two loudspeakers to two ears, and L' and R' denote the actual loudspeaker signals. The reproduced binaural pressures are:

$$\begin{bmatrix} P_L' \\ P_R' \end{bmatrix} = \begin{bmatrix} H_{LL} & H_{LR} \\ H_{RL} & H_{RR} \end{bmatrix} \begin{bmatrix} L' \\ R' \end{bmatrix}. \tag{10.15}$$

Let the binaural pressures in Equation (10.15) be equal to those in Equation (10.14), yielding:

$$\begin{bmatrix} L' \\ R' \end{bmatrix} = \begin{bmatrix} H_{LL} & H_{LR} \\ H_{RL} & H_{RR} \end{bmatrix}^{-1} \begin{bmatrix} H_{LL}^1 & H_{LR}^1 \\ H_{RL}^1 & H_{RR}^1 \end{bmatrix} \begin{bmatrix} L \\ R \end{bmatrix}. \tag{10.16}$$

Equation (10.16) converts the original stereophonic signals L and R for a standard loudspeaker configuration into the precorrected signals L' and R'. Therefore, rendering the precorrected signals L' and R' through a pair of loudspeakers with a nonstandard configuration can generate the same binaural pressures and, consequently, the same effect as in standard stereophonic reproduction. In other words, the nonstandard loudspeaker configuration is corrected. This method is not applicable to all nonstandard stereophonic loudspeaker configurations. For some configurations, the acoustic transfer matrix from two loudspeakers to two ears is singular and, therefore, non-invertible. In this case, Equation (10.16) is invalid.

An important example of correcting nonstandard stereophonic loudspeaker configurations is *stereophonic expansion* (also called stereo spreader, or stereo-based widening). As shown in Figure 10.6, the span angle between two stereophonic loudspeakers in some practical applications (such as TV or multimedia computers) is less than the 60° standard. According to the law of sines [Equation (1.25)], this angle narrows the baseline width for frontal virtual sources. Therefore, algorithms for correcting or retaining baseline width are required.

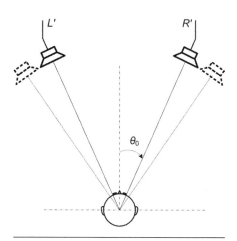

Figure 10.6 Stereophonic expansion

In a left-right symmetrical case, let

$$H_{LL} = H_{RR} = H_\alpha \quad H_{LR} = H_{RL} = H_\beta \quad H_{LL}^1 = H_{RR}^1 = H_{\alpha 1} \quad H_{LR}^1 = H_{RL}^1 = H_{\beta 1}. \quad (10.17)$$

Then, Equation (10.16) becomes

$$L' = G_L(\theta_L, f)L + G_L(\theta_R, f)R \quad R' = G_R(\theta_L, f)L + G_R(\theta_R, f)R, \quad (10.18)$$

where $\theta_L = 330°$ and $\theta_R = 30°$ are loudspeaker azimuths in conventional stereophonic configurations. Furthermore,

$$G_L(\theta_L, f) = G_R(\theta_R, f) = \frac{H_\alpha H_{\alpha 1} - H_\beta H_{\beta 1}}{H_\alpha^2 - H_\beta^2}$$

$$G_R(\theta_L, f) = G_L(\theta_R, f) = \frac{H_\alpha H_{\beta 1} - H_\beta H_{\alpha 1}}{H_\alpha^2 - H_\beta^2}. \quad (10.19)$$

Therefore, rendering signals L' and R' through a pair of loudspeakers with a narrow span angle yields the same baseline width for frontal virtual sources as that for standard stereophonic reproduction. This scheme can be interpreted as creating a pair of standard virtual stereophonic loudspeakers by a pair of narrow-spanned actual loudspeakers. The Equation (10.19) can also be obtained by letting $\theta = 330°$, $E_0(f) = L$ and $\theta = 30°$, $E_0(f) = R$ in Equations (9.11) and (9.12), respectively, and then summing the results.

At low frequencies, the signal processing provided by Equation (10.18) is equivalent to the case of out-of-phase stereophonic loudspeaker signals, in which the summing virtual source can be positioned outside the span of two actual loudspeakers (Section 1.6.1). However, Equation (10.18) is a more general algorithm for stereophonic expansion. The artificial head or human HRTFs at the entire audible frequency range are applicable to stereophonic expansion processing. Nevertheless, the pinna-related peaks and notches in an artificial head or human HRTFs make signal processing difficult. The pinna-related spectral cue is vital to front-back discrimination and vertical localization, but contributes little to frontal stereophonic localization. Moreover, the interaural phase delay difference below 1.5 kHz, caused by the interchannel level difference of loudspeaker signals, is a dominant localization cue in stereophonic reproduction. As frequency increases above 1.5 kHz, the summing localization principle in intensity stereophonic reproduction gradually becomes invalid. Therefore, the HRTFs of an artificial head without pinnae (Section 9.5.2) or spherical head model can be used for stereophonic expansion processing. Even a bandlimited implementation of stereophonic expansion below 1.5 kHz can result in satisfactory subjective performance (Xie and Zhang, 1999). In addition, taking advantage of the left-right symmetry of loudspeaker configurations [Equation (10.17)], we can simply implement the stereophonic expansion algorithm by a shuffler structure similar to Equation (6.106). The constant-power equalization provided by Equations (9.69) and (9.70) can also be incorporated into processing, thereby reducing timbre coloration in reproduction. Walsh and Jot (2006) used similar methods in stereophonic expansion (as well as virtual 5.1-channel surround sound expansion), such as that implemented when using spherical head HRTFs, processing below 8 kHz, and symmetrical shuffler structure. Walsh and Jot, however, used a different equalization scheme. Some other stereophonic expansion algorithms based on similar principles are available (Kim et al., 2004). The stereophonic expansion algorithm suggested by Aarts (2000) uses the simplest shadowless head model, in which two ears are approximated by two points separated by $2a$ in free space. The resultant algorithm is straightforward and can be implemented by analog circuit.

In practical applications, stereophonic expansion can be implemented, either in the preprocessing stage of program recording, or in the post-processing stage of reproduction, which may sometimes prompt erroneous repeated use of stereophonic expansion in both preprocessing and post-processing stages. Ven et al. (2007) proposed a method of blind cancellation for stereophonic expansion to detect and resolve the errors caused by repeated stereophonic expansions. This method avoids the degradation of subjective performance.

Another example is the correction of asymmetrical stereophonic loudspeaker configurations. In some cases (e.g., in a car), stereophonic loudspeakers may be arranged left-right asymmetrically with respect to the listener. In such cases, the signal processing in Equation (10.16) is a remedial measure (Xie, 2003).

Two real loudspeakers are assumed located at ear level. In some practical applications, real loudspeakers may be arranged at elevated (or low) locations. Some works suggested correcting elevated loudspeaker arrangement by the transaural synthesis processing in Equation (10.16) (Tan et al., 2000; Gan et al., 2001; Yang et al., 2002). Similar to the case in Section 9.2.1, however, head turning fails to bring correct dynamics cues to static binaural reproduction through two loudspeakers. The high-frequency spectral cue caused by pinnae is critical to vertical localization, but it is visibly individual dependent and sensitive to listener position. Therefore, stably delivering high-frequency spectral cues for elevated correction in loudspeaker reproduction is difficult to achieve. The actual performance of the elevated correction technique requires further validation.

10.3 Stereophonic Enhancement Algorithms

Stereophonic enhancement algorithms are a class of techniques for consumer electronics applications; these algorithms manipulate conventional stereophonic signals and then reproduce them through a pair of frontal loudspeakers to enhance certain spatial perception effects (Maher, 1997). Various stereophonic enhancement algorithms, including SRS 3D stereo by SRS Labs (Sound Retrieval System, 3-dimensional) and QXpander by Qsound Labs (http://www.srslabs.com, http://www.qsound.com) are available. These algorithms may differ in terms of implementation, but share similar principles. Some of these algorithms have been applied in commercial products or software-based plug-ins in multimedia computers.

Figure 10.7 shows a block diagram of a typical stereophonic enhancement algorithm. The consideration implemented with the algorithm is described as follows. A pair of conventional

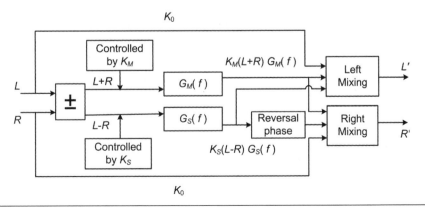

Figure 10.7 Block diagram of a typical stereophonic enhancement algorithm

stereophonic signals can be equally represented by their sum signal $M = L + R$ and difference signal $S = L - R$. The sum signal contains more frontal information (such as direct sounds), and the difference signal includes more side ambient information (such as lateral reflections). Accordingly, a selected enhancement of the difference signal exaggerates the side information. The stereophonic signals are first converted into the sum and difference signals by a mid-side (MS) matrix (or circuit; Figure 10.7). The amplitudes of the sum and difference signals are controlled by two weighted or gain coefficients K_M and K_S, respectively. The weighted sum and difference signals are then processed by a pair of filters $G_M(f)$ and $G_S(f)$, respectively. These filters are designed similar to the transaural principle to widen the perceived baseline or spatial distribution of reproduced sound. Filter $G_M(f)$ may be disregarded, corresponding to a unit response. The outputs of the two filters are given by:

$$G_M K_M M = K_M (L + R) G_M(f), \qquad (10.20)$$

and:

$$G_S K_S S = K_S (L - R) G_S(f). \qquad (10.21)$$

Finally, the signals in the filter outputs are remixed with weighted (K_0) original L and R signals to form loudspeaker signals thus:

$$L' = K_0 L + K_M (L + R) G_M(f) + K_S (L - R) G_S(f) \\ R' = K_0 R + K_M (L + R) G_M(f) - K_S (L - R) G_S(f) \qquad (10.22)$$

On this procedure, the stereophonic enhancement algorithm exaggerates the side information and simultaneously widens the perceived spatial distribution of reproduced sound (to two sides, at most). The principle of stereophonic enhancement algorithms is somewhat similar to that of the stereophonic expansion discussed in Section 10.2. The difference is that the latter expands the spatial distribution of reproduced sound to an appropriate extent, thereby compensating for a narrow stereophonic loudspeaker configuration; the former both exaggerates side information in stereophonic signals and expands the spatial distribution of reproduced sound to two sides. As stated in Section 10.2, at low frequencies, these signal-processing algorithms are equivalent to the case of out-of-phase stereophonic loudspeaker reproduction.

Nonetheless, exaggerating side information and excessive expansion in stereophonic enhancement algorithms may cause a hole in frontal (central) stereophonic virtual sources and degrade the perceived definition of virtual sources. Therefore, stereophonic enhancement algorithms cannot improve virtual source localization. The functions of stereophonic enhancement algorithms are to exaggerate and expand the lateral information in stereophonic reproduction, so as to enhance subjective ambience and immersion. The algorithm is not always valid for all stereophonic signals. For example, it is invalid for stereophonic signals with interchannel time difference, such as those recorded by the spaced microphone (A-B) technique, because of comb filtering in sum and difference signals. Certain stereophonic enhancement algorithms also alter timbre, thereby causing perceivable coloration in reproduction. Olive (2001) conducted a subjective experiment to evaluate five software-based plug-ins of stereophonic enhancement algorithms. The results indicate that three of the five algorithms are inferior to conventional stereophonic reproduction.

The stereophonic enhancement system is often called a "3-D stereo system" or "virtual surround sound system." In contrast to binaural presentation through headphones, the stereophonic enhancement system with two frontal loudspeakers cannot recreate full three-dimensional auditory experiences. Therefore, assigning the 3-D label to the stereophonic enhancement system is inappropriate. The misuse of "virtual surround sound system" originates from advertisements, an issue to be addressed in Section 10.4.4.

10.4 Virtual Reproduction of Multi-channel Surround Sound through Loudspeakers

10.4.1 Virtual Reproduction of 5.1-Channel Surround Sound

Multi-channel surround sound reproduction necessitates multiple loudspeakers, which is complex and inconvenient in some practical applications, such as those for TV or multimedia computers. To solve these problems, researchers introduced some virtual loudspeaker-based approaches for multi-channel surround sound reproduction (called *virtual surround sound*; Bauck and Cooper, 1996; Davis and Fellers, 1997; Kawano et al., 1998; Toh and Gan, 1999; Hawksford, 2002; Bai and Shih, 2007a). Through transaural processing, for example, the central and two surround loudspeakers in 5.1-channel surround sound can be simulated through a pair of actual stereophonic loudspeakers. During the past decades, this approach has been patented, and some commercial products have been introduced (e.g., TruSurround by SRS Labs, Qsurround by Qsound Labs, Dolby Virtual Speaker or Virtual Surround by Dolby Laboratories, etc.). Some of these techniques are outlined on web pages (http://www.srslabs.com/, http://www.qsound.com, http://www.dolby.com).

Existing virtual 5.1-channel surround sounds may differ in terms of implementation, but have similar basic principles. Figure 10.8 shows a block diagram of the system (the LFE channel, which is processed similar to the central channel, is excluded). The five channel signals are processed and then mixed into signals L' and R', which are reproduced through a pair of frontal loudspeakers at $\theta_L = 330°$ and $\theta_R = 30°$, respectively. The two loudspeaker signals can be written in the frequency domain as:

$$L' = L + 0.707C + G_L(\theta_{LS}, f)LS + G_L(\theta_{RS}, f)RS$$
$$R' = R + 0.707C + G_R(\theta_{LS}, f)LS + G_R(\theta_{RS}, f)RS \quad , \tag{10.23}$$

where θ_{LS} and θ_{RS} are the target azimuths for virtual left and right surround loudspeakers, respectively. In Equation (10.23), signals L and R are directly fed into left and right loudspeakers, respectively, to create a summing virtual source within the span of the two loudspeakers. Signal C is attenuated by –3 dB and then simultaneously fed into the left and right loudspeakers. According to Equation (10.23), when other signals satisfy $L = R = LS = RS = 0$, we derive $L' = R' = 0.707\ C$. In this case, signal C is reproduced by the phantom center channel method, leading to a summing virtual source at the front $\theta = 0°$. The surround signals LS and RS are filtered by transaural synthesis filters and then fed into the loudspeakers. For the left surround signal LS, when other signals satisfy $L = R = C = RS = 0$, Equation (10.23) becomes:

$$L' = G_L(\theta_{LS}, f)LS \qquad R' = G_R(\theta_{LS}, f)LS, \tag{10.24}$$

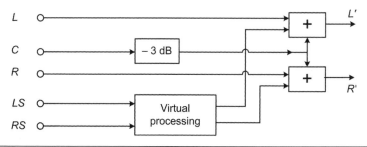

Figure 10.8 Block diagram of virtual 5.1-channel surround sound

According to Equation (9.12b), for a symmetrical loudspeaker configuration, the responses of transaural filters are given by:

$$G_L(\theta_{LS}, f) = \frac{H_\alpha H_L(\theta_{LS}, f) - H_\beta H_R(\theta_{LS}, f)}{H_\alpha^2 - H_\beta^2}$$

$$G_R(\theta_{LS}, f) = \frac{-H_\beta H_L(\theta_{LS}, f) + H_\alpha H_R(\theta_{LS}, f)}{H_\alpha^2 - H_\beta^2},$$

(10.25)

where $H_L(\theta_{LS}, f)$ and $H_R(\theta_{LS}, f)$ are the transfer functions (HRTFs) from virtual left surround loudspeakers to two ears. H_α and H_β are transfer functions from an actual left or right loudspeaker to the ipsilateral and contralateral ears, respectively.

The right surround signal RS is processed similarly to Equation (10.24). The responses of transaural filters are given by:

$$G_L(\theta_{RS}, f) = \frac{H_\alpha H_L(\theta_{RS}, f) - H_\beta H_R(\theta_{RS}, f)}{H_\alpha^2 - H_\beta^2}$$

$$G_R(\theta_{RS}, f) = \frac{-H_\beta H_L(\theta_{RS}, f) + H_\alpha H_R(\theta_{RS}, f)}{H_\alpha^2 - H_\beta^2},$$

(10.26)

where $H_L(\theta_{RS}, f)$ and $H_R(\theta_{RS}, f)$ are the transfer functions from virtual right surround loudspeakers to two ears.

The configuration of a pair of virtual surround loudspeakers is also left-right symmetric. Let $H_L(\theta_{LS}, f) = H_R(\theta_{RS}, f) = H_{\alpha 2}$ and $H_R(\theta_{LS}, f) = H_L(\theta_{RS}, f) = H_{\beta 2}$, yielding:

$$G_L(\theta_{LS}, f) = G_R(\theta_{RS}, f) = \frac{H_\alpha H_{\alpha 2} - H_\beta H_{\beta 2}}{H_\alpha^2 - H_\beta^2}$$

$$G_R(\theta_{LS}, f) = G_L(\theta_{RS}, f) = \frac{H_\alpha H_{\beta 2} - H_\beta H_{\alpha 2}}{H_\alpha^2 - H_\beta^2},$$

(10.27)

Through a pair of actual frontal loudspeakers, therefore, the transaural synthesis in Equation (10.24) creates a pair of virtual surround loudspeakers, through which the surround signals LS and RS are reproduced. Equations (10.24–10.27) can be directly derived using a procedure similar to that for deriving Equations (10.18) and (10.19). Given the left-right symmetry in Equation (10.27), virtual surround loudspeaker processing can also be simply implemented by the shuffler structure similar to Equation (6.106).

10.4.2 Improvement of Virtual 5.1-Channel Surround Sound Reproduction through Stereophonic Loudspeakers

Virtual 5.1-channel surround sound is based on transaural synthesis. All the problems encountered in transaural reproduction, including narrow listening regions, localization errors, and coloration in reproduction (Lorho and Zacharov, 2004), also occur in virtual 5.1-channel surround sound. The factors that cause such problems were analyzed in Chapter 9.

The following schemes for improving the performance of virtual 5.1-channel surround sound are suggested (Xie et al., 2005c; 2005f):

1. Improving virtual source stability

In the virtual 5.1-channel surround sound system shown in Figure 10.8, a pair of actual stereophonic loudspeakers is arranged with a conventional span angle of 60°. Alternatively, a loudspeaker configuration can be chosen, so that loudspeaker-to-ear transfer functions (matrix) and binaural pressures are stable against head translation. In this case, the size of the listening region is expanded at a specific tolerable error in binaural pressures.

As stated in Section 9.3.1, a narrow span angle between a left and right loudspeaker pair improves the stability of virtual sources against lateral head translation, thereby expanding the listening region. A straightforward method is to use the stereo dipole in virtual 5.1-channel surround sound reproduction, but stereo dipole requires increased low-frequency components of signals, leading to instability at low frequencies. As a trade-off between robustness and better low-frequency performance in reproduction, a 20°–30° angle span [i.e., the left and right loudspeakers are arranged at azimuths 345° (350°) and 15° (10°), respectively] is a better option. Unlike the conventional 60° span angle, such a loudspeaker configuration improves the stability of virtual sources and, simultaneously, avoids a large increase in the low-frequency components of signals. It also facilitates loudspeaker arrangement in TV and multimedia computers. To retain the baseline width for frontal virtual sources, however, this loudspeaker configuration requires transaural synthesis filters, other than those for virtual surround loudspeakers.

2. Improving the distribution of virtual source directions

As stated in Section 9.2.1, the perceived virtual source positions in two-front loudspeaker reproduction are usually restricted in the region of frontal-horizontal quadrants. The virtual source intended for rear-horizontal quadrants is often perceived at the mirror position in frontal-horizontal quadrants. In some virtual 5.1-channel surround sound algorithms, the target azimuth for virtual right surround loudspeakers is $\theta_{RS} = 110°$, as specified by Equation (10.3). Therefore, the actual perceived azimuth for virtual right surround loudspeakers is often located at a mirror azimuth $\theta = 70°$. A similar problem occurs in virtual left surround loudspeakers. This problem narrows the distribution region of perceived virtual sources. As a trade-off, target azimuths of $\theta_{RS} = 90°$ and $\theta_{LS} = 270°$ can be selected for two virtual surround loudspeakers. This surround loudspeaker configuration is a compromise in actual 5.1-channel sound reproduction for when a listening room is unsuitable for the ITU-recommended configuration.

3. Timbre equalization

Reducing coloration in reproduction necessitates incorporating the constant-power equalization algorithm discussed in Section 9.5.1 into signal processing. The loudspeaker signals in the frequency domain then become:

$$L' = G'_L(\theta_L,f)L + G'_L(\theta_R,f)R + 0.707C + G'_L(\theta_{LS},f)LS + G'_L(\theta_{RS},f)RS$$
$$R' = G'_R(\theta_L,f)L + G'_R(\theta_R,f)R + 0.707C + G'_R(\theta_{LS},f)LS + G'_R(\theta_{RS},f)RS \ , \quad (10.28)$$

where signal C is processed in the same manner as that in Equation (10.23). It is attenuated by −3 dB and then simultaneously fed into left and right loudspeakers to create the summing virtual source at the frontal direction $\theta = 0°$.

After incorporating the equalization algorithm, the $G_L(\theta_{LS},f)$, $G_R(\theta_{LS},f)$, $G_L(\theta_{RS},f)$, and $G_R(\theta_{RS},f)$ in Equation (10.23) are replaced by the $G'_L(\theta_{LS},f)$, $G'_R(\theta_{LS},f)$, $G'_L(\theta_{RS},f)$, and $G'_R(\theta_{RS},f)$ defined by Equations (9.69) and (9.70). For a left-right symmetrical configuration, let the transfer function from two virtual surround loudspeakers to two ears be $H_L(\theta_{LS}, f) = H_R(\theta_{RS}, f) = H_{\alpha 2}$ and $H_R(\theta_{LS}, f) = H_L(\theta_{RS}, f) = H_{\beta 2}$, yielding:

$$
\begin{aligned}
G'_L(\theta_{LS}, f) = G'_R(\theta_{RS}, f) &= \frac{H_\alpha H_{\alpha 2} - H_\beta H_{\beta 2}}{\sqrt{|H_\alpha H_{\alpha 2} - H_\beta H_{\beta 2}|^2 + |H_\alpha H_{\beta 2} - H_\beta H_{\alpha 2}|^2}} \cdot \frac{|H_\alpha^2 - H_\beta^2|}{H_\alpha^2 - H_\beta^2} \\
G'_R(\theta_{LS}, f) = G'_L(\theta_{RS}, f) &= \frac{H_\alpha H_{\beta 2} - H_\beta H_{\alpha 2}}{\sqrt{|H_\alpha H_{\alpha 2} - H_\beta H_{\beta 2}|^2 + |H_\alpha H_{\beta 2} - H_\beta H_{\alpha 2}|^2}} \cdot \frac{|H_\alpha^2 - H_\beta^2|}{H_\alpha^2 - H_\beta^2}
\end{aligned}
. \quad (10.29)
$$

In contrast to the H_α and H_β in Equation (10.27), those in Equation (10.29) denote the transfer functions from two loudspeakers at azimuths 15° and 345° (rather than 30° and 330°) to the ipsilateral and contralateral ears, respectively.

In contrast to the implementation in Equation (10.23), the original signals L and R should be virtually processed to retain the baseline width of the frontal virtual source. That is, a pair of frontal loudspeakers with a 30° azimuthal span is used to create a pair of virtual front loudspeakers with a 60° azimuthal span. Let $\theta_L = 330°$ and $\theta_R = 30°$ be the azimuths of virtual left and right loudspeakers, respectively. For a left-right symmetrical configuration, let the transfer functions from two virtual front loudspeakers to two ears be $H_L(\theta_L, f) = H_R(\theta_R, f) = H_{\alpha 1}$ and $H_R(\theta_L, f) = H_L(\theta_R, f) = H_{\beta 1}$, similar to Equation (10.29), yielding:

$$
\begin{aligned}
G'_L(\theta_L, f) = G'_R(\theta_R, f) &= \frac{H_\alpha H_{\alpha 1} - H_\beta H_{\beta 1}}{\sqrt{|H_\alpha H_{\alpha 1} - H_\beta H_{\beta 1}|^2 + |H_\alpha H_{\beta 1} - H_\beta H_{\alpha 1}|^2}} \cdot \frac{|H_\alpha^2 - H_\beta^2|}{H_\alpha^2 - H_\beta^2} \\
G'_R(\theta_L, f) = G'_L(\theta_R, f) &= \frac{H_\alpha H_{\beta 1} - H_\beta H_{\alpha 1}}{\sqrt{|H_\alpha H_{\alpha 1} - H_\beta H_{\beta 1}|^2 + |H_\alpha H_{\beta 1} - H_\beta H_{\alpha 1}|^2}} \cdot \frac{|H_\alpha^2 - H_\beta^2|}{H_\alpha^2 - H_\beta^2}
\end{aligned}
. \quad (10.30)
$$

A direct implementation of the algorithm given by Equation (10.28) necessitates eight transaural synthesis filters. Similar to the case of Equation (6.106), left-right symmetry yields simplified shuffler structure implementation. Equation (10.28) is equivalent to:

$$
\begin{bmatrix} L' \\ R' \end{bmatrix} = \begin{bmatrix} 0.707 & 0.707 \\ 0.707 & -0.707 \end{bmatrix} \left\{ \begin{bmatrix} 1 \\ 0 \end{bmatrix} C + \begin{bmatrix} \Sigma_1 & 0 \\ 0 & \Delta_1 \end{bmatrix} \begin{bmatrix} 1 & 1 \\ 1 & -1 \end{bmatrix} \begin{bmatrix} L \\ R \end{bmatrix} + \begin{bmatrix} \Sigma_2 & 0 \\ 0 & \Delta_2 \end{bmatrix} \begin{bmatrix} 1 & 1 \\ 1 & -1 \end{bmatrix} \begin{bmatrix} LS \\ RS \end{bmatrix} \right\}, \quad (10.31)
$$

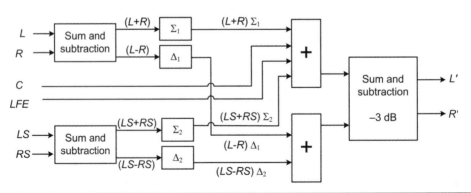

Figure 10.9 Block diagram for shuffler structure implementation of signal processing in Equation (10.31)

where:

$$\Sigma_1 = 0.707[G'_L(\theta_L, f) + G'_L(\theta_R, f)] \qquad \Delta_1 = 0.707[G'_L(\theta_L, f) - G'_L(\theta_R, f)]$$
$$\Sigma_2 = 0.707[G'_L(\theta_{LS}, f) + G'_L(\theta_{RS}, f)] \qquad \Delta_2 = 0.707[G'_L(\theta_{LS}, f) - G'_L(\theta_{RS}, f)]$$ (10.32)

Four filters are required to implement the algorithm given by Equation (10.31), thereby simplifying signal processing. Figure 10.9 shows a block diagram for the signal processing designed according to Equation (10.31), where the LFE channel has been supplemented. The filters in the block diagram are implemented by a finite impulse response structure with a length 1.3–2.7 ms; the corresponding perceived performance is validated by psychoacoustic experiments (Xie et al., 2006b). As expected, the algorithm can recreate the summing virtual source in frontal-horizontal quadrants, but the virtual source intended for rear-horizontal quadrants is often perceived at the mirror position in frontal-horizontal quadrants.

10.4.3 Virtual 5.1-Channel Surround Sound Reproduction through More than Two Loudspeakers

In addition to conventional two-front loudspeaker reproduction, some algorithms for virtual 5.1-channel surround through more than two ($M \geq 3$) loudspeakers have been suggested for various purposes.

As stated in Sections 9.3.1 and 9.3.2, a narrow span angle between a left and right loudspeaker pair improves the stability of virtual sources against head translation, thereby expanding the listening region. Nevertheless, a loudspeaker configuration with a narrow span angle (such as the stereo dipole) requires a considerable increase in the low-frequency components of signals. This increase complicates signal processing. Thus, this configuration improves the performance of reproduction only at medium and high frequencies. The three-front loudspeaker configuration depicted in Figure 9.4 is another option for improving stability in virtual 5.1-channel surround sound reproduction. In some practical applications, however, arranging a central loudspeaker system is inconvenient because of front-located screens. In this case, the central loudspeaker system is usually arranged above the screen, which causes a deviation between the heights of the central and left/right loudspeaker systems, as well as a deviation of the tweeter and woofer units of the central loudspeaker system from the front. All these aspects degrade virtual source performance in reproduction.

Bauck (2001) proposed a four-loudspeaker configuration and crossover filters for reproduction. The low-frequency components of signals are reproduced through a pair of woofers arranged at a large span angle (such as on two sides of a screen). The high-frequency components of signals are reproduced through a pair of small tweeters arranged at a narrow span angle (such as above the screen). This method improves the stability of virtual source in reproduction, but the loudspeaker configuration used is incompatible with conventional stereophonic sound. The loudspeaker configuration and signal processing also present slight complexity.

To improve the stability in virtual 5.1-channel surround sound reproduction, Tong and Xie (2005) proposed a three-loudspeaker configuration and system, whose basic idea is described as follows. The low-frequency sound waves possess long wavelength, thereby enabling the relative stability of virtual sources against head translation. By contrast, the short wavelength of high-frequency sound waves causes instability in high-frequency virtual sources. Therefore, the high-frequency components of signals are reproduced through three (left, center, and right) frontal loudspeakers and corresponding transaural synthesis to improve stability. Meanwhile, the low-frequency components of signals are reproduced through a pair of left and right loudspeakers because of the inherent stability of such a pair. In practical applications, this reproduction is implemented through a pair of two-way loudspeaker systems with full audio bandwidths and a central tweeter. The two-way loudspeaker systems are arranged with a 60° span angle, as in conventional stereophonic sound. The central tweeter is arranged at the front $\theta = 0°$. The small size of the tweeter enables its easy arrangement (e.g., above the screen). The signal processing for low-frequency components is also easily realizable because the span angle between left and right loudspeakers is not as narrow as the stereo dipole. In addition, this loudspeaker configuration is compatible with conventional stereophonic sound. Figure 10.10 shows a block diagram of this system. The 5.1-channel signals are filtered by crossover filters and then processed by transaural synthesis for three- and two-loudspeaker reproduction.

Two-front loudspeaker reproduction cannot create stable perceived virtual sources in rear-horizontal quadrants. To improve virtual source localization in rear-horizontal quadrants, researchers suggested a method for virtual 5.1-channel surround sound reproduction through four loudspeakers, that is, a pair in front and a pair in the rear (Shi, 2007a; Shi and Xie, 2007b). Although four-loudspeaker reproduction discards only a central loudspeaker, as compared with standard 5.1-channel reproduction, the arrangement of the loudspeakers is relatively flexible and unrestricted by the ITU recommendation (Figure 1.23)—when certain algorithms (similar to those in Section 10.2) for correcting loudspeaker locations are incorporated. Another advantage of four-loudspeaker reproduction is that it enhances subjective envelopment, when decorrelation processing and virtual multiple surround loudspeakers (Section 10.1.3) are incorporated. Multiple virtual surround loudspeakers are also applicable to 5.1-channel surround sound reproduction with five full audio bandwidth loudspeakers. This approach improves subjective envelopment. The multiple virtual surround loudspeaker method is also suitable for virtual 5.1-channel surround sound reproduction through a pair of frontal loudspeakers, but the virtual surround loudspeakers are perceived in frontal-horizontal quadrants. Similar algorithms are applicable to converting conventional two channel stereophonic signals for three-front loudspeaker reproduction (such as in 5.1-channel surround sound), or vice versa (Bauck 2000a; 2000b).

10.4.4 Notes on Virtual Surround Sound

The preceding sections focused on virtual 5.1-channel surround sound. Similar algorithms are suitable for other surround sound formats, such as Dolby Pro-Logic and 7.1-channel surround

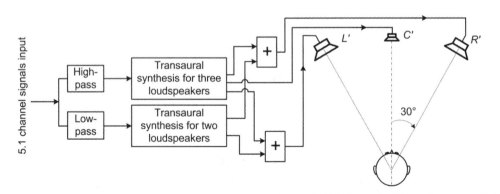

Figure 10.10 Virtual 5.1-channel surround sound reproduction through a pair of full audio bandwidth loudspeaker systems along with a central tweeter

sound. However, some commercially exaggerated claims have caused confusion in understanding virtual surround sound (Xie, 2000).

Virtual surround sound is simple and useful because few (usually two) loudspeakers are required in reproduction. It can be regarded as a special application of transaural reproduction through loudspeakers. Accordingly, the defects in transaural reproduction (Chapter 9)—such as errors in virtual source localization, narrow listening regions, and timbre coloration—also exist in virtual surround sound reproduction. Some methods for partially reducing such errors have been suggested, but these techniques do not completely eliminate the aforementioned defects. In particular, two frontal loudspeakers can create a stable virtual source only in frontal-horizontal quadrants and not in all horizontal directions. A localization error for lateral virtual sources also occurs when mismatched HRTFs are used in processing, or when the head rotates around vertical axes in reproduction (Section 9.4). Numerous commercial advertisements assert, without scientific proof, that virtual surround sound reproduction through two frontal loudspeakers can recreate stable virtual sources in the entire horizontal plane.

Strictly speaking, "virtual surround sound" refers to the virtual reproduction of multi-channel surround sound through (fewer) loudspeakers. Less strictly and at most, this term also pertains to binaurally reproducing multi-channel surround sound through headphones. In some commercial advertisements, almost all algorithms for HRTF-processing are mislabeled as "virtual surround sound," "3-D sound," or "3-D stereo." In particular, the stereophonic enhancement algorithms discussed in Section 10.3 should not be categorized as virtual surround sound or 3-D stereo algorithms because these algorithms manipulate only stereophonic signals and cannot create full three-dimensional spatial auditory perception. Certain commercial advertisements make the following claims: "3-D stereo restores the spatial cues lost in conventional stereophonic record/playback processing and corrects them in terms of human auditory characteristics, so that the spatial information is reproduced at appropriate directions;" "it improves the localization and is effective over a wide area so that listener is no longer restricted to sitting at a narrow sweet point;" and "it is effective at various listening positions." These misleading claims confuse the subjective attributes of virtual source localization and ambience or immersion. As stated in Section 10.3, stereophonic enhancement algorithms enhance subjective ambience and immersion, rather than improve virtual source localization in reproduction or expand the listening region.

In practical applications, virtual surround sound algorithms can be exactly implemented by various digital signal-processing chips or software-based plug-ins for multimedia computers. For example, some chips can simultaneously decode Dolby digital (AC-3—5.1-channel audio) data streams and implement virtual 5.1-channel surround sound algorithms. The resultant two-channel digital signals are converted into analog signals by digital-to-analog conversion. These chips are relatively expensive.

Some chips for virtual surround sound consist of analog processing circuits, along with some external resistors and capacitors. These features translate to reduced cost. The performance of these chips is usually inferior because of the low accuracy of analog circuit implementation for HRTF-based or transaural processing. Some chips or products that are claimed to be capable of manipulating 5.1-channel signals can actually only receive and process two-channel downmixing of 5.1-channel signals. Because original surround sound information is lost in the downmixing stage, these chips or products use stereophonic enhancement algorithms that only create certain subjective effects.

There are two opposing views on the application prospects of virtual surround sound. The first originates from some commercial advertisements, which assert that virtual surround sound with only two loudspeakers can recreate the same auditory experience as that generated by multi-channel surround sound. These advertisements also claim that such an approach is simpler and more suitable for users than multi-channel surround sound, which is evaluated in the advertisement as too complex for wide application. The second view completely rejects virtual surround sound because of its defects, as well as the deviation between advertisements and actual performance. Misled by some advertisements, some people mistake the stereophonic enhancement algorithms discussed in Section 10.3 as virtual surround sound or as inclusive of all VAD contents. Such misunderstanding drives people to reject all VADs and HRTF-related techniques.

The absence of definitive scientific support for these advertisements prevents people from comprehensively understanding the concepts and state-of-the-art advantages of HRTFs and VADs. As will be shown in Chapter 14, VADs are applicable to various fields of scientific research, engineering, and consumer electronics; virtual surround sound is one such popular application. Virtual surround sound suffers from certain defects so that its perceived performance is generally no match for that of multi-channel surround sound. However, the former is simple and suitable for certain applications (such as TV and multimedia computers). With reasonable design, the perceived spatial performance of virtual surround sound is usually superior to the direct reproduction of two-channel downmixing of multi-channel surround sound signals. Therefore, both multi-channel and virtual surround sound techniques prove valuable to different types of applications. The key point is that the performance and functions of the two techniques should be scientifically evaluated to accurately reveal their essence and to eliminate the negative influence caused by unscientific claims. As for stereophonic enhancement algorithms, they are a class of post-processing schemes for stereophonic sound. Their performance (i.e., success or failure) is unrelated to the virtual reproduction of multi-channel surround sound or to VAD techniques.

10.5 Summary

This chapter discusses a category of application of virtual auditory display, namely, applying the principles of binaural or transaural synthesis to stereophonic and multi-channel surround sound signals to accommodate headphone presentation or various loudspeaker configurations.

Stereophonic and multi-channel sound signals are originally intended for loudspeaker reproduction rather than headphone reproduction. When they are presented through a pair of headphones, unnatural spatial auditory perception (i.e., in-head localization or lateralization) occurs. Binaural synthesis is applicable to improving stereophonic or surround sound reproduction through headphones. That is, HRTF-based filtering is used to simulate acoustic transmission, from loudspeakers to two ears, so that stereophonic or surround sound signals are reproduced through virtual loudspeakers (sources). This processing presents correct stereophonic information to two ears.

Incorporating reflections and reverberations into binaural signals also relatively improves externalization. This approach has been applied to the binaural processing of 5.1-channel surround sound algorithms. However, reflections should be carefully introduced in accordance with auditory scenarios to avoid creating unnatural auditory perception. Surround signal decorrelation and the use of multiple virtual surround loudspeakers are suitable for enhancing subjective performance in the headphone presentation of 5.1-channel surround sound. These methods facilitate externalization and enhance subjective envelopment without introducing unnatural auditory perception.

In some applications, the standard configuration or arrangement of stereophonic loudspeakers may be infeasible. In these cases, transaural synthesis is applicable to correcting nonstandard loudspeaker configurations. For television and multimedia computer applications in particular, the span angle between stereophonic loudspeakers is less than 60°. In this case, stereophonic expansion algorithms are helpful. Stereophonic enhancement algorithms cannot improve virtual source localization. Instead, they exaggerate and expand lateral information in stereophonic reproduction, thereby enhancing subjective ambience and immersion.

Multiple loudspeakers with full audio bandwidths are needed in standard multi-channel (such as 5.1-channel) surround sound reproduction. In some practical applications, however, arranging multiple loudspeakers is infeasible. Virtual surround sound is designed in accordance with transaural principles, in which more virtual loudspeakers are created by few actual loudspeakers (such as a pair of frontal loudspeakers), thereby simplifying the system. A limited listening region, localization errors, and coloration in reproduction are common problems encountered in virtual surround sound. These shortcomings can be alleviated by using a narrow-spanned frontal loudspeaker configuration with an appropriate timbre equalization algorithm.

Virtual surround sound originally refers to the virtual reproduction of multi-channel surround sound through (few) loudspeakers. The misuse of this term, especially in relation to stereophonic enhancement algorithms, as well as some exaggerated claims, has caused confusion. Scientific assessments are needed to eliminate the negative influence of unscientific advertisements and clearly elucidate the essence of virtual surround sound.

11

Binaural Room Modeling

This chapter addresses the problems in binaural room modeling—that is, binaural modeling and rendering a room, or environmental reflections. Basic physics-based methods for binaural room modeling, such as the image-source method and the ray-tracing method, are outlined. Some perception-based methods for binaural room modeling, including various artificial delay and reverberation algorithms are also discussed.

Heretofore, room or environment reflection has been disregarded in free-field virtual source synthesis. As stated in Section 1.7, however, reflections exist in most actual rooms and are crucial to spatial auditory perception. Therefore, a complete virtual auditory display should include a reflection-modeling component, hereafter called *virtual auditory or acoustic environment* (*VAE*). Incorporating reflections into VAE processing presents the following advantages: it (1) recreates the spatial auditory perception in a room or reflective environment; (2) eliminates or reduces in-head localization in headphone presentation; and (3) enables the control of the perceived virtual source distance. This chapter addresses the problem of reflection rendering in VAE. Two basic methods are commonly available for rendering room or environment reflections. *Physics-based methods* simulate the physical propagation of sound from source to receiver inside a room, or, equally, binaural room impulse responses (BRIRs), and then synthesize the binaural signals by convoluting input stimuli with the BRIRs. *Perception-based methods* recreate the desired auditory perception of reflections by signal processing algorithms, from a perceptual rather than a physical viewpoint. Section 11.1 discusses the physics-based methods, including the concepts and methods for room acoustics and BRIR modeling. Section 11.2 discusses the perception-based methods, including various artificial delay and reverberation algorithms.

11.1 Physics-based Methods for Room Acoustics and Binaural Room Impulse Response Modeling

11.1.1 BRIR and Room Acoustics Modeling

The binaural pressures in a room are contributed from the direct sound caused by sources and reflections from boundary surfaces. In the static case, temporal and spatial information on direct sound and reflections are encoded in BRIRs (Section 1.7.1). BRIRs are defined as the impulse response of a linear-time-invariant system that is composed of a source, room, and listener (two ears). Therefore, a straightforward scheme for rendering direct sounds and reflections involves convoluting mono anechoic recording, or a synthesized stimulus, with a pair of BRIRs. This procedure is known as *binaural auralization*.

BRIRs can be measured from an artificial head or human subject, an issue to be discussed in Section 14.2.1. Alternatively, BRIRs can be obtained by physical modeling, that is, by modeling sound transmission from a source to two ears. Complete physical modeling involves source modeling (such as radiation pattern modeling), transmission or room acoustics modeling (e.g., surface reflection, scattering and absorption, air absorption modeling, etc.), and listener modeling (modeling the scattering and diffraction imposed by human anatomical structures). Given the geometrical and physical data of a room, source, and listener, the sound pressure inside the room and BRIRs can, in principle, be calculated by solving the wave equation—subject to certain boundary conditions. However, even in the case of point sources and with the absence of a listener, the analytical solution of the wave equation can be obtained only in rare cases, such as that achieved for a rectangular room with rigid surfaces. In general cases, the acoustic field in a room is simulated by approximate methods or simplified physical models.

Computers have been used for room acoustics modeling since the 1960s. Because of the rapid development of computer techniques in recent decades, computational modeling has become the most important method for analyzing room acoustics. According to physical principles, the methods for room acoustics modeling are classified into two categories: *geometrical acoustics-based methods* and *wave acoustics-based methods*. Geometrical acoustics yields the asymptotic solution of the acoustic field in a room. The approximation of geometrical acoustics is valid only when the sound wavelength is considerably smaller than the boundary surface dimension, while much larger than boundary surface roughness. That is, geometrical acoustics-based methods are reasonable for high frequencies and smooth boundary surfaces. In geometrical acoustics approximation, sound (with its wave nature disregarded) is treated as similar to a ray, which obeys the rule of reflections. *Image-source methods* and *ray-tracing methods* are two common examples of geometrical acoustics-based methods.

Wave acoustics-based methods, which take the wave nature of sound into account, solve the wave equation of pressure inside a room and yield more accurate results. Usually, the wave equation and boundary problem of a room with complex geometry are solved by numerical methods, such as the finite element method (FEM), boundary element method (BEM), finite-difference time-domain (FDTD) method, and digital waveguide mesh method. Limited to extensive computational workloads, these numerical methods are suitable only for low-frequency and small-room modeling.

The succeeding sections outline the methods commonly used for room acoustics modeling. The research and application of room acoustics modeling have developed substantially during recent decades. The subsequent sections discuss only the issues related to VAEs because the details of room acoustics modeling are beyond the scope of this book. For in-depth discussion, consult other references (Vorländer, 2008; Lehnert and Blauert, 1992; Kleiner et al., 1993; Svensson and Kristiansen, 2002).

11.1.2 Image-source Methods for Room Acoustics Modeling

Image-source methods decompose the reflected sound field into the radiations of multiple image sources in free space. Figure 11.1 depicts the simplest case of image-source methods. A real point source S with intensity Q_S is located in front of a rigid and smooth surface (e.g., a wall) with an infinite area. The pressure at the receiving point A is the sum of those pressures caused by direct sound

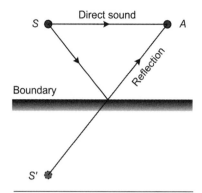

Figure 11.1 Specular reflection and image source caused by a rigid and smooth surface with an infinite area

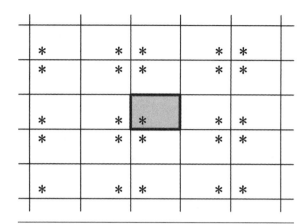

Figure 11.2 Array of image sources in a two-dimensional rectangular room

and reflection from the surface. The reflection is regarded as though it originated from an image source S' with intensity $Q_{S'} = Q_S$. The position of the image source is determined by reflecting the real source off the surface (specular reflection), thereby revealing the reflection path. Accordingly, the sound at the receiving point A is calculated by summing the sounds from the real source S and image source S'. For an impulse source, the pressure at the receiving point A is just the overall acoustic impulse response of transmission.

In an enclosed space, such as a room with multiple reflective surfaces, the reflection from each surface is re-reflected by other surfaces, and so on, resulting in second-, third-, and even higher-order reflections. This process can be equivalently modeled by corresponding high-order image sources. Figure 11.2 shows image sources in a two-dimensional rectangular room, which is equivalent to a three-dimensional rectangular room with four rigid walls and a fully absorptive floor and ceiling. The distribution of image sources forms a two-dimensional array, and the number of image sources rapidly grows with increasing order of reflection. Similarly, the reflections in a three-dimensional rectangular room can be modeled by a three-dimensional array of image sources. The Lth-order reflections have $6 \times 5^{(L-1)}$ image sources and reflections up to the Lth-order have $3 \times (5^L - 1)/2$ image sources. Image-source methods also yield the exact solution for a rectangular room with rigid surfaces. Image-source calculation for a room with complex geometry is more intricate than that for a rectangular room.

Visibility checking is generally needed in image-source calculation because not all image sources are contributable. As an example, Figure 11.3 shows the section of a simplified hall and its ceiling, floor, back wall, and balcony (Savioja et al., 1999). The

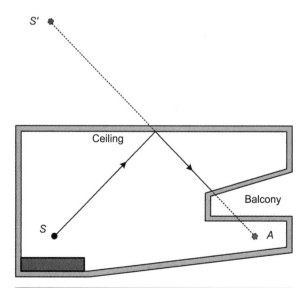

Figure 11.3 Visibility checking in a simplified hall (adapted from Savioja et al., 1999)

sound from the real source S is reflected by the ceiling, which is modeled by the first-order image source S'. Occluded by the balcony, the image source S' is not contributable to the response in the receiving point A. An image source that is contributable to the response in the receiving point should satisfy the following conditions: (a) no obstacle should intersect between the reflection path from the image source to the receiving point, and (b) the reflection point should be located on the finite reflective surface. Visibility testing discards irrelevant images, thereby effectively reducing the number of potential image sources to be handled.

A practical boundary surface is not always rigid. For a uniform, non-rigid, and locally-reacting boundary surface with a specific acoustic impedance $Z(f)$, the pressure reflection coefficient of the plane wave is generally given by (Morse and Ingrad, 1968):

$$R_p(f,\beta) = \frac{\xi(f)\cos\beta - 1}{\xi(f)\cos\beta + 1} \quad \text{with} \quad \xi(f) = \frac{Z(f)}{\rho_0 c}, \tag{11.1}$$

where ρ_0 and c are the density and sound speed of air, respectively; and β denotes the incident angle (equal to the reflection angle). Accordingly, for an approximate plane wave incidence, from a distant point source with intensity Q_S, the reflection can be modeled by an image source with intensity:

$$Q_{S'}(f,\beta) = Q_S R_p(f,\beta). \tag{11.2}$$

The pressure reflection coefficient $R_P(f, \beta)$ varies as a function of frequency and incident angle, so that the equivalent intensity of the image source depends upon frequency and incident angle. In Equation (11.2), different $R_P(f, \beta)$ should be selected for surfaces with different specific acoustic impedance. Usually, however, measurement (in a reverberation room) yields the diffuse field (energy) absorption coefficient $\alpha(f)$, which is related to the mean $R_P(f, \beta)$ over incident angles by:

$$|\overline{R}_P(f)| = \sqrt{1 - \alpha(f)}. \tag{11.3}$$

In image-source calculation, $R_P(f, \beta)$ is usually approximated by its mean values in Equation (11.3) and represented in octave (or one-third octave) bands. This approximation inevitably causes some errors. Equation (11.3) provides only the magnitude of the pressure reflection coefficient. The complex-valued pressure response of a surface material is derived by minimum-phase reconstruction. Although the real response of the surface material is non-minimum phase, minimum-phase reconstruction usually does not generate perceivable artifacts in VAEs because of the insensitivity of human hearing to the phase information in reflections.

$Q_{S'}(f, \beta)$ describes the direction-dependent reflection characteristics of the surface, and can be represented by the complex frequency response of a filter. $Q_{S'}(f, \beta)$ is the frequency response of the corresponding surface filter for a first-order image source, or the cascaded frequency response of two or more surface filters for a second- or higher-order image source. Accordingly, the impulse response of each reflection is calculated by the inverse Fourier transform of $Q_{S'}(f, \beta)$ cascading with propagation delay and $1/r$ distance attenuation for the corresponding image source, in which the last two terms are evaluated on the basis of the relative distance between the image source and receiver. The entire room impulse response (RIR) includes the sum of the impulse responses of all reflections.

Image-source methods can, in principle, accurately determine all reflection paths, from which the intensity, arrival time, and direction of each reflection can be evaluated, and from which the overall RIR can be calculated. However, the basic image-source method is applicable to specular reflections and not to diffuse reflections and diffraction. The method becomes complex for a room with complex geometry. In addition, the number of image sources, and therefore, the computational load grow exponentially with increasing order of reflections. Thus, the basic image-source method is typically used to simulate several preceding orders of early reflections. Fortunately, an increase in the number of reflections or image sources generates a more diffuse sound field. At the same time, the relative contribution of each high-order image source to the entire RIR gradually declines because of increasing accumulated propagation distance (1/r attenuation) and boundary material absorption. As a result, late reflections can be modeled by some other computationally efficient, but less accurate, methods.

11.1.3 Ray-tracing Methods for Room Acoustics Modeling

Ray tracing is another method commonly used for room acoustics modeling, in which sound radiation is supposed to act similarly to a number of rays. Each ray carries a certain amount of energy and propagates at sound speed. When the ray comes into contact with the boundary surface, part of its energy is reflected and the rest is absorbed. Ray-tracing methods simulate the radiations from a source with a number of rays, and then trace the propagation of each ray in terms of the geometrical and physical characteristics of a room surface. Figure 11.4 shows a sketch of ray-tracing in a room. The response in a receiving point is contributed from all the rays that pass through it. Different variations of ray-tracing methods are available, but only the basic approach is discussed here.

The first step is determining the reflection point. Starting from the sound source, the equation of the ray (spatial straight line) of a given direction is formulated. The reflection point, which is the intersecting point between the ray and the surface of a room, is determined by solving the ray equation, along with the spatial plane equation, of the surface.

The second step is determining the direction of the reflected ray at the reflection point, according to the specular or diffuse reflection rule. Specular reflection occurs for an even and smooth surface. The direction of the specular reflected ray is uniquely determined by that of incidence, that is, the reflected angle (the angle between the reflected ray and the normal direction of the surface) is always equal to the incident angle. Diffuse reflection occurs for a rough or nonuniform surface. The directional distribution of the diffuse reflected ray or energy is determined according to certain laws, which vary according to surface material. The *uniform diffuse reflection law* states that reflected intensity (energy) is evenly distributed in semi-spherical directions. *Lambert's law* states that reflected intensity is distributed as a cosine function of reflected angle β':

$$I(\beta') = I_0 \cos\beta'. \tag{11.4}$$

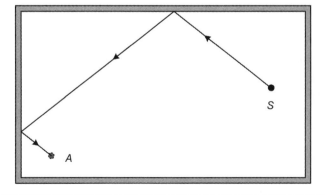

Figure 11.4　Sketch of ray tracing in a room

Diffuse reflection laws, such as Equation (11.4), describe the statistical directional distribution of reflected rays or energy.

Practical materials may have mixed properties: part specular reflection, part diffuse reflection, and part absorption. Let $\alpha = \alpha(f)$ be the energy absorption coefficient of a surface material, and d, which is defined as the ratio of diffuse reflected energy to overall reflected energy, be the diffuse coefficient of the surface material. Then, energy conservation requires:

$$\alpha + d(1-\alpha) + (1-\alpha)(1-d) = 1 \tag{11.5}$$

The first, second, and third terms on the left side of Equation (11.5) denote the proportion of the absorbed, diffuse reflected, and specular reflected energy, respectively. Various approaches to using ray-tracing methods in modeling surfaces with mixed properties are available. The most common is to let each incident ray be reflected either specularly or diffusively, which is implemented by generating a random number δ between 0 and 1. The reflected ray is treated as the diffuse reflection for $\delta < d$. Otherwise, it is regarded as specular reflection, in which the reflected ray direction is uniquely determined by the incident direction. In diffuse reflection, the reflected ray direction is randomly selected according to a desired statistical distribution function, such as Equation (11.4).

The third step is tracing a reflected ray. The reflected ray, whose equation is determined by its direction, serves as the incident ray of sequential reflection. The ray is continuously traced as the first and second steps until the ray energy satisfies certain conditions. Some of the energy of a ray is absorbed when a ray comes into contact with a surface. After being reflected/absorbed N times, the energy of a ray becomes:

$$E(N) = E(0) \prod_{i=1}^{N} (1 - \alpha_i), \tag{11.6}$$

where $E(0)$ is the initial energy and α_i denotes the energy absorption coefficient of the ith surface. The tracing of the ray is terminated when the ray energy is below a preset threshold E_{min}—that is, $E(N) < E_{min}$.

The fourth step involves tracing different rays. Each ray radiated from the source is traced, through the first to third steps, until all the rays have been traced.

The fifth step entails detecting and recording the ray-tracing results. Recording the rays at a given receiving point yields the arrival time (or accumulated propagation distance) and the direction of reflected energy. Recording the rays in various receiving points generates the temporal and spatial distributions of reflection energy. Effectively detecting rays necessitates that the receiver be defined as a target centered on the receiving point with finite volume, rather than a point. For example, a spherical target is often used, given its symmetry. All the rays that pass through the target are recorded. In binaural room modeling, the energies, arrival times, and directions of rays should be recorded.

The selection of the number of rays radiated from a source and the target dimension are important issues in ray-tracing. The former determines the spatial density of rays, through which calculation accuracy is achieved. It is also closely related to the latter. An inappropriate combination of ray density and target dimension may result in missing or repeatedly detecting certain rays. Details on this issue can be found in the literature (Lehnert and Blauert, 1992).

Unlike image-source methods, ray-tracing methods are computationally efficient. They are also applicable to diffuse reflection modeling. However, the ray-tracing methods are usually less accurate than image-source methods. Some improved ray-tracing methods, such as the cone-tracing method, have been proposed. Ray-tracing methods are typically used for late reflection modeling.

11.1.4 Other Methods for Room Acoustics Modeling

Edge diffraction is disregarded in the basic image-source method discussed in Section 11.1.2. When a real source is located within a connection of two semi-infinite rigid surfaces with interior corner angles of 90°, 60°, 45°, 36° …, the basic image-source method for reflection calculation is valid. For other corner angles, however, edge diffraction should be considered to satisfy boundary conditions. The basic image-source method also yields a discontinuous acoustic response when the receiver passes through a regional boundary of the visibility of an image source. In this case, the image source or reflection suddenly emerges or disappears. Edge diffraction eliminates this discontinuity. The complicated calculation involved in edge diffraction is beyond the scope of this book.

In addition to image-source and ray-tracing methods, the radiosity method, which is often used in lighting, heat radiation, and computer graphics, is also applicable to room acoustics modeling. The radiosity method is also based on geometrical acoustics, in which boundary surfaces are divided into a number of small elements. Each element radiates rays and receives rays from all other elements and from the original source. Sound energy is absorbed and attenuated when it is exchanged among the elements. The room response at a receiving point is contributed from all these elements, as well as by the original source. The radiosity method is applicable to partially diffuse reflection.

As mentioned in Section 11.1.3, ray-tracing is typically used for late reflection modeling. As the order of reflection increases, however, the computational cost of ray-tracing remains large. Thus, more efficient perception-based methods are proposed for late reflection modeling in VAEs (Section 11.2).

Geometrical acoustics-based methods largely disregard the wave nature of sound, making these approaches suitable only for high frequencies. Numerical methods based on wave acoustics are needed for low-frequency and small-room modeling, in which wave nature should be taken into account. Some numerical methods for solving the boundary problem of the wave equation have been applied to room acoustics modeling.

FEM is a type of numerical method that differentiates space into a number of small volume elements, and converts the wave equation and boundary problem into a set of linear algebra equations. The various connections and combinations of elements with different shapes approximate complex room geometry. FEM is applicable to room response calculations and eigenmodels.

The BEM discussed in Section 4.3 is also applicable to room acoustics modeling. In contrast to FEM, BEM converts the boundary problem of the wave equation into the boundary surface integral, after which the boundary surface is differentiated into a mesh of elements, resulting in a set of linear algebra equations. Both BEM and FEM usually yield the frequency-domain response.

FDTD is another method that has been applied to room acoustics modeling, in which the wave equation and boundary problem are both spatially and temporally differentiated, resulting in a finite-difference algorithm. In essence, the problem of solving partial differential equations is converted into an issue of iteration calculation. FDTD is efficient because it is carried out in the time domain, resulting in the easy acquisition of the time-domain response. However, boundary processing is complex.

These numerical methods for the wave equation require the differentiation of spatial or temporal variables into elements. The large number of small elements needed for high-frequency modeling incurs heavy computational workloads that may exceed computer capacity. These methods are therefore suitable only for low frequencies and small rooms.

11.1.5 Source Directivity and Air Absorption

Simulating an omnidirectional point source is the simplest case of source modeling. Sound radiation from some practical sources, such as the human head (with torso) and musical instruments, exhibits directivity. The RIR and BRIRs are related to the directivity pattern of sound sources. The directivity factor $D(\Omega', f)$, which is defined as the magnitude of radiation pressure at direction Ω' relative to that of main-axis of radiation, is typically used to describe the directivity pattern of a sound source:

$$D(\Omega', f) = \left| \frac{P(\Omega', f)}{P(\Omega'_{ref}, f)} \right|. \tag{11.7}$$

Generally, $D(\Omega', f)$ varies as a function of direction and frequency. It is obtained by measurement, or, sometimes, by calculation from a physical model of sound sources. $D(\Omega', f)$ is complex for certain practical sound sources (Giron, 1996). In room acoustics modeling and other applications, the continuous and parameterized representation of $D(\Omega', f)$ can be derived by the spatial basis functions (such as spherical harmonic functions) decomposition of the measured $D(\Omega', f)$. $D(\Omega', f)$ and HRTFs represent the frequency-dependent directivity patterns of radiation and receivers, respectively. As indicated by the acoustic principle of reciprocity, $D(\Omega', f)$ decomposition is similar to the HRTF decomposition in Section 6.3.

Source directivity can be incorporated into room acoustics modeling. In image-source methods, source directivity is simulated by weighting the source intensity Q_S with $D(\Omega', f)$, which is equivalent to filtering the source signals with the direction-dependent filters $D(\Omega', f)$. In ray-tracing methods, two techniques can be used to incorporate source directivity. One is selecting directionally uniform ray density (as for an omnidirectional source) and weighted ray intensity according to source directivity. The other is selecting direction-independent ray intensity and weighted ray density according to source directivity.

As stated in Section 1.4.6, the high-frequency attenuation caused by air absorption should be considered for a distant source. Although the air absorption of direct sound is negligible in an ordinary-sized room, it should be taken into account for reflections caused by a large accumulated propagation distance, especially in a large hall.

The pressure magnitude attenuation caused by air absorption is evaluated by (Morse and Ingrad, 1968):

$$P(r_{total}) = P_0 \exp[-\alpha_{air}(f) r_{total}], \tag{11.8}$$

where $\alpha_{air}(f)$ is the air absorption coefficient related to temperature and humidity (among other factors); and r_{total} is the accumulated propagation distance from source to receiver. Air absorption can be modeled by a low-pass filter according to Equation (11.8), and the analytical expressions of $\alpha_{air}(f)$ are specified by the International Organization for Standardization (Huopaniemi et al., 1997b). In image-source methods, air absorption can be incorporated by cascading the response of air absorption filters with the response from each image source to the receiver.

11.1.6 Calculation of Binaural Room Impulse Responses

A pair of BRIRs describes the physical course, from a source to two ears, in a reflective room or environment (Section 11.1.1). Complete binaural room simulation involves source, room, and listener modeling. Source and room modeling have been separately discussed in preceding sections. BRIRs are obtained by combining source and room modeling with the acoustic transfer characteristics (head-related transfer functions, HRTFs, or head-related impulse responses, HRIRs) of a listener. A complete scheme for BRIR modeling is outlined as follows.

1. The acoustic field of a room is simulated by appropriate methods that are selected on the basis of previous knowledge of a source and room. In principle, the methods in preceding sections or some hybrid methods are applicable to room acoustics modeling. For example, image-source methods are used for early reflection modeling (e.g., up to second- or third-order reflections) because of their accuracy, and ray-tracing methods are employed for late reflection modeling because of their computational efficiency. Edge diffraction modeling by special methods is sometimes required. Generally, geometrical acoustics-based modeling yields the intensity, spectrum, arrival time, and direction of the direct sound and reflections of a given source and receiver position. Geometrical acoustics-based methods are valid only at high frequencies. If necessary, therefore, some wave acoustics-based methods are used as a supplement to low-frequency modeling.

2. Room acoustics modeling yields the monophonic (mono) responses of direct sound and each reflection. The entire mono room response is obtained by combining the responses of direct sound and all reflection paths. To convert the responses into binaural responses, the mono response of each direct or reflection path is filtered with a pair of HRTFs (or convoluted with a pair of HRIRs) at corresponding directions. The results are then combined to form complete BRIRs.

Some room acoustics modeling methods (especially ray-tracing methods) usually yield the energy-time responses (square pressure impulse response) of a room, with phase information excluded. Moreover, the response is often provided on each octave band (or one-third octave band). To apply the responses to binaural synthesis, the energy-time responses should be converted into pressure impulse responses. A common scheme is to first convert the energy-time responses on each frequency band to frequency-domain responses. Then, the pressure magnitude responses are obtained by square root manipulation and interpolation in the frequency domain. Finally, the pressure impulse responses are derived by minimum-phase reconstruction and inverse Fourier transform (Lehnert and Blauert, 1992; Kuttruff, 1993). An alternative method for obtaining pressure impulse responses is first generating white noise of an appropriate length. The noise is then modulated with an amplitude envelope obtained by the square root of the energy-time response in each frequency band. Finally, each modulated signal is band-pass filtered at the corresponding central frequency, and all the filtered signals are summed (Farina, 1995). The impulse response obtained in this manner retains the same energy decay pattern as the original energy-time response, but possesses a random phase. Both methods exclude the phase information on RIR. Some experimental results indicate that the phase information on RIR is insignificant to auditory perception (Kuttruff, 1991).

Binaural reflection rendering in VAEs can be directly implemented by convoluting an input stimulus with a pair of known HRIRs. Some efficient convolution algorithms have been developed for this purpose (Gardner, 1995b). The direct rendering technique is convenient for non-real time implementation, such as the static room acoustic auralization to be discussed in Section 14.2. Alternatively, in real-time dynamic VAEs (Chapter 12), binaural reflection rendering is often implemented by using parametric and decomposed structures. That is, source radiation patterns, sound propagation, surface materials, and air absorption, as well as HRTF-based filtering, are modeled in real time and then separately realized by corresponding low-order or simplified filters and delay lines. This approach indicates that, overall, BRIRs are approximated by appropriate series-parallel filters and delay line connections (Walker, 2000).

11.2 Artificial Delay and Reverberation Algorithms

As stated in preceding sections, a straightforward scheme for binaural reflection rendering is convoluting input stimuli with BRIRs. However, modeling a pair of complete BRIRs is a complex task. Moreover, because the length of an actual BRIR is on the order of room reverberation time (typically from hundreds of milliseconds to several seconds), convolution is complex, especially under real-time conditions. As an alternative, artificial delay and reverberation algorithms are appropriate for some applications (such as consumer electronics), in which precisely simulating reflections is unnecessary (Blesser, 2001). According to some measured or precalculated room acoustic attributes or parameters (such as reverberation time), these algorithms render reflections from the perceptual, rather than physical, point of view. They are classified as perception-based methods in order to distinguish them from the previously-discussed physics-based methods.

Artificial delay and reverberation have long been applied to audio signal processing for recording mono, stereophonic, and multichannel surround sound programs. In early years, the spring reverberation unit, plate reverberation unit, reverberation room, circulation magnetic recording sound, BBD (bucket brigade device) electric delay and reverberation unit were adopted. These components have been replaced by digital delay and reverberation unit during the past two decades. However, some signal processing structures in traditional methods remain applicable to digital delay and reverberation unit. This section discusses the common delay and reverberation algorithms for VAEs. An overview of these issues is found in Gardner and Dattorro (1997).

11.2.1 Artificial Delay and Discrete Reflection Modeling

As stated in Section 1.7.1, room reflections comprise discrete early reflections and late reverberation. Artificial delay algorithms are designed to model discrete early reflections.

The output $y(n)$ of a digital delay line with a length of m samples is related to the input signal $x(n)$ by:

$$y(n) = x(n-m). \tag{11.9}$$

The length of delay in seconds is calculated as:

$$\tau_D = \frac{m}{f_s}, \tag{11.10}$$

where f_s is the sampling frequency.

According to Equation (2.19), the Z transform of Equation (11.9) is given by:

$$Y(z) = z^{-m} X(z), \tag{11.11}$$

Therefore, multiplying a factor of z^{-m} to the signal in the Z-domain is equivalent to the delay of the corresponding signal in the time domain by m samples. The system function of a delay line is:

$$H_D(z) = \frac{Y(z)}{X(z)} = z^{-m}. \tag{11.12}$$

The corresponding system impulse response is the unit sampling sequence in Equation (2.17):

$$h_D(n) = \delta(n-m). \tag{11.13}$$

If the weighted output $x(n - m)$ with gain g of the delay line is mixed with the original signal $x(n)$, the resultant signal is given by:

$$y(n) = x(n) + g\,x(n-m). \tag{11.14}$$

Equation (11.14) simulates the case of direct sound along with single reflection. A gain (attenuation) with $|g| < 1$ simulates surface material absorption. Gain g can be regarded as the pressure reflection coefficient. The impulse response and system function that correspond to Equation (11.14) are respectively provided by:

$$h_D(n) = \delta(n) + g\delta(n-m), \tag{11.15}$$

$$H_D(z) = 1 + gz^{-m}. \tag{11.16}$$

The algorithm in Equation (11.14) is implemented by the finite infinite response (FIR) filter structure shown in Figure 11.5a. This filter consists of a parallel connection between a direct line and a delayed line with gain g. Let $z = \exp(j\omega)$, with ω denoting the digital angular frequency in Equation (2.20). Then, Equation (11.16) becomes:

$$H_D(\omega) = 1 + g\exp(-jm\omega) \qquad |H_D(\omega)| = \sqrt{1 + 2g\cos(m\omega) + g^2}. \tag{11.17}$$

Figure 11.5b shows the variations in the magnitude response $|H_D(\omega)|$, with ω within the range of $0 \le \omega < 2\pi$ for $m = 10$ and $g = 0.7$. Magnitude $|H_D(\omega)|$ exhibits a comb filtering characteristic, with notch $(1 - g)$ at $\omega = (2q + 1)\pi/m$ and peak $(1 + g)$ at $\omega = 2q\pi/m$ for $q = 0,12 \ldots (m - 1)$. This characteristic stems from the interference caused by the modeled direct sound and reflection. Therefore, Figure 11.5a is an FIR comb filter structure.

To simulate multiple discrete early reflections with various delays caused by the different surfaces in a room, the outputs of Q delay lines with different lengths m_q and gains g_q are combined with the original signal $x(n)$, yielding:

$$y(n) = x(n) + \sum_{q=1}^{Q} g_q x(n - m_q). \tag{11.18}$$

Correspondingly, the impulse response and system function are respectively given by:

$$h_D(n) = \delta(n) + \sum_{i=1}^{Q} g_q \delta(n - m_q), \tag{11.19}$$

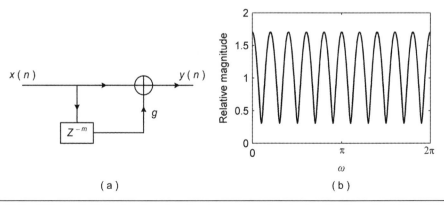

(a)   (b)

Figure 11.5 FIR comb filter structure for direct sound and a single reflection simulation: (a) FIR filter structure; (b) comb filtering

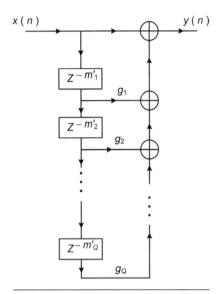

Figure 11.6 FIR filter structure for simulating multiple discrete early reflections

and:

$$H_D(z) = 1 + \sum_{q=1}^{Q} g_q z^{-m_q}. \tag{11.20}$$

The algorithms for simulating multiple discrete early reflections are implemented by the FIR filter structure shown in Figure 11.6, with $m_1 = m'_1$, $m_2 = m'_1 + m'_2$, ... $m_Q = m'_1 + m'_2 + ... + m'_Q$. The lengths of the delay lines in the algorithms are chosen according to the arrival times of early reflections in a modeled room or environment. In some practical cases, the lengths of delay lines can also be selected according to the subjective preferred delay times for early reflections (Ando, 1985).

All the gains g_q in the delay lines of Equation (11.20) or Figure 11.6 are independent of frequency. Accordingly, the spectrum of each modeled reflection is identical to that of the input signal (direct sound), except that the former is attenuated by a constant factor. The gain in each delay line can be replaced by a low-pass filter unit with the system function $G_{LOW,q}(z)$ to simulate the frequency-dependent surface and air absorption. In this case, the overall impulse response and system function become:

$$h_D(n) = \delta(n) + \sum_{q=1}^{Q} g_{LOW,q}(n - m_q), \tag{11.21}$$

$$H_D(z) = 1 + \sum_{q=1}^{Q} G_{LOW,q}(z) z^{-m_q}, \tag{11.22}$$

where $g_{LOW}(n)$ is the impulse response of $G_{LOW,q}(z)$.

The low-pass filter unit can be implemented by an FIR or infinite impulse response (IIR) structure. Well-designed filters can simulate various surface and air absorptions. Figure 11.7 depicts an example of a 1-order IIR low-pass filter, whose system function is provided by:

$$G_{LOW}(z) = \frac{b_0 + b_1 z^{-1}}{1 + a_1 z^{-1}}. \tag{11.23}$$

11.2.2 Late Reflection Modeling and Plain Reverberation Algorithm

The number of reflections from different surfaces rapidly grows with increasing order of room reflections. At the same time, reflection energy density decays with time because of surface

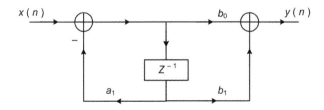

Figure 11.7 1-order IIR low-pass filters

materials and air absorptions. In many cases, the late reflection in rooms is approximated as a diffuse reverberation field, in which numerous reflections arrive at a receiver from various directions at every instant. The exact arrival time and direction of each reflection are perceptually insignificant, and the exponential decay of reverberation energy can be calculated by statistical acoustical methods. When these features are exploited, the late reflection in a VAE can be simply simulated by various artificial reverberation algorithms.

The plain reverberation algorithm simulates the successive reflections and decay in a room, which is equivalent to combining the outputs of an infinite number of delay lines with lengths of $m, 2m, 3m$ ... samples and gains of g, g^2, g^3 ... relative to the original signal $x(n)$:

$$y(n)=x(n)+\sum_{q=1}^{\infty}g^q x(n-qm). \tag{11.24}$$

Equation (11.24) can be written in recursive form thus:

$$y(n)=x(n)+g\,y(n-m). \tag{11.25}$$

Correspondingly, the impulse response and system function are:

$$h_{REV}(n)=\delta(n)+\sum_{q=1}^{\infty}g^q\delta(n-qm), \tag{11.26}$$

and:

$$H_{REV}(z)=1+\sum_{q=1}^{\infty}g^q z^{-qm}=\frac{1}{1-gz^{-m}}. \tag{11.27}$$

Therefore, the impulse response $h_{REV}(n)$ in Equation (11.26) consists of an infinite series of unit impulses weighted with an infinite power series of gain g. The algorithm given by Equation (11.25) is called the *plain reverberation algorithm*, which is implemented by the IIR filter structure shown in Figure 11.8. The structure comprises a feedback loop with an m-sample delay line and gain g. That the input and output of Figure 11.8 satisfy Equation (11.25) is easy to prove.

Equation (11.26) shows that each modeled reflection is delayed by m samples and attenuated g times, compared with the preceding reflection. The magnitude of the qth reflection is g^q times that of direct sound. As stated in Section 1.7.1, the reverberation time of the plain reverberation algorithm is evaluated by assuming $20\log_{10}g^q=-60$ (dB):

$$T_{60}=\frac{-3m}{f_s\log_{10}g}. \tag{11.28}$$

For a given sampling frequency and delay line length m, T_{60} increases with feedback gain g.

The plain reverberation algorithm is simple with adjustable reverberation time, and accommodates the exponential energy decay of natural reverberation. However, it suffers from the following drawbacks:

1. The reverberation time is frequency independent, preventing it from simulating frequency-dependent reverberation time, which usually decreases with increasing frequency in real rooms.
2. The equal time interval between two successive reflections tends to create fluttering. Moreover, the modeled reflection density f_s/m is invariable

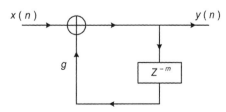

Figure 11.8   IIR filter structure for the plain reverberation algorithm

with time, which contradicts the phenomenon of increasing reflection density with time in real rooms.

3. The system function $H_{REV}(z)$ in Equation (11.27) includes m poles that are located at an equal interval in a circle with radius $g^{1/m}$ in the Z-plane:

$$z_p = g^{1/m} \exp\left(j\frac{2\pi p}{m}\right) \qquad p = 0,1,2...(m-1). \tag{11.29}$$

These poles cause the system function magnitude $|H_{REV}[\exp(j\omega)]|$ to vary with frequency, resulting in comb filtering characteristics that are similar to those shown in Figure 11.5b. These characteristics cause subjective coloration in timbre. The peaks in $|H_{REV}[\exp(j\omega)]|$ correspond to the poles of function $H_{REV}(z)$. The digital angular frequencies of peaks are evaluated by substituting $z = \exp(j\omega)$ in Equation (11.29) as:

$$\omega_p = \frac{2\pi p}{m} \qquad p = 0,1...(m-1). \tag{11.30}$$

Because of these drawbacks, using a single plain reverberation unit for late reflection simulation often causes perceivable artifacts. This problem prompts the improvement of the plain reverberation algorithm (Section 11.2.3).

11.2.3 Improvements on Reverberation Algorithm

The feedback gain g in Equation (11.28) controls reverberation time. To simulate the surface and air absorption-induced decrease in reverberation time at high frequencies, the *low-pass reverberation algorithm* shown in Figure 11.9 is implemented, in which the feedback gain g in Figure 11.8 is replaced with a low-pass filter unit $G_{LOW}(z)$. The impulse response and system function of the low-pass reverberation algorithm are given by:

$$h_{REV}(n) = \delta(n) + g_{LOW}(n-m) + g_{LOW}(n) * g_{LOW}(n-2m) \tag{11.31}$$
$$+ g_{LOW}(n) * g_{LOW}(n) * g_{LOW}(n-3m) + ...,$$

and:

$$H_{REV}(z) = \frac{1}{1 - G_{LOW}(z)z^{-m}}, \tag{11.32}$$

where $g_{LOW}(n)$ is the impulse response related to the low-pass filter $G_{LOW}(z)$, and "*" denotes convolution. Similar to the case in Equation (11.21), the low-pass filter unit can be implemented by an FIR or IIR structure (Figure 11.7).

In addition to the decrease in reverberation time at high frequencies, the reflection density in the low-pass reverberation algorithm increases with time. This behavior is consistent with the

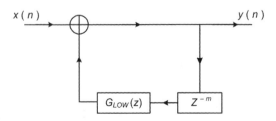

Figure 11.9 Low-pass reverberation algorithm

characteristics of late reflections in real rooms. The high-order reflections in the low-pass rever-beration algorithm are modeled by multiple convolutions with $g_{LOW}(n)$, which increases reflection density.

To produce a flat static magnitude response and reduce perceived coloration, Schroeder (1962) proposed the well-known *all-pass reverberation algorithm*. The impulse response and system function of this algorithm are given by

$$h_{REV}(n) = (A+B)\delta(n) + B\sum_{q=1}^{\infty} g^q \delta(n-qm), \tag{11.33}$$

and:

$$H_{REV}(z) = \frac{-g+z^{-m}}{1-gz^{-m}} = A + \frac{B}{1-gz^{-m}} = (A+B) + B\sum_{q=1}^{\infty} g^q z^{-qm}, \tag{11.34}$$

$$A = -\frac{1}{g} \qquad B = \frac{(1-g^2)}{g}.$$

The impulse response $h_{REV}(n)$ comprises an infinite series of unit impulses with time-decaying gains. The magnitude of the $(q+1)$th reflected impulse is g times that of the qth reflected impulse. The system function satisfies the equation:

$$|H_{REV}[\exp(j\omega)]| = 1. \tag{11.35}$$

Therefore, the system function magnitude is frequency independent. The all-pass reverberation algorithm can be implemented by the all-pass IIR filter structure shown in Figure 11.10. Equation (11.34) corresponds to the following input-output equation:

$$y(n) = -gx(n) + x(n-m) + g\,y(n-m). \tag{11.36}$$

The all-pass reverberation algorithm cannot completely eliminate timbre coloration because of the short time frequency analysis of human hearing. On the other hand, the flat static magnitude response of the algorithm is the consequence of long-term Fourier analysis.

To increase reflection density, several all-pass reverberation units can be connected in series. Different delays can be chosen for each all-pass reverberation unit, so that the reflections are in-consistent at the same instant. In practice, the *Schroeder reverberation algorithm* or the structure shown in Figure 11.11 is often adopted. The algorithm consists of several parallel plain reverbera-tion units and a series connection of all-pass reverberation units. The plain reverberation units are suitable for modeling discrete early reflections, and the all-pass reverberation units are intended to increase reflection density. Reducing the coloration caused by comb filters necessitates using plain reverberation units with incommensurate delays. However, the modeled reflection density in

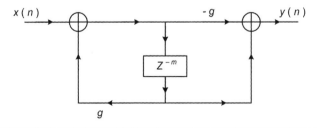

Figure 11.10   IIR filter structure of the all-pass reverberation algorithm

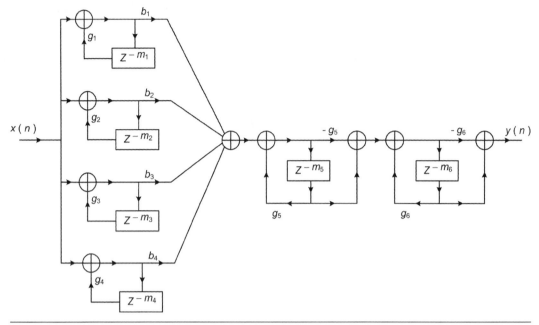

Figure 11.11 Schroeder reverberation algorithm

Figure 11.11 is constant over time. The low-pass reverberation units in Figure 11.9 can be used to replace the plain reverberation units in Figure 11.11, thereby enabling the simulation of increasing reflection density with time in real rooms.

Jot and Chaigne (1991) proposed a *feedback delay network* (FDN) structure for reverberation algorithms. As shown in Figure 11.12, the outputs of N parallel delay lines with different lengths are connected to all N inputs by an $N \times N$ feedback matrix $[A]$. Designing the feedback matrix and delay in each delay line is possible, so that the outputs of the delay lines are uncorrelated. Therefore, the FDN structure is applicable to creating multi-channel uncorrelated reverberations. To

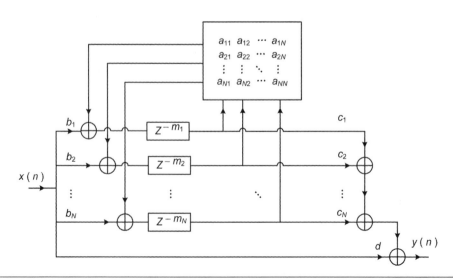

Figure 11.12 Feedback delay network structure for reverberation algorithms

simulate frequency-dependent reverberation time, appropriate filters (such as low-pass filters) can be inserted before/after each delay line, which also cause reflection density to increase with time.

11.2.4 Application of Delay and Reverberation Algorithms to Virtual Auditory Environments

The mono delay and reverberation algorithms were presented in Sections 11.2.1, 11.2.2, and 11.2.3 with the spatial information on reflections disregarded. However, the spatial information on reflections should be taken into account in binaural VAEs (Gardner, 1998).

As stated in Section 1.7.2, the spatial information on early reflections is perceptually important. For example, early lateral reflections are crucial to generating auditory source width in a hall. To simulate the directional characteristics of direct sound and discrete early reflections, the input stimulus and the outputs of each delay line in Figure 11.6 are filtered with a pair of corresponding HRTFs. The outputs of all left and right HRTF filters are separately combined to synthesize binaural signals. The block diagram of direct sound and reflection modeling for binaural signals is shown in Figure 11.13. To simulate surface and air absorptions, low-pass filter units can also be inserted between the output of each delay line and the input of HRTF-based filters.

The algorithm in Figure 11.13 precisely simulates the directional characteristics of multiple discrete reflections. However, the simulation is complex because each discrete reflection requires processing with a pair of HRTF-based filters. In practical applications, only a few early reflections (usually up to the first or second order) are processed in this manner. In some applications, the delayed signals of vicinity-reflected directions are grouped and then filtered with a pair of HRTFs at the center of these directions. This approach is equivalent to simulating vicinity reflections that are centralized in one direction, simplifying signal processing but inevitably degrading perceived performance. As a rougher approximation, binaural reflections are simply simulated by

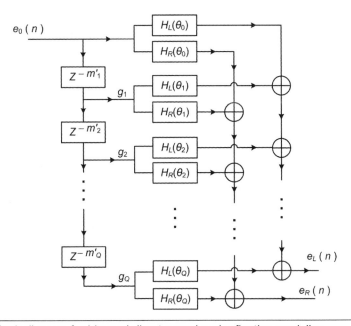

Figure 11.13   Block diagram for binaural direct sound and reflection modeling

the appropriate delay and attenuation of reflections to form correct binaural interaural time difference and interaural level difference information.

Under the approximation of the diffuse reverberation field, numerous late reflections arrive at two ears from various directions at every instant. Individually simulating each late reflection is often difficult and unnecessary. Alternatively, modeling late reflections with two uncorrelated reverberation signals is perceptually efficient for many practical applications. Figure 11.14 shows a block diagram of the creation of two low-correlation reverberation signals, which are used in the digital interactive virtual acoustics dynamic VAE system (Section 12.4; Saviojia et al., 1999). This artificial reverberation structure contains four parallel feedback loops. Each feedback loop consists of a delay line τ, a low-pass filter $H_A(z)$, and a comb- or all-pass filter $H_B(z)$. The delay in each delay line is chosen to generate inconsistent reflections at the same instant. In addition, the decorrelated reverberation algorithm using reciprocal maximal length sequence pairs is also applicable to binaurally modeling late reflection (Trivedi et al., 2009a; 2009b).

If necessary, a pair of partially uncorrelated reverberation signals is created by weighted mixing of two uncorrelated reverberation signals $y_L(n)$ and $y_R(n)$ as:

$$e_L(n) = \cos\gamma\, y_L(n) + \sin\gamma\, y_R(n)$$
$$e_R(n) = \sin\gamma\, y_L(n) + \cos\gamma\, y_R(n) \tag{11.37}$$

where γ is related to the interaural cross-correlation coefficient (IACC) in Equation (1.34) thus:

$$\gamma = \frac{\arcsin(IACC)}{2}. \tag{11.38}$$

Therefore, the interaural correlation of reverberation signals is controlled by γ.

The reflection modeling methods discussed in this section are not as precise as physical-based modeling and BRIR convolution-based methods. However, the former can be simply implemented using various filter structures, thereby yielding appropriate perceptual performance.

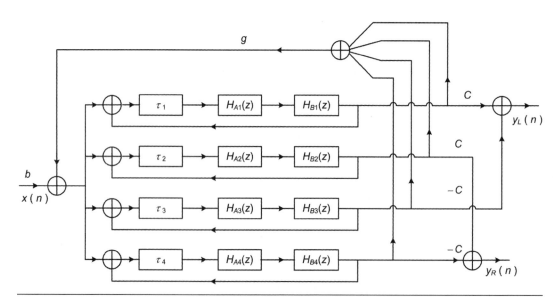

Figure 11.14 Block diagram of the creation of two low-correlation reverberation signals (adapted from Saviojia et al., 1999)

11.3 Summary

A complete virtual auditory display, which includes room reflection rendering, is also called a virtual auditory environment. A straightforward scheme for binaural reflection rendering is convoluting a mono anechoic recording stimulus with a pair of binaural room impulse responses obtained by measurement or physical modeling.

Complete physical modeling involves source, room, and listener modeling. In past decades, computers have been an important tool for room acoustics modeling, which is classified into two categories: wave acoustics-based methods and geometrical acoustics-based methods. Geometrical acoustics-based methods, which disregard the wave nature of sound, are suitable for high frequencies and smooth boundary surfaces.

Image-source and ray-tracing methods are two main types of geometrical acoustics-based methods. The basic image-source method simulates each reflection from a boundary surface by an image source, from which the intensity, arrival time, and direction of each reflection can be evaluated. However, the method is suitable for specular reflections, rather than for diffuse reflections or diffraction. This method becomes complex for high-order reflection modeling or for a room with complex geometry. Therefore, it is normally used to simulate preceding-order early reflections.

Ray tracing simulates the radiations from a source with a number of rays, and then traces the propagation of each ray according to the geometrical and physical characteristics of a room surface. The responses in a receiving point are contributed from all the rays that pass through it. Ray-tracing methods are more efficient than image-source methods and applicable to diffuse reflection modeling. However, the basic ray-tracing method is less accurate than image-source methods.

The wave acoustics-based methods should be used when the wave nature of sound is considered. Common numerical methods include finite element method, boundary element method, finite-difference time-domain method, and digital waveguide mesh method. These approaches are feasible only at low frequencies and in small rooms.

For some applications, in which the physical accuracy of reflection modeling is not a crucial requirement, some perception-based methods, such as artificial delay and reverberation algorithms, are applicable to reflection rendering. They usually yield reasonable perceptual performance.

Some artificial delay and reverberation algorithms—including plain reverberation, low-pass reverberation, all-pass reverberation, Schroeder reverberation, and feedback delay network reverberation—have been proposed and applied to audio processing. These algorithms are relatively simple and can be implemented by various finite impulse response and infinite impulse response filter structures. In many applications of virtual auditory environments, the directional characteristics of direct sound and a few early reflections are individually modeled, and late reflections are often modeled by uncorrelated reverberation signals.

12

Rendering System for Dynamic and Real-time Virtual Auditory Environments (VAEs)

This chapter discusses the dynamic and real-time virtual auditory environment (VAE). The basic principle and structure of a dynamic VAE are presented. The methods for modeling and rendering dynamic information in a VAE are addressed. The dynamic characteristics of VAE are discussed. Some examples of dynamic VAE systems are also outlined.

The static VAD (virtual auditory display) or VAE (virtual auditory environment) has been discussed in previous chapters, in which both virtual sources and listeners are assumed to be in fixed locations and real-time processing is not always required. In a real acoustic environment, however, either source or listener movement alters binaural pressures and provides dynamic acoustic information. This dynamic information should be incorporated into VAD or VAE processing because it is significant, for both source localization and for recreating convincing auditory perceptions of acoustic environments. Therefore, in addition to modeling sound sources, room (environment), and listener, a sophisticated VAE should be able to constantly detect the position and orientation of a listener's head, on whose basis signal processing is updated in real time. In other words, a realistic VAE should be an interactive, dynamic, and real-time rendering system, hereafter called a *dynamic and real-time VAE system* or *dynamic VAE system*.

This chapter discusses the dynamic VAE. Section 12.1 outlines the basic principle and structure of a dynamic VAE. Section 12.2 discusses the modeling and rendering of dynamic information that stems from head movement—including head tracking, dynamic free-field virtual sources, environment reflection synthesis, the dynamic characteristics of VAE, and dynamic crosstalk cancellation for loudspeaker reproduction. The simulation of moving virtual sources is addressed in Section 12.3. Finally, some examples of dynamic VAE systems are outlined in Section 12.4.

12.1 Basic Structure of Dynamic VAE Systems

In real environments, the sound waves radiated by sources arrive at a listener by direct and reflected/scattered propagations off boundary surfaces. The scattering, diffraction, and reflection from human anatomical structures further modify the pressures received by both ears. A static VAE, which involves source, room (environment), and listener modeling, simulates the static acoustic course, from source to two ears, under various physical conditions. According to the prior knowledge of geometrical and physical data on sources and environments, a static VAE first simulates the radiation of sources, as well as the direct and reflected/scattered propagations of sound waves, to acquire temporal and spatial information on sound fields. Then, the static VAE simulates the overall effects of human anatomical structures by filters based on HRTF (head-related transfer function) to encode the information on sound fields into binaural signals.

A listener's head movement (including translation and turning) in real environments alters the location of two ears and the overall scattering/diffraction/reflection effects of anatomical structures. It ultimately alters binaural pressures and associated information. Accordingly, a dynamic and real-time VAE should constantly adjust source, environment (propagation), and listener modeling according to the temporary position and orientation of the listener's head in order to accommodate the dynamic variations in binaural signals. The idea of dynamic binaural presentation dates back to the 1970's (Boerger, et al., 1977). Various applications call for different emphases and detail structures in dynamic VAEs, but have similar basic principles and general structures. Figure 12.1 shows the basic structure of a dynamic VAE system, which consists of three parts:

1. *Information input and definition:* This part inputs the prior information and data for the dynamic VAE through a user interface. These information and data are classified into three categories: source information, environment information, and listener information. Source information includes types of source stimuli, the number, spatial positions, orientations, directivities (radiation patterns), and levels of sources, or the predetermined trajectory for a moving source. Environment information includes room or environment geometry and the absorption coefficients of surface materials and air. Listener information includes the initial spatial position, orientation, and individual data on a listener (such as HRTFs). A head-tracking device detects the position and orientation of the listener's head and then provides the information to the system.

2. *Dynamic VAE signal processing:* According to the prior information and data in part (1), this component simulates sound sources, as well as both the direct and reflected/scattered propagations from sound sources to two ears, using certain physical algorithms. On the basis of the temporary position of the head that is detected by head-tracking devices, the HRTFs for binaural synthesis are constantly updated to obtain dynamic binaural signals.

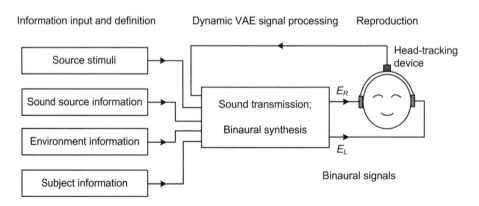

Figure 12.1 Structure of a typical dynamic VAE system

3. *Reproduction:* The resultant binaural signals are reproduced through headphones after headphone-to-ear canal transmission equalization, or through loudspeakers after cross-talk cancellation.

The input stimuli can be prerecorded in an anechoic chamber (dry signals) or synthesized by computer (such as musical instrument digital interface). The basic principles and schemes for binaural synthesis and static room or environmental acoustics modeling have been discussed in previous chapters. In dynamic and real-time rendering, however, processing becomes complicated, as will be discussed in the next section. In practice, multiple virtual sources at different positions may need to be rendered. Moreover, a dynamic VAE may be an interactive system with more than one user, in which the input information from each user is simultaneously processed. All these aspects incur enormous computational cost for a dynamic VAE.

The dynamic VAE systems that were developed early on were implemented on special digital signal processors (DSPs). With the development of PCs (personal computers), the recently developed systems are usually implemented on PC platforms, along with software. The advantage of PC-based implementation is flexibility in system updates. For multiple users, a dynamic VAE is implemented as a distributed system with a server platform and multiple terminals. In addition to auditory information, visual and other sensory information are required in multimedia and virtual reality applications. All these require a complex hardware structure and signal-processing scheme. This chapter focuses on the acoustic problems and signal-processing scheme in dynamic VAEs, with less emphasis on hardware structure.

12.2 Simulation of Dynamic Auditory Information

12.2.1 Head Tracking and Simulation of Dynamic Auditory Information

In contrast to a static VAE, a dynamic VAE needs constant updating of signal processing, according to the detected position and orientation of the head. The head is able to move in six degrees of freedom in space, with three being translation movements and the rest being turning movements. Accordingly, six coordinate parameters are needed to fully describe the position and orientation of the head. The position of the head center is specified by three Cartesian coordinate parameters. As shown in Figure 12.2, the head is initially located at the coordinate origin ($x = 0$, $y = 0$, $z = 0$) with x, y, and z pointing to the right, front, and top, respectively. The translation of the head is specified by (Δx, Δy, Δz).

The head is able to turn in three degrees of freedom, including turning around the x or left-right axes (pitch), around the y or front-back axes (roll) and around the z or up-down (vertical) axes (yaw or rotation). Accordingly, three angular parameters are needed to describe head turning.

An electromagnetic head tracker or other device (such as an infrared or laser-based device, computer vision instrument, or ultrasonic instrument) is used to detect the head position and orientation in a dynamic

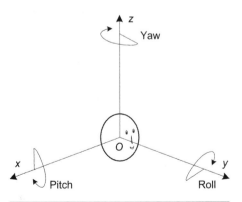

Figure 12.2 Head turning around three coordinate axes

VAE (Georgiou et al., 2000; Mannerheim and Nelson, 2008; Hess, 2012). Some commercial head trackers provide appropriate precision and resolution for a dynamic VAE. For example, the Polhemus Fastrak electromagnetic head tracker is capable of detecting six degrees of freedom of head translation and turning, with a distance precision of 0.08 cm, distance resolution of 0.0005 cm, angular precision of 0.15°, and angular resolution of 0.025°. When one receiver alone is active, the update rate is 120 Hz and the delay is 4 ms. These properties are beneficial to efficient VAE performance, as will be demonstrated in Section 12.2.4.

Head movement in any degree of freedom alters the spatial relationship between ears and sources or between ears and environments in real or virtual spaces, thereby altering binaural pressures. Ideally, head movement in all six degrees of freedom should be detected, on the basis of which dynamic signal processing can be implemented. However, incorporating the dynamic information on all six degrees of freedom in movement may require a computational capacity that exceeds that of the system. In this case, only the parts of the complete degrees of freedom that are closely related to important auditory perceptions are adopted in signal processing. For example, only the dynamic information related to horizontal translation and turning around vertical axes is retained.

Incorporating head tracking and corresponding dynamic processing improves the performance of VAE in terms of the following aspects:

1. *Improvement of virtual source localization:*   As stated in Section 1.4.3, the dynamic cue introduced by head turning is crucial for resolving front-back ambiguity and vertical localization. Section 8.3 also indicated that front-back confusion and error in vertical localization commonly occur in headphone reproduction because of the lack of dynamic cues in a static VAD. Numerous psychoacoustic experiments have confirmed that incorporating the dynamic information introduced by head rotation around vertical axes in VAD or VAE visibly reduces the percentage of front-back confusion in localization, especially in nonindividualized HRTF binaural synthesis (Wenzel et al., 1993a, 1995, 1996; Sandvad, 1996; Horbach et al., 1999; Wightman and Kistler, 1999). Using speech stimuli, Begault et al. (2001) compared the effects of head tracking, reverberation, and individualized HRTFs on front-back confusion in virtual source localization. The results indicate that head tracking is considerably more significant than the other two factors for resolving front-back (or back-front) confusion, with the confusion rate decreasing from 50% to 28%. Therefore, incorporating dynamic information is the most effective method for resolving front-back ambiguity in headphone reproduction. The aforementioned results also indicate that auditory localization cues are somewhat redundant (Section 1.4.5) because the auditory system can still localize virtual sources—even with partially incorrect cues, such as the incorrect spectral cues caused by the use of nonindividualized HRTFs rather than a subject's own HRTFs.

   Mackensen et al. (2000) investigated the effect of incorporating the dynamic cues caused by head rolling on elevation localization errors, and obtained negative results. These findings are inconsistent with some previous results, including those of Rao and Xie (2005) and Zhang and Xie (2012). Thus, in-depth investigations are needed.

2. *Enhancement of the subjective sense of realism:*   In real environments, head and source movements alter both direct and environment-reflected/scattered transmissions from sources to two ears, thereby altering the information encoded in binaural pressures and the resultant perceived sense of the surrounding acoustic environment. Incorporating

dynamic information in VAE enhances the subjective sense of realism, or presence, in virtual acoustic environments (Wenzel, 1991; Saviojia et.al, 1999).

3. *Dynamic crosstalk cancellation in loudspeaker reproduction:* Crosstalk cancellation is needed in binaural or transaural reproduction through loudspeakers. As stated in Sections 9.2 and 9.3, the performance of crosstalk cancellation depends on head position and orientation in relation to loudspeakers. Given a loudspeaker configuration and processing, crosstalk cancellation is effective at the default head position. Head movement (including translation and turning) disables crosstalk cancellation. In addition to dynamic binaural synthesis, therefore, dynamic crosstalk cancellation, according to the temporary head position, is also needed (see Section 12.2.5).

12.2.2 Dynamic Information in Free-field Virtual Source Synthesis

Dynamic free-field virtual source synthesis is the simplest case of dynamic VAE synthesis. The head is assumed initially located at the origin ($x = 0$, $y = 0$, $z = 0$) in the coordinate system shown in Figure 12.2, and the virtual source is assumed located at (r_0, θ_0, ϕ_0) relative to the head. Then, the parameters (r_0, θ_0, ϕ_0) for the source position and the corresponding HRTFs are selected in binaural synthesis. When the head moves from the origin with a translation (Δx, Δy, Δz), the position (r, θ, ϕ) of a fixed virtual source relative to the moved head center can be evaluated by (r_0, θ_0, ϕ_0) and (Δx, Δy, Δz) as:

$$\begin{aligned} r_0 \cos\phi_0 \sin\theta_0 - \Delta x &= r\cos\phi\sin\theta \\ r_0 \cos\phi_0 \cos\theta_0 - \Delta y &= r\cos\phi\cos\theta. \\ r_0 \sin\phi_0 - \Delta z &= r\sin\phi \end{aligned} \tag{12.1}$$

Therefore, head translation alters the direction and distance of virtual sources relative to the head.

Head turning alters the direction, but not the distance, of virtual sources relative to the head. For example, when the head rotates around the vertical axis with an azimuth $\Delta\theta$ (as shown in Figure 12.2, a positive $\Delta\theta$ denotes clockwise rotation), the elevation of the virtual source relative to the head remains unchanged, but the azimuth becomes:

$$\theta = \theta_0 - \Delta\theta. \tag{12.2}$$

Head rolling is slightly complex. When the head turns around the y-axis with an angle $\Delta\gamma$ (along the direction shown in Figure 12.2), the direction (θ, ϕ) of the virtual source relative to the head becomes:

$$\begin{aligned} \cos\phi\sin\theta &= \cos\phi_0 \sin\theta_0 \cos(\Delta\gamma) + \sin\phi_0 \sin(\Delta\gamma) \\ \cos\phi\cos\theta &= \cos\phi_0 \cos\theta_0. \\ \sin\phi &= \sin\phi_0 \cos(\Delta\gamma) - \cos\phi_0 \sin\theta_0 \sin(\Delta\gamma) \end{aligned} \tag{12.3}$$

Using the interaural polar coordinate in Figure 1.3 to describe head pitch is a convenient approach. The virtual source is assumed initially located at (Θ_0, Φ_0) relative to the head. When the head turns around the x-axis with an angle $\Delta\Phi$ (along the direction shown in Figure 12.2), the interaural polar azimuth of the virtual source relative to the head remains unchanged with $\Theta = \Theta_0$, but the interaural polar elevation becomes:

$$\Phi = \Phi_0 + \Delta\Phi. \tag{12.4}$$

The interaural polar coordinate can be converted into the default coordinate in this book with Equation (1.2).

In practice, the head may move (including both translation and turning) in a more complicated manner. Similar to the case of the turning of a rigid body in mechanics, head turning can be conveniently described by three Eular angles (Goldstein, 1980). In a general case, the virtual source position relative to the head can be evaluated according to geometry.

Dynamic binaural synthesis is implemented according to the temporary virtual source position (r, θ, ϕ) relative to the head. A straightforward scheme is to constantly update HRTFs in response to the temporary virtual source position (relative to the head) from a premeasured HRTF database for all possible source positions (Jin et al., 2005). This scheme is called *direct rendering* or *data-based rendering*.

HRTF varies as a continuous function of source position, but measurement usually yields only the HRTFs at discrete and finite source positions. The HRTFs at unmeasured positions should be interpolated from measured positions, as stated in Section 6.1. HRTF interpolation can be accomplished offline. This approach is equivalent to preparing an HRTF database with sufficient spatial resolution. Provided that the spatial resolution of the interpolated HRTFs is finer than the auditory resolution, the discontinuity in synthesized virtual source positions is inaudible. However, such a full set of HRTFs requires enormous memory for storage and a high hard disk access rate for real-time processing.

Alternatively, parametric schemes can be introduced to reduce data volume. For example, HRTFs with moderate spatial resolution are prepared and saved in advance. Then, the HRTFs at arbitrary unmeasured positions are obtained by real-time interpolation. In addition, because of the nearly distance-independent feature of far-field HRTFs, the distance dependence of a far-field virtual source can be simply simulated by filtering with far-field HRTFs at a fixed source distance. The filter is then cascaded with a variable propagation delay $\tau = r/c$ and a variable scaling (gain) factor $1/r$ for distance attenuation [refer to the discussion after Equation (1.37)]. Therefore, the parametric scheme reduces the required memory for HRTF storage, but at the cost of increasing computational burden.

The various HRTF interpolation schemes discussed in Chapter 6 are, in principle, applicable to dynamic VAE processing. Adjacent linear interpolation or bilinear interpolation is often used in real-time processing because of its relatively low computational cost. Under minimum-phase approximation, minimum-phase HRTFs (or head-related impulse responses, HRIRs) and pure delay can be separately interpolated. Some parametric models can also be used to calculate the ITD (interaural time difference) contributed by the interaural difference in the linear phase of a pair of HRTFs (see the equations provided in Section 7.2). HRTF interpolation can be implemented on various parametric HRTF models (such as the common-acoustical pole and zero-based interpolation discussed in Section 6.1.4). All the aforementioned methods reduce the amount of HRTF data in terms of length or number, or reduce the computational cost of interpolation.

As stated in Section 11.1.5, the directivity of a sound source is modeled by a direction-dependent filter. When the virtual source or head moves, the direction-dependent filter should be updated. This update can be implemented by loading the prepared directional filter coefficient (if necessary, directional interpolation is implemented) or using source directivity models (Wenzel et al., 2000).

12.2.3 Dynamic Information in Room Reflection Modeling

The dynamic binaural information caused by head movement is more complex in a reflective environment than in the free field because source or head movement alters both direct and reflected propagation paths. Ideally, a dynamic VAE should be capable of simulating all such information.

Similar to the case of dynamic free-field virtual source synthesis, direct (data-based) rendering is, in principle, applicable to dynamic room or environment reflection synthesis, in which BRIRs (binaural room impulse responses) are precalculated or premeasured, and then loaded and updated according to the temporary position and orientation of heads and sources (Horbach et al., 1999). The advantage of direct rendering is that it considerably decreases real-time computational cost by directly loading premeasured or precalculated BRIRs. However, direct rendering requires a database of spatially dense BRIRs for different head/source positions and orientations. Given the actual length of each BRIR (in the order of room reverberation time), the amount of data is enormous, requiring vast storage and making real-time access difficult. Long BRIRs cause difficulty in dynamic and real-time binaural synthesis, even when the hybrid frequency-time domain convolution algorithm is adopted. Therefore, some simplification schemes for BRIRs, such as truncating BRIRs with a time window (which only accounts for early reflections) and restricting the degrees of freedom of head movement, should be adopted.

Alternatively, the *parametric rendering scheme* is often adopted for dynamic room or environment reflection simulation. In contrast to direct rendering, parametric rendering simulates propagation from sources to two ears and synthesizes binaural signals in real-time. It does not, therefore, require a large memory for BRIR storage. However, most of the methods for binaural room modeling discussed in Section 11.1 are unsuitable for real-time rendering because of their high computational cost. Some simplifications and compromises should be employed in practical application.

The image-source method presented in Section 11.1.2 is often used for reflection simulation in parametric rendering, in which each reflection is modeled by an image source and all BRIRs are represented by summing the responses of direct and image sources. Correspondingly, binaural room reflection synthesis is equivalently implemented by the binaural synthesis of multiple free-field sources with different propagation delays, distance attenuations, and frequency-dependent absorption spectra, if necessary (Zotkin et al., 2002). Provided that the position of the direct source that corresponds to the direct sound is unchanged, the spatial positions of image sources that correspond to the reflections are fixed and independent of a listener's head position. This feature is characteristic of image-source methods. A listener's head movement alters the geometrical relationship between the head and sources. Head turning alters both direct and image source directions relative to the head, but leaves the source distances and visibility intact. Head translation, on the other hand, alters both the directions and distances of direct and image sources, and may also alter the visibility of image sources. Therefore, repeated visibility checking on image sources is required after head translation (Section 11.1.2). In practice, instead of convoluting stimuli with a pair of entire BRIRs, the overall transmission from a source to two ears is often decomposed into a series of direct and reflected propagation paths (direct and image sources). Each path is modeled by a signal-processing module that comprises filters, delay lines, and scaling factors to account for source directivity, propagation delay, distance attenuation, boundary and air absorption, and HRTFs, among other factors. Figure 12.3 shows a block diagram of this module (surface material absorption is unnecessary for direct paths). A listener's head movement alters the paths from sources to two ears, so that the parameters of filters, delay lines, and scaling factors should be constantly updated. A set of parallel modules, as shown in Figure 12.3, is needed for synthesizing multiple virtual sources, including direct and reflection image sources.

The highest order of reflection achievable for the image source method depends on available hardware/software capacity and the extent of complexity of environment (room) boundaries. Overall, the number of image sources exponentially grows with increasing order of reflections, thereby rapidly increasing the number of computational modules (Figure 12.3). In practice, some

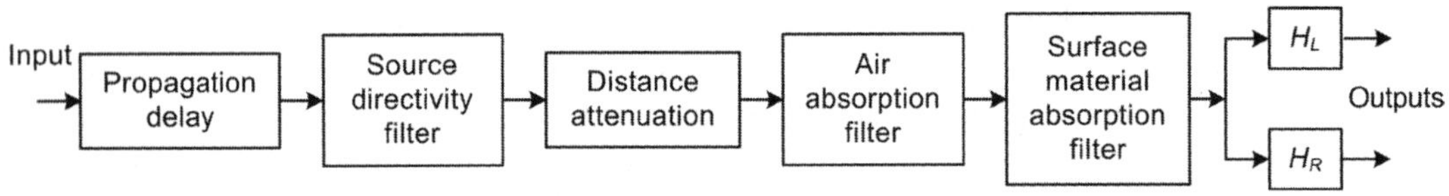

Figure 12.3 Signal processing module for modeling a propagation path

psychoacoustic rules are often used to reduce computational cost. For example, human hearing is less sensitive to reflections than to direct sound; as a benefit of this feature, simplified (low-order) HRTF filters can be used to synthesize virtual image sources, as opposed to using fine (high-order) HRTF filters to synthesize direct sound sources (Saviojia et al., 1999). Even if reflection synthesis has been simplified, most existing real-time systems can individually model and render direct sound and early reflections up to the first or second order. The aforementioned case is therefore a compromise solution. A large proportion of computational cost in parametric rendering is consumed in visibility checking and HRTF-based synthesis of multiple image sources.

Figure 12.3 shows that each virtual source synthesis, either for direct or reflection image sources, needs an HRTF-based filter for each ear. For M virtual source synthesis, M HRTF-based filters or HRIR-based convolution manipulations are required for each ear. The computational cost is directly proportional to the virtual source number M and ultimately exceeds available computational capacity. The virtual loudspeaker-based algorithms discussed in Section 6.5.1 are applicable to simplifying multiple source synthesis, in which binaural synthesis is implemented by commonly available and invariable filters for a few virtual loudspeakers. Accordingly, dynamic virtual source rendering is accomplished by adjusting the gains of input stimuli to virtual loudspeakers according to some signal mixing methods or panning laws, without requiring HRTF interpolation or updating (Noisternig et al., 2003; Jot et al., 2006a). Similarly, the basis function decomposition-based algorithms are also applicable to simplifying the synthesis of dynamic multiple sources, in which dynamic virtual source rendering is accomplished by adjusting the weights $w_q(\theta, \phi)$ and scaling factor $1/r$ for distance attenuation, and $\tau = r/c$ for the propagation delay in Figure 6.12.

Late reflections are usually treated as ideal diffuse fields with uniform energy density, in which binaural pressures remain static against head movement. The various artificial reverberation algorithms discussed in Section 11.2 are applicable to simulating the late reflections in dynamic VAEs. The simulation remains static in response to head or source movement. Diffuse-field approximation substantially simplifies late reflection rendering, thereby enabling real-time processing. The parameters (e.g., reverberation time) for artificial reverberation algorithms are obtained by premeasurement from a real room or precalculation from room acoustics simulation. Alternatively, late reflections can be rendered from the latter part of premeasured or precalculated BRIRs, which are assumed invariant against head movement. Thus, this approach is a hybrid scheme of data-based and parametric rendering (Takala et al., 1996; Pellegrini, 1999; Horbach et al., 1999; Saviojia et al., 1999).

Takane et al. (2003a, 2003b) introduced a different scheme for VAE processing in developing a VAE system called ADVISE (acoustic display based on the virtual sphere model). This method defines a spherical surface around a listener, and this surface is then differentiated by a dense grid. The acoustic transmission from a source outside the spherical surface to a listener's ears is divided into two components. One is transmission from the source to the points on the virtual spherical surface (including direct and reflected paths); this transmission is independent of the listener's head position and orientation, and can therefore be predetermined. The other is transmission from each point on the spherical surface to the ears. According to Huygens's principle and the Kirchhoff-Helmholtz boundary integral equation (Section 4.22), the sound field samples on the spherical boundary are modeled by a series of monopole sources, with intensity equal to the pressure derivative at surface and dipole sources, which have an intensity equal to the pressure at the surface. The pressure at each ear caused by each monopole or dipole source is calculated by HRTF filtering or HRIR convolution. The overall pressure at each ear is the sum of the pressures contributed by all monopole and dipole sources. When the listener moves inside the virtual sphere, the dynamic change in binaural pressures can be modeled by updating HRTF filters, as well as propagation delay and attenuation for the transmission from each monopole and dipole source to each ear alone. This method is similar to the virtual loudspeaker-based algorithms for binaural synthesis in Section 6.5.1. The wave field synthesis mentioned in Section 1.8.4 is a sound reproduction method based on the Kirchhoff-Helmholtz boundary integral equation. Similar to the binaural reproduction of multi-channel sound discussed in Sections 6.5 and 10.1, the aforementioned scheme is, in nature, equivalent to the dynamic binaural rendering of wave field synthesis signals. It also illustrates the relationship between sound fields and binaural reproduction. According to the spatial sampling theorem, however, the aforementioned scheme requires the synthesis of numerous virtual sources on the spherical surface, making it a highly complex method. Moreover, when a source position changes (such as a moving source), transmission from the source to all field points on the spherical surface changes accordingly. Under this condition, the pressures and their normal derivatives for each field point on the spherical surface should be recalculated and updated. These tasks entail intensive computational cost, making practical application difficult.

In the foregoing discussions, the dynamic binaural signals are obtained by real-time HRTF-based synthesis associated with head tracking. However, dynamic binaural recording cannot be realized with an artificial head. Algazi et al. (2004) proposed an alternative method called motion-tracked binaural sound (MTB) for dynamic binaural recording. As shown in Figure 12.4, MTB uses an array with M microphones arranged on the rigid surface of a sphere (or cylinder) to capture the sound field. The sphere radius is similar to the head radius. This approach is equivalent to sampling the sound field in the vicinity of a spherical head. Suppose that head turning is allowed but translation is forbidden, and a head tracker is used to detect the listener's head orientation in reproduction. When the concerned ear is located at the orientation of only one microphone, the

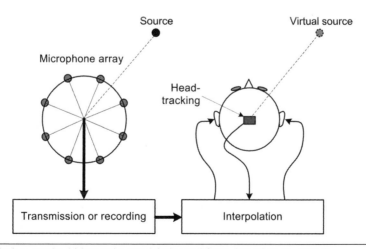

Figure 12.4 Motion-tracked binaural sound (adapted from Algazi et al., 2004)

recorded signal from that microphone is used as the ear signal. When the concerned ear is located at the orientation between the microphone arrangements, the weighted sum of some adjacent microphone signals is used as the ear signal. That is, the ear signal is derived by spatially interpolating the adjacent microphone signals. If only the head rotation around the vertical axis is taken into account (Figure 12.2), the M microphones can be arranged on the equator of a rigid sphere at equal intervals. If the head orientation is restricted within a small frontal azimuthal region of $\pm \Delta\theta$, the M microphones can be arranged in the azimuthal regions $90° \pm \Delta\theta$ and $270° \pm \Delta\theta$. Algazi also proposed some interpolation schemes for microphone signals. The results indicate that, for a rigid sphere with radius $a_0 = 0.0875$ m and $M = 8$ microphones arranged on the equator, dynamic binaural signals up to 1.25 kHz can be accurately derived by interpolation. This basic MTB method was enhanced in succeeding works. Melick et al. (2004) incorporated customized HRTFs into MTB processing. Hom et al. (2004) introduced a method for the continuous high-frequency interpolation of the spectral magnitudes of adjacent microphones to avoid the spectral discontinuities caused by head rotation. On the basis of MTB, Algazi et al. (2005) also proposed a scheme for the binaural reproduction of multi-channel surround sound.

Additionally, other microphone array and signal-processing schemes, especially the spherical or hemispherical array technique and the spherical harmonics (SH)-based beamforming technique discussed in Section 6.6, are applicable to deriving dynamic binaural signals (Li and Duraiswami, 2006). The advantage of MTB or other similar array techniques for binaural recording is that dynamic information can be recovered by sound field recording, making these approaches individual independent. However, limited by the spatial sampling theorem discussed in Chapter 6, numerous microphones are required to accurately recover high-frequency binaural signals in array recording. MTB, ADVISE, and direct rendering are similar in terms of spatial sampling.

12.2.4 Dynamic Behaviors in Real-time Rendering Systems

Ideally, the binaural signals or auditory scenarios created by a dynamic VAE should synchronously vary with head movement, as in real environments. Therefore, an ideal dynamic VAE should be a linear-time-variable system. However, the signal processing schemes in dynamic VAEs are deduced from the static scheme, in which a series of short "static states" are used to approximate the transient states. Thus, the dynamic behaviors of VAEs should be considered.

The first problem of interest in dynamic rendering is the *auditory continuity of data*, or scenario update. When the head movement detected by the head tracker exceeds a certain preset threshold, the system updates the data or parameters (such as HRTFs) in processing. If the updating is instantly accomplished, the auditory scenario abruptly switches, rather than smoothly transitions, from one scenario to the next, a situation that may generate audible artifacts. For free-field virtual source rendering, audible artifacts are caused by two factors. One is the auditory discontinuity caused by the physical differences between two adjacent HRTFs. This problem is related to the static audible difference between two adjacent HRTFs. The other factor pertains to the artifacts (such as clicks) caused by abruptly switching signals. Using the HRTFs of a VALDEMAR artificial head, Hoffmann and Møller (2005a, 2005b, 2005c, 2008a, 2008b, 2008c) carried out a series of psychoacoustic experiments to evaluate the audibility of adjacent HRTFs under both static and dynamic conditions. The results indicate that the audible directional thresholds for direct switching between adjacent HRTF magnitudes vary from 4.1° to 48.2° and the audible thresholds for direct switching ITD range from 5 to 9.4 μs, which is smaller than 10 μs (the just-noticeable difference of static ITD).

For smooth transition from one auditory scenario to the next, a spatially dense HRTF database is needed, or alternatively, some cross-fading (commutation) methods are incorporated into signal processing. Two cross-fading methods are available: output cross-fading and parameter cross-fading (Wenzel et al., 2000).

Dynamic far- and free-field virtual source synthesis is used as an example to illustrate cross-fading. Let θ_1 denote past scenario parameters, such as past source position relative to the head, and $h(\theta_1, t)$ denote the corresponding HRIR. The notation for the left or right ear is excluded from the denotation of HRIRs for simplicity. According to Equation (1.37) and succeeding discussions, the synthesized binaural signals of the past virtual source position for the input stimulus $e_0(t)$ are given by:

$$e_1(t) = \frac{1}{r_1} h(\theta_1, t) * e_0(t - \tau_1),$$
(12.5)

where the scaling factor $1/r_1$ and $\tau_1 = r_1/c$ account for the distance attenuation and propagation delay for the virtual source at distance r_1, respectively. Similarly, let θ_2 and $h(\theta_2, t)$ denote the current parameter and corresponding HRIR, respectively. The synthesized binaural signals of the current virtual source position are provided by:

$$e_2(t) = \frac{1}{r_2} h(\theta_2, t) * e_0(t - \tau_2),$$
(12.6)

where r_2 is the current virtual source distance and $\tau_2 = r_2/c$ is the propagation delay.

In output cross-fading, the system output is a time-variable weighted sum of the two-parallel processing of Equations (12.5) and (12.6) thus:

$$e(t) = \lambda(t) e_2(t) + [1 - \lambda(t)] e_1(t).$$
(12.7)

When weight $\lambda(t)$ continuously varies from zero to a unit, the system output cross-fades from $e_1(t)$ to $e_2(t)$. Output cross-fading is straightforward, but it entails twice the computational cost during transition. Moreover, the cross-faded output $e(t)$ does not strictly correspond with cross-faded scenario parameters.

In parameter cross-fading, the signal processing parameter θ (including source direction, distance, and propagation delay) is constantly updated. The system output is given by:

$$e(t) = \frac{1}{r(t)} h[\theta(t), t] * e_0[t - \tau(t)]$$
$$\theta(t) = \lambda(t)\theta_2 + [1 - \lambda(t)]\theta_1 \quad r(t) = \lambda(t)r_2 + [1 - \lambda(t)]r_1.$$
$$\tau(t) = \lambda(t)\tau_2 + [1 - \lambda(t)]\tau_1$$
(12.8)

When weight $\lambda(t)$ continuously varies from zero to a unit, the parameters and system output cross-fade from the past to the current one.

If parameter θ denotes the horizontal azimuth of the virtual source relative to the listener, then using Equation (12.8) is equivalent to changing the horizontal azimuth of HRIRs. Because measurement usually yields HRIRs at discrete directions, the interpolation or basis function decomposition discussed in Chapter 6 is required to derive the parametric representation of HRIRs at arbitrary azimuth θ. From this point of view, the basis function decomposition-based algorithms for binaural synthesis are convenient for parameter updating. The output cross-fading scheme provided by Equation (12.7) is analogous to pairwise signal mixing (panning) for multi-channel sound, which is also equivalent to the adjacent linear interpolation of HRTFs (Section 6.4).

The update rate of cross-fading parameters should be high enough that the artifacts generated by parameter updating are inaudible. Different parameters require different update rates. Some parameters, such as the propagation delay, require the updating of every sample to avoid audible artifacts, particularly to maintain the Doppler frequency shift for a fast-moving source (Section 12.3). In addition, the appropriate time length for cross-fading from one scenario to another is usually larger than several milliseconds, depending on the stimuli and the movements of a listener.

These cross-fading methods for free-field virtual sources can be extended to cross-fading of image sources for reflections. As stated in Section 12.2.3, image-source methods simulate the reflections by multiple free-field sources with different propagation delays and distance attenuations.

The second problem of interest in dynamic rendering is scenario update rate. In real environments, binaural pressures and auditory scenarios synchronously and smoothly vary with head movement. A dynamic VAE updates the binaural signals and auditory scenarios at certain time intervals. The *scenario update rate* of a VAE refers to the number of update scenario manipulations per second, which should not be confused with the parameter update rate in parameter cross-fading. The higher the scenario update rate, the closer the auditory perception to the real environment, but a higher computational cost is incurred. Given the limitations in the available capacity of a system, some trade-offs have been made in existing dynamic VAE systems on the basis of psychoacoustic rules. The psychoacoustic experiment conducted by Sandvad (1996) indicates that a scenario update rate of 10 Hz or less degrades the localization speed; a scenario update rate of 20 Hz hardly influences the localization speed, although audible artifacts may occur during moderate to fast head movements.

The third problem of interest in dynamic rendering is system latency time. When the head moves, the synthesized binaural signals in existing VAEs do not change in a synchronous manner, but with delays. *System latency time* is defined as the time from which a listener's head moves to the time at which the corresponding change in the synthesized binaural signal output occurs. System latency is contributed from a series of factors, including latency in the response of the head tracker, data transmission and communication, time required for update and signal processing, and data buffer. System latency is determined by hardware and software structures, and appears to be an inevitable occurrence. In practice, system latency time should be reduced with appropriate hardware and software designs to generate a perceptually tolerable time.

Psychoacoustic experiments have been conducted to evaluate the effect of system latency time on the perceptual performance of dynamic VAEs from two perspectives. The first is the influence on virtual source localization, including the accuracy and the listener's response time in localization. The second is the audible artifacts caused by system latency. These experiments yielded varied results. Some results indicate that a system latency time of 150 ms (Bronkhorst, 1995) or 500 ms (Wenzel, 1997, 1999; Yairi et al., 2008) slightly influences source localization. Other findings (Sandvad, 1996) demonstrate that a system latency time of 93 ms causes obvious localization errors and degrades localization speed. Results typically depend on the time length of stimuli. For a short stimulus with a large latency time, the listener may not have time to move his/her head to obtain dynamic information. Brungart et al. (2006) indicate that a system latency time larger than 73 ms increases localization errors for short stimuli and increases a listener's localization response time for continuous stimuli. The perceivable threshold of system latency time for isolated stimuli and best listening is 60–70 ms. The threshold decreases to 25 ms when accompanied with a low-latency reference tone. Brungart also suggested that a system latency time lower than 60 ms is adequate for most applications, and a latency time less than 30 ms is difficult to perceive in highly demanding VAEs.

12.2.5 Dynamic Crosstalk Cancellation in Loudspeaker Reproduction

Dynamic binaural signals can also be reproduced through loudspeakers with appropriate crosstalk cancellation. As shown in Chapter 9, static crosstalk cancellation is sensitive to head position. For a given loudspeaker configuration, crosstalk cancellation is effective only at the default listening position (sweet point). In addition to dynamic binaural synthesis, therefore, dynamic crosstalk cancellation, in accordance with the temporary position of a listener's head, is required in a dynamic loudspeaker-based VAE.

The two-front loudspeaker reproduction in Figure 9.1 provides an illustrative example. Let H_{LL} = $H_{LL}(\theta_L, f)$, $H_{RL} = H_{RL}(\theta_L, f)$, $H_{RR} = H_{RR}(\theta_R, f)$, and $H_{LR} = H_{LR}(\theta_R, f)$ denote the transfer functions from loudspeakers to ears in the default head position and orientation. The crosstalk cancellation matrix is given by Equation (9.6). For simplicity, the head movement is restricted to within the horizontal plane. In this case, the head can move in three degrees of freedom, including rotation around the vertical axis and translation in the horizontal plane.

When the head rotates around the vertical axis with an azimuth $\Delta\theta$ (a positive $\Delta\theta$ denotes clockwise rotation), the distance of two loudspeakers relative to the head center is fixed, but the directions of two loudspeakers relative to the head change. In this case, the transfer functions from loudspeakers to two ears become $H'_{LL} = H_{LL}(\theta_L - \Delta\theta, f)$, $H'_{RL} = H_{RL}(\theta_L - \Delta\theta, f)$, $H'_{RR} = H_{RR}(\theta_R - \Delta\theta, f)$, and $H'_{LR} = H_{LR}(\theta_R - \Delta\theta, f)$. Therefore, dynamic crosstalk cancellation requires the updating of HRTFs in Equation (9.6), that is, substituting the HRTFs in Equation (9.6) with H'_{LL}, H'_{RL}, H'_{RR}, and H'_{LR}.

Head translation alters both the directions and distances of two loudspeakers relative to the head. The case of lateral translation is depicted in Figure 9.6. After head translation, the directions and distances of two loudspeakers relative to the head become (θ'_L, r'_L) and (θ'_R, r'_R), which are calculated similarly to Equation (12.1). Dynamic crosstalk cancellation requires the updating of the HRTFs in Equation (9.6) in terms of the parameters of new loudspeaker positions relative to the head. Dynamic crosstalk cancellation is complex in near-field loudspeaker distances with r'_L or r'_R < 1.0 m because the distance dependence of near-field HRTFs requires consideration. In far-field loudspeaker distances with r'_L and r'_R > 1.0 m, dynamic crosstalk cancellation can be implemented using distance-independent far-field HRTFs, but appropriate delay and attenuation should be incorporated to account for the variations in the loudspeaker distances r_L and r_R. Recursive structures similar to those discussed in Section 9.6.3 are convenient for the practical implementation of dynamic crosstalk cancellation. Especially for the interaural transfer function (ITF) elements in a crosstalk canceller [Figure 9.13 and Equation (9.81)], the ITD included in ITF can be approximated by some of the parameter equations discussed in Section 7.2.1. Accordingly, updating the ITD included in ITF is implemented by updating the parameter in the ITD equation. As with the problems encountered in dynamic binaural synthesis, the auditory continuity of scenario updates, scenario update rates, and system latency times should be considered in dynamic crosstalk cancellation (Mannerheim et al., 2006).

Gardner (1997) comprehensively investigated dynamic crosstalk cancellation for two-loudspeaker reproduction and proposed some implementation schemes. The results of psychoacoustic experiments confirm that dynamic binaural synthesis and crosstalk cancellation can recreate virtual sources that are stable against head translation and rotation. Moreover, the back-front confusion rate in localization for dynamic processing is 50.7%, in contrast to 91.4% for conventional static processing. Therefore, dynamic processing decreases the back-front confusion in reproduction with two frontal loudspeakers.

As indicated in Equation (9.6), crosstalk cancellation is stable and realizable only when the acoustic transfer matrix from loudspeakers to two ears is non-singular and therefore invertible. Crosstalk cancellation should also be causal. For two-front loudspeaker configuration, crosstalk cancellation is realizable (at least approximately) for a frontal head orientation. When the head rotates around the vertical axis with an azimuth $\Delta\theta$, the acoustic transfer matrix from loudspeakers to two ears changes and crosstalk cancellation is not always realizable. Gardner (1997) confirmed that crosstalk cancellation is impossible for a rotation with $|\Delta\theta| = 90°$. The HRTFs are front-back symmetric for a spherical head with two ears located diametrically opposite the head surface. Even human HRTFs are approximately front-back symmetric below 1 kHz. The symmetric HRTFs generate the singular and, therefore, invertible acoustic transfer matrix $[H]$ in Equation (9.6) when $|\Delta\theta| = 90°$. The analysis here is closely related to the discussion in Section 6.4.2. When a pair of loudspeakers is front-back symmetrically located at the side, pairwise signal mixing in conventional multichannel sound cannot recreate lateral virtual sources, according to the summing localization law.

Lentz and Schmitz (2002) also confirmed that, for frontal loudspeaker configuration with an azimuthal span of $2\theta_0 = 90°$, dynamic crosstalk cancellation deteriorates when $|\Delta\theta|$ reaches $45°$ (i.e., the listener faces one loudspeaker). To enable dynamic crosstalk cancellation for a horizontal 360°-rotation listener, Lentz (2006b) proposed a four horizontal loudspeaker configuration with equal azimuthal intervals. According to the head orientation, the loudspeaker pair, toward which a listener faces, is selected as the active loudspeakers for obtaining optimal crosstalk cancellation. The measurement indicates that crosstalk cancellation exhibits a mean channel separation of 20 dB across a frequency range of 100 Hz to 12 kHz (Lentz et al., 2002, 2005). Four horizontal loudspeaker configuration has been previously proposed (Krebber et al., 2000), but the study lacked a detailed analysis or sufficient experimental results.

Similar to dynamic binaural synthesis and crosstalk cancellation, adaptive binaural synthesis and crosstalk cancellation are applicable to virtual 5.1-channel surround sound reproduction through two frontal loudspeakers. As stated in Section 10.4, the listening region for the virtual reproduction of 5.1-channel surround sound is limited. Kim et al. (2008) proposed an adaptive virtual surround sound system for wide screen TVs. When a listener moves to a new position, the remote-controlled infrared and ultrasonic tracking system detects the listener's position, and then the optimal listening point is steered toward the listener's new position. However, the processing only adapts to the new position when the listener provides instructions via remote control. This scheme expands the listening region, but it is inapplicable for the simultaneous treatment of multiple listeners.

12.3 Simulation of Moving Virtual Sources

In the preceding discussions, the virtual sources are assumed static. To simulate real environments with moving sources, however, the case of moving virtual sources (sometimes accompanied with a moving listener) should be taken into account in a complete dynamic VAE.

In the free field, source movement alters the source distance and direction relative to the listener. In a reflective environment, source movement also alters reflections. If the reflections are modeled by image-source methods, the position and visibility of image sources relative to the listener vary with the position of the direct source. Accordingly, moving virtual source simulation should constantly update signal-processing parameters. For example, the HRTF filters, propagation delays, and scaling factors for distance attenuation (among others) in the signal-processing module of Figure 12.3 should be constantly updated, according to prior knowledge of the moving source trajectory.

Similar to the case of a moving listener in Section 12.2, a spatially continuous HRTF is, in principle, needed for moving virtual source simulation. In practice, a moving virtual source is usually simulated by constantly loading and updating the HRTFs at discrete positions. If the difference between two switched HRTFs is sufficiently small, a perceptually continuous moving virtual source is rendered. Otherwise, some audible artifacts may be created.

Therefore, a database of spatially dense measured or interpolated HRTFs is needed for moving virtual source simulation. Alternatively, a scheme similar to the output cross-fading in Section 12.2.4 is applicable to creating moving virtual sources. This scheme is equivalent to HRTF interpolation, or panning, in conventional multi-channel sound, but it increases the computational cost in moving virtual source simulation. Some schemes similar to partition convolution were also proposed for moving virtual source simulation (Matsumoto and Tohyama, 2003).

Similar to the case of the dynamic binaural synthesis for a moving listener, the various basis function decomposition-based algorithms of HRTFs discussed in Section 6.5.2 are applicable to moving virtual source simulation. Signal processing, similar to that shown in Figure 6.12, is implemented. Evans (1998b) proposed an SH decomposition-based algorithm for moving virtual source simulation. In far-field cases (a similar algorithm is valid for near-field cases), HRTFs can be decomposed as a weighted sum of the continuous basis functions [Equations (6.14) and (6.57)]. The spatial trajectory of a moving virtual source is described by the parameter equation:

$$r = r(t_1) \qquad \theta = \theta(t_1) \qquad \phi = \phi(t_1). \tag{12.9}$$

Substituting Equation (12.9) into Equation (6.14) or Equation (6.57) yields the parametric representation of decomposed HRTFs. Then, the binaural signals for a moving virtual source in the free field are synthesized by filtering the input stimulus E_0 with decomposed HRTFs and by incorporating the time-variable distance attenuation $r(t_1)$ and propagation delay $\tau = r(t_1)/c$:

$$E = \frac{E_0}{r(t_1)} H \left[\theta(t_1), \phi(t_1), f \right] \exp\left[-j2\pi f \frac{r(t_1)}{c} \right]. \tag{12.10}$$

The notation for the left or right ear is excluded in Equation (12.10). Transmission from a moving source to an ear is a linear-time-variable course. Equation (12.10) is interpreted as the approximation of the short-period Fourier transform of binaural signals in the time domain, with t_1 as a time parameter.

The Doppler frequency shift should be considered in simulating a fast-moving virtual source (more strictly, a fast relative movement between source and listener; Krebber et al., 2000). As shown in Figure 12.5, source S moves at velocity v_S, and listener L moves at velocity v_L. The projections of vectors v_S and v_L to the line that connects the source and listener are v_{S1} and v_{L1}, respectively. For a sound radiation with the static frequency f_0, the relative movement between the source and listener creates a received frequency at the ears as:

$$f = \frac{c + v_{L1}}{c - v_{S1}} f_0 = K f_0 \qquad K = \frac{c + v_{L1}}{c - v_{S1}} \qquad c > v_{S1}. \tag{12.11}$$

The Doppler frequency shift can be implemented by changing (up/down) the sampling frequency of input stimuli, a task equivalent to applying instantaneous linear frequency modulation to input stimuli. Therefore, a complete scheme for rendering

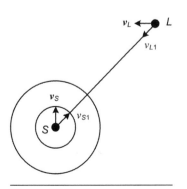

Figure 12.5 Source movement relative to the listener

fast-moving virtual sources involves two steps. The first is simulating the Doppler frequency shift, and the second is dynamic HRTF-based synthesis. The Doppler frequency shift should be calculated prior to HRTF-based synthesis.

The propagation delay in Equation (12.5) should be constantly updated according to the temporary source position relative to the listener's head. The update rate of propagation delay (as well as ITD) should be high enough to maintain the effect of the Doppler frequency shift embedded in the change in time delays. In a digital system with sampling frequency 44.1 kHz (sampling interval, 22.7 μs), the propagation delay should be updated for every sample. Manipulating the propagation delay update and the Doppler frequency shift are often combined in practical processing. Updating the propagation delay at high rates, according to the distance between the source and listener, is equivalent to changing the sampling frequency of input stimuli, which automatically realizes the Doppler frequency shift. Iwaya and Suzuki (2007) divided a pair of HRIRs into a pair of main responses and initial delays, and then simulated the Doppler frequency shift by separately changing the delays for the left and right ears.

Moreover, the image sources for reflections move at different directions relative to the listener. Thus, the Doppler frequency shift should be individually processed for each direct and image source. Here, only the simplest case of the Doppler frequency shift is presented. More complex cases can be found in the literature (Strauss, 1998; Jenison et al., 1998).

12.4 Some Examples of Dynamic VAE Systems

Some dynamic VAE systems have been developed for various purposes and applications. In these systems, early reflections are usually modeled by image-source methods and realized by the signal processing modules similar to that shown in Figure 12.3. Late reflections are usually simulated by various artificial reverberation algorithms.

The Sound Laboratory (SLAB) developed by NASA (the National Aeronautics and Space Administration) is intended to provide a platform for psychoacoustic study (Wenzel et al., 2000; Miller and Wenzel, 2002; Begault et al., 2010). It is a software-based system written in VC++ language, and is implemented on a PC or server using Microsoft Windows. Through an application-programming interface, SLAB provides access to different psychoacoustic studies. In SLAB, the dynamic virtual auditory environment caused by moving sources can be rendered in real time. The source radiation pattern, sound propagation, environment reflection and absorption, and air absorption can also be simulated. Six of the first-order reflections are modeled by image-source methods. The maximum achievable number of virtual sources depends upon the available computational ability of a system (typically four CPUs, or central processing units). The typical scenario update rate is 120 Hz. Excluding the external latency caused by a head tracker, the internal system latency time is 24 ms. The binaural signals are reproduced through headphones. The SLAB system has been updated several times (http://humanfactors.arc.nasa.gov/SLAB). The latest version also supports the use of individualized HRTFs (Begault et al., 2006).

The Digital Interactive Virtual Acoustics (DIVA) system developed by Helsinki University of Technology (which is now merged as Aalto University) is intended for research on multimedia and virtual reality (Saviojia, et al., 1999; 2002). It is implemented in workstations with a UNIX operating system and software written in C++ language. DIVA can render the virtual auditory environment caused by static or moving sources in real time, including the simulation of source radiation patterns, sound propagation, environment reflection and absorption, and air absorption.

The number of virtual sources depends upon the available computational ability of a system. In first- and second-order reflection modeling, image-source methods are employed, and artificial reverberation algorithms are used in late reflection modeling. The typical scenario update rate is 20 Hz, and the typical system latency time varies from 110–160 ms. Finally, the binaural signals are reproduced through headphones.

Ruhr-Universität Bochum in Germany initiated VAE research in the mid-1980s. Within the scope of the European ESPRIT Research Project, the first-generation system (named SCATIS, Spatially Co-ordinated Auditory/Tactile Interactive Scenario) was developed from 1993 to 1996 (Blauert et al., 2000; Djelani et al., 2000). The system is intended for psychoacoustic and virtual reality research, and is implemented in UNIX workstations. The signal-processing scheme is similar to aforementioned systems, except that 40 24-bit fixed-point digital signal processors, or DSPs (Motorola DSP56002), are used for real-time auralization processing. Individualized HRTFs are incorporated into processing. Each virtual source is synthesized by two 80-point FIR filters. The system can simultaneously render 32 virtual sources. The scenario update rate is 60 Hz, and the system latency time is 66 ms.

Ruhr-Universität Bochum began research on the second-generation system (named IKA-SIM) in 1998 (Silzle et al., 2004). This system is software-based and uses Microsoft Windows and software written in C++ language. No additional DSPs are needed. The system provides three selections—low, middle, or high—for FIR filter length. For example, the length of HRTF filters can be 64, 96, or 128 points. Under standard computer specifications (Intel P4 CPU with 2.6 GHz) and using the middle filter length, the system can simultaneously render 40–60 virtual sources (direct or image sources). The scenario update rate is 60 Hz, and the average system latency time is 80 ms.

The system developed by the Acoustic Lab at South China University of Technology is intended for psychoacoustic research. It is also a PC and C++ language-based system (Zhang and Xie, 2012). The system detects and simulates dynamic information for six degrees of freedom of head movement, and can synthesize both far- and near-field virtual sources at various distances and directions, as well as early reflections. Two alternative algorithms are used for dynamic binaural synthesis. The first is the conventional HRTF-based algorithm, and the second is the PCA-based algorithm, in which HRTFs at various directions and distances are decomposed by PCA (principal component analysis), and binaural synthesis is implemented by a parallel bank of 16 filters (Section 6.5.2; Xie and Zhang, 2012a). The system can simultaneously render up to 280 virtual sources (direct or image) using the conventional algorithm, or 4500 virtual sources using the PCA-based algorithm. The scenario update rate is 120 Hz and the system latency time is 25.4 ms. The improved system performance is attributed to improvements in both PC performance and signal-processing algorithms.

Distinguished from most other systems, the system developed by RWTH (Rheinisch-Westfaelische Technische Hochschule) Aachen Universität in Germany is designed for loudspeaker reproduction (Lentz et al., 2006a). Dynamic crosstalk cancellation (Section 12.2.5) is incorporated into the processing. Distance-dependent near-field HRTFs are used in processing to render near-field virtual sources. The system latency time is 31.7 ms.

The structures of most other dynamic VAE systems for research purposes are similar to those of the aforementioned systems, except in a few cases (e.g., ADVISE, discussed in Section 12.2.3). Some dynamic VAE systems are also based on special hardware platforms. These systems have considerable power for VAE real-time processing and are especially suitable for rendering multiple virtual sources, but they are expensive. Other PC-based or DSP-based commercial dynamic VAE products are intended for consumer electronic applications, such as 3-D (three-dimensional) games. The basic principle of these systems is similar to that of the systems designed for research purposes, but the structures and algorithms of the former are considerably simplified, given the relatively large allowable errors in consumer electronic applications.

12.5 Summary

A realistic virtual auditory environment should be an interactive, dynamic, and real-time rendering system. Aside from synthesizing auditory information on sources and environments, it also renders the dynamic auditory information generated by source or listener movement.

Most dynamic virtual auditory environment systems are similar in basic principle and structure, and generally consist of three parts: information input and definition, dynamic virtual auditory environment signal processing, and reproduction. A listener's head movement alters the virtual source position relative to the head. A head tracker detects the listener's head position and orientation in real time, on whose basis the system implements corresponding dynamic signal processing. Incorporating head tracking and dynamic processing improves the performance of virtual auditory environments in relation to three aspects: virtual source localization, subjective sense of realism, and listening region of loudspeaker reproduction.

Direct rendering is applicable to free-field virtual sources, in which HRTFs are loaded and updated according to the temporary virtual source position relative to the listener's head. Parametric schemes can be introduced to alleviate computational load, in which the distance dependence of far-field virtual sources can be simply modeled by distance-dependent propagation delay and attenuation. Direct rendering is, in principle, also applicable to dynamic room or environment reflection synthesis, in which the binaural room impulse responses for binaural synthesis are constantly updated, according to the temporary position and orientation of heads and sources from a prepared database. However, direct reflection rendering requires enormous memory for the storage of binaural room impulse responses. Alternatively, parametric rendering schemes are often adopted for dynamic environment or room reflection synthesis. The issues concerning implementing dynamic crosstalk cancellation are similar to those for implementing virtual source rendering.

For a smooth transition from one auditory scenario to the next, cross-fading methods are incorporated into signal processing. The two kinds of cross-fading methods are output cross-fading and parameter cross-fading.

Scenario update rate and system latency time are two other important system parameters in system performance evaluation. Although psychoacoustic experiments yield various results, a scenario update rate larger than 20 Hz and a latency time lower than 60 ms are adequate for most applications.

The case of moving virtual sources, or both moving virtual sources and moving listeners, should be taken into account for a complete dynamic virtual auditory environment. The Doppler frequency shift should also be simulated in rendering fast-moving virtual sources. Dynamic virtual auditory environment systems have been developed for various purposes and applications.

13

Psychoacoustic Evaluation and Validation of Virtual Auditory Displays (VADs)

This chapter presents the methods for psychoacoustic evaluation and validation of virtual auditory display (VAD). The conditions required in psychoacoustic experiments for VAD evaluation are discussed. Some commonly used experimental methods, including auditory comparison and discrimination experiments, virtual source localization experiments, and quantitative evaluation experiments on grading certain subjective attributes, are addressed. The statistical methods for analyzing psychoacoustic experimental results are outlined. Binaural auditory models used to objectively assess VAD are also discussed.

Validating the actual effects of VADs and related signal processing schemes necessitates the performance evaluation of the schemes. Similar to the evaluation of conventional electroacoustic systems, objective and subjective assessments are two major methods of evaluation. Because a VAD creates a spatial auditory event by synthesizing and rendering binaural signals, the most straightforward method for objective assessment is to compare the synthesized and baseline/reference binaural signals, as well as related localization cues, and subsequently analyze errors. Error analysis for the HRTF (head-related transfer function) approximations in Section 5.1 falls into this category. However, given that the performance of VADs is closely related to the psychoacoustic and physiological acoustic aspects of human hearing, a purely physical or mathematical evaluation method cannot fully reflect actual auditory perception. A possible solution is to introduce the psychoacoustic and physiological acoustics factors into the objective assessment of binaural signals. Some preliminary attempts have been undertaken, but further efforts are needed. Because the psychology and physiology of hearing are highly complex issues that are influenced by many factors, state-of-the-art objective assessment is insufficient to accurately reflect actual auditory perception. At the current stage, therefore, the subjective assessment and validation of VADs by psychoacoustic experiments are extraordinarily important. A variety of psychoacoustic experiment methods that correspond to different contexts have been proposed.

This chapter discusses the psychoacoustic evaluation and validation of VADs. Section 13.1 focuses on the conditions required in psychoacoustic experiments for VAD evaluation. Section 13.2 presents the principle of auditory comparison and discrimination experiments, along with representative examples. Section 13.3 discusses the important principles of virtual source localization experiments and relevant results. Quantitative evaluation experiments on grading certain subjective attributes are presented in Section 13.4. Section 13.5 outlines the statistical methods used to analyze psychoacoustic experimental data, with some typical results provided. Finally, Section 13.6 presents an objective assessment method for VADs, in which binaural auditory models based on psychoacoustic and physiological acoustic principles are incorporated.

13.1 Experimental Conditions for the Psychoacoustic Evaluation of VADs

The psychoacoustic evaluation of VADs involves assessing the performance of VADs in accordance with a subject's auditory perception. Assessing the subjective auditory perception of sound is part of experimental psychology. Reliable results can only be obtained with strict psychoacoustic experimental methods and conditions, as well as appropriate statistical analyses. Commercial promotion and amateurish application result in considerable misunderstanding of the subjective evaluation of sound reproduction. For example, conclusions are often drawn from informal listening tests, which cannot be applied to scientific experimental studies on VADs.

No consensus or standard for the assessment of VADs—with respect to evaluation contents, methods, experimental conditions, as well as data processing—has been reached. The International Telecommunication Union (ITU) developed a subjective assessment standard for multi-channel surround sound (ITU-R, BS.1116-1, 1997), but the standard focuses only on the timbre impairment that occurs in the low-bit rate (compressive) coding of multi-channel signals, rather than on VADs. Nevertheless, the standard can be referred to in some VAD evaluations, particularly in evaluating the virtual reproduction of stereophonic and multi-channel surround sound (Chapter 10).

Formulating clear aims and objectives is important in designing psychoacoustic experiments for evaluating VADs. Two schemes are typically used for evaluation. One is comparing VADs with a given reference (e.g., a free-field real sound source) to evaluate the realism, or similitude, of the auditory perceptions generated by VADs. The other scheme involves directly evaluating a subject's preferences for a specific auditory perception created by VADs. For example, the senses of immersion in virtual acoustical environments, which may not necessarily exist in natural environments, are assessed. Thus, experimental design and content vary depending on application and object. Commonly used evaluation contents include overall perceived similitude to specific references, the perceived direction and distance of virtual sources, and the naturalness and realism of auditory events, timbre, and sense of immersion. Evaluation content is selected on the basis of practical requirements.

Psychoacoustic experiments for evaluating VADs should be conducted within certain acoustical environments. For example, headphone-based experiments can be conducted in an ordinary room because reflections from surroundings hardly influence auditory perception. However, an anechoic chamber is needed when the auditory perceptions of free-field virtual sources and real sound sources are compared in an experiment. For loudspeaker reproduction, experiments should, in principle, be conducted in an anechoic chamber to avoid room reflections. In some cases, especially in evaluating virtual multi-channel surround sound (see Section 10.4), experiments can also be carried out in listening rooms with appropriate surface absorption, as in the case of multi-channel surround sound reproduction. The recommended listening room in ITU-R BS.1116-1 for multi-channel sound reproduction has an area of 30–70 m^2 and a mean reverberation time across 0.2–4 kHz, expressed as:

$$\overline{T}_{60} = 0.25\left(\frac{V}{V_0}\right)^{1/3},\tag{13.1}$$

where $V_0 = 100 \text{ m}^3$ and V is the volume of the listening room. The standard also allows for the acceptable deviation of reverberation time at different frequency ranges.

Whether reproduction is conducted in an anechoic chamber or listening room, less background noise is desired to avoid the disturbance caused by such noise. As recommend by ITU-R BS.1116-1, an indoor background noise of no more than NR (noise rating) 10 is preferred. The recommended upper limit is NR 15.

The stimuli selected for psychoacoustic experiments are also related to evaluation contents, methods, objects, and applications, among other factors. For example, in evaluating the performance of VAD systems with respect to rendering virtual sources at different spatial directions, broadband stimuli—such as white noise, pink noise, and pulse—are often employed. These stimuli enable the inclusion of a variety of localization cues at different frequency ranges. In evaluating room auralization, anechoic-recorded music and speech stimuli are frequently used. For application fields, such as speech communication, speech should be the natural choice. Although no specific standards have been established for stimuli selection, chosen stimuli should comprehensively reflect the characteristics, or reveal the defects, of evaluated objects.

In psychoacoustic experiments, the sound pressure (level) presented by VADs should be modest, to avoid excessively weak pressures that affect assessment or excessively strong pressures that cause auditory fatigue and discomfort. The types of sound pressures commonly measured include the free-field pressure at the head center with the head absent, the pressure at the entrance to the blocked or open ear canal, and the pressure at the eardrum.

Given the diffraction and scattering of the head and transmission characteristics of the ear canal, the sound pressures measured at different reference points vary. In ear canal and eardrum measurements, the sound pressures of the left and right ears are also different and vary with target virtual source positions, frequencies, and characteristics of individual subjects. In some studies, the measured binaural sound pressures P_L and P_R are converted into the free-field pressure P_0 at the head center by dividing corresponding HRTFs [Equation (1.15)]. Various weighted sound pressure levels, such as linear, or A-weights, are also used in different studies. No clearly established standards for the presented sound pressure levels in evaluating VADs currently exist, but a P_0 (or P_L and P_R) of 60–70 dB or dB(A) is usually selected in the majority of extant literature.

The subjects recruited for psychoacoustic experiments should have normal hearing. Young people are typically chosen, to avoid the high-frequency hearing loss associated with increasing age. More strictly, each subject should be assessed by hearing screening prior to psychoacoustic experiments. For example, some researchers used a hearing loss of no more than 10 and 15 dB within a frequency range of 250 Hz to 4 kHz or 8 kHz, respectively, as a criterion for subject selection (Hoffmann and Møller, 2006a).

A sufficient number of subjects are needed to guarantee valid statistical results. In theory, the higher the number of subjects, the more reliable the results obtained. However, the increase in the number of subjects results in complex experiments. No clear standards have been established for the number of subjects needed for psychoacoustic experiments on VADs. ITU-R BS.1116-1 (for evaluating multi-channel sound) indicates that 20 subjects are sufficient, but this standard is made on the basis of experience. In the literature, most experiments involve as few as 4 to 12 subjects. Strictly speaking, an equal number of male and female subjects should be chosen because of statistical gender differences in anatomical shape and size, as well as resultant HRTFs and related sound localization cues. This uniformity enables the acquisition of unbiased results, an issue that has, nonetheless, been neglected in most studies. To expand experimental data samples, repeated experiments on each condition and each subject are often conducted. Experiments with more contents should be carried out in separate stages to avoid the hearing fatigue caused by lengthy experiments. Moreover, each trial should not be less than half an hour with break intervals longer than the experimental period. Prior to formal experiments, training is necessary to familiarize subjects with evaluation methods and contents.

The raw data for psychoacoustic experiments are collected from all subjects and repeated experiments. From the perspective of statistics, therefore, such experiments are subject to certain statistical rules on random variables. The experimental data should be analyzed further by appropriate statistical methods to yield results of a certain confidence level. This crucial issue is discussed in Section 13.5.

13.2 Evaluation by Auditory Comparison and Discrimination Experiment

13.2.1 Auditory Comparison and Discrimination Experiment

Auditory comparison and discrimination experiments are a typical approach to evaluating VADs. These are types of qualitative evaluations, in which VAD performance is assessed through the overall subjective comparison of target reproduction signals (or reproduction methods) and reference signals under controlled experimental conditions. Some commonly employed experimental design methods are described in what follows.

The A/B comparison experiment is the simplest, in which a subject evaluates whether a difference exists between samples A and B, which are randomly rendered by turns. More strictly, one of the following auditory discrimination and forced-choice paradigms are used:

1. The two-interval, two-alternative forced-choice (**2I/2AFC** for short) paradigm consists of two signal segments, in which signals A and B are arranged in random order, leading to two different combined signal sequences AB and BA. Subjects determine which segment in the combined signal contains target (known) perception attributes; if the subjects are unable to identify the segment, forced choices that are presented in a random manner are needed. Repeated evaluations by each subject with equal repetitions of sequences AB and BA are often necessary.

2. The three-interval, two-alternative forced-choice (**3I/2AFC**) paradigm includes three signal segments, in which the first segment is always reference signal A, followed by signal A and target signal B in random order. This order yields two different combined signal sequences, AAB and ABA. Subjects determine whether the second or third segment is different from (or identical to) the first segment; if the subject is unable to identify the segment, randomly presented forced choices are needed. Repeated evaluations by each subject with equal repetitions of sequences AAB and ABA are often necessary.

3. The three-interval, three-alternative forced-choice (**3I/3AFC**) paradigm includes three signal segments, in which two segments are classified as signal A and the remaining segment is signal B, generating three different combined signal sequences AAB, ABA, and BAA. Subjects determine which segment in the combined signal differs from the other two segments; should the subjects be unable to distinguish the required segment,

randomly presented forced choices are needed. Frequently, each subject is asked to repeatedly evaluate the signal segments at equal repetitions of each combined sequence.

4. The four-interval, two-alternative forced-choice (**4I/2AFC**) paradigm includes four signal segments, in which the first and fourth segments are always denoted as signal A and the second and third segments are A or B in random order, generating two different combined signal sequences AABA and ABAA. Subjects determine whether the second or third segment is different from (or identical to) the first (or fourth) segment; forced choices are randomly presented upon failure to identify the required segment. Each subject is also often asked to provide repeated assessments at equal repetitions of each combined sequence.

5. The four-interval, three-alternative forced-choice (**4I/3AFC**) paradigm includes four signal segments, in which the first segment is always signal A followed by B and two repetitions of A in random order, producing three different combined signal sequences AAAB, AABA, and ABAA. Subjects determine which of the second, third, or fourth segment of the combined signal differs from the first; if the subjects cannot identify the required segment, forced choices are randomly presented. Each subject is asked to perform repeated evaluations at equal repetitions of each combined sequence.

6. The four-interval, one oddball, two-alternative forced-choice (**4I/1O/2AFC**) paradigm is a deformation of the 4I/2AFC paradigm. 4I/1O/2AFC includes four signal segments with combined sequences AABA, ABAA, BABB, and BBAB, in which three out of the four segments are identical, and one (either the second or third) differs from the other three. Subjects determine which of the second or third segment in the combined signal varies from the other three; if the subjects are unable to identify the required segment, forced choices are randomly presented. Each subject performs repeated assessments at equal repetitions of each combined sequence.

Other methods, with principles similar to those underlying these six, are available. Given the similarity in principles, such methods can be regarded as variants.

The proportion of correct discriminations is usually calculated to analyze experimental data. Calculation is carried out, either over the repetitions of all subjects, or the repetitions of each subject, depending on experimental purpose and requirement. In the two- and three-alternative forced-choice paradigms, if all the subjects cannot distinguish the target signal and provide random responses, the expected statistical values of the proportion of correct discriminations are 0.5 and 0.33, respectively.

Statistical methods are needed to further analyze and validate experimental data. The most commonly used approach is to test the hypothesis that the proportion of correct discriminations is equal to, or greater than, a specific value under certain significance levels. Let us assume that the result of discrimination is a random variable x, with $x = 1$ representing a correct evaluation and $x = 0$ representing failure. Thus, x obeys $(0,1)$ distribution, where the probability of $x = 0$ is $1 - p$ and the probability of $x = 1$ is p. The probability distribution function is:

$$P(x = k) = p^k (1 - p)^{1-k} \quad k = 0, 1. \tag{13.2}$$

The corresponding expected value and variance are:

$$\mu_0 = p \qquad \sigma^2 = p(1 - p). \tag{13.3}$$

In random selection, we have $p = p_0 = 0.5$ and 0.33 for the two- and three-alternative forced-choice paradigms, respectively.

Suppose that $(x_1, x_2, \ldots x_N)$ denotes N independent evaluations (samples) of the random variable x. Such evaluations may come from N different subjects, N repetitions of a specific subject, or K repetitions for each L subject with $N = L \times K$. The corresponding mean value and standard deviation of the samples are calculated by:

$$\bar{x} = \frac{1}{N}\sum_{n=1}^{N} x_n, \quad \sigma_x = \sqrt{\frac{1}{N-1}\sum_{n=1}^{N}(x_n - \bar{x})^2}. \tag{13.4}$$

Taking advantage of statistical hypothesis testing methods (Marques, 2007) facilitates the evaluation of the discrimination between target B and reference A from actual samples (subjects' evaluations). Usually, hypothesis p as equal to a specific value p_0 is used as the criterion for categorizing evaluation as nondiscrimination, whereas p as larger than another specific value p_1 is the criterion used to classify assessment as discrimination. In the two-alternative forced-choice paradigm, $p_0 = 0.5$ and $p_1 = 0.75$ are usually selected. In the three-alternative forced-choice paradigm, $p_0 = 0.33$ and $p_1 = 0.67$ are typically used. Thus, two statistical hypotheses for the proportion of correct discrimination p are tested at a significance level of α:

1. Testing two-sided hypothesis H_0: $p = p_0$ and alternative hypothesis H_1: $p \neq p_0$. Accepting $p = p_0$ indicates that no audible difference (discrimination) between target B and reference A is detected.
2. Testing one-sided hypothesis H_0: $p \geq p_1$ and alternative hypothesis H_1: $p < p_1$. Accepting $p \geq p_1$ while rejecting $p = p_0$ indicates that audible difference (discrimination) between target B and reference A is detected.

For N independent samples of the random variable x, the probability of q correct evaluations is:

$$P(N,q) = \frac{N!}{q!(N-q)!}p^q(1-p)^{N-q}. \tag{13.5}$$

For the two-alternative forced-choice paradigm with $p = p_0 = 0.5$ in two-sided hypothesis testing, we suppose that N is even for convenience. The overall probability of the correct evaluation number within $(N/2 - n) \leq q \leq (N/2 + n)$ is:

$$P_{all} = \sum_{q=(N/2-n)}^{(N/2+n)} P(N,q). \tag{13.6}$$

Lower and upper bounds are determined by increasing n in Equation (13.6) from $n = 0$ in the range $0 \leq n \leq N/2$ by turns until $n = n_0$, where P_{all} is equal to or greater than $1 - \alpha$. Then, the hypothesis of $p = p_0 = 0.5$ is accepted when the mean correct rate of the samples is within:

$$\bar{x}_{low} < \bar{x} < \bar{x}_{upper}, \quad \bar{x}_{low} = \frac{1}{N}\left(\frac{N}{2} - n_0\right) \quad \bar{x}_{upper} = \frac{1}{N}\left(\frac{N}{2} + n_0\right). \tag{13.7}$$

Failure occurs in the experiment when the mean correct rate of the samples is lower than $\bar{x}_{low}$.

As to the one-sided hypothesis testing of $p \geq p_1$, the overall probability of the correct evaluation number within $n \leq q \leq N$ is:

$$P_{all} = \sum_{q=n}^{N} P(N,q). \tag{13.8}$$

The lower bound is determined by decreasing n in Equation (13.8) from round $(p_1 N)$ in the range $0 \leq n \leq$ round $(p_1 N)$ by turns until $n = n_0$, where P_{all} is equal to or greater than $1 - \alpha$. The one-sided hypothesis of $p \geq p_1$ is then rejected when the mean correct rate of the samples satisfies:

$$\overline{x} \leq \overline{x}'_{low} = \frac{n_0}{N}. \tag{13.9}$$

Moreover, $\overline{x}$ and:

$$u = \frac{\overline{x} - p_0}{\sigma / \sqrt{N}}, \tag{13.10}$$

are random variables. For a large N, variable u approximately obeys a normal (Gaussian) distribution of $N(0,1)$ with zero mean and unit variance. Under this condition, the hypothesis test for the distribution of u at a significance level of α is more convenient.

Let $u_{1-\alpha}$ denote the quantile of $1-\alpha$ of normal distribution $N(0,1)$, which is defined by the integral on probability distribution function $f(\xi)$ $(-\infty < \xi < +\infty)$:

$$\int_{-\infty}^{u_{1-\alpha}} f(\xi)d\xi = 1 - \alpha \quad or \quad \int_{u_{1-\alpha}}^{+\infty} f(\xi)d\xi = \alpha \quad 0 \leq \alpha \leq 1. \tag{13.11}$$

For the two-sided hypothesis test H_0: $p = p_0$, when:

$$u = \left| \frac{\overline{x} - p_0}{\sigma / \sqrt{N}} \right| < u_{1-\alpha/2}, \tag{13.12}$$

hypothesis $p = p_0$ is accepted. For a normal distribution of $N(0,1)$, $u_{1-\alpha/2} = 1.96$ at a significance level of $\alpha = 0.05$.

$$u = \frac{\overline{x} - p_0}{\sigma / \sqrt{N}} < u_{\alpha/2}, \tag{13.13}$$

means failure in the experiment.

For the one-sided hypothesis test H_0: $p \geq p_1$, when:

$$\frac{\overline{x} - p_1}{\sigma / \sqrt{N}} \leq u_{\alpha}, \tag{13.14}$$

hypothesis $p \geq p_1$ is rejected. For a normal distribution of $N(0,1)$, $u_{\alpha} = -1.64$ at a significance level of $\alpha = 0.05$. These calculations can be implemented by statistical analysis software.

Some issues concerning auditory comparison and discrimination experiments are worth noting.

1. The experiments aim to determine whether audible difference exists between reference and target signals, and any perceived difference can be used as a discrimination criterion. In this sense, as many artificial artifacts (introduced in signal processing) should be reduced as possible. For example, the overall level difference between reference and target signals should be avoided.

2. These types of experiments only qualitatively analyze the perceptual difference between reference and target signals. Should further quantificational analysis of specific perceived attributes be necessary, the methods to be discussed in Sections 13.3 and 13.5 would be helpful.

3. From the perspective of mathematical statistics, the number of samples N should be sufficiently large, to enable the derivation of reliable statistical results. A value greater than 50 or 100 is usually preferred, with a low limit of $N \geq 30$. For a limited number of subjects, repetitions are necessary.

13.2.2 Results of Auditory Discrimination Experiments

The methods in Section 13.2.1 can be used to evaluate the perceptual discrimination of head-phone-rendered virtual sources and free-field real sources, with the former classified as target B and the latter as reference A.

Using the 2I/2AFC paradigm, Zahorik et al. (1996) investigated the perceptual equivalence of headphone-rendered virtual sound sources and free-field real sources. The authors conducted an experiment in an anechoic chamber, with small loudspeakers as free-field real sources. The distance of the loudspeakers relative to the subject's head center was 1.4 m. Individualized HRIRs (head-related impulse responses) and HpTFs (headphone-to-ear canal transfer functions), which were measured by a probe microphone positioned at the eardrums under the same experimental layout as in free-field real sources, were used in binaural synthesis and headphone equalization. Target signal B was reproduced through a small cushion headphone to reduce headphone-to-real source hearing interference and to avoid the influence of headphone on microphone position in HpTF measurement. The stimulus was broadband noise within 300–12000 Hz. The virtual sources synthesized from HRIRs at 15 different spatial directions (θ = 180°, 225°, 270°, 315°, and 0° at elevations ϕ = −30°, 0°, and 30° in the default coordinate) and with different HRIR lengths were evaluated. In the experiment, reference and target signals were rendered in random sequence, and the subjects were asked to determine which signal was reproduced from the real loudspeaker. Four subjects participated in the experiment. For each subject and each condition (HRIR length and virtual source direction), 64 repetitions were conducted. The statistical analysis shows that the subjects could not discriminate between reference and target signals when HRIR length was 20.48 ms. Langendijk and Bronkhorst (2000) conducted similar experiments, in which both 2AFC and 4I/1O/2AFC paradigms were employed. The authors drew similar conclusions.

The aforementioned experimental results demonstrate the subjective perception equivalence between headphone-rendered virtual sound sources and free-field real sound sources under strictly controlled experimental conditions. These findings are important experimental validations of headphone-based VADs. However, most of the actual conditions in various applications do not satisfy stringent experimental requirements, resulting in a variety of drawbacks to the headphone-based VADs (Sections 8.3 and 8.4).

Auditory comparison and discrimination experiments can also be used to evaluate various simplifications or approximations of measured HRTFs, if the measured HRTFs are used as reference A and the simplified HRTFs are used as target B. These experiments are extensively used in evaluating the temporal windowing, smoothing, filter design, and spatial interpolation of HRTFs. For example, Kulkarni and Colburn (2004) used the 4I/2AFC paradigm to validate their infinite impulse response (IIR) filter design (Section 5.3.4).

13.3 Virtual Source Localization Experiment

13.3.1 Basic Methods for Virtual Source Localization Experiments

Sound source position is one of the subjective attributes of sound. One of the primary purposes of a VAD is to generate virtual sound sources at different spatial positions in terms of direction and distance. Thus, the ability to generate differentially positioned virtual sources is an important measure of VAD performance. This ability is typically evaluated through virtual source localization experiments (Wightman and Kistler, 2005), the most commonly used methods for evaluating and validating VADs.

A localization experiment involves an appropriate number of subjects determining the perceived spatial position (direction and distance) of headphone- or loudspeaker-rendered virtual sources under certain physical conditions. This method is classified as absolute evaluation, which often requires further statistical analysis of raw results.

In some experiments, in which the static virtual sources in VADs are considered, a subject's head position should be fixed. This requirement is particularly demanding in static loudspeaker-based reproduction, wherein the perceived virtual source position strongly depends on head position. In practice, head position can be monitored by various optic systems, cameras, or head-tracking systems. Stimuli of short durations (e.g., hundreds of milliseconds) are also often employed to avoid possible head movements during presentation. Conversely, in certain experiments, in which the virtual sources in dynamic VADs are considered, head movements are allowed and encouraged. Stimuli of long durations (a few seconds or more) are frequently used to provide subjects enough time to move their heads, through which they can fully utilize the resultant dynamic localization cues.

In a spherical coordinate system (Figure 1.1) or interaural polar coordinate system (Figure 1.3), (r, θ, ϕ) or (r, Θ, Φ) is used to specify source position. To help a subject determine the perceived virtual source position, an appropriate spatial coordinate system is often established in the listening room. Given the limited accuracy of human auditory distance perception, as well as the difficulty of accurately controlling perceived virtual source distances in VADs, most studies emphasize only the determination of the source direction (θ, ϕ) or (Θ, Φ), with a qualitative assessment of source distance, such as within/outside the head. For a virtual source in the horizontal plane, determining its azimuth along the plane is sufficient. For a virtual source in the median or lateral plane, the interaural polar coordinate system is convenient because determining the interaural polar azimuth Θ (for sources located in the upper-half space) or interaural polar elevation Φ of the virtual source is sufficient.

Various methods are used to identify virtual source positions in localization experiments (Wightman and Kistler, 2005). The most straightforward is oral identification (or writing down the perceived source position). Other methods include pointing by hand or laser pointer, clicking by mouse in a graphic interface representation of three-dimensional space in a computer, touching a three-dimensional spherical model, and turning the head toward the perceived source direction in dynamic localization associated with a head-tracking system (Wightman and Kistler, 1989b; Bronkhorst, 1995; Martin et al., 2001; Pernaux et al., 2002). Although some studies have noted that a certain identification method may be more accurate under certain conditions (Pernaux et al., 2003), the more popular view is that every method has advantages and disadvantages, and no good reason drives the choice of preferred method. Alternatively, some researchers predefine a certain number of spatial directions and ask subjects to select the single direction closest to the perceived direction (Møller et al., 1996b). The effectiveness of this method depends upon the interval between predefined directions.

13.3.2 Preliminary Analysis of the Results of Virtual Source Localization Experiments

Obtaining statistical results necessitates preliminary statistical analysis of raw data. The simplest method is to calculate the mean and standard deviation of a perceived source direction in terms of azimuth and elevation. Let us assume that, for a target virtual source direction (θ_S, ϕ_S), N experimental observations (samples) $[\theta_I(n), \phi_I(n)]$ exist; $n = 1,2 \ldots N$, which may be from N different subjects, N repetitions of a subject, or K repetitions for each L subject with $N = L \times K$. Thus, the mean and standard deviation of these experimental samples are:

$$\overline{\theta}_I = \frac{1}{N}\sum_{n=1}^{N}\theta_I(n) \qquad \sigma_\theta = \sqrt{\frac{1}{N-1}\sum_{n=1}^{N}[\theta_I(n)-\overline{\theta}_I]^2}\,,$$

$$\overline{\phi}_I = \frac{1}{N}\sum_{n=1}^{N}\phi_I(n) \qquad \sigma_\varphi = \sqrt{\frac{1}{-1}\sum_{n=1}^{N}[\phi_I(\ )-\overline{\phi}_I]^2}\,,$$

(13.15)

where $\overline{\theta}_I$ and $\overline{\phi}_I$ represent the mean or centric-perceived azimuth and elevation, respectively; and σ_θ and σ_ϕ reflect the dispersion of the sample. Ideally, $\overline{\theta}_I$ and $\overline{\phi}_I$ should equal the target values of θ_S and ϕ_S, respectively. This equivalence means that the samples should be distributed in a diagonal straight line on the plot of function $\overline{\theta}_I$ (or $\overline{\phi}_I$) against θ_S (or ϕ_S). Otherwise, evaluation errors occur. We define:

$$\Delta\theta_1 = \frac{1}{N}|\sum_{n=1}^{N}[\theta_I(n)-\theta_S]| = |\overline{\theta}_I - \theta_S|,$$

$$\Delta\phi_1 = \frac{1}{N}|\sum_{n=1}^{N}[\phi_I(n)-\phi_S]| = |\overline{\phi}_I - \phi_S|,$$

(13.16)

as the deviations (errors) between mean perceived value (azimuth or elevation) and target value, and:

$$\Delta\theta_2 = \frac{1}{N}\sum_{n=1}^{N}|\theta_I(n)-\theta_S| \qquad \Delta\phi_2 = \frac{1}{N}\sum_{n=1}^{N}|\phi_I(n)-\phi_S|,$$

(13.17)

as the mean unassigned (absolute) errors between the perceived value (azimuth or elevation) and target value across samples.

For virtual sources within the horizontal plane, such as in the case of sound reproduction with two frontal loudspeakers, source direction is determined by azimuth θ alone, so that only the error on azimuth θ is analyzed. Similarly, for sound sources within the lateral or median plane, only the error on the interaural polar azimuth Θ or interaural polar elevation Φ is analyzed.

For source distance perception, if merely qualitative evaluations are provided, calculating the ratio or percentage of the perceived virtual sources located within/on the surface or outside the head is sufficient. If quantitative assessments are collected, a statistical analysis similar to that carried out for direction evaluations is needed. Given N samples $r_I(n)$, $n = 1,2 \ldots N$ of perceived source distances that correspond to a target distance r_S, the mean and standard deviation are:

$$\overline{r}_I = \frac{1}{N}\sum_{n=1}^{N}r_I(n) \qquad \sigma_r = \sqrt{\frac{1}{N-1}\sum_{n=1}^{N}[r_I(n)-\overline{r}_I]^2}\,.$$

(13.18)

When the perceived virtual source is close to the horizontal plane, these statistical methods for azimuth and elevation errors are reasonable. In three-dimensional space, however, the arithmetic mean and standard deviation of $[\theta_I(n), \phi_I(n)]$, $n = 1,2 \ldots N$ cannot be directly used to describe the directional dispersion of data because the source directions tested are distributed on the spherical surface whose center is consistent with that of the subject's head. In the horizontal plane of $\phi = 0°$, for example, an actual deviation of $\Delta\theta_1$ or $\Delta\theta_2 = 30°$ is considerably greater from the perspective of absolute direction than at high elevations, such as $\phi = 60°$. This particular problem, coupled with the fact that azimuth and elevation errors are almost certainly not independent of each other, complicates the statistics of perceived source directions.

Wightman et al. (1989b) used spherical statistics to analyze three-dimensional perceived source directions. First, the target direction is represented by a unit vector r_S from the origin to (θ_S, ϕ_S); then, the result of the nth perceived sample is represented by a unit vector $r_I(n)$ from the origin to the perceived direction $[\theta_I(n), \phi_I(n)]$. Similar to the case in Equation (13.17), the mean angular error is defined as the unassigned mean of the angular difference between target and perceived directions:

$$\Delta_2 = \frac{1}{N} \sum_{n=1}^{N} |\arccos[r_I(n) \bullet r_S]|, \tag{13.19}$$

where the dot denotes the scalar multiplication of two vectors. The mean perceived source direction $\bar{r}_I$ is obtained by the sum of all perceived direction vectors thus:

$$\bar{r}_I = \sum_{n=1}^{N} r_I(n). \tag{13.20}$$

The angle between $\bar{r}_I$ and r_S, expressed as Δ_1, reflects the absolute directional deviation between the target and the mean perceived direction, with similar meanings as those indicated in Equation (13.16):

$$\Delta_1 = \arccos\left(\frac{\bar{r}_I \bullet r_S}{|\bar{r}_I|}\right). \tag{13.21}$$

The length of $\bar{r}_I$, that is, $R = |\bar{r}_I|$, indicates the dispersion of perceived source directions. Ideally, N samples of perceived directions should be identical to the target direction; thus, $R = N$. The shorter the length of R, the greater the dispersion observed in the results on perceived directions. We can define κ^{-1} to describe this dispersion. Under small samples with $N < 16$, an unbiased estimate of κ is given by:

$$\kappa = \frac{(N-1)^2}{N(N-R)}. \tag{13.22}$$

Thus, the lower the value of κ^{-1} is, the less the dispersion. When $R = N$, $\kappa^{-1} = 0$.

Conversely, reversal errors (front-back and up-down confusion) may arise for some subjects in virtual source localization experiments, as analyzed in Section 8.3. In this case, the mean perceived source direction in Equation (13.15) or Equation (13.20) is meaningless, and the mean angular error provided by Equation (13.17) or Equation (13.19) may be considerable, or potentially misleading. In the case of a low confusion rate, two approaches are commonly used to treat confusion:

1. prior to processing, confusion is excluded from raw data; and
2. prior to processing, reversal is resolved (i.e., through spatial reflection, the response is coded as though it indicates the correct hemisphere).

After the application of these treatment approaches, counting the confusion rate is necessary to faithfully reflect the characteristics of experimental results. Spatial reflection for sound sources near the mirror plane imposes a low effect on the Δ_2 in Equation (13.19) while increasing confusion rate. For example, for a sound source with a target azimuth $\theta_S = 95°$ and perceived azimuth $\theta_I = 85°$, the angular error is $10°$ without spatial reflection process. With spatial reflection process, the angular error becomes zero but adds the account of front-back confusion. In this case, spatial reflection process is insignificant because an angular error of 10° may be attributed to a slight error in localization.

Finally, if the experimental results on different subjects differ considerably (such as in the case of the signal processing of nonindividualized HRTFs), calculating the mean perceived source direction across subjects is of little significance. Under this situation, many researchers simply calculate the mean perceived source direction across the repetitions for each subject and present results for each subject. As a simple alternative, researchers provide the distribution of perceived samples without averaging.

13.3.3 Results of Virtual Source Localization Experiments

Many studies on VADs include virtual source localization experiments. The current section provides only representative examples.

Wightman et al. (1989b) used localization experiments to compare the localization performance of free-field real sources and headphone-rendered virtual sources. The experiments were conducted in an anechoic chamber with small loudspeakers as free-field real sources. Individualized HRTFs and HpTFs were also incorporated into binaural synthesis and equalization to precisely duplicate the at-eardrum sound pressures in headphone presentation as those pressures of free-field real sources.

Target source positions were uniformly distributed at 72 spatial directions, including six different elevation planes; that is, $\phi = -36°, -18°, 0°, 18°, 36°$, and 54°. The stimulus in this experiment was a train of eight 250-ms bursts of band-pass Gaussian noise (200 Hz–14 kHz). To prevent subjects from becoming familiar with specific stimuli after several trials, the authors shaped the energy spectrum of noise according to an algorithm that divides the spectrum into critical bands. Random intensity (uniform distribution in a 20-dB range) was then assigned to the noise within each critical band.

Eight adults ($L = 8$, four males and four females) participated in the experiment. All had normal hearing, as verified by audiometric screening within 15 dB hearing loss, and had no history of hearing problems of any kind. Within the 72 target source directions under free-field real source conditions, each subject performed $K = 12$ repeated evaluations of each of the initial 36 target source directions, and $K = 6$ repeated assessments of each of the remaining 36 target source directions. Within the 72 target source directions in headphone presentation, each subject carried out $K = 10$ repeated evaluations of each of the initial 36 target source directions, and $K = 6$ repeated assessments of each of the remaining 36 target source directions. Finally, the experimental results for each subject and for all subjects were separately analyzed.

The study (Wightman et al., 1989b) shows the scatterplots of perceived source azimuth and elevation (in the insets) versus target source azimuth and elevation for all subjects in both free-field real source and headphone presentation. Figure 13.1 shows the results for a subject (identified by SDO). The definition of elevation in the aforementioned study is identical to the default coordinate in this book, but the azimuthal range in the former is $-180° < \theta \le 180°$. In the horizontal plane $\phi = 0°$, $\theta = 0°, 90°, 180°, -90°$ represent directly front, right, rear, and left directions, respectively. In the left hemispherical space, therefore, the definition of θ differs from that in this book. To avoid confusion, in addition to the azimuth defined in Wightman's study, the bracketed azimuths in Figure 13.1 are transformed into the default coordinate of this book. As shown in the figure, after spatial reflection is performed to resolve the reversal in raw data, the perceived directions of

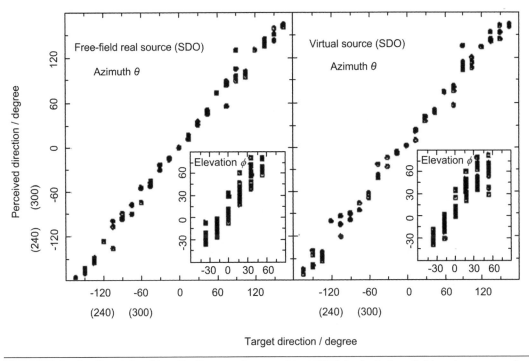

Figure 13.1   Scatterplots of localization results on subject SDO for both free-field real source and headphone presentation (from Wightman et al., 1989b, by permission of J. Acoust. Soc. Am.)

both free-field real sources and headphone-rendered virtual sources are nearly distributed along a diagonal straight line. This result suggests that the perceived directions are close to the target directions. Results also vary among subjects. The localization results for subject SDO are the best among those for the eight subjects.

Additional calculations yield the mean angular error Δ_2 in Equation (13.19), κ^{-1} in Equation (13.22), and mirror confusion rate. Wightman's study shows the statistical results for each subject and the overall statistical results across eight subjects. For simplicity, Table 13.1 lists only the overall statistical results across eight subjects and for nine separate regions of auditory space: front, side (left and right combined), and rear quadrants; at high, medium, and low elevations. The detailed divisions with transformations into the default coordinate of this book are listed as follows.

1.   Low elevation, $\phi = -36°, -18°$;
2.   Middle elevation, $\phi = 0°, 18°$;
3.   High elevation, $\phi = 36°, 54°$;
4.   Front region, $\theta = 0°, 15°, 30°, 45°, 315°, 330°, 345°$;
5.   Side region, $\theta = 60°, 75°, 90°, 105°, 120°, 240°, 255°, 270°, 285°, 300°$;
6.   Back region, $\theta = 135°, 150°, 165°, 180°, 195°, 210°, 225°$.

Table 13.1 shows that the mean angular errors Δ_2 of the free-field real source and headphone-rendered virtual source are similar, but the confusion rate in headphone presentation is higher than that in the free-field real source, especially in the front region. The averaging of the confusion rate over all the spatial regions yields about 11% for headphone presentation and 6% for the free-field real source. The reasoning behind these results was analyzed in Section 8.3.

Table 13.1 Regional measures of localization performance (adapted from Wightman et al., 1989b)

Azimuths		Front		Side		Back	
		Free-field source	Headphone presentation	Free-field source	Headphone presentation	Free-field source	Headphone presentation
Low elevation	Average Δ_2/degree	20.4	21.0	17.7	18.4	19.4	17.8
	Average κ^{-1}	0.06	0.04	0.04	0.03	0.03	0.03
	Average confusion rate/%	4	10	6	11	1	1
Middle elevation	Average Δ_2/degree	17.9	21.5	16.1	15.1	21.1	19.7
	Average κ^{-1}	0.04	0.05	0.04	0.03	0.05	0.05
	Average confusion rate/%	1	7	8	8	1	2
High elevation	Average Δ_2/degree	25.2	29.3	22.6	26.7	29.8	31.2
	Average κ^{-1}	0.06	0.10	0.07	0.09	0.09	0.10
	Average confusion rate/%	16	38	18	17	4	13

Wenzel et al. (1993a) further studied the effect of nonindividualized HRTFs on virtual source localization. The experimental conditions and methods are basically similar to those adopted by Wightman et al. (1989b), but in the latter, both the free-field real source and virtual source included 24 spatial directions, and headphone stimuli were synthesized using HRTFs from a representative subject SDO. Sixteen subjects (2 males and 14 females) participated in the experiment. In each of the 24 directions, every subject performed $K = 9$ repeated evaluations. Statistical calculations were conducted in the sub-region space, as in Wightman's study, and $M = 24$ directions (θ, ϕ) were divided into the following sub-regions (transformed into the default coordinate of this book):

1. Front-low: (315°, −36°), (45°, −18°), (345°, −18°), (0°, 0°);
2. Front-high: (315°, 18°), (30°, 36°), (345°, 18°), (15°, 54°), (0°, 36°), (315°, 54°);
3. Side-low: (285°, 0°), (90°, −36°), (270°, 0°), (105°, −18°);
4. Side-high: (255°, 36°); (120°, 36°), (255°, 54°);
5. Back-low: (210°, −36°), (180°, −36°), (225°, −18°),(165°, 0°);
6. Back-high: (135°, 18°); (150°, 18°), (180°, 54°).

Figure 13.2 shows the regional average of the mean angular error Δ_2, and κ^{-1} for the free-field real source and headphone-rendered virtual source provided by Wenzel. Figure 13.3 illustrates the statistical distribution of mirror confusion rate. Each open square in the figure represents the regional average result over M_i spatial directions and $K = 9$ repeated evaluations of each of the 16 subjects. Here, each M_i denotes the number of target directions in a specific subregion, with $\sum_{i=1}^{6} M_i = M = 24$. That is, these perceived directions are statistical distributions of different subjects, with averaging over $K \times M_i$. The filled squares represent the overall regional average results for the 16 subjects (i.e., averaged over M_i directions $\times$ L subjects $\times$ K repetitions). For comparison, the average results over $M_i \times L \times K$ in nominally equivalent regions (open circles) in Wightman's study (obtained using individualized HRTFs) are also given. In Wightman's study, the total spatial

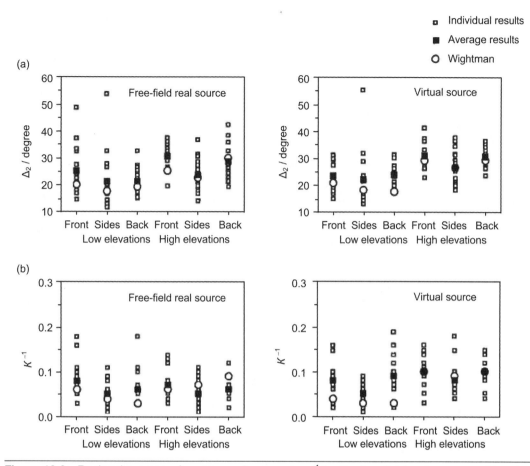

Figure 13.2 Regional average of mean angular error Δ_2, κ^{-1} for the free-field real source and head-phone-rendered virtual source (from Wenzel et al., 1993a, by permission of J. Acoust. Soc. Am.)

directions $M = 72$; moreover, the K, M_i, and L values differ from those in Wenzel's study. For comparison, the results for part of the spatial directions in Wightman's study are excluded from Figures 13.2 and 13.3.

As indicated in Figures 13.2 and 13.3, obvious individual differences in localization performance are observed when nonindividualized HRTFs are used. The detailed analysis indicates that after spatial reflection is performed to resolve the reversal in raw data, the localization of virtual sources is accurate and comparable to that of the free-field real sources for 12 of 16 subjects; 2 subjects exhibit virtual source localization that is poorer than for free-field real sources (mainly in terms of elevation), and 2 other subjects show poor elevation accuracy for both free-field real sources and virtual sources. The main problem caused by the use of nonindividualized HRTFs is increased confusion rate (especially front-back confusion). For virtual sources, the front-back and up-down confusion rates are 31% and 18%, respectively. For free-field real sources, the front-back and up-down confusion rates are 19% and 6%, respectively. These results have been frequently referred to in previous sections (such as Section 8.3).

In addition to Wightman and Wenzel, many other researchers conducted localization experiments on headphone-rendered virtual sources with similar principles. Bronkhorst (1995) synthesized

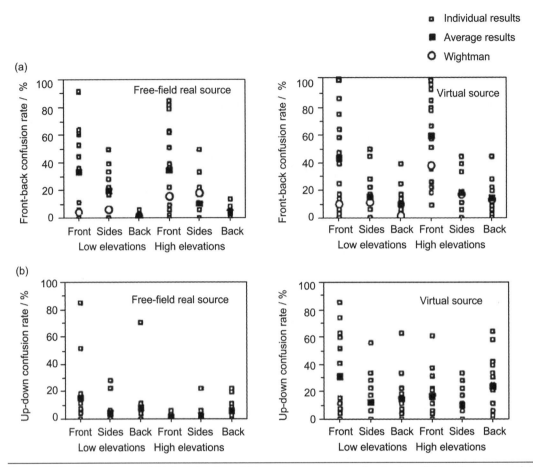

Figure 13.3 Percentages of (a) front-back and (b) up-down confusion for the free-field real source and headphone-rendered virtual source (from Wenzel et al., 1993a, by permission of J. Acoust. Soc. Am.)

virtual sources with individualized HRTF filtering and individualized HpTF equalizing (measured at a position 3 mm to the eardrum with a probe microphone). Localization experiments were conducted for both dynamic and static virtual sources (with and without head tracking). The results, including mean angular errors and confusion rates, were subjected to analysis of variance, or ANOVA (see Section 13.5.1). The conclusions drawn by Bronkhorst are similar to those of Wightman and Wenzel.

Martin et al. (2001) used 354 target directions, including elevations from −40° to 70°, and Gaussian noise as stimulus. Three subjects were recruited, and the individualized HRTFs and HpTFs obtained by blocked ear canal measurement were incorporated into binaural signal synthesis. The ANOVA of mean angular errors and mirror confusion rates indicates that headphone-rendered virtual sources can exhibit a localization performance equivalent to that of free-field real sources. A similar conclusion was discussed in Section 13.2.2. However, this finding slightly differs from Wightman's partly because of the strictly controlled conditions in Martin's experiment. In addition, Martin used few participants.

Pedersen and Minnaar (2006) used similar experimental methods for virtual source localization to evaluate the performance of real-time and dynamic VADs with head tracking. They reported a perceived offset of elevation that strongly depends on the stimulus elevation.

Results of localization experiments are naturally distributed in three-dimensional auditory space, which can be conveniently described using relevant spherical statistical and graphical methods (Leong and Carlile, 1998). Figure 13.4 shows an example of the localization results for free-field virtual sources rendered by a headphone-based real-time and dynamic virtual auditory environment (VAE) (Section 12.4; and Zhang and Xie, 2012). White noise was used as stimulus and MIT-KEMAR HRTFs were used in dynamic binaural synthesis. Twenty-eight target virtual directions, which were distributed in the right-hemispherical space at four elevations ($\phi = -30°$, 0°, 30°, 60°) with seven azimuths ($\theta = 0°$, 30°, 60°, 90°, 120°, 150°, and 180°) at each elevation, were chosen for the experiment. Six teachers or postgraduate students in acoustics (3 males and 3 females) participated in the experiment. They were asked to evaluate the perceived virtual source positions in terms of direction and distance. Each subject assessed each target virtual source position 3 times, yielding 18 evaluations of each target position. Figure 13.4 shows the statistical results for the perceived virtual source directions in dynamic synthesis after front-back and up-down confusion were resolved. The results are demonstrated on the surface of a sphere and viewed from the front, right, and rear directions, respectively. Notation "+" represents the target virtual source direction. The black points at the center of the ellipses are the average perceived directions across subjects and repetitions. The ellipses are the confidence regions at a significance level $\alpha = 0.05$. The results also indicate overall front-back and up-down confusion rates of 2.5% and 13.7%, respectively; the average perceived virtual source distances across subjects and repetitions range from 0.86–1.23 m. Therefore, dynamic binaural synthesis visibly reduces front-back and up-down confusion, and improves externalization of headphone presentation—even when nonindividualized HRTFs are used.

As an example of loudspeaker-based virtual source reproduction, Figure 13.5 shows the localization results for the lateral virtual source in the improved virtual 5.1-channel surround sound described in Section 10.4.2 (Xie et al., 2005c). The principle of signal processing is depicted in Figure 10.9. In the experiment, MIT-KEMAR HRTFs with a sampling frequency of 44.1 kHz, 16-bit quantification, and a length of 512-points were used (Table 2.1). The inputs were 5.1-channel surround sound signals created by pairwise mixing (Section 6.4.2), then 0 input signals, except for the L (left) and LS (left surround) channel signals, were used to generate left lateral virtual sources.

A localization experiment was conducted in a listening room with a reverberation time of 0.3 s. Loudspeakers were arranged at $\theta_R = 15°$ and $\theta_L = 345°$ on a circle with a radius of 2.0 m and a height of 1.2 m from the ground. Eight subjects, aged 18 to 35, were recruited and asked to evaluate the perceived azimuths of virtual sources. The final result was obtained by averaging across all the subjects using Equation (13.15). The standard deviation σ_θ was also calculated.

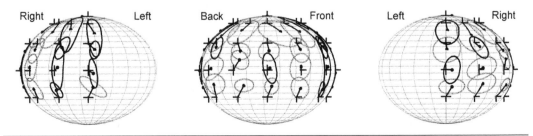

Figure 13.4   Graphical representation of localization results of dynamic binaural synthesis

Figure 13.5 shows the localization results for speech stimuli. The coordinate system used in the original article (Xie et al., 2005c) is different from the default coordinate of this book; thus, the virtual source in the left horizontal region is denoted by $360° - \bar{\theta}_I$. The figure shows that virtual sources continuously move from $\bar{\theta}_I \approx 330°$ to $\bar{\theta}_I \approx 280°$ (i.e., $360° - \bar{\theta}_I$ from 30° to 80°) when the interchannel level difference $20\log_{10}$ (LS/L) changes from −24 to +24 dB. This movement suggests that virtual 5.1-channel surround sound reproduction with two frontal loudspeakers can recreate lateral virtual sources. Therefore, lateral virtual source localization is improved relative to conventional 5.1-channel surround sound reproduction through ITU (International Telecommunication Union) loudspeaker configuration and pairwise signal mixing (Section 6.4.2; Xie, 2001a). This result is similar to the case of conventional 5.1-channel surround sound reproduction, in which two surround loudspeakers are moved forward to two sides ($\theta_{RS} = 90°$ and $\theta_{LS} = 270°$).

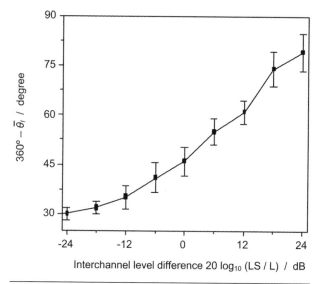

Figure 13.5   Localization results for lateral virtual sources in improved virtual 5.1-channel surround sound with two frontal loudspeakers

However, the localization experiments show (not plotted in Figure 13.5) that the virtual sources intended for the rear appear at the frontal direction in virtual 5.1-channel surround sound reproduction with two frontal loudspeakers. This is a common problem in two-front loudspeaker reproduction.

13.4 Quantitative Evaluation Methods for Subjective Attributes

In addition to virtual source position, some other auditory perceptual attributes require assessment. These attributes include realism or naturalness, timbre, and sense of immersion, depending upon different applications and research purposes. An important point in subjective assessment experiments is that subjects should fully understand the implications of these attributes; otherwise, confusion may arise.

In various quantitative evaluation methods, a subject usually rates an evaluated object in comparison to a reference object. Thus, the score is expressed in a relative sense, corresponding to a certain degree of subjective perception. For example, the auditory experience of a free-field real source can be used as the reference for the naturalness and timbre of a virtual source. In some cases, such as in exploring the influence of simplified signal processing on timbre, an alternative approach is to use original processing without simplification as reference. For a virtual auditory environment, in-situation recording with an artificial head is often considered as a reference (Lokki et al., 2001a; Lokki and Jarvelainen, 2001b).

Different studies adopt different rating scales, such as five-grade, seven-grade, and ten-grade scales. A commonly used scale is the five-grade type, with scales ranging from 1–5 or 0–4. Sometimes, several subjective attributes are simultaneously rated, but such an approach increases experimental difficulty.

No universal standards currently exist for rating the subjective attributes of VADs and related signal processing schemes. The ITU-recommended standard (ITU-R, BS.1116-1, 1997) is designed for the subjective assessment of small impairments in multi-channel sound rather than for VADs. Nevertheless, some methods in this standard can be introduced into the subjective assessment of VADs, especially for virtual surround sound reproduction.

The ITU standard recommends a five-grade scale ranging from 1.0–5.0 (with a precision of one decimal place) to quantitatively evaluate the degree of impairment in perceived quality against a reference object. The elevation scale is detailed as follows:

5.0: imperceptible;
4.0: perceptible but not annoying;
3.0: slightly annoying;
2.0: annoying; and
1.0: very annoying.

Various evaluation methods are presented in the literature. The ITU standard recommends the "double-blind triple-stimulus with hidden reference" method. It consists of three stimuli (A, B, C), and the known reference is always stimulus A. The hidden reference and object are randomly assigned to B and C, respectively. Therefore, two kinds of combinations are generated: AAB and ABA. A subject rates the second and third stimuli, with the first stimulus as the reference stimulus. Repetitions are often employed, with the two combinations presented at equal numbers. Reference A is hidden in either the second or third stimuli to validate the reliability of the quantitative evaluation of perceived quality. Ideally, the score of reference A, hidden in either the second or third stimulus, should be 5.0 from a statistical perspective; otherwise, the subject's assessment data are unreliable.

A preliminary statistical processing of raw data from subjective assessment experiments is necessary. Let x_n, $n = 1,2 \dots, N$ denote the scores of N subjective evaluations, which may come from N different subjects, N repetitions of a subject, or L subjects with K repetitions ($N = L \times K$). Correspondingly, the mean and standard deviation are:

$$\bar{x} = \frac{1}{N} \sum_{n=1}^{N} x_n \qquad \sigma_x = \sqrt{\frac{1}{N-1} \sum_{n=1}^{N} (x_n - \bar{x})^2}, \qquad (13.23)$$

which represent the average score and dispersion of the score data.

13.5 Further Statistical Analysis of Psychoacoustic Experimental Results

13.5.1 Statistical Analysis Methods

The data collected from either the localization experiments in Section 13.3 or the subjective evaluation experiments in Section 13.4 can be regarded as samples of random variables that obey certain statistical rules. Previous analyses provide only preliminary statistical results, such as mean and standard deviation. Strictly speaking, only when appropriate statistical methods are used can a conclusion of a certain confidence level be drawn. Commonly used statistical methods include hypothesis testing and ANOVA. The complete analysis of experimental data by statistical techniques is beyond the scope of this book. Some extensively applied statistical methods associated with previous experiments are briefly introduced here. Refer to Marques (2007) or specialized statistical software, such as statistical product and service solutions and statistics analysis system, for more details on statistical methods and their applications.

(1) Mean value test for a set of experimental data:

Let x_n with $n = 1, 2, \ldots, N$ denote a set of observed samples under certain experimental conditions. Generally, x_n can be either the score from quantitative evaluation experiments or the deviation between the perceived and target source direction $x_n = |\theta_I(n) - \theta_S|$. The mean $\bar{x}$ and standard deviation σ_x of the samples can be calculated using Equation (13.23).

In practice, verifying whether the mean $\bar{x}$ of a set of data is equal to, greater than, or less than a specific value μ_0 is often necessary. An example is validating the $\Delta\theta_2$ and $\Delta\phi_2$ in Equation (13.17) to determine whether they are less than a specific value. Again, in the double-blind triple-stimulus with hidden reference method recommended by the ITU, an essential task is to determine whether the mean score of the reference stimulus hidden in the second or third segment is 5.0.

The test for the mean value of N observed sample data is a hypothesis test on one population. Suppose that sample data are from a normally distributed population $N(\mu, \sigma^2)$ (this should not be confused with the number N of observed samples). μ and σ^2 are expectation and variance, respectively. The t-test in Table 13.2 should be used in testing the hypothesis under a significance level α (usually, $\alpha = 0.05$). Here, $\bar{x}$ and σ_x are calculated from Equation (13.23); $t_{1-\alpha}(N-1)$ and $t_{1-\alpha/2}(N-1)$ are the $(1 - \alpha)$ and $(1-\alpha/2)$ quantiles of t-distribution, respectively:

(2) Mean value test for two sets of experimental data

Let x_n ($n = 1, 2, \ldots, N_1$) and y_n ($n = 1, 2, \ldots, N_2$) denote two sets of observed samples under different conditions. The mean $\bar{x}, \bar{y}$ and the standard deviation σ_x, σ_y of each set of observed samples can be calculated by Equation (13.23).

Verifying whether the mean of these two sets of data are equal (or whether one is greater than the other) is necessary. For example, to compare the localization accuracy of the free-field real source and virtual source in Section 13.3, the κ^{-1} of Equation (13.22) and the confusion rate in the mirror direction should be compared. In quantitatively assessing subjective attributes, comparing the mean score under two different conditions (such as two different kinds of signal processing schemes) is also necessary.

The mean value test for two sets of observed values from different sample data (two-mean test) is a hypothesis test on two populations. Suppose that the two sets of observed values x_n ($n = 1, 2, \ldots,$

Table 13.2 Mean value test on a single normal distribution

	Null hypothesis	Alternative hypothesis	Reject the null hypothesis at significance level α if
(1)	$\mu = \mu_0$	$\mu > \mu_0$	$\bar{x} \geq \mu_0 + \dfrac{\sigma_x}{\sqrt{N}} t_{1-\alpha}(N-1)$
(2)	$\mu = \mu_0$	$\mu < \mu_0$	$\bar{x} \leq \mu_0 - \dfrac{\sigma_x}{\sqrt{N}} t_{1-\alpha}(N-1)$
(3)	$\mu = \mu_0$	$\mu \neq \mu_0$	$|\bar{x} - \mu_0| \geq \dfrac{\sigma_x}{\sqrt{N}} t_{1-\alpha/2}(N-1)$

N_1) and y_n ($n = 1, 2, ..., N_2$) are from two normally distributed populations $N(\mu_1, \sigma_1^2)$ and $N(\mu_2, \sigma_2^2)$, respectively. μ_1, μ_2 and σ_1^2, σ_2^2 are corresponding expectations and variances. To verify the mean value, the homogeneity of variances should be tested beforehand; that is, testing hypothesis $\sigma_1^2 = \sigma_2^2$ at a significance level α. When:

$$\frac{\sigma_x^2}{\sigma_y^2} \geq F_{1-\alpha/2}(N_1 - 1, N_2 - 1) \quad or \quad \frac{\sigma_y^2}{\sigma_x^2} \geq F_{1-\alpha/2}(N_2 - 1, N_1 - 1), \tag{13.24}$$

then the hypothesis should be rejected. $F_{1-\alpha/2}$ is the $(1-\alpha/2)$ quantile of F distribution. Only when hypothesis $\sigma_1^2 = \sigma_2^2$ is true can the mean values of the two sets of observed values be tested. Then, the t-test is used to validate the two sets of observed values $x_n(n = 1, 2, ..., N_1)$ and $y_n(n = 1, 2, ..., N_2)$ at a significance level α, as outlined in Table 13.3. The definitions of the symbols in Table 13.3 are the same as those in Table 13.2, with:

$$\sigma_w^2 = \frac{(N_1 - 1)\sigma_x^2 + (N_2 - 1)\sigma_y^2}{N_1 + N_2 - 2}. \tag{13.25}$$

(3) Analysis of variance

The mean value test is for two sets of experimental results under two different conditions. As a more general case, in localization and quantitative assessment experiments, validating the mean values of three or more sets of data under different controlled conditions is a common approach. Examples are: comparing the confusion rate in localization, or comparing the mean scores from three different signal-processing methods. This is an issue of multi-population mean testing in statistics.

Many factors cause changes in experimental conditions. In free-field virtual source synthesis, for instance, the length of an HRTF filter, individualized or nonindividualized HRTFs, and/or the presence or absence of head tracking may affect results. If only one factor changes, then the experiment is classified as a single-factor experiment. If more than one factor changes, then it is categorized as a multi-factor experiment. The different conditions of a factor are called levels. For example, the length of an HRTF filter can be divided into three levels: 128, 256, and 512 points; the choice of HRTF can be divided into two levels: individualized and nonindividualized.

As a most general case, mean value testing should be carried out on multiple populations under multiple factors; more than one factor and their interaction effects will likely influence results. This kind of problem, including evaluating statistically significant differences among multiple mean values and testing the significant effects of multiple factors, is addressed by ANOVA in statistics. Numerous conclusions on the psychoacoustic experiments on VADs are drawn from ANOVA. The

Table 13.3 Mean value test on two normally distributed populations

	Null hypothesis	Alternative hypothesis	Reject the null hypothesis at significance level α if
(1)	$\mu_1 = \mu_2$	$\mu_1 > \mu_2$	$\bar{x} - \bar{y} \geq t_{1-\alpha}(N_1 + N_2 - 2)\sigma_w\sqrt{\dfrac{1}{N_1} + \dfrac{1}{N_2}}$
(2)	$\mu_1 = \mu_2$	$\mu_1 < \mu_2$	$\bar{x} - \bar{y} \leq -t_{1-\alpha}(N_1 + N_2 - 2)\sigma_w\sqrt{\dfrac{1}{N_1} + \dfrac{1}{N_2}}$
(3)	$\mu_1 = \mu_2$	$\mu_1 \neq \mu_2$	$\|\bar{x} - \bar{y}\| \geq t_{1-\alpha/2}(N_1 + N_2 - 2)\sigma_w\sqrt{\dfrac{1}{N_1} + \dfrac{1}{N_2}}$

mathematical methods of ANOVA are discussed in relevant textbooks (Marques, 2007). Details are not discussed here given space limitations.

13.5.2 Statistical Analysis Results

As stated in Section 5.5.2, Huopaniemi and Zacharov (1999b) conducted an assessment experiment on the subjective attributes of different types of HRTF filters in headphone-based binaural synthesis. The attributes assessed included localization and timbre impairment. A five-grade scale similar to that recommended by the ITU (ITU-R, BS.1116-1, 1997) was adopted. The individualized HRTFs were measured using the blocked ear canal technique. The 256-order (257-point) FIR (finite impulse response) filters at a sampling frequency of 48 kHz were used as hidden references. The authors evaluated three types of filters with various orders: FIR filters designed with the rectangular time windowing method (Section 5.3.1), IIR (infinite impulse response) filters designed by Prony's method (Section 5.3.2), and WIIR (warped infinite impulse response) filters with warping coefficient $\lambda = 0.65$ (Section 5.5.2). Each type of filter included five different orders. Four target directions at $\theta = 0°, 40°, 90°$, and $210°$ in the horizontal plane were tested, and 10 subjects participated in the experiment. Thus, the multiple factors included filter structure, filter order, target direction, and subject. The multi-factor ANOVA of the experimental results indicates that the order and structure of the filters significantly affect reproduction performance. The $(Q, P) = (24, 24)$-order WIIR filters achieve a mean score of more than 4.5. However, 96-order (97-point) traditional FIRs or $(Q, P) = (48, 48)$-order IIR filters are needed to achieve a similar score.

Begault et al. (2001) directly compared the influence of head tracking, room reflection, and individualized HRTFs on the spatial perception of a virtual speech source. Specifically, the authors used two levels or states (tracking on and tracking off) for head tracking; three levels (anechoic, early reflection, and full auralization) for room reflection; and two levels (individualized and generic, from an artificial head) for HRTF selection. Thus, the total number of different conditions was $2 \times 3 \times 2 = 12$. In the study, "anechoic" indicates that only direct sounds (free-field virtual sources) were synthesized. Both "early reflection" and "full auralization" were synthesized by a model of a rectangular room obtained from a room prediction–auralization software (CATT, see Section 14.2). In the former, HRTF-filtered early reflections up to the preceding 80 ms were added to direct sound synthesis, and, in the latter, the late reverberation field from 80 ms to 2.2 s was incorporated into early reflection. Head tracking was used to incorporate the head movement-induced dynamic cue to the direct sound and early reflections, with an update rate of 33 Hz and an average system latency time of 45.3 ms.

Virtual sources at six horizontal azimuths $\theta = 0°, 45°, 135°, 180°, 225°, 315°$ (converted into the default coordinate of this book) were evaluated under 12 different combinations of experimental conditions. The subjects were asked to localize and assess the perceived attributes of virtual sources in each case. Five measures were used in the assessment process: the mean unassigned (absolute) error $\Delta\theta_2$ and $\Delta\phi_2$ of azimuth and elevation in Equation (13.17), the confusion rate in mirror direction, externalization (in- or out-of-head), and perceived realism (five-grade scale from 0–4). $L = 9$ subjects participated, and for each combination of experimental conditions, each subject performed evaluations $K = 5$. Consequently, a statistical result of $L \times K = 45$ was derived.

For each measure in the assessment (e.g., confusion rate), multi-factor ANOVA should be used to analyze the mean value. The analysis shows that only room reflections significantly influence azimuth and elevation error (after reversal is resolved); head tracking significantly reduces the confusion rate from 59% to 28%. Room reflections pose a significant effect on externalization, with the mean rate of subjects' externalized evaluations increasing from 40% under the anechoic condition to 79% under the early reflection and full auralization conditions (no significant differences were found between early reflections and full auralization conditions). Perceived realism exhibits no significant difference among 12 combined conditions, a result that may be due to the subjects having no common understanding of the meaning of perceived realism.

13.6 Binaural Auditory Model and Objective Evaluation of VADs

Similar to the case of a conventional electroacoustic system, objective and subjective assessments are two classes of methods for evaluating the performance of VADs and related signal-processing schemes. As indicated in the preceding discussions, subjective assessment by psychoacoustic experiment effectively reveals actual perceived performance, but is complicated and time-consuming. Elaborately designed experiments with sufficient samples and appropriate statistical methods are needed to obtain reliable conclusions.

Objectively assessing VADs is a relatively simple task. The essence of VADs lies in synthesizing binaural signals by HRTF-based signal processing. Thus, a straightforward approach to evaluating VADs is to compare synthesized and ideal/target binaural signals, and then to analyze the differences. A variety of error criteria for the HRTF approximations in Section 5.1 follow this ideal and can be used to evaluate the performance of free-field virtual source synthesis in VADs. However, these analyses neglect the psychoacoustic/physiological factors in binaural hearing, thereby generating resultant physical and/or mathematical errors that do not correspond to perceived errors. Therefore, some psychoacoustic/physiological factors should be incorporated into the objective analysis of VADs.

A binaural auditory model describes the physical, physiological, and psychoacoustic processing of the hearing system on received sound waves. Since Jeffress (1948) proposed the basic principle

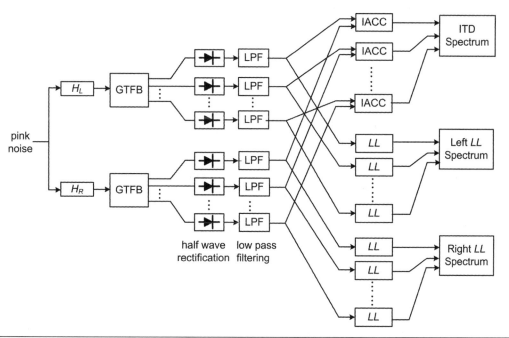

Figure 13.6   Block diagram of a binaural auditory model (adapted from Huopaniemi and Zacharov, 1999b)

of a binaural auditory model, it has been improved and developed by many researchers, as well as applied to the objective assessment of sound quality and spatial sound reproduction (including multi-channel sound and VADs; Macpherson, 1991; Pulkki et al., 1999; Huopaniemi and Zacharov, 1999b). Figure 13.6 shows the block diagram of a binaural auditory model used to evaluate different HRTF filter designs.

Pink noise represents a desired mean power spectrum of a sound source. A pair of HRTFs (with the eardrum as the reference point) simulates the effect of the head, torso, and external ear on the incident sound wave, which transforms the free-field pressure of a point source into pressure signals at the eardrums.

Strictly speaking, a complete auditory model should include the function of the middle ear. In a left-right symmetric model, however, the middle ear has no effect on interaural cues, such as ITD (interaural time difference) and ILD (interaural level difference). For simplicity, therefore, the binaural model in Figure 13.5 disregards the effect of the middle ear.

A series of parallel band-pass filter banks are analogous to the frequency analysis function of the inner ear; that is, the simulation of auditory filters. When a band-pass filter is represented by a GammaTone filter, its impulse response is:

$$g(t) = \frac{at^{N-1}\cos(2\pi f_c t + \phi)}{\exp(Bt)} \qquad t \geq 0, \tag{13.26}$$

where a is a parameter for filter gain, f_c is the central frequency (in units of Hz) of the filter, φ denotes the initial phase, N represents the order of the filter, and B is a parameter that characterizes the filter bandwidth. When $N = 4$, $B = 2\pi$ ERB, the GammaTone filter bank is an approximation of auditory filters. In Equation (1.6), the equivalent rectangular bandwidth (ERB) of the auditory filter is given in units of Hz, whereas the unit of central frequency is kHz. Thirty-two parallel GammaTone filters were used by Huopaniemi et al. (1999b).

Following each ERB band-pass filtering, half-wave rectification and low-pass filtering with a cutoff frequency of 1 kHz are used to simulate the behavior of hair cells and auditory nerves in each band-pass channel. For simplicity, the adaptive process in the auditory system is excluded.

ITD as a function of frequency [transformed into the units of equivalent rectangular bandwith number (ERBN) given in Equation (1.7)] is then evaluated by applying the cross-correlation calculation described in Section 3.2.1 to the binaural outputs generated after rectification and low-pass filtering. The ITD calculated here can represent the frequency dependence of ITD because it is calculated in each frequency bank in accordance with the resolution of the auditory system. Below the cutoff frequency of low-pass filtering, filtering slightly influences the signals. Thus, the normalized cross-correlation function and ITD are relevant to the fine structures (phases) of binaural signals. Above the cutoff frequency, low-pass filtering smoothes the signals. Therefore, the normalized cross-correlation function gradually becomes relevant to the envelope of the binaural signals rather than to the fine structures. These signals appropriately indicate that the phase delay difference of interaural sound pressures is a localization cue for low frequencies, whereas the envelope delay difference is a localization cue for middle and high frequencies.

Loudness spectra in units of Sones/ERB for each ERB bank can be calculated for either of the two ears using the following equation:

$$L_L = (\overline{E_L^2})^{1/4} \qquad L_R = (\overline{E_R^2})^{1/4}, \tag{13.27}$$

where $(\overline{E_L^2})$ and $(\overline{E_R^2})$ are the average power over time for the left and right ears, respectively. Equation (13.27) is an approximation of the loudness formula provided by Zwicker. In Zwicker's formula, however, the exponential factor is 0.23 (Zwicker and Fastl, 1999), rather than 1/4, as in Equation (13.27).

The loudness level spectra LL in units of Phons/ERB for each ERB bank are:

$$LL_L = 40 + 10\log_2 L_L \qquad LL_R = 40 + 10\log_2 L_R. \tag{13.28}$$

From this equation, an evaluation of ILD as a function of frequency (in units of ERBN) can be calculated from the difference in loudness level between the left and right ears as:

$$\Delta LL = LL_R - LL_L. \tag{13.29}$$

In addition, the binaural loudness level spectrum as a function of frequency (in units of ERBN) is calculated by summing the loudness spectra of both ears and then transforming them to loudness levels, which can be applied to the analysis of timbre in reproduction.

Evaluating VADs and related signal-processing schemes by a binaural auditory model is implemented by feeding the evaluated binaural signals into the model, and then comparing the resultant loudness level spectra of the left and right ears, as well as the ITD, with an ideal or referenced condition. Binaural signals for evaluation can be obtained in two ways:

1. Simulation. For example, mono stimulus is processed by the scheme under evaluation.
2. Measurement. The signals obtained from the scheme under evaluation are reproduced through headphones or loudspeakers. The binaural signals are then recorded in reproduction by using a pair of microphones positioned at the ears of an artificial head or a human subject.

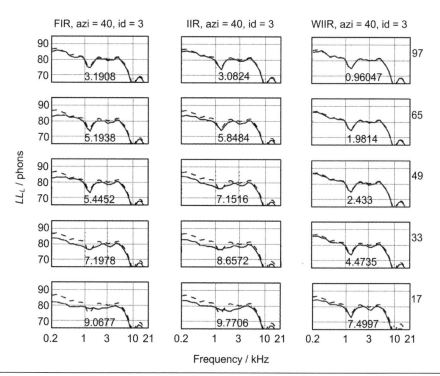

Figure 13.7 Left-ear loudness level spectra for different HRTF filter structures (from Huopaniemi and Zacharov, 1999b, by permission of AES)

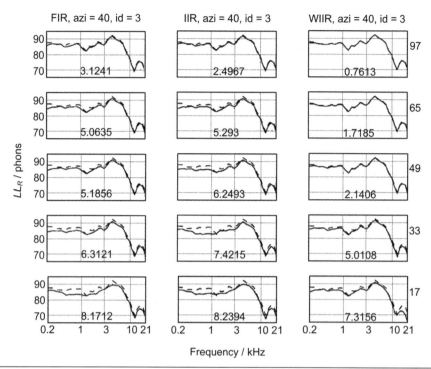

Figure 13.8 Right-ear loudness level spectra for different HRTF filter structures (from Huopaniemi and Zacharov, 1999b, by permission of AES)

As stated in Section 5.5.2, Huopaniemi and Zacharov (1999) also used the binaural auditory model to analyze the performance of different types of HRTF filters in headphone presentation. Similar to the case of subjective assessment in Section 13.5.2, three types of filters (i.e., FIR, IIR, and WIIR filters) of various orders were analyzed. The minimum-phase approximation of HRTFs was implemented prior to filter design. Figures 13.7 and 13.8 plot the simulated results for the loudness level spectra of the left and right ears, respectively (subject no. 3). The input stimulus was pink noise and the target direction of the HRTFs was at horizontal azimuth $\theta = 40°$. The number on the right side of the figures denotes different filter lengths or orders. For example, 97 on the right side of the top pane denotes a 96-order (97-point) FIR or $(Q, P) = (48, 48)$-order IIR filter at a sampling frequency of 48 kHz. The solid lines are the results for the filters under evaluation and the dashed lines are those for a 256-order (257-point) reference FIR filter. As indicated in Figures 13.7 and 13.8, with the same filter order, the WIIR filter has a minimum error compared with the reference. This result also holds for ITD. Thus, frequency warping is a good choice for filter design from the aspect of auditory perception (Section 5.5.2).

Only the errors in loudness level spectra and ITD are analyzed in the binaural auditory model. On this basis, further steps can be taken to predict subjective attributes, such as perceived virtual source direction and timbre. Such prediction requires a more elaborate model, which includes signal-processing procedures in a high-level neural system. Some work (Park et al., 2006) has been done on this matter, but more improvement is necessary.

Section 5.2.4 discussed a study on the audibility of spectral details of HRTFs at high frequencies in a psychoacoustic experiment (Xie and Zhang, 2010). Further analysis, using the binaural loudness model proposed by Moore et al. (1997, 2008), shows that the audibility of a smoothed HRTF magnitude spectrum is relevant to the loudness spectrum differences (monaural and binaural) between original and smoothed HRTFs in each auditory filter bank. Such audibility is also, therefore, related to the direction of the sound source and the individuality of each subject.

The traditional binaural auditory models mentioned here consist of a bottom-up (signal-driven) architecture. As pointed out and suggested by Blauert (2012), these traditional models are sufficient for a number of applications, including sound source localization. For evaluating some other perceived attributes of VAD that concern auditory cognition, auditory models that have inherent knowledge bases and employ top-down (hypothesis-driven) strategies are needed. This kind of model is an important direction for future research.

13.7 Summary

Objective and subjective assessments are two methods for evaluating virtual auditory displays and related signal-processing schemes, in which subjective assessment by psychoacoustic experiments is the major experimental approach for evaluating and validating perceived performance of virtual auditory display.

Conditions should be stringently controlled and appropriate statistical methods should be applied in subjective evaluation experiments. No universal standards of contents, methods, conditions, and data analyses for subjective evaluation experiments are currently in place. Thus, these factors may differ depending on the study. In practical applications, the design of a subjective evaluation experiment should be based on the specific situations and contents to be evaluated. Experimental conditions—such as acoustical environment, background noise, stimuli, pressure levels in reproduction, as well as hearing conditions and number of subjects—should also be meticulously designed and chosen.

Auditory comparison and discrimination experiments are applicable to qualitatively evaluating the overall performance of virtual auditory displays through a comparison of the subjective differences between targets and references. These experiments are usually designed to examine the perceptual similarity between a free-field real source and a virtual sound source, or to evaluate the effect of various simplified HRTF filters.

Virtual source localization experiments are extensively used to quantitatively assess and validate performance of a virtual auditory display. Many important conclusions discussed in the preceding chapters are also drawn from the data analysis of localization experiments. Moreover, the quantitative evaluation of other subjective attributes, such as naturalness, timbre, and sense of immersion in virtual auditory displays, are often conducted through subject rating by grade scale.

The raw data obtained from localization experiments or other quantitative evaluations should be analyzed by statistical methods to draw conclusions of a certain confidence level. The hypothesis test and ANOVA are two commonly used approaches.

Developed objective evaluation methods currently incorporate psychoacoustic/physiological factors in assessments. Binaural auditory models have been applied to the objective evaluation of virtual auditory displays, which is an issue that may be a worthwhile direction for future research.

14

Applications of Virtual Auditory Displays (VADs)

This chapter overviews a variety of applications of virtual auditory displays, including application to scientific research experiments, binaural auralization, sound recording and reproduction, virtual reality, multimedia, communication and information systems, as well as possible application to clinical diagnosis.

In recent decades, VADs (virtual auditory displays) have been widely employed in a variety of fields, such as scientific research, engineering, communication, multimedia, consumer electronic products, and entertainment. Although the basic principles of VADs and HRTFs (head-related transfer functions) have been presented in the preceding chapters, in which Chapter 10 partially addresses a specific application of VADs (domestic sound reproduction), systematic and summative discussion is needed. This chapter summarizes various VAD applications. Covering all possible applications is difficult; thus, only some representative applications, which sufficiently demonstrate the application values of VADs, are presented (Xie and Guan, 2004).

Section 14.1 provides an overview of the application of VADs in scientific research experiments that are related to binaural hearing. Section 14.2 focuses on the application of VADs to binaural auralization, as well as on existing problems—with emphasis on the aspects of room acoustic design and subjective evaluation. The application of VADs to sound reproduction and relevant sound program recording is presented in Section 14.3. The application of VADs to virtual reality, communication and information systems, and multimedia is discussed in Section 14.4. Finally, Section 14.5 outlines the possible application of VADs to clinical diagnosis.

14.1 VADs in Scientific Research Experiments

Binaural pressure signals contain numerous information or cues on hearing perception, which are comprehensively analyzed by high-level neural systems to form auditory events. For years, researchers have devoted efforts to psycho- and physioacoustical phenomena through various binaural listening experiments. In such experiments, explicitly distinguishing the function of a particular cue in binaural hearing under specific conditions necessitates separately examining all the information or cues contained in binaural signals. This examination requires artificially controlling (e.g., preserving or discarding) certain auditory information and then collecting the resultant subjective or physiological responses. In the study of localization cues, for example, different ILDs (interaural level differences) or ITDs (interaural time differences) are stimulated in headphone presentation by changing the amplitude and time delay of left- and right-channel headphone signals (Blauert, 1997). However, this simple manipulation of binaural signals is insufficient to precisely simulate complex localization information, thereby resulting in unnatural auditory perceptions, such as in-head localization.

HRTF-based VADs allow for complete and precise control over binaural signals under different physical conditions, achieving an auditory perception close to natural hearing. Thus, HRTF-based VADs have become an important experimental tool in binaural scientific research. For scientific research in general, accuracy is of the ultimate importance. Headphone presentation is popular because it is immune to the influence of environmental noise and reflections, and is easy to implement. As indicated in Sections 8.3 and 8.4, however, headphone presentation also suffers from disadvantages. Alleviating the drawbacks of headphone presentation necessitates the careful consideration of each step of signal processing; for instance, adopting individualized HRTFs and headphone equalization should be considered, according to the context.

Numerous binaural experimental studies were conducted with the use of VADs. Wightman and Kistler (1992a) investigated the relative salience of conflicting ITD and ILD cues for localization. In the experiments, the authors synthesized the baseline signals that correspond to the binaural signals generated by real sound sources by filtering Gaussian noise with individualized HRTFs and inverse headphone-to-ear canal transfer functions. They generated the experimental signals with conflicting ITD and ILD cues by manipulating the magnitude or phase characteristics of the digital filters used in binaural synthesis. The authors then compared the baseline and experimental signals, yielding results concerning the relative dominance of low-frequency ITD to ILD in various conditions (Section 1.4.5). Using dynamic VADs associated with head tracking, Wightman and Kistler (1999) explored the influence of the dynamic localization cues introduced by head movement on resolving front-back ambiguity (Section 1.4.3).

In the study of high-frequency spectral cues caused by pinnae to localization, some researchers modified high-frequency spectral cues by filling pinna cavities with certain materials (Gardner and Gardner, 1973; Hofman et al., 1998; Morimoto, 2001), such as soft-rubbers. However, quantitative control over spectral cues is difficult to achieve through this modification method. This problem can be overcome by using VADs.

Kulkarni and Colburn (1998) used VADs to investigate the influence of the fine structures of HRTF magnitudes on localization (Section 5.2.4). Langendijk and Bronkhorst (2002) also employed VADs to study the contribution of spectral cues to human sound localization by eliminating the cues in 1/2-, 1-, or 2-octave bands at a frequency range above 4 kHz. The results show that removing the cues in 1/2-octave bands do not affect localization, whereas removing the cues in 2-octave bands makes correct localization virtually impossible.

Using VADs, Jin et al. (2004) investigated the relative contribution of monaural versus interaural spectral cues to resolving directions within the cone of confusion. In the experiments, the natural values of overall ITD and ILD were maintained. An artificial flat spectrum was presented at the left eardrum, and the right ear sound spectrum was adjusted to preserve either the true right monaural spectrum or the true interaural spectrum. The localization experiments indicate that neither the preserved interaural spectral difference cue nor the preserved right monaural spectral cue sufficiently maintains accurate elevation in the presence of a flat monaural spectrum at the left eardrum. Macpherson and Sabin (2007) quantized the weights assigned by listeners to the spectral cues from each ear through VADs (Section 1.4.4).

VADs can be also used as an important experimental tool in research on distance perception, as stated in Section 8.4.2 (Zahorik, 2002a). With VADs, Bronkhorst and Houtgast (1999) predicted the source distance perception in a room from the direct-to-reverberant energy ratio (Section 1.7.2). Other than localization, VADs have also been used in studies relevant to binaural hearing. Drullman and Bronkhorst (2000) investigated speech intelligibility and talker recognition under different conditions. The results show that, compared with conventional monaural and binaural presentation, directional presentation through VADs yields better speech intelligibility when two or more competing talkers are present. This finding is useful in communication and teleconference applications (to be shown in Section 14.4.2). Moreover, VADs were employed in the study of spatial masking (Kopco and Shinn-Cunningham, 2003), as well as psychoacoustics and neurophysiology (Hartung et al., 1999b).

Although only representative examples of the use of VADs as a research tool are given here, they are sufficient to demonstrate the scientific value of such methods.

14.2 Applications of Binaural Auralization

14.2.1 Application of Binaural Auralization in Room Acoustics

The sound field in a room consists of direct sounds and reflections (Section 1.7). The subjective perception of room acoustic quality is closely related to the physical properties of the sound field in a room. The existing physical parameters, or indices, used to describe the sound field, however, are currently insufficient to fully represent the subjective perception of the quality of room acoustics because these parameters or indices are incomplete. Thus, subjective assessment is necessary in evaluating the acoustic performance of rooms, such as concert halls, theatres, and multifunction halls.

Onsite listening is the most straightforward method of subjective assessment, but is impractical in most cases. For example, employing onsite listening methods to compare the acoustic quality of different music halls located in different countries or cities incurs expensive travel costs and suffers from the short auditory memory of listeners.

In the early developmental stages of room acoustics, a variety of sound recording and reproduction technologies were used in research on room acoustic quality. Among these technologies, binaural recording and reproduction with an artificial head or a real human subject is commonly performed. By reproducing recorded binaural signals, listeners can compare and evaluate the perceived acoustic quality of different rooms. In practice, however, onsite recording is often inconvenient.

Auralization aims to regenerate/evoke auditory perceptions at a specific listening position by reproducing the information in the sound field that is obtained by mathematical or physical simulation. Various methods are used to implement auralization; details can be found in a review (Kleiner et al., 1993). BRIRs (binaural room impulse responses) contain core information on direct sounds and reflections. Binaural auralization is achieved by convoluting a mono "dry" signal with mathematically or physically obtained BRIRs, and reproducing the synthesized binaural signals through headphones or loudspeakers with appropriate crosstalk cancellation. Binaural auralization has become an important tool in the research and design of room acoustic quality. It was applied to the evaluation of various subjective attributes, such as spaciousness, envelopment, and speech intelligibility.

For an already-built room (such as an auditorium), the most accurate way to obtain BRIRs is by measurement, which can be conducted on artificial heads or human subjects. In the early stages, sparks and starting guns were frequently used as impulse sound sources. These have recently been replaced by loudspeakers and various test signals. The principle that underlies the measurement of BRIRs is similar to that of the HRTF measurement discussed in Chapter 2. Nevertheless, some differences should be noted:

1. The length of BRIRs is in the order of room reverberation time. This length is considerably longer than that of free-field HRIRs (head-related impulse responses). To avoid time aliasing, therefore, the length (or periodical length) of the numerical sequences (such as maximal length sequence) used in BRIR measurement should be longer than the length of room impulse responses.

2. In room acoustic measurements, dodecahedral loudspeaker systems are often adopted as omnidirectional sound sources. Most of the currently available dodecahedral loudspeaker systems are omnidirectional only below certain frequencies (e.g., 8 kHz). When they are used at high frequencies, directivity occurs. Under this situation, a necessary approach may be to design an improved dodecahedral loudspeaker system with omnidirectional characteristics within a wide frequency range. Conversely, actual sound sources (e.g., musical instruments) have certain complex and frequency-dependent directivities, leading to differences between actual and measured BRIRs. Subjective results show that binaural auralization with different sound source directivities result in different subjective perceptions for various aspects (Rao and Xie, 2007). Thus, a sound source with directivity similar to that in actual situations is preferred in BRIR measurement or simulation. Some studies used a monitor loudspeaker system with certain directivities as the sound sources in BRIR measurement.

3. Binaural auralization usually requires BRIRs in the entire audio frequency range. However, commonly used sound sources (such as dodecahedral loudspeakers) work only at a limited frequency range and sometimes generate un-flat magnitude responses. At the working frequency range, frequency equalization effectively compensates for the non-ideal magnitude responses of sound sources. Beyond the working frequency range, this method is invalid.

BRIRs contain primary temporal and spatial information on direct and reflected sounds; aside from binaural auralization, therefore, BRIRs are useful in analyzing binaurally relevant measures in room acoustics [such as the IACC (interaural cross-correlation coefficient) in Equation (1.34)]. BRIR measurement becomes a built-in function in some specialized software for room acoustic measurements [e.g., DIRAC (B & K) 7841]. The University of Parma in Italy conducted numerous studies on BRIR measurement and auralization. In 2004, a large-scale research plan, intended to facilitate research on famous cultural heritages, was launched; the plan included the room acoustic measurements (such as BRIR measurement) of 20 world-famous opera houses in Italy (Farina et al., 2004). Given the structures and purposes of opera houses, a dodecahedral loudspeaker with a subwoofer was used as the sound source for measurements in the orchestra pit, and a monitoring loudspeaker with directivity was used for measurements on the stage.

These auralization methods for already-built rooms are invalid for unbuilt rooms. Because resolving acoustic defects for an already-built room is difficult and costly, the predictive evaluation of the objective and subjective properties of a room in the design stage is necessary. In conventional room acoustic design, an acoustic scale model is used to evaluate the acoustic properties of the room being designed. In detail, a modeled room with N times size reduction, relative to a real room, is first constructed. Then, the acoustic properties of the room being designed are predicted by rescaling the acoustic properties of the modeled room $1/N$ times in frequency. Xiang and Blauert (1991, 1993) incorporated binaural auralization into the acoustic scale model, in which the BRIRs for auralization were obtained by acoustic scale.

With the advent of computer technology, auralization with room acoustics modeling has become an important means of room acoustic research and design, and an especially useful tool for the predictive evaluation of room acoustic properties in the design stage (Møller, 1992). The principles of room acoustics modeling and binaural auralization are discussed in Chapter 11 [Lehnert and Blauert (1992); Kleiner et al (1993)]. The required modeling accuracy depends on application. For room acoustic research and design, modeling accuracy is a comparatively more important concern than calculation time.

Some software products for room acoustic design that enable binaural auralization have been developed. Such software products usually provide an interface for defining room geometry, as well as the positions of sound sources and listeners. These software products also provide a significant library of surface materials and physical data on sound sources for users. Some libraries are open to users, enabling later supplementation of materials and sound source data. On the basis of input information, these software products simulate acoustic transmission in a room, calculate various objective acoustic parameters and BRIRs, and realize auralization.

ODEON, developed by the Technical University of Denmark (Naylor, 1993), is a type of software for room acoustic design. It is still under development, and its latest version (v. 11.0) was released in 2011. The hybrid method used in ODEON combines image source modeling and ray tracing, while considering boundary diffraction and scattering in simulating early reflections. An appropriate diffuse reflection model is incorporated into the simulation of third- and higher-order reflections. This software not only predicts the parameters of conventional room acoustics, but also realizes auralization. Various anechoic recording stimuli are included in the software, and further extension of stimuli is allowed (http://www.odeon.dk).

Another software product for room acoustics and electroacoustic design is EASE/EARS, developed by SDA Software Design Ahnert GmbH, Germany (Ahnert, 1993). The software is constantly updated and its latest version (v. 4.3) was released in April 2011. It can simulate interior rooms and open-air spaces for a single sound source or for multiple sound sources (http://www.ada-acoustic design.de/). Similar simulation software products include CATT-Acoustic (http://www.catt.se/), LMS RAYNOISE (http://www.lmsintl.com/RAYNOISE), and RAMSETE (http://pcfarina.eng.unipr.it/Ramsete_Ultimo/index.htmRAMSETE).

International round-robin tests on room acoustic computation were undertaken three times from 1995 to 2005 (Vorländer, 1995; Bork, 2000, 2005a, 2005b), with a focus on comparing calculated and measured objective acoustic parameters. The future plan of round-robin tests is to include the comparison of auralization schemes.

14.2.2 Existing Problems in Room Acoustic Binaural Auralization

Although great progress has been achieved in the computational simulation of sound fields and binaural auralization (with many successful applications in room acoustic design), some problems remain. In particular, perceptible differences exist between auralization and onsite listening conditions, which adversely affect the application and acceptability of auralization. This phenomenon may result from two sources:

1. Constrained by model complexity and computational capability, certain simplifications were implemented in the computational simulation of room acoustics, which yields approximate results. For example, geometric acoustics-based methods, which are reasonable only at high frequencies, are generally applied in room acoustics modeling. Therefore, wave acoustics-based methods are needed to improve the accuracy of room acoustics modeling at low frequencies. In most cases, simulations of sound source directivities, boundary absorptions, and reflections may also be less accurate. These factors adversely affect BRIR calculation, and, therefore, auralization quality.

2. Binaural auralization is related to VAD technology, which nevertheless suffers from inherent drawbacks, such as directional error and the in-head localization of rendered virtual sources (Sections 8.3 and 8.4). These drawbacks inevitably affect the performance of auralization. As observed, perceptual artifacts remain—even when measured BRIRs are used, or when binaural signals are directly recorded in auralization.

Improving auralization quality generally necessitates further enhancing the accuracy of room acoustics modeling and binaural reproduction. Improving room acoustics modeling entails more accurate analysis models and calculation methods, and enhancing the quality of binaural reproduction entails more accurate simulation of binaural signals, including dynamic information. As indicated in Chapters 8 and 12, other than appropriate equalization of headphone-to-ear canal transmission, the dynamic information generated by head movement and individualized HRTF processing is needed in authentic headphone-rendered VADs.

However, the limitations presented by current computer capability constrain real-time simulations. A possible approach involves performing real-time simulations of the dynamic information generated by head movements for direct sounds and preceding early reflections; this information is simulated by the parametric and decomposed structures of binaural reflection rendering (Section 12.2.3). As a result, most of the current binaural auralization software products for room acoustic design are based on static simulations, and some other dynamic VAEs (virtual auditory environments) may include dynamic binaural modeling for, at most, direct sounds and preceding early reflections (Noisternig et al., 2008). Alternatively, incorporating the dynamic information on early reflections in auralization is possible by a data-based direct rendering technique, in which premeasured BRIRs (truncated by time window) for various head orientations are used according to temporal head position. This approach is termed *binaural room scanning* (Horbach et al., 1999; Karamustafaoglu and Spikofski, 2001; Spikofski and Fruhmann, 2001; Moldrzyk et al., 2004).

The effect of individualized HRTFs on auralization quality remains an open issue. Li et al. (2006) recorded human subjects' BRIRs at the concert hall in the Sydney Opera House, and conducted a psychoacoustic experiment to investigate the listeners' abilities to discriminate between synthesized binaural signals using their own and others' BRIRs; music, white noise, and speech were used as mono input stimuli. The results indicate a discrimination performance greater than 90%. Rao and Wu (2008b) investigated the perceptual differences in auralization quality using individualized BRIRs from human subjects and generic BRIRs from KEMAR recorded in a multimedia classroom. The psychoacoustic experimental results show a discrimination performance greater than 93% for four test stimuli (two types of music, speech, and white noise), and that timbre difference is the dominant perceptible cue.

These findings support the need for individualized HRTFs in binaural auralization, but further subjective validation is required. Once this need is reconfirmed, the use of population-matched

HRTFs should be more comprehensively considered. Commonly used room acoustic design software products currently employ HRTFs from Western populations, although some allow for loading user-defined HRTFs. These types of software may be unsuitable for Eastern (such as Chinese) users. Thus, a meaningful endeavor is to investigate how the individualized HRTFs of Chinese subjects, or at least the representative HRTFs of Chinese subjects, are incorporated into auralization.

14.2.3 Other Applications of Binaural Auralization

Binaural recording and reproduction is an extensively used method for sound event recording and reproduction. Aside from room acoustics research, this method has been generally used for sound archives, subjective assessment—such as noise evaluation (Gierlich, 1992; Song et al., 2008)—and the subjective assessment of sound reproduction systems (Toole, 1991).

In-car sound quality, designed to provide a comfortable sound environment, has recently received considerable attention from acousticians, automobile designers, and automobile manufacturers. In-car sound quality is related to motor- and tire-related noise reduction, the speech intelligibility of interior colloquy as well as related communication equipment, and in-car sound reproduction quality. As one application of the auralization technique, automotive acoustic quality auralization often uses artificial head-recorded or computer-simulated binaural impulse responses for car cabins (similar to BRIRs). A car cabin is a small and enclosed space; thus, numerical computation based on wave acoustics, rather than on geometrical acoustics, is required for acoustic simulation (Granier et al., 1996).

Researchers in Parama University, Italy, conducted numerous studies on this issue (Farina and Ugolotti, 1997, 1998; Farina and Bozzoli, 2002). Some researchers also applied headphone-based auralization to the study of the preference ratings for loudspeakers in different listening rooms (Hiekkanen et al., 2009), or applied binaural room scanning to assess and design in-car sound reproduction systems (Olive and Nind, 2007).

14.3 Applications in Sound Reproduction and Program Recording

The binaural or virtual reproduction of stereophonic and multi-channel surround sound discussed in Chapter 10 can be regarded as a domestic application of VAD techniques. Among these, the headphone-based rendering in Section 10.1 is widely used in various types of portable sound playback devices, such as MP3 players. The loudspeaker-based rendering in Sections 10.2 and 10.4 is also extensively used in televisions and other domestic sound reproduction systems, wherein dedicated hardware, such as a digital signal processor, is often needed to process the signals from conventional stereophonic or surround sound programs in real time.

Another application of VAD or binaural techniques is sound program recording. The raw programs are prerecorded using VAD or binaural techniques, stored and delivered through various media (such as tapes, CDs, DVDs, hard disks, and the Internet), and then directly replayed through headphone or stereophonic loudspeakers.

Some sound programs with the label "recorded from artificial head" are created from direct binaural recording on artificial (or human) heads, and are suitable for headphone presentation. Although this type of sound program is much less popular than those created from conventional stereo, it still captures a certain share of the market.

HRTF-based processing is also applicable to program recording. Figure 14.1 shows an example of a binaural mixing console. M independent input signals e_{01}, e_{02} ... e_{0M} are separately convolved with M pairs of HRIRs for different spatial positions. The convolution results are then combined to form the binaural signals e_L and e_R. This combination corresponds to synthesizing M virtual sound sources at different spatial positions. An artificial reverberator is also used to simulate reflections.

Aside from the binaural simulation of direct sound sources, binaural synthesis is also applicable to simulating reflections from environments in program recording. For example, convolving anechoic (dry) stimuli with premeasured BRIRs simulates the binaural recording in a hall without onsite recording. Using physical-based modeling methods enables binaural recording on a virtual, rather than an actual, hall. Such methods not only simplify the procedure for program recording, but also reduce costs.

The sound programs created with these methods are intended for headphone presentation. They can also be reproduced through loudspeakers with appropriate crosstalk cancellation (Chapter 9). If crosstalk cancellation is incorporated in the recording stage, the resultant program is then directly suitable for loudspeaker reproduction. This scheme for signal synthesis is called transaural pan-pot, which generates natural and authentic virtual sources in the frontal-horizontal quadrants in stereophonic loudspeaker reproduction. Being compatible with conventional stereophonic signal recording or mixing techniques, transaural pan-pot synthesis is an alternative method for two-channel stereophonic program recording.

Some commercial software products for sound recording can realize these functions. An example is Panorama, a DirectX-based plug-in module developed by Wave Arts (www.wavearts.com). It is compatible with other commonly used sound editing and recording software, such as Adobe Audition. In addition to the basic signal-processing function in Figure 14.1, the simulation of the sound field in a rectangular room, up to first-order early reflections and diffuse reverberations, is performed by Panorama. It allows for user-defined parameters, such as virtual source direction and distance, room size, surface materials with various absorption capacities, HRTF data, and optional reproduction patterns for either headphone or loudspeaker layouts. Panorama originally contains the HRTFs of KEMAR, as well as those of nine human subjects, taken from the CIPIC database.

Binaural or transaural synthesis is applicable to converting multi-channel (such as the 5.1-channel) surround sound signals for headphone or stereophonic loudspeaker reproduction (Sections 10.1.2, 10.1.3, 10.1.4, and 10.4). If binaural or transaural conversion is pre-implemented, the resultant signals can be stored or delivered by two-channel media (such as CDs) and then directly reproduced through headphones or a pair of stereophonic loudspeakers without additional signal processing.

Transaural synthesis can also be applied to 5.1-channel surround sound program recording. As mentioned in Section 6.4.2, 5.1-channel loudspeaker configuration, along with the

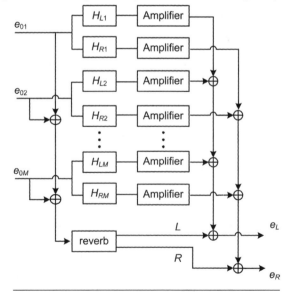

Figure 14.1 Block diagram of a binaural mixing console (adapted from Gardner, 1995c)

conventional pairwise signal mixing method, always results in a hole for virtual sources in lateral directions. Alternatively, lateral virtual sources can be recreated using two or three frontal loudspeakers in 5.1-channel configuration with signal mixing on the basis of the transaural principle presented in Section 9.1 (Horbach, 1997).

In summary, VAD or binaural techniques are successfully applied to sound program recording in various ways and with different purposes. The key point is application with flexibility.

14.4 Applications in Virtual Reality, Communication, and Multimedia

14.4.1 Applications in Virtual Reality

Virtual environments or virtual reality systems provide users the feeling of being present in natural environments through computer-controlled artificial surroundings (Blauert et al., 2000). Virtual reality includes virtual visual, auditory, and tactile stimuli. The interaction and complementarity of multiple information on these aspects strengthen the sense of reality and immersion.

An important application of virtual reality is virtual training. Unlike actual training, virtual reality offers a safe and low-cost task-training environment. A typical example is driving training simulation (Krebber et al., 2000). A virtual car acoustic environment, which is part of the virtual driving environment, requires the following components:

1. an external moving sound source with respect to the driver (e.g., traffic flow with Doppler shift);
2. fixed engine sound, which depends on engine speed and torque;
3. fixed tire sound, which depends on speed and road conditions;
4. fixed wind noise, which depends on speed;
5. background noises, commands to the driver, and other related elements.

Virtual acoustic environment systems dynamically synthesize or read sound signals from prerecorded sound databases according to a driver's control maneuvers, then reproduce the sound signals through headphones or loudspeakers with appropriate signal processing.

Similar methods can be applied to special training environments, such as virtual aviation, aerospace, and submarine environments. Doeer et al. (2007) designed a VAE system for training pilots to fly civil aircraft (Airbus A340). VAE can also be used to simulate battle environments or other types of military operational scenarios (Jones et al., 2005).

Moreover, VAEs benefit hearing training for musicians and sound engineers. An experimental study indicates that virtual acoustic game training improves participants' abilities to perform sound localization for real sound sources (Honda et al., 2007). Virtual reality can also be used in scene displays, including different types of exhibitions and entertainment fields (Kan et al., 2005).

The binaural auralization discussed in Section 14.2 can be regarded as a special application of virtual reality because auralization simulates a particular reflective (room) acoustic environment. More generally, the application of binaural synthesis to sound program recording in the preceding sections is also a special application of VAEs, an indication that VAEs are applicable to a broad range of fields.

14.4.2 Applications in Communication

An important purpose of applying VADs to speech communication is to improve speech intelligibility. In real life, conversation usually occurs in environments with background noise and simultaneously competing speech sources. When target speech sources and other competing speech sources are spatially separated, the hearing system can use the cocktail party effect (Section 1.6.4) to obtain expected information and guarantee speech intelligibility. This ability is attributed to binaural hearing.

However, mono signal transmission is dominantly used in currently available communication systems, in which the inability to spatially separate targets and competing sources degrades speech intelligibility. As mentioned in Section 14.1, the spatial distribution information in sound sources facilitates the improvement of speech intelligibility, and can be incorporated into communication systems by VADs (Begault and Erbe, 1994b; Drullman and Bronkhorst, 2000). Psychoacoustic experimental results indicate that spatially separating multiple speech sources by VADs improves speech intelligibility for either full-bandwidth or 4 kHz-low pass (phone quality) speech signals (Begault, 1999).

In addition, commonly used headphone presentation suffers from unnatural in-head localization and hearing fatigue after headphones have been worn for long periods. In this sense, the other significant purpose of applying VADs to speech communication is to produce natural auditory perception, and, consequently, reduce hearing fatigue.

Teleconferencing is another important application of VADs in speech communication, in which multiple speakers are simultaneously present (Kang and Kim, 1996; Evans et al., 1997). In remote conferencing, if participants are distributed in two or more separate meeting rooms, the direct approach to preserving the spatial information and improving the intelligibility of transmitted speech is to combine and reproduce the binaural sound signals obtained by artificial-head recording in each meeting room. Telecommunications industries provide face-to-face communication services; thus, VADs can be incorporated to virtually manipulate recorded sound signals toward particular spatial distributions and acoustic environments. Using VADs eliminates spatial distance, making participants feel as though discussions are held in the same (virtual) meeting room. A similar method can be used in situations that require the simultaneous monitoring of multiple speeches, such as those occurring in military command centers and emergency telephone systems.

VADs also contribute to aeronautical communication, and numerous investigations on this application were undertaken by the NASA Ames Research Center (Begault, 1998). The projects are categorized as a combined application of VADs on speech communication and information orientation. Given that civil aircraft cockpits are characterized by high environmental noise, headphones (aside from speech communication) are used to reproduce air traffic warnings, on whose basis pilots determine target (e.g., other aircraft) directions or identify corresponding visual targets (e.g., radar display) and, accordingly, take appropriate measures. Applying VADs to aeronautical communication improves speech intelligibility and reduces the search or reaction time of pilots with the help of spatialized auditory warnings. The latter is important to flight safety. Additionally, headphone presentation may be combined with active noise control to reduce pilot exposure to binaural noise.

The VAD applications in aeronautical communication include auditory-based information display and orientation. In some cases, vision is often superior to hearing in terms of target identification and orientation. However, acoustic information becomes particularly important when a target is out of visual range (e.g., behind the listener) or when overloading of visual information occurs (such as multiple visual targets). In real life, auditory information often guides visual orientation (Bolia et al., 1999), and goals can be localized through hearing even without visual help (Lokki and Grohn, 2005). Revealing target information and orientation is therefore another important application of VADs.

Audio navigation systems, which combine global positioning systems with VADs, reproduce sounds as they are emitted from target directions. These systems are applied primarily in civil or military rescue searches (Kan et al., 2004). A similar method can be used to present various types of spatial auditory information (Dobrucki et al., 2010), such as that contained in guidance and information systems for the blind (Loomis et al., 1998; Bujacz, et al., 2012), or in tourism and museum applications (Gonot et al., 2006).

Monitoring multiple targets (such as different instruments and meters) is often necessary in practice; such targets cause visual overload. Under this situation, VADs are used to alleviate the visual burden by transforming part of the visual presentation of spatial information into auditory presentation (i.e., nonvisual orientation). Actually, various forms of sound design to provide useful information are called *sonification* (Barrass, 2012).

14.4.3 Applications in Multimedia and Mobile Products

In professional applications, various functions, such as communication and virtual reality, may be implemented by corresponding equipment. In consumer applications, however, users may prefer multifunctional and integrated equipment.

Characterized by integration and interaction, multimedia PCs can comprehensively handle various types of information, including audio, video, image, text, and data. Information exchange between computers is also possible through the Internet. Even standard PCs possess these functions, making them ideal platforms for communication, information processing, and virtual reality.

VADs are widely incorporated into the entertainment functions of multimedia PCs. Currently, a multimedia PC connected to a pair of headphones or small loudspeakers located on both sides of the monitor is often used to reproduce diverse audio-video programs (e.g., DVDs, CDs, and MP3s). Most of these programs are recorded in conventional stereophonic or 5.1-channel surround sound format, making the binaural or virtual processing in Chapter 10 applicable to improving the reproduction effect. A variety of 3-D computer games also use VADs to produce distinct spatial audio-visual effects. Sound cards for multimedia PCs and application software under Windows platforms can synthesize spatial virtual sources or process binaural/virtual stereophonic and multi-channel surround sound signals. Moreover, interactive and dynamic binaural synthesis on multimedia computer platforms is possible, through the addition of head-tracking devices (Lopez and Gonzalez, 1999; Kyriakakis, 1998).

Various applications of VADs to virtual reality, as well as communication and information systems, can be implemented on multimedia computer platforms. These applications include digital multimedia broadcasting and virtual online shopping. Multimedia computers are also used as teleconferencing terminals.

MPEG-4, a low-bit-rate coding standard for multimedia video and audio, was formulated by the Moving Picture Experts Group under ISO and IEC in 1995, with the first and second editions released in 1999 and 2000, respectively. MPEG-4 differs from earlier MPEG-1 and MPEG-2 formats, in that the former not only specifies the coding scheme for low-bit-rate audio signals, but also enables object-oriented coding. It considers every sound source's signal (natural or synthesized sound) as an independent transmitted object or element, and resynthesizes it into a complete auditory (or, more precisely, audio-visual) scene in a user terminal. To achieve sound scene combination, MPEG-4 adopts *Audio Binary Format for Scene Description* (audio BIFS) as a tool to describe sound scene parameters, while retaining flexibility in defining combination methods. Users can flexibly compile and combine these objects, and local interaction is allowable for synthesized scenes from different viewing (listening) positions and angles. MPEG-4 supports virtual auditory environment applications, which have become part of MPEG-4. Substantial research has been devoted to such applications (Scheirer et al., 1999; Vaananen, 2000; Vaananen and Huopaniemi, 2004; Jot and Trivi, 2006b; Dantele et al., 2003; Seo et al., 2003; Lee et al., 2005).

The second edition of MPEG-4 provides parameters that describe three-dimensional acoustic environments in advanced audio BIFS, which includes the parameters of rectangular rooms (e.g., room size and frequency-dependent reverberation time), the parameters of sound source characteristics (e.g., frequency-dependent directivity, position, and intensity), and the acoustic parameters of surface materials (e.g., frequency-dependent reflection or absorption coefficients). Auditory scenes are synthesized at a user's terminal in terms of these parameters. A listener's movement in virtual space causes changes in binaural signals. Interactive signal processing is supported to simulate the dynamic behavior of binaural signals in accordance with a listener's temporal head orientation. Because MPEG-4 does not specify sound synthesis and reproduction methods, many types of sound synthesis and reproduction technologies can be adopted, depending on application requirements and hardware performance at a user's terminal. A real-time and dynamic virtual auditory environment system (Chapter 12) is usually an appropriate choice.

Another possible application of VADs related to PCs is audio interface for the blind (Wersenyi, 2012). The typical graphical user interfaces of a PC are inapplicable to blind people, driving the need for alternative interfaces with auditory displays.

Mobile communication and handheld sound reproduction products have seen rapid development in recent years. From a practical perspective, using VADs in these types of products may be a promising direction. Some corporations and research institutes have already launched relevant studies (AES Staff Technical Writer, 2006; Yasuda et al., 2003; Paavola et al., 2005; Choi et al., 2006). Mobile products are characterized by a combination of functions, such as speech communication, interactive VAEs, teleconferencing, spatial auditory information presentation (e.g., traffic directions), and entertainment (e.g., video-audio reproduction and 3-D games). Therefore, such products can be regarded as an application of multimedia technology. The improved speed and bandwidth of wireless communication networks enhance the possibility of realizing these functions.

The manner by which VADs are applied in mobile products is similar to that by which VADs are used in other applications. However, two problems require discussion. One is the appropriate simplification of data and algorithms, given the limited processing and storage ability of mobile products. The other problem is that headphones are better reproduction devices than loudspeakers because the former consume relatively low power; headphones also avoid the poor low-frequency performance associated with small loudspeakers and difficulty in crosstalk cancellation for a pair of close-span loudspeaker configuration. These problems are not absolute because situations may change with the development of technology. Some researchers have proposed crosstalk cancellation and speaker reproduction methods for mobile products (Bai et al., 2007). A number of mobile phone brands are already equipped with these functions. With the development of inexpensive head trackers, dynamic VADs are also expected to be applied to mobile products (Porschmann, 2007; Algazi and Duda, 2011).

Augmented reality is a new application that aims to both realistically and virtually reproduce environments. Augmented auditory or acoustic information can be regarded as complementary to natural auditory information. In some cases, for example, VAD listeners need to simultaneously listen to environmental sounds, such as the user of some mobile products. This requirement can be satisfied by arranging a pair of miniature microphones outside the in-ear headphone (Lokki et al., 2004). This miniature binaural microphone functions as a head tracker, in addition to recording environmental sounds (Tikander et al., 2004), and makes dynamic virtual auditory signal processing possible. In this application, the headphone reproduces the combined virtual acoustic signal and real sound signal, in which virtual auditory information is reproduced for communication and information services.

With the support of the European Commission in the Context of the Information Society Technology, a few European research institutes have undertaken the LISTEN plan since 2001 (Eckel, 2001; Delerue, 2002). The plan aims to study the manner by which real environments are augmented by reproduction systems for virtual auditory information.

14.5 Applications in Clinical Auditory Evaluations

As stated in Sections 1.6.4 and 14.4.2, the hearing system uses the cocktail party effect to obtain targeted speech information and guarantee speech intelligibility in noisy environments, which are characterized by competing sound sources. This ability is attributed to binaural hearing. Therefore, binaural hearing is crucial to information identification in noisy environments for listeners of normal hearing. For individuals with impaired hearing, however, the ability to identify binaural signals declines or disappears, resulting in a phenomenon of "hearing without understanding." Hence, the binaural perception of spatial sound can be used as a means of clinical diagnosis (Shinn-Cunningham, 1998c). One approach is to use VADs to diagnose patient hearing ability by sound localization test or speech intelligibility evaluation in (virtual) noisy environments. Hearing aids designed in accordance with binaural principles partially improve patients' binaural spatial hearing and identification abilities.

14.6 Summary

Virtual auditory displays (VADs) present significant application values in many fields and this chapter discussed only some typical applications. In binaural scientific experiments, researchers adopt VADs to artificially control binaural signals for the purpose of investigating the relative contributions of various acoustic information and cues under different conditions. Binaural auralization evaluates and predicts the subjective attributes of room sound quality; it is, therefore, a major tool for room acoustic design and sound quality research. Many types of room acoustic design software are currently endowed with binaural auralization functions. Binaural auralization also applies to other acoustic designs and subjective evaluations. Some defects in binaural auralization require resolution. The principle that underlies binaural and transaural syntheses in VADs can be used in sound recording with various methods and purposes. Virtual reality is an important application of VADs, especially in a variety of civil and military training programs. VADs are applied in communication and information systems for three reasons: to improve speech intelligibility, enhance sound orientation, and alleviate visual burden. Multimedia platforms are ideal environments for virtual reality, as well as for information communication and processing. The multimedia video and audio compression standard, MPEG-4, supports the application of virtual auditory environments. Mobile communication and handheld sound reproduction products will be the most promising applications of VADs in the future. VADs are also applicable to hearing diagnoses in medical science.

Appendix A
Spherical Harmonic Functions

Spherical harmonic functions (SHFs) are an infinite series of functions of the angular variables $0° \leq \alpha \leq 180°$ and $0° \leq \beta < 360°$, which specify spatial direction. In the default coordinate in Figure 1.1, angle $\alpha = 90° - \phi$ denotes the elevation and $\beta = \theta$ denotes the azimuth. Alternatively, in the interaural polar coordinate in Figure 1.3, $\alpha = \Gamma = 90° - \Theta$ denotes the complementary of interaural polar azimuth and $\beta = \Phi$ represents the interaural polar elevation. Although variables (α, β) denote different angles in two coordinate systems, the equations for calculating SHFs are consistent. In some mathematical and physical textbooks on SHFs, variable θ usually denotes elevation and ϕ represents azimuth, which conflicts with most literature on HRTFs (head-related transfer functions) and VADs (virtual auditory displays) and may, therefore, cause confusion. Therefore, notations (α, β) rather than (θ, ϕ) are used in this work to denote the variables in SHFs. For simplicity, symbol Ω is used to denote angles (α, β), i.e., $\Omega = (\alpha, \beta)$.

Two equivalent forms of SHFs—real-valued and complex-valued SHFs—are available. Both are often applied to acoustical (including HRTF) analysis. Normalized *complex-valued SHFs* are defined as:

$$Y_{lm}(\Omega) = Y_{lm}(\alpha, \beta) = N_{lm} P_l^{|m|}(\cos\alpha)\exp(jm\beta)$$
$$l = 0,1.2\dots; m = 0,\pm 1,\pm 2,\dots \pm l \tag{A.1}$$

where l is the order of SHFs and N_{lm} is the normalized factor, expressed as:

$$N_{lm} = \sqrt{\frac{(l-|m|)!\ (2l+1)}{(l+|m|)!\ 4\pi}}. \tag{A.2}$$

$P_l^{|m|}$ is the associated Legendre polynomial and is defined by:

$$P_l^{|m|}(x) = (1-x^2)^{\frac{|m|}{2}} \frac{d^{|m|}}{dx^{|m|}} P_l(x). \tag{A.3}$$

$P_l(x)$ is the l-degree Legendre polynomial, which is the solution to the Legendre equation:

$$(1-x^2)\frac{d^2 y}{dx^2} - 2x\frac{dy}{dx} + l(l+1)y = 0 \qquad -1 \leq x \leq 1. \tag{A.4}$$

The zero- and first-order complex-valued SHFs are:

$$Y_{00}(\Omega) = \frac{1}{\sqrt{4\pi}} \qquad Y_{10}(\Omega) = \sqrt{\frac{3}{4\pi}}\cos\alpha \qquad Y_{1\pm1}(\Omega) = \sqrt{\frac{3}{8\pi}}\sin\alpha\exp(\pm j\beta). \tag{A.5}$$

According to Equation (A.1) and Euler equation $\exp(\pm jm\beta) = \cos(m\beta) \pm j\sin(m\beta)$, normalized *real-valued SHFs* are defined as:

$$Y_{lm}^1(\Omega) = Y_{lm}^1(\alpha,\beta) = N_{lm}^1 P_l^m(\cos\alpha)\cos(m\beta)$$
$$Y_{lm}^2(\Omega) = Y_{lm}^2(\alpha,\beta) = N_{lm}^2 P_l^m(\cos\alpha)\sin(m\beta).$$
$$l = 0,1,2.....;\quad m = 0,1,2....l$$

(A.6)

Because $m \geq 0$ in Equation (A.6), the absolute value sign for m is unnecessary in the associated Legendre polynomial. The normalized factor is provided by:

$$N_{lm}^1 = N_{lm}^2 = \sqrt{\frac{(l-m)!\ (2l+1)}{(l+m)!\ 2\pi\Delta_m}} \qquad \Delta_m = \begin{cases} 2 & m=0 \\ 1 & m\neq 0 \end{cases}.$$

(A.7)

The zero- and first-order real-valued SHFs are:

$$Y_{00}^1(\Omega) = \frac{1}{\sqrt{4\pi}} \qquad Y_{10}^1(\Omega) = \sqrt{\frac{3}{4\pi}}\cos\alpha$$

$$Y_{11}^1(\Omega) = \sqrt{\frac{3}{4\pi}}\sin\alpha\cos\beta \qquad Y_{11}^2(\Omega) = \sqrt{\frac{3}{4\pi}}\sin\alpha\sin\beta$$

(A.8)

The complex and real-valued SHFs are related by the equations:

$$Y_{l0}(\Omega) = Y_{l0}^1(\Omega)$$

$$Y_{lm}(\Omega) = \begin{cases} \dfrac{\sqrt{2}}{2}[Y_{lm}^1(\Omega) + jY_{lm}^2(\Omega)] & m>0 \\[2mm] \dfrac{\sqrt{2}}{2}[Y_{l,-m}^1(\Omega) - jY_{l,-m}^2(\Omega)] & m<0 \end{cases}.$$

(A.9)

The complex-valued SHFs satisfy orthonormality thus:

$$\int Y_{l'm'}^*(\Omega)Y_{lm}(\Omega)d\Omega = \int_{\beta=0}^{2\pi}\int_{\alpha=0}^{\pi} Y_{l'm'}^*(\alpha,\beta)Y_{lm}(\alpha,\beta)\sin\alpha\,d\alpha\,d\beta = \delta_{ll'}\delta_{mm'},$$

(A.10)

where superscript notation "*" denotes complex conjugation.

The real-valued SHFs also satisfy orthonormality as:

$$\int Y_{l'm'}^{\sigma'}(\Omega)Y_{lm}^{\sigma}(\Omega)d\Omega = \int_{\beta=0}^{2\pi}\int_{\alpha=0}^{\pi} Y_{l'm'}^{\sigma'}(\alpha,\beta)Y_{lm}^{\sigma}(\alpha,\beta)\sin\alpha\,d\alpha\,d\beta = \delta_{ll'}\delta_{mm'}\delta_{\sigma\sigma'}\quad \sigma,\sigma'=1,2.$$

(A.11)

Because of the orthonormality and completeness of SHFs, any square integrable function $F(\Omega) = F(\alpha,\beta)$ can be decomposed by complex-valued or real-valued SHFs thus:

$$F(\Omega) = \sum_{l=0}^{\infty}\sum_{m=-l}^{l} d_{lm}Y_{lm}(\Omega) = a_{00}Y_{00}^1(\Omega) + \sum_{l=1}^{\infty}\sum_{m=0}^{l}[a_{lm}Y_{lm}^1(\Omega) + b_{lm}Y_{lm}^2(\Omega)].$$

(A.12)

Note that $Y_{l0}^2(\alpha,\beta) = 0$, which is retained in Equation (A.12) only for convenience. With the orthonormality of SHFs given by Equations (A.10) and (A.11), the coefficients of complex-valued SH (spherical harmonic) decomposition are evaluated by:

$$d_{lm} = \int F(\Omega)Y_{lm}^*(\Omega)d\Omega = \int_{\beta=0}^{2\pi}\int_{\alpha=0}^{\pi} F(\alpha,\beta)Y_{lm}^*(\alpha,\beta)\sin\alpha\,d\alpha\,d\beta.$$

(A.13)

The coefficients of real-valued SH decomposition are evaluated by:

$$a_{lm} = \int F(\Omega)Y_{lm}^1(\Omega)d\Omega = \int_{\beta=0}^{2\pi} \int_{\alpha=0}^{\pi} F(\alpha,\beta)Y_{lm}^1(\alpha,\beta)\sin\alpha \, d\alpha \, d\beta$$

$$b_{lm} = \int F(\Omega)Y_{lm}^2(\Omega)d\Omega = \int_{\beta=0}^{2\pi} \int_{\alpha=0}^{\pi} F(\alpha,\beta)Y_{lm}^2(\alpha,\beta)\sin\alpha \, d\alpha \, d\beta \qquad (A.14)$$

The coefficients of complex-valued and real-valued SH decomposition are related by the equations:

$$a_{l0} = d_{l0} \qquad d_{lm} = \begin{cases} \dfrac{\sqrt{2}}{2}[a_{lm} - jb_{lm}] & m > 0 \\[2mm] \dfrac{\sqrt{2}}{2}[a_{l,-m} + jb_{l,-m}] & m < 0 \end{cases}, \qquad (A.15)$$

or:

$$a_{lm} = \frac{\sqrt{2}}{2}[d_{lm} + d_{l,-m}] \qquad b_{lm} = \frac{\sqrt{2}}{2}j[d_{lm} - d_{l,-m}] \qquad 1 \le m \le l, \qquad (A.16)$$

and:

$$\sum_{m=-l}^{l} |d_{lm}|^2 = \sum_{m=0}^{l} [|a_{lm}|^2 + |b_{lm}|^2]. \qquad (A.17)$$

As a special case, the Dirac delta function can be decomposed by SHFs thus:

$$\delta(\Omega - \Omega') = \frac{1}{\sin\alpha}\delta(\alpha - \alpha')\delta(\beta - \beta') = \delta(\cos\alpha - \cos\alpha')\delta(\beta - \beta')$$

$$= \sum_{l=0}^{\infty}\sum_{m=-l}^{l} Y_{lm}^*(\Omega')Y_{lm}(\Omega). \qquad (A.18)$$

The SH decomposition yields the addition equation:

$$\frac{2l+1}{4\pi}P_l(\cos\Gamma) = \sum_{m=-l}^{l} Y_{lm}^*(\Omega')Y_{lm}(\Omega), \qquad (A.19)$$

where Γ is the angle between directions Ω and Ω'.

A directional-continuous function $F(\Omega) = F(\alpha, \beta)$ is required for calculating SH coefficients with Equation (A.13) or Equation (A.14). Sometimes, however, only the discrete samples of the $F(\Omega)$ measured at M directions—$F(\Omega_i) = F(\alpha_i, \beta_i)$, $i = 0,1 \ldots (M-1)$—are available. Using appropriate directional sampling schemes, the integral in Equation (A.13) or Equation (A.14) can be replaced by a weighted sum of M directional samples of $F(\Omega)$, so that SH coefficients up to a finite order $(L-1)$ or L^2 SH coefficients are calculated. Accordingly, the truncated or spatial bandlimited SH representation of $F(\Omega)$ is obtained. Directional sampling (Shannon-Nyquist) theorem requires $L^2 \le M$. That is, SH coefficients up to the $\left(\sqrt{M} - 1\right)$ order, at most, can be evaluated from the M directional sampling of $F(\Omega)$.

Function decomposition by complex-valued SHFs provides an example. The truncated SH representation of $F(\Omega)$, as well as the associated SH coefficients, are provided by [the coefficients of real-valued SH decomposition are related to those of complex-valued SH decomposition by Equation (A.16)]:

$$F(\Omega) = \sum_{l=0}^{L-1} d_{lm} Y_{lm}(\Omega), \tag{A.20}$$

$$d_{lm} = \sum_{i=0}^{M-1} \lambda_i F(\Omega_i) Y_{lm}^*(\Omega_i), \tag{A.21}$$

where λ_i is the quadrature weight that determines the relative contribution of the $F(\Omega)$ sample at direction Ω_i to the summation. Weight set $\{\lambda_i, i = 0,1 \dots (M-1)\}$ depends on the directional sampling scheme. The directional sampling scheme and associated weight set should be appropriately selected to ensure that the M directional samples of SHFs satisfy the discrete orthonormality:

$$\sum_{i=0}^{M-1} \lambda_i Y_{l'm'}^*(\Omega_i) Y_{lm}(\Omega_i) = \delta_{ll'}\delta_{mm'}. \tag{A.22}$$

Some directional sampling schemes can be used to effectively calculate SH coefficients from the discrete directional samples of function $F(\Omega)$. Three common schemes are outlined here [Rafaely (2005)].

Scheme 1, *Equiangle sampling.* Elevation $\alpha = 90° - \phi$ and azimuth $\beta = \theta$ are, respectively, uniformly sampled at $2L$ angles, resulting in $M = 4L^2$ directional samples denoted by $(\alpha_q, \beta_{q'})$, $q, q' = 0,1, \dots (2L-1)$. The SH coefficients in Equation (A.21) are evaluated by:

$$d_{lm} = \sum_{q=0}^{2L-1} \sum_{q'=0}^{2L-1} \lambda_q F(\alpha_q, \beta_{q'}) Y_{lm}^*(\alpha_q, \beta_{q'}) \quad l \le (L-1), |m| \le l. \tag{A.23}$$

Weights λ_q satisfy the relationship:

$$\sum_{q=0}^{2L-1} \lambda_q P_l(\cos\beta_q) = \sqrt{2}\delta_{l0}. \tag{A.24}$$

The equiangle sampling scheme is convenient for practical measurement, but requires $M = 4L^2$ directional samples to calculate SH coefficients up to the $(L-1)$ order, which is fourfold the lower bound imposed by the Shannon-Nyquist theorem. Therefore, the equiangle sampling scheme exhibits low efficiency.

Scheme 2, *Gauss-Legendre sampling.* Elevation is first sampled at L angles $\alpha_q = 90° - \phi_q$ ($q = 0,1, \dots L-1$), so that $\{\cos(\alpha_q)\}$ are the Gauss-Legendre nodes (the zeros of the L-degree Legendre polynomial) and $\{\gamma_{q'}\}$ are the weights within the range $[-1,1]$. Then, the azimuth in each elevation are sampled at $2L$ angles $\beta_{q'} = \theta_{q'}$ ($q' = 0,1,2 \dots, 2L-1$) with even intervals. Therefore, $M = 2L*L = 2L^2$ directional samples are derived and denoted by $(\alpha_q, \beta_{q'})$, $q = 0, 1,\dots (L-1)$, $q' = 0,1,\dots (2L-1)$. The SH coefficients in Equation (A.21) are calculated by:

$$d_{lm} = \sum_{q=0}^{L-1} \sum_{q'=0}^{2L-1} \lambda_q F(\alpha_q, \beta_{q'}) Y_{lm}^*(\alpha_q, \beta_{q'}). \tag{A.25}$$

The Gauss-Legendre sampling scheme requires $M = 2L^2$ directional samples to calculate SH coefficients up to the $(L-1)$ order, which is twice the lower bound imposed by the Shannon-Nyquist theorem. Gauss-Legendre sampling is more efficient than equiangle sampling in terms of the number of directional samples required, but it is usually inconvenient for practical HRTF measurement.

Scheme 3, *Nearly uniform sampling.* Directional samples are uniformly or nearly uniformly distributed on the surface of a spatial sphere, so that the distance between neighboring samples is constant or nearly constant. A uniform sampling scheme requires at least $M = L^2$ directional samples to calculate SH coefficients of up to the $(L-1)$ order, which is the lower bound imposed by the Shannon-Nyquist theorem. In practice, 1.3 to 1.5 times the lower bound is required, and the latter yields a set of equal weights λ_i in the calculation of Equation (A.21). Nearly uniform sampling is the most efficient in terms of the number of directional samples required, but it is also inconvenient for practical measurement.

Appendix B

Multipole Re-expansions for Calculating the HRTFs of the Snowman Model

As indicated in the discussion in Section 4.2.1, the acoustic principle of reciprocity is incorporated into the calculation of HRTFs for the snowman model. As shown by Equation (4.18), the overall pressure at the spatial field point r caused by a point source located at the ear position r' consists of three parts: the free-field pressure $P_0(r, r', f)$ generated by the point source, as well as the scattered pressures $P_A(r, r', f)$ and $P_B(r, r', f)$ caused by spherical head A and spherical torso B, respectively.

$$P(r,r',f) = P_0(r,r',f) + P_A(r,r',f) + P_B(r,r',f). \tag{B.1}$$

On the surfaces of spheres A and B, $P(r, r', f)$ satisfies the rigid boundary condition given by Equations (4.16) and (4.17) as:

$$\frac{\partial P(r,r',f)}{\partial n}\bigg|_{SA} = 0, \tag{B.2a}$$

$$\frac{\partial P(r,r',f)}{\partial n}\bigg|_{SB} = 0. \tag{B.2b}$$

The free-field pressure $P_0(r, r', f)$ is given by Equation (4.19) as:

$$P_0(r,r',f) = j\frac{k\rho_0 cQ_0}{4\pi|r-r'|}\exp(-jk|r-r'|). \tag{B.3}$$

Equation (4.20) indicates that, in the local coordinate centered at spherical head A (short for head-centered coordinate), $P_A(r, r', f)$ can be expanded as a series of complex-valued spherical harmonic functions (SHFs):

$$P_A(r,r',f) = P_A(r_A,r'_A,f) = \sum_{l=0}^{\infty}\sum_{m=-l}^{l} A_{lm}^{A} h_l(kr_A)Y_{lm}(\Omega_A). \tag{B.4}$$

As shown by Equation (4.21), in the local coordinate centered at spherical torso B (short for torso-centered coordinate), $P_B(r, r', f)$ can also be expanded as a series of complex-valued SHFs:

$$P_B(r,r',f) = P_B(r_B,r'_B,f) = \sum_{l'=0}^{\infty}\sum_{m'=-l'}^{l'} A_{l'm'}^{B} h_{l'}(kr_B)Y_{l'm'}(\Omega_B). \tag{B.5}$$

The boundary conditions given by Equations (B.2a) and (B.2b) are applied to the overall pressure provided by Equation (B.1) to obtain the coefficients A_{lm}^{A} and $A_{l'm'}^{B}$. Applying the boundary conditions on the surface of spherical head A necessitates representing pressures $P_0(\boldsymbol{r}, \boldsymbol{r}', f)$ and $P_B(\boldsymbol{r}, \boldsymbol{r}', f)$ in the head-centered coordinate. We use the expansion:

$$\frac{\exp(-jk|\boldsymbol{r}-\boldsymbol{r}'|)}{4\pi|\boldsymbol{r}-\boldsymbol{r}'|} = \begin{cases} -jk\sum_{l=0}^{\infty}\sum_{m=-l}^{l} j_l(kr')h_l(kr)Y_{lm}^*(\Omega')Y_{lm}(\Omega) & r>r' \\[3mm] -jk\sum_{l=0}^{\infty}\sum_{m=-l}^{l} h_l(kr')j_l(kr)Y_{lm}^*(\Omega')Y_{lm}(\Omega) & r<r' \end{cases}, \tag{B.6}$$

where superscript notation "*" denotes complex conjugation; and $j_l(kr)$ and $h_l(kr)$ are the l-order spherical Bessel function and spherical Hankel function of the second kind, respectively. In the head-centered coordinate, the source is located at the ear position with $r'_A = a_A$. For the arbitrary field point outside the head, Equation (B.3) can be expanded by complex-valued SHFs thus:

$$P_0(\boldsymbol{r},\boldsymbol{r}',f)=P_0(\boldsymbol{r}_A,\boldsymbol{r}'_A,f)=\sum_{l=0}^{\infty}\sum_{m=-l}^{l} C_{lm}^A h_l(kr_A)Y_{lm}(\Omega_A) \quad r_A>r'_A. \tag{B.7}$$

The SH (spherical harmonic) coefficients are given by:

$$C_{lm}^A = C_{lm}^A(r'_A,\Omega'_A,f)=k^2\rho_0 cQ_0\, j_l(kr'_A)Y_{lm}^*(\Omega'_A). \tag{B.8}$$

The SH coefficients are determined when the source (ear) position (r'_A, Ω'_A) is given.

To represent the $P_B(\boldsymbol{r}, \boldsymbol{r}', f)$ given by Equation (B.5) in the head-centered coordinate, the following transform equation is applied:

$$h_{l'}(kr_B)Y_{l'm'}(\Omega_B)=\sum_{l=0}^{\infty}\sum_{m=-l}^{l} D_{l'm',lm}^{BA}\, j_l(kr_A)Y_{lm}(\Omega_A), \tag{B.9}$$

where $D_{l'm',lm}^{BA}$ are the coefficients for transforming the SH representation from the torso-centered coordinate to the head-centered coordinate, which are determined by the geometrical relationship between two coordinates. The details for calculating $D_{l'm',lm}^{BA}$ are complicated and therefore excluded here. A more comprehensive explanation can be found in the literature (Gumerov and Duraiswami, 2002a; Gumerov et al., 2002b).

Substituting Equations (B.4), (B.5), (B.7), and (B.9) into Equation (B.1) yields the coordinate A-based representation of overall pressure at the field point $\boldsymbol{r}_A$:

$$P(\boldsymbol{r}_A,\boldsymbol{r}'_A,f)=\sum_{l=0}^{\infty}\sum_{m=-l}^{l}\left\{\left[\sum_{l'=0}^{\infty}\sum_{m'=-l'}^{l'} D_{l'm',lm}^{BA} A_{l'm'}^B\right]j_l(kr_A)+(C_{lm}^A+A_{lm}^A)h_l(kr_A)\right\}Y_{lm}(\Omega_A). \tag{B.10}$$

In the torso-centered coordinate, for sources located on the ear position $\boldsymbol{r}'_B$ and field point $\boldsymbol{r}_B$ near the torso surface, $r_B < r_B'$ is always derived. Similar to that derivation, the overall pressure at the field point $\boldsymbol{r}_B$ is given by:

$$P(\boldsymbol{r}_B,\boldsymbol{r}'_B,f)=\sum_{l=0}^{\infty}\sum_{m=-l}^{l}\left\{\left[\sum_{l'=0}^{\infty}\sum_{m'=-l'}^{l'} D_{l'm',lm}^{AB} A_{l'm'}^A+C_{lm}^B\right]j_l(kr_B)+A_{lm}^B h_l(kr_B)\right\}Y_{lm}(\Omega_B), \tag{B.11}$$

where C_{lm}^B is a set of coefficients for the SH expansion of Equation (B.3) in the torso-centered coordinate representation:

$$C_{lm}^B = C_{lm}^B(r'_B,\Omega'_B,f)=k^2\rho_0 cQ_0\, h_l(kr'_B)Y_{lm}^*(\Omega'_B). \tag{B.12}$$

Coefficients C_{lm}^B are determined when the source position (r'_B, Ω'_B) is given. $D_{l'm',lm}^{AB}$ are the coefficients for transforming the SH representation from coordinates A to B, which are determined by the geometrical relationship between two coordinates or transforms.

The boundary conditions provided by Equations (B.2a) and (B.2b) are applied in Equations (B.10) and (B.11), respectively. That is,

$$\left.\frac{\partial P(\mathbf{r}_A, \mathbf{r}'_A, f)}{\partial r_A}\right|_{r_A = a_A} = 0 \qquad \left.\frac{\partial P(\mathbf{r}_B, \mathbf{r}'_B, f)}{\partial r_B}\right|_{r_B = a_B} = 0$$

The orthonormality of complex-valued SHFs yields:

$$h_l'(ka_A)A_{lm}^A + j_l'(ka_A)\sum_{l'=0}^{\infty}\sum_{m'=-l'}^{l'} D_{l'm',lm}^{BA}A_{l'm'}^B = -h_l'(ka_A)C_{lm}^A, \qquad (B.13a)$$

$$h_l'(ka_B)A_{lm}^B + j_l'(ka_B)\sum_{l'=0}^{\infty}\sum_{m'=-l'}^{l'} D_{l'm',lm}^{AB}A_{l'm'}^A = -j_l'(ka_B)C_{lm}^B, \qquad (B.13b)$$

$$l = 0,1,2.....\infty \quad m = -l,.....+l.$$

These equations are an infinite set of linear algebraic equations, with an infinite set of coefficients A_{lm}^A and A_{lm}^B as unknowns. In practical calculation, Equations (B.13a) and (B.13b) can be truncated to order $(L-1)$—that is, we let $A_{lm}^A = A_{lm}^B = 0$ when $l \geq L$. Then, Equations (B.13a) and (B.13b) become a finite set of linear algebraic equations, from which A_{lm}^A and A_{lm}^B can be solved. The truncation order $(L-1)$ is selected as (Gumerov and Duraiswami, 2002a):

$$(L-1) = \text{int}eger\left(\frac{ekr_{AB}}{2}\right), \qquad (B.14)$$

where r_{AB} is the distance between the centers of the spherical head and torso. Substituting the resultant A_{lm}^A and A_{lm}^B into Equations (B.4) and (B.5) and using Equation (B.1) yield the overall pressure $P(\mathbf{r}, \mathbf{r}', f)$. In addition, the coefficients C_{lm}^A and C_{lm}^B in Equations (B.13a) and (B.13b) are source position dependent, but only one source position (ear) is of concern. Therefore, Equations (B.13a) and (B.13b) are solved only once for every frequency or wave number. This feature is the advantage of reciprocal calculation.

The first and second terms on the left side of Equation (B.13a) respectively correspond to the contributions of the scattering of spherical head A and the scattering of spherical torso B to the overall pressure at the surface of spherical head A. The second term is 0 in the conventional rigid-spherical head model because the torso is disregarded. Similarly, the first and second terms on the left side of Equation (B.13b) respectively correspond to the contributions of the scattering of spherical torso B and the scattering of spherical head A to the overall pressure at the surface of spherical torso B. As the first-order approximation, the first-order scattering from the torso is considered and the higher-order (multiple) head-torso scattering is disregarded. These conditions are realized by excluding the second term on the left side of Equation (B.13b) and solving the A_{lm}^B from the equation. A_{lm}^A can, in turn, be solved by substituting the resultant A_{lm}^B into Equation (B.13a). Finally, the overall pressure at the field point $\mathbf{r}$ is derived by substituting A_{lm}^B and A_{lm}^A into Equations (B.4) and (B.5), and then truncating them to the $(L-1)$ order. Equation (B.1) is subsequently used. The calculation indicates that, for far-field snowman-HRTFs (head-related transfer functions), the first-order approximate yields results with adequate accuracy. Compared with the first-order approximation, the higher order approximation results in only 1–2 dB improvement in the calculated HRTF magnitudes (Gumerov et al., 2002b). Because only the first-order torso scattering is significant to the far-field snowman HRTFs, the model and calculation can be simplified. Even a geometrical acoustic model for the torso is adequate. However, calculating the near-field snowman HRTFs necessitates the consideration of second- and third-order approximations (Yu et al., 2010b). For a source no less than 0.2 m and an elevation higher than −45°, the magnitude error for the third-order approximation is within ± 1.0 dB up to a frequency of 20 kHz. Therefore, the contributions of the fourth- and higher-order scattering between the head and torso can be disregarded.

This calculation can be extended to nonrigid and locally reacting head and torso surfaces with a uniformly distributed impedance boundary. The symmetrical nature of the snowman model is helpful for simplifying calculation. Refer to the literature (Algazi et al., 2002a; Gumerov and Duraiswami, 2002a, 2004, 2005; Gumerov et al., 2002b) for details on the snowman model and multipole re-expansions. This analysis is noticeably similar to the angular momentum coupling and multiple scattering theorem in quantum mechanics. The spherical head HRTF is similar to the solution to the scattering problem of a spherical potential barrier in quantum mechanics. Therefore, some mathematical methods for quantum mechanics are applicable to HRTF calculation. The discussion here shows that the methods in two branches of physics are interchangeable. For a discussion of quantum mechanics, refer to Schiff (1968).

Methods similar to the aforementioned techniques are applicable to analyzing the effect of nearby boundaries on HRTFs (Gumerov and Duraiswami, 2001). As stated in Section 4.1, the head is simplified as a rigid sphere with radius a and the pinnae and torso are excluded. When a point source and spherical head are near an infinite rigid plane (such as a wall), multiple scattering between the head and plane occurs. According to the acoustic principle of mirror images, the effect of a plane can be modeled on an image sphere and image source. Therefore, multipole re-expansions can be used to solve this problem.

References

Aarts R.M. (**2000**). "Phantom sources applied to stereo-base widening," J. Audio Eng. Soc. **48(3)**, 181-189.

Adams N., and Wakefield G. (**2007**). "Efficient binaural display using MIMO state-space systems," in Proceeding of IEEE 2007 International Conference on Acoustics, Speech and Signal Processing, Honolulu, HI, USA, Vol (1), 169-172.

Adams N., and Wakefield G. (**2008**). "State-space synthesis of virtual auditory space," IEEE Trans. Audio, Speech, and Language Processing **16(5)**, 881-890.

Adams N., and Wakefield G. (**2009**). "State-space models of head-related transfer functions for virtual auditory scene synthesis," J. Acoust. Soc. Am. **125(6)**, 3894-3902.

AES Staff Technical Writer (**2006**). "Binaural technology for mobile applications," J. Audio Eng. Soc. **54 (10)**, 990-995.

Ahnert W., and Feistel R. (**1993**). "EARS auralization software," J. Audio Eng. Soc. **41(11)**, 894-904.

Ajdler T., Faller C., and Sbaiz L., et al. (**2005**). "Interpolation of head related transfer functions considering acoustics," in AES 118th Convention, Barcelona, Spain, Preprint: 6327.

Ajdler T., Sbaiz L., and Vetterli M. (**2007**). "Dynamic measurement of room impulse responses using a moving microphone," J. Acoust. Soc. Am. **122(3)**, 1636-1645.

Ajdler T., Faller C., and Sbaiz L., et al. (**2008**). "Sound field analysis along a circle and its applications to HRTF interpolation," J. Audio. Eng. Soc. **56(3)**, 156-175.

Akeroyd M.A., Chambers J., and Bullock D., et al. (**2007**). "The binaural performance of a cross-talk cancellation system with matched or mismatched setup and playback acoustics," J. Acoust. Soc. Am. **121(2)**, 1056-1069.

Algazi V.R., Avendano C., and Thompson D. (**1999**). "Dependence of subject and measurement position in binaural signal acquisition," J. Audio Eng. Soc. **47(11)**, 937-947.

Algazi V.R., Avendano C., and Duda R.O. (**2001a**). "Elevation localization and head-related transfer function analysis at low frequencies," J. Acoust. Soc. Am. **109(3)**, 1110-1122.

Algazi V.R., Duda R.O., and Thompson D.M., et al. (**2001b**). "The CIPIC HRTF database," in Proceeding of 2001 IEEE Workshop on the Applications of Signal Processing to Audio and Acoustics, New Platz, NY, USA, 99-102.

Algazi V.R., Avendano C., and Duda R.O. (**2001c**). "Estimation of a spherical-head model from anthropometry," J. Audio. Eng. Soc. **49(6)**, 472-479.

Algazi V.R., Duda R.O., and Morrison R.P.,et al. (**2001d**). "Structural composition and decomposition of HRTFs," in Proceeding of 2001 IEEE Workshop on Applications of Signal Processing to Audio and Acoustics, New Platz, NY, 103-106.

Algazi V.R., Duda R.O., and Duraiswami R., et al. (**2002a**). "Approximating the head-related transfer function using simple geometric models of the head and torso," J. Acoust. Soc. Am.**112(5)**, 2053-2064.

Algazi V.R., Duda R.O., and Thompson D.M. (**2002b**). "The use of head-and-torso models for improved spatial sound synthesis," in AES 113rd Convention, Los Angeles, CA, USA, Preprint: 5712.

Algazi V.R., Duda R.O., and Thompson D.M. (**2004**). "Motion-tracked binaural sound," J. Audio. Eng. Soc. **52(11)**,1142-1156.

Algazi V.R., Dalton R.J., and Duda R.O. (**2005**). "Motion-tracked binaural sound for personal music players," in AES 119th Convention, New York, USA, Preprint: 6557.

Algazi V.R., and Duda R.O. (**2011**). "Headphone-based spatial sound, exploiting head motion for immersive communication," IEEE Signal Processing Magazine **28(1)**, 33-42.

Ando Y. (**1985**). *Concert hall acoustics* (Springer-Verlag, New York).

Ando Y. (**1998**). *Architectural acoustics, blending sound sources, sound fields, and listeners* (Springer-Verlag, New York).

Andreopoulou A., and Roginska A. (**2011**). "Towards the creation of a standardized HRTF repository," in AES 131st Convention, New York, NY, USA, Paper: 8571.

ANSI (**1985**). *ANSI S3.36-1985, Manikin for simulated in-situ airborne acoustic measurements, American National Standard* (American National Standards Institute, New York).

ANSI (**1989**). *ANSI S3.25/ASA80-1989, Occluded ear simulator* (American National Standard, American National Standards Institute, New York).

ANSI (**2007**). *ANSI S3.4-2007, Procedure for the computation of loudness of steady sounds, American National Standard* (American National Standards Institute, New York).

Aoshima N. (**1981**). "Computer-generated pulse signal applied for sound measurement," J. Acoust. Soc. Am. **69(5)**, 1484-1488.

Asano F., Suzuki Y., and Sone T. (**1990**). "Role of spectral cues in median plane localization," J. Acoust. Soc. Am. **88(1)**, 159–168.

Avendano C., Algazi V.R., and Duda R.O. (**1999**). "A head-and-torso model for low-frequency binaural elevation effects," in Proceeding of 1999 IEEE Workshop on Applications of Signal Processing to Audio and Acoustics, New Paltz, New York, USA, 179-182.

Bai M.R., Zeung P. (**2004**). "Fast convolution technique using a nonuniform sampling scheme: algorithm and applications in audio signal processing," J. Audio. Eng. Soc. **52(5)**, 496-505.

Bai M.R., and Ou K.Y. (**2005a**). "Head-related transfer function (HRTF) synthesis based on a three-dimensional array model and singular value decomposition," J. Sound and Vibration **281(3/5)**, 1093-1115.

Bai M.R., Tung C.W., and Lee C.C. (**2005b**). "Optimal design of loudspeaker arrays for robust cross-talk cancellation using the Taguchi method and the genetic algorithm," J. Acoust. Soc. Am. **117(5)**, 2802-2813.

Bai M.R., and Lin C. (**2005c**). "Microphone array signal processing with application in three-dimensional spatial hearing," J. Acoust. Soc. Am. **117(4)**, 2112-2121.

Bai M.R., and Lee C.C. (**2006a**). "Development and implementation of cross-talk cancellation system in spatial audio reproduction based on subband filtering," J.Sound and Vibration **290(3/5)**, 1269-1289.

Bai M.R., and Lee C.C. (**2006b**). "Objective and subjective analysis of effects of listening angle on crosstalk cancellation in spatial sound reproduction," J. Acoust. Soc. Am. **120(4)**, 1976-1989.

Bai M.R., and Shih G.Y. (**2007a**). "Upmixing and downmixing two-channel stereo audio for consumer electronics," IEEE Trans.Consumer Electronics **53(3)**, 1011-1019.

Bai M.R., Shih G.Y., and Lee C.C. (**2007b**). "Comparative study of audio spatializers for dual-loudspeaker mobile phones," J. Acoust. Soc. Am. **121(1)**, 298-309.

Balmages I., and Rafaely B. (**2007**). "Open sphere designs for spherical microphone arrays," IEEE Trans. Audio,Speech and Language Processing **15(2)**, 727-732.

Barrass S. (**2012**). "Digital fabrication of acoustic sonifications," J. Audio. Eng. Soc. **60(9)**, 709-715.

Barron M. (**1971**). "The subject effects of first reflections in concert halls—the need of lateral reflections," J.Sound and Vibration **15(4)**, 475-494.

Batteau D.W. (**1967**). "The role of the pinna in human localization," Proc.Royal.Soc., London, **168 (Ser, B)**, 158-180.

Bauck J., and Cooper D.H. (**1996**). "Generalization transaural stereo and applications," J. Audio. Eng. Soc. **44(9)**, 683-705.

Bauck J. (**2000a**). "Conversion of two-channel stereo for presentation by three frontal loudspeakers," in AES 109th Convention, Los Angeles, CA, USA, Preprint: 5239.

Bauck J. (**2000b**). "Equalization for central phantom images and dependence on loudspeaker spacing: reformatting from three loudspeakers to two loudspeakers," in AES 109th Convention, Los Angeles, CA, USA, Preprint: 5240.

Bauck J. (**2001**). "A simple loudspeaker array and associated crosstalk canceler for improved 3D audio," J. Audio. Eng. Soc. **49(1/2)**, 3-13.

Bauer B.B. (**1961**). "Stereophonic earphones and binaural loudspeakers," J. Audio. Eng. Soc. **9(2)**, 148-151.

Begault D.R. (**1991**). "Challenges to the successful implementation of 3-D sound," J. Audio Eng. Soc. **39(11)**, 864-870.

Begault D.R. (**1992**). "Perceptual effects of synthetic reverberation on three-dimensional audio systems," J. Audio Eng. Soc. **40(11)**, 895-904.

Begault D.R., and Wenzel E. M. (**1993**). "Headphone localization of speech," Hum. Factors **35(2)**, 361–376.

Begault D.R. (**1994a**). *3-D Sound for Virtual Reality and Multimedia* (Academic Press Professional Cambridge, MA.).

Begault D.R., and Erbe T. (**1994b**). "Multichannel spatial auditory display for speech communications," J. Audio. Eng. Soc. **42(10)**, 819-826

Begault, D.R. (**1998**). "Virtual acoustics, aeronautics, and communications," J. Audio. Eng. Soc. **46(6)**, 520-530.

Begault D.R. (**1999**). "Virtual acoustic displays for teleconferencing: Intelligibility advantage for 'telephone-grade' audio" J. Audio. Eng. Soc. **47(10)**, 824-828.

Begault D.R., Wenzel E.M., and Anderson M.R. (**2001**). "Direct comparison of the impact of head tracking, reverberation, and individualized head-related transfer functions on the spatial perception of a virtual speech source," J. Audio. Eng. Soc. **49(10)**, 904-916.

Begault D.R.. Godfroy M., and Miller J.D., et al. (**2006**). "Design and verification of HeadZap, a semiautomated HRIR measurement system," in AES 120th Convention, Paris, France, Preprint: 6655.

Begault D.R., Wenzel E M., and Godfroy M, et al. (**2010**). "Applying spatial audio to human interfaces: 25 years of NASA experience," in AES 40th International Conference, Tokyo, Japan.

Bekesy G. (**1960**). *Experiments in hearing* (Mcgraw-Hill, New York, USA).

Beliczynski B., Kale I., and Cain G.D. (**1992**). "Approximation of FIR by IIR digital filters: an algorithm based on balanced model reduction," IEEE Trans. Signal Processing **40(3)**, 532–542.

Beranek L. (**1996**). *Concert Halls and Opera Houses* (Acoustical Society of America, USA).

Berkhout A.J., Vries D., and Vogel P. (**1993**). "Acoustic control by wave field synthesis," J. Acoust. Soc. Am. **93(5)**, 2764-2778.

Bernfeld B. (**1975**). "Simple equations for multichannel stereophonic sound localization," J. Audio. Eng. Soc. **23(7)**, 553-557.

Bernhard U. S., and Hugo F. (**2003**). "Subjective selection of non-individual head-related transfer functions," in Proceedings of the 2003 International Conference on Auditory Display, Boston, MA, USA, 259-262.

Blauert J. (1997). *Spatial Hearing* (Revised edition, MIT Press, Cambridge, MA, England).

Blauert J., Brueggen M., and Bronkhorst A.W., et al. (**1998**). "The AUDIS catalog of human HRTFs," J. Acoust. Soc. Am. **103(5)**, 3082.

Blauert J., Lehnert H., and Sahrhage J., et al. (**2000**). "An interactive virtual-environment generator for psychoacoustic research I: architecture and implementation," Acustica united with Acta Acustica **86(1)**, 94–102.

Blauert J. (**2012**). "Modeling binaural processing: What next? (abstract)," J. Acoust. Soc. Am. **132(3, Pt2)**, 1911.

Blesser B. (**2001**). "An interdisciplinary synthesis of reverberation viewpoints," J. Audio Eng. Soc. **49(10)**, 867-903.

Blommer M.A., and Wakefield G.H. (**1997**). "Pole-zero approximations for head-related transfer functions using a logarithmic error criterion," IEEE Trans. on Speech and Audio Processing **5(3)**, 278-287.

Bloom P.J. (**1977**). "Determination of monaural sensitivity changes due to the pinna by use of minimum-audible-field measurements in the lateral vertical plane," J. Acoust. Soc. Am.**61(3)**, 820-828.

Boer K. de (**1940**). "Stereophonic sound reproduction," Philips Tech. Rev. **1940(5)**, 107-114.

Boerger G., Laws P., and Blauert J. (**1977**). "Stereophonic reproduction through headphones with control of special transfer functions by head movements," Acustica **39(1)**, 22–26.

Bolia R.S., D'Angelo W.R., and McKinley R.L. (**1999**). "Aurally aided visual search in three-dimensional space," Human Factors **41(4)**, 664–669.

Boone M.M., Verheijen E.N.G., and Van Tol P.F. (**1995**). "Spatial sound field reproduction by wave field synthesis," J. Audio. Eng. Soc. **43(12)**, 1003-1012.

Bork I. (**2000**). "A comparison of room simulation software–The 2nd round robin on room acoustical computer simulation," Acta Acustica united with Acustica **86(6)**, 943–956.

Bork I. (**2005a**). "Report on the 3rd round robin on room acoustical computer simulation–Part I: measurements," Acta Acustica united with Acustica **91(4)**, 740–752.

Bork I. (**2005b**). "Report on the 3rd round robin on room acoustical computer simulation–Part II: calculations," Acta Acustica united with Acustica **91(4)**, 753-763.

Bouéri M., and Kyirakakis C. (**2004**). "Audio signal decorrelation based on a critical band approach," in AES 117th Convention, San Francisco, CA, U.S.A, Preprint: 6291.

Bovbjerg B.P., Christensen F., and Minnaar P., et al. (**2000**). "Measuring the head-related transfer functions of an artificial head with high directional resolution," in AES 109th Convention, Los Angeles, USA, Preprint: 5264.

Breebaart J., and Kohlrausch A. (**2001**). "The perceptual (ir)relevance of HRTF amplitude and phase spectra," in AES 110th Convention, Amsterdam, The Netherlands, Preprint: 5406.

Bronkhorst A. W. (**1995**). "Localization of real and virtual sound sources," J. Acoust. Soc. Am. **98(5)**, 2542-2553.

Bronkhorst A.W., and Houtgast T. (**1999**). "Auditory distance perception in rooms," Nature **397**, 517-520.

Bronkhorst A. W. (**2000**). "The cocktail party phenomenon: a review of research on speech intelligibility in multiple-talker conditions," Acta Acustica united with Acustica **86(1)**, 117-128.

Brookes T., and Treble C. (**2005**). "The effect of non-symmetrical left/right recording pinnae on the perceived externalisation of binaural recordings," in AES 118th Convention, Barcelona, Spain, Preprint: 6439.

Brown C.P., and Duda R.O. (**1998**). "A structural model for binaural sound synthesis," IEEE Trans.Speech and Audio Processing **6(5)**, 476-488.

Brungart D.S. (**1998**). "Control of perceived distance in virtual audio displays," in Proceedings of the 20th Annual International Conference of the IEEE Engineering in Medicine and Biology Society, Hong Kong, China, Vol (3), 1101-1104.

Brungart D.S., and Rabinowitz W.M. (**1999a**). "Auditory localization of nearby sources. Head-related transfer functions," J. Acoust. Soc. Am. **106(3)**, 1465-1479.

Brungart D.S., Durlach N.I., and Rabinowitz W.M. (**1999b**). "Auditory localization of nearby sources. II. Localization of a broadband source," J. Acoust. Soc. Am. **106(4)**, 1956-1968.

Brungart D.S. (**1999c**). "Auditory localization of nearby sources. III. Stimulus effects," J. Acoust. Soc. Am. **106 (6)**, 3589-3602.

Brungart D.S. (**1999d**). "Auditory parallax effects in the HRTF for nearby sources," in Proceeding of 1999 IEEE Workshop on Applications of Signal Processing to Audio and Acoustics, New Paltz, NY, USA, 171-174.

Brungart D.S., and Simpson B.D. (**2002**). "The effects of spatial separation in distance on the informational and energetic masking of a nearby speech signal," J. Acoust. Soc. Am. **112(2)**, 664-676.

Brungart D.S., Kordik A.J., and Simpson B.D. (**2006**). "Effects of headtracker latency in virtual audio displays," J. Audio. Eng. Soc. **54(1/2)**, 32-44.

Brungart D.S., and Romigh G.D. (**2009**). "Spectral HRTF enhancement for improved vertical-polar auditory localization," in Proceeding of 2009 IEEE Workshop on Applications of Signal Processing to Audio and Acoustics, New Paltz, NY, USA, 305-308.

Bujacz M., Skulimowski P., and Strumillo P. (**2012**). "Naviton—a prototype mobility aid for auditory presentation of three-dimensional scenes to the visually impaired," J. Audio. Eng. Soc. **60(9)**, 696-708.

Burkhard M.D., and Sachs R.M. (**1975**). "Anthropometric Manikin for acoustic research," J. Acoust. Soc. Am. **58(1)**, 214-222.

Busson S., Nicol R., and Katz B.F.G. (**2005**). "Subjective investigations of the interaural time difference in the horizontal plane," in AES 118th Convention, Barcelona, Spain, Preprint: 6324.

Bustamante F.O., Lopez J.J., and Gonzalez A. (**2001**). "Prediction and measurement of acoustic crosstalk cancellation robustness," in Proceedings of 2001 IEEE International Conference on Acoustics, Speech and Signal Processing, Salt Lake City, UT, USA, **Vol (5)**, 3349-3352.

Butler R.A., and Belendiuk K. (**1977**). "Spectral cues utilized in the localization of sound in the median sagittal plane," J. Acoust. Soc. Am. **61(5)**, 1264-1269.

Candes E.J., and Wakin M.B. (**2008**). "An introduction to compressive sampling," IEEE Signal Processing Magazine **25(2)**, 21-30.

Carlile S., and Pralong D. (**1994**). "The location-dependent nature of perceptually salient features of the human head-related transfer functions," J. Acoust. Soc. Am. **95(6)**, 3445–3459.

Carlile S., Jin C., and Harvey V. (**1998**). "The generation and validation of high fidelity virtual auditory space," in Proceedings of the 20th Annual International Conference of the IEEE Engineering in Medicine and Biology Society, Hong Kong, China, Vol (3), 1090-1095.

Carlile S., Jin C., and Van Raad V. (**2000**). "Continuous virtual auditory space using HRTF interpolation: acoustic and psychophysical errors," in Proceeding of 2000 International Symposium on Multimedia Information Processing, Sydney, Australia, 220-223.

Chanda P.S., and Park S. (**2005**). "Low order modeling for multiple moving sound synthesis using head-related transfer functions' principal basis vectors," in Proceedings of 2005 IEEE International Joint Conference on Neural Networks, Montreal, Canada, **Vol.(4)**, 2036-2040.

Chanda P.S., Park S., and Kang T.I. (**2006**). "A binaural synthesis with multiple sound sources based on spatial features of head-related transfer functions," in Proceeding of 2006 IEEE International Joint Conference on Neural Networks, Vancouver, Canada, 1726-1730.

Chen J., Van Veen B.D., and Hecox K.E. (**1992**). "External ear transfer function modeling: a beamforming approach," J. Acoust. Soc. Am. **92(4)**, 1933-1944.

Chen J., Van Veen B.D., and Hecox K.E. (**1993**). "Synthesis of 3D virtual auditory space via a spatial feature extraction and regularization model," in Proceeding of 1993 IEEE Virtual Reality Annual International Symposium, Seattle, WA , USA, 188-193.

Chen J., Van Veen B.D., and Hecox K.E. (**1995**). "A spatial feature extraction and regularization model of the head-related transfer function," J. Acoust. Soc. Am. **97(1)**, 439-452.

Chen Z.W., Yu G.Z., and Xie B.S., et al. (**2012**). "Calculation and analysis of near-field head-related transfer functions from a simplified head-neck-torso model," Chinese Physics Letters **29(3)**, 034302.

Cheng C.I., and Wakefield G.H. (**1999**). "Spatial frequency response surfaces (SFRS'S): an alternative visualization and interpolation technique for head relation transfer functions (HRTF's)," in AES 16th International Conference, Rovaniemi, Finland.

Cheng C.I. (**2001a**). "Visualization, measurement, and interpolation of head-related transfer functions (HRTFs) with applications in electro-acoustic music," Dissertation of Doctor degree, The University of Michigan, USA.

Cheng C.I., and Wakefield G.H. (**2001b**). "Introduction to head-related transfer functions (HRTFs): representations of HRTFs in time, frequency, and space," J. Audio. Eng. Soc. **49(4)**, 231-249.

Cherry E.C. (**1953**). "Some experiments on the recognition of speech, with one and with two ears," J. Acoust. Soc. Am. **25(5)**, 975-979.

Cheung N.M., Trautman S., and Horner A. (**1998**). "Head-related transfer function modeling in 3-D sound systems with genetic algorithms," J. Audio. Eng. Soc. **46(6)**, 531-539.

Chi S.L., Xie B.S., and Rao D. (**2008**). "Improvement of virtual sound image by loudspeaker equalization," Technical Acoustics (in Chinese) **27(5, Pt.2)**, 370-371.

Chi S.L., Xie B.S., and Rao D. (**2009**). "Effect of mismatched loudspeaker pair on virtual sound image," Applied Acoustics (in Chinese) **28(4)**, 291-299.

Choi T., Park Y.C., and Youn D.H. (**2006**). "Efficient out of head localization system for mobile applications," in AES 120th Convention, Paris, France, Preprint: 6758.

Chong U.P., Kim I.H., and Kim K.N. (**2001**). "Improved 3D sound using wavelets," in Proceeding of The IEEE Fifth Russian-Korean International Symposium on Science and Technology, Tomsk, Russia, **Vol. (1)**, 54-56.

Christensen F., Møller H., and Minnaar P., et al. (**1999**). "Interpolating between head-related transfer functions measured with low-directional resolution," in AES 107th Convention, New York, USA, Preprint: 5047.

Christensen F., Jensen C.B., and Møller H. (**2000**). "The design of VALDEMAR–an artificial head for binaural recording purposes," in AES 109th Convention, Los Angeles, CA, USA, Preprint: 5253.

Christensen F., Martin G., and Minnaar P., et al. (**2005**). "A listening test system for automotive audio Part 1: system description," in AES 118th Convention, Barcelona, Spain, Preprint: 6358.

Ciskowski R.D., and Brebbia C.A. (**1991**). *Boundary Element Methods in Acoustics* (Computational Mechanics Publications, Southampton, U.K.).

Clack H.A.M., Dutton G.F., and Vanderlyn P.B. (**1957**). "The 'Stereosonic' recording and reproduction system," IRE Trans. on Audio **AU-5(4)**, 96-111.

Cooper D.H. (**1982**). "Calculator program for head-related transfer function," J. Audio. Eng. Soc. **30(1/2)**, 34-38.

Cooper D.H. (**1987**). "Problem with shadowless stereo theory: asymptotic spectral status," J. Audio. Eng. Soc. **35(9)**, 629-642.

Cooper D.H., and Bauck J.L. (**1989**). "Prospects for transaural recording," J. Audio. Eng. Soc. **37(1/2)**, 3-19.

Cooper D.H., and Bauck J.L. (**1990**). "Head diffraction compensated stereo system," U.S.Patent 4,893,342.

Daigle J. N., and Xiang N. (**2006**). "A specialized fast cross-correlation for acoustical measurements using coded sequences," J. Acous. Soc. Am. **119(1)**, 330-335.

Damaske P. (**1969/1970**). "Directional dependence of spectrum and correlation functions of the signals received at the ears," Acustica **22(4)**, 191-204.

Damaske P. (**1971**). "Head-related two-channel stereophony with loudspeaker reproduction," J. Acoust. Soc. Am. **50 (4B)**, 1109–1115.

Damaske P., and Ando Y. (**1972**). "Interchannel crosscorrelation for multichannel loudspeakers reproduction," Acoustica **27(4)**, 233-238.

Daniel J. (**2003**). "Spatial sound encoding including near field effect: introducing distance coding filters and a viable, new ambisonic format," in AES 23rd International Conference, Copenhagen, Denmark.

Dantele A., Reiter U., and Schuldt M., et al. (**2003**). "Implementation of MPEG-4 audio nodes in an interactive virtual 3D environment," in AES 114th Convention, Amsterdam, The Netherlands, Preprint: 5820.

Dattorro J. (**1997**). "Effect design: part 1: reverberator and other filters," J. Audio Eng. Soc. **45(9)**, 660-684.

Davis M.F., and Fellers M.C. (**1997**). "Virtual surround presentation of Dolby AC-3 and Pro Logic signal," in AES 103rd Convention, New York, USA., Preprint: 4542.

Deif A.S. (**1982**). *Advanced matrix theory for scientists and engineers* (Abacus Press, Tunbridge Wells, Kent, England).

Delerue O., and Warusfel O. (**2002**). "Authoring of virtual sound scenes in the context of the LISTEN project," in AES 22nd International Conference, Espoo, Finland.

Djelani T., Pörschmann C., and Sahrhage J., et al. (**2000**). "An interactive virtual-environment generator for psychoacoustic research II: collection of head-related impulse responses and evaluation of auditory localization," Acta Acustica united with Acustica **86 (6)**, 1046–1053.

Dobrucki A., Plaskota P., and Pruchnicki P., et al. (**2010**). "Measurement system for personalized head-related transfer functions and its verification by virtual source localization trials with visually impaired and sighted individuals," J. Audio Eng. Soc. **58(9)**, 724-738.

Doerr K.U., Rademacher H., and Huesgen S., et al. (**2007**). "Evaluation of a low-cost 3D sound system for immersive virtual reality training systems," IEEE Trans. On Visualization and Computer Graphics **13(2)**, 204-212.

Drullman R., and Bronkhorst A.W. (**2000**). "Multichannel speech intelligibility and talker recognition using monaural, binaural, and three-dimensional auditory presentation," J. Acoust. Soc. Am. **107(4)**, 2224-2235.

Duda R.O., and Martens W.L. (**1998**). "Range dependence of the response of a spherical head model," J. Acous. Soc. Am. **104(5)**, 3048-3058.

Duda R.O., Avendano C., and Algazi V.R. (1999). "An adaptable ellipsoidal head model for the interaural time difference," in Proceedings of 1999 IEEE International Conference on Acoustics, Speech, and Signal Processing, Phoenix, AZ, USA, **Vol.(2)**, 965-968.

Duraiswami R., Zotkin D.N., and Gumerov N.A. (**2004**). "Interpolation and range extrapolation of HRTFs," in Proceedings of 2004 IEEE International Conference on Acoustics, Speech, and Signal Processing, Montreal, Quebec, Canada,**Vol.(4)**, 45-48.

Duraiswami R., Zotkin D.N., and Li Z.Y., et al. (**2005**). "High order spatial audio capture and its binaural head-tracked playback over headphones with HRTF cues," in AES 119th Convention, New York, USA., Preprint: 6540.

Durant E.A., and Wakefield G.H. (**2002**). "Efficient model fitting using a genetic algorithm: pole-zero approximations of HRTFs," IEEE Trans. on Speech and Audio Processing **10(1)**, 18-27.

Durlach N.I, and Colburn S. (**1978**). "Binaural Phenomena," in *Handbook of perception*, Vol. IV, (edited by Carterette E.C. and Friedman M.P., Academic Press, New York, USA).

Durlach N.I., Rigopulos A., and Pang X.D., et al. (**1992**). "On the externalization of auditory image," Presence **1(2)**, 251–257.

Eckel G. (**2001**). "Immersive audio-augmented environments, the LISTEN project," in Proceedings of the IEEE 5th International Conference on Information Visualisation, London, UK, 571-573.

Enzner G. (**2008**). "Analysis and optimal control of LMS-type adaptive filtering for continuous-azimuth acquisition of head related impulse responses," in Proceeding of 2008 IEEE International Conference on Acoustics, Speech and Signal Processing, Las Vegas, NV, USA., 393-396

Enzner G. (**2009**). "3D-continuous-azimuth acquisition of head-related impulse responses using multi-channel adaptive filtering," in Proceeding of 2009 IEEE Workshop on Applications of Signal Processing to Audio and Acoustics, New Paltz, NY, USA, 325-328.

Evans M.J., Tew A.I., and Angus J.A.S., et al. (**1997**). "Spatial audio teleconferencing–which way is better?," in Proceedings of the Fourth International Conference on Auditory Displays (ICAD 97), Palo Alto, CA, USA, 29-37.

Evans M.J., Angus J.A.S., and Tew A.I. (**1998a**). "Analyzing head-related transfer function measurements using surface spherical harmonics," J. Acoust. Soc. Am. **104 4**), 2400-2411.

Evans M.J. (**1998b**). "Synthesizing moving sounds," in The Fifth International Conference on Auditory Display, Glasgow, UK.

Fahn C.S., and Lo Y.C. (**2003**). "On the clustering of head-related transfer functions used for 3D sound localization," J. Information Science and Engineering **19(1)**, 141-157.

Faller K.J., Barreto A., and Gupta N., et al. (**2005**). "Enhanced modeling of head-related impulse responses towards the development of customizable sound spatialization," in Proceeding of the 2005 WSEAS/IASME International Conference on Computational Intelligence, Man-Machine Systems and Cybernetics, Miami, Florida, USA, 82-87.

Faller K.J., Barreto A., and Adjouadi M. (**2010**). "Augmented Hankel total least-squares decomposition of head-related transfer functions," J. Audio. Eng. Soc. **58(1/2)**, 3-21.

Farina A. (**1995**). "Auralization software for the evaluation of the results obtained by a pyramid tracing code: results of subjective listening tests," in 15th International Conference on Acoustics, Trondheim, Norway.

Farina A., and Ugolotti E. (**1997**). "Subjective comparison of different car audio systems by the auralization technique," AES 103rd Convention, New York, USA, Preprint: 4587.

Farina A., and Ugolotti E. (**1998**). "Numerical model of the sound field inside cars for the creation of virtual audible reconstructions," in 1st COST-G6 Workshop on Digital Audio Effects (DAFX98), Barcelona, Spain.

Farina A., and Bozzoli F. (**2002**). "Measurement of the speech intelligibility inside cars," in AES 113th Convention, Los Angeles, CA, USA, Preprint: 5702.

Farina A., Armelloni E., and Martignon P. (**2004**). "An experimental comparative study of 20 Italian opera houses: measurement techniques," J. Acoust. Soc. Am. **115(5)**, 2475-2476.

Fels J., Buthmann P., and Vorländer M. (**2004**). "Head-related transfer functions of children," Acta Acustica united with Acustica **90(5)**, 918-927.

Fels J., and Vorländer M. (**2009**). "Anthropometric parameters influencing head-related transfer functions," Acta Acustica united with Acustica **95(2)**, 331-342.

Fletcher H. (**1940**). "Auditory patterns," Rev.Mod.Psys. **12(1)**, 47-65.

Fontana S., Farina A., and Grenier Y. (**2006**). "A system for rapid measurement and direct customization of head related impulse responses," in AES 120th Convention, Paris, France, Preprint: 6851.

Freeland F.P., Biscainho L.W.P., and Diniz P.S.R. (**2004**). "Interpositional transfer function for 3D-sound generation," J. Audio. Eng. Soc. **52(9)**, 915-930.

Fujinami Y. (**1998**). "Improvement of sound image localization for headphone listening by wavelet analysis," IEEE Trans. Consumer Electronics **44(3)**, 1183-1188.

Fukudome K., Suetsugu T., and Ueshin T.,et al. (**2007**). "The fast measurement of head related impulse responses for all azimuthal directions using the continuous measurement method with a servo-swiveled chair," Applied Acoustics **68(8)**, 864-884.

Gan W.S., Tan S.E., and Er M.H., et al. (**2001**). "Elevated speaker projection for digital home entertainment system," IEEE Trans. Consumer Electronics **47(3)**, 631-637.

Gardner M.B., and Gardner R.S. (**1973**). "Problem of localization in the median plane: effect of pinnae cavity occlusion," J. Acoust. Soc. Am. **53(2)**, 400-408.

Gardner W.G., and Martin K.D. (**1995a**). "HRTF measurements of a KEMAR," J. Acoust. Soc. Am. **97(6)**, 3907-3908.

Gardner W.G. (**1995b**). "Efficient convolution without input-output delay," J. Audio. Eng. Soc. **43(3)**, 127-136.

Gardner W.G. (**1995c**). "Transaural 3D audio," Technical Report No. 342, MIT Media Laboratory, Perceptual Computing Section.

Gardner W.G. (**1997**). "3-D Audio using loudspeakers," Doctor thesis of Massachusetts Institute of Technology, Massachusetts, USA.

Gardner W.G. (**1998**). "Reverberation algorithms," in *Applications of signal processing to audio and acoustics*, Edited by Kahrs M. and Brandenburg K. (Kluwer Academic Publishers, USA).

Gardner W.G. (**1999**). "Reduced-rank modeling of head-related impulse responses using subset selection," in Proceeding of 1999 IEEE Workshop on Applications of Signal Processing to Audio and Acoustics, New Paltz, New York, USA, 175-178.

Geisler C.D. (**1998**). *From sound to synapse: physiology of the mammalian ear* (Oxford University Press, New York, USA).

Gelfand S.A. (**2010**). *Hearing: An Introduction to psychological and physiological acoustics* (5th Edition, Informa Healthcare, London, UK).

Genuit K. (**1984**). "Ein modell zur beschreibung von aussenohr ubertragungseigenschaften," Dissertation of Doctor of Philosophy, Rheinisch-Westfälischen Technichen Hoch-schule Aachen, Germany.

Genuit K., Xiang, N. (**1995**). "Measurements of artificial head transfer functions for auralization and virtual auditory environment," in Proceeding of 15th International Congress on Acoustics (invited paper), Trondheim, Norway, II 469-472.

Genuit K., and Fiebig A. (**2007**). "Do we need new artifical head?," in Proceeding of the 19th International Congress on Acoustics, Madrid, Spain.

Georgiou P., and Kyriakakis C. (**1999**). "Modeling of head related transfer functions for immersive audio using a state-space method," in Conference Record of IEEE Thirty-Third Asilomar Conference on Signals, Systems, and Computers, Pacific Grove, CA, USA, **Vol.(1)**, 720-724.

Georgiou P.G., Mouchtaris A., and Roumeliotis S.I., et al. (**2000**). "Immersive sound rendering using laser-based tracking," in AES 109th Convention, Los Angeles, CA, USA, Preprint: 5227.

Gerzon M.A. (**1985**). "Ambisonics in multichannel broadcasting and video," J. Audio. Eng. Soc. **33(11)**, 859-871.

Gierlich H.W. (**1992**). "The application of binaural technology," Applied Acoustics **36(3/4)**, 219-243.

Giron F. (**1996**). "Investigation about the directivity of sound sources," Dissertation of Doctor Degree of Philosophy, Faculty of Electrotechnical Eng., Ruhr-University, Bochum, Germany.

Goldstein H. (1980). Classical mechanics (second edition, Addison-Wesley Publishing Company Inc., Massachusetts, USA).

Gong M., Xiao Z., and Qu T.S., et al. (**2007**). "Measurement and analysis of near-field head-related transfer function," Applied Acoustics (in Chinese) **26(6)**, 326-334.

Gonot A., Chateau N., and Emerit M. (**2006**). "Usability of 3D-sound for navigation in a constrained virtual environment," in AES 120th Convention, Paris, France, Preprint: 6800.

Granier E., Kleiner M., and Dalenbäck B.I., et al. (**1996**). "Experimental auralization of car audio installations," J. Audio. Eng. Soc. **44(10)**, 835-849.

Grantham D.W., Willhite J.A., and Frampton K.D., et al. (**2005**). "Reduced order modeling of head related impulse responses for virtual acoustic displays," J. Acoust. Soc. Am. **117(5)**, 3116-3125.

Grassi E., Tulsi J., and Shamma S. (**2003**). "Measurement of head-related transfer functions based on the empirical transfer function estimate," in Proceedings of the 2003 International Conference on Auditory Display, Boston, MA, USA, 119-122.

Greff R., and Katz B.F.G. (**2007**). "Round robin comparison of HRTF simulation systems: preliminary results," in AES 123rd Convention, New York, USA., Preprint: 7188.

Griesinger D. (**1989**). "Equalization and spatial equalization of dummy-head recordings for loudspeaker reproduction," J. Audio. Eng. Soc. **37(1/2)**, 20-29.

Grindlay G., and Vasilescu M.A.O. (2007). "A multilinear (tensor) framework for HRTF analysis and synthesis," in Proceeding of IEEE 2007 International Conference on Acoustics, Speech and Signal Processing, Honolulu, HI, USA, **Vol.(1)**, 161-164.

Guan S.Q. (**1995**). "Some thoughts on Stereophonic," Applied Acoustics (in Chinese) **14(6)**,6-11.

Guillon P., Guignard T., and Nicol R. (**2008a**). "Head-related transfer function customization by frequency scaling and rotation shift based on a new morphological matching method," in AES 125th Convention, San Francisco, CA, USA, Preprint: 7550.

Guillon P., and Nicol R., and Simon L. (**2008b**). "Head-related transfer functions reconstruction from sparse measurements considering a priori knowledge from database analysis: a pattern recognition approach," in AES 125th Convention, San Francisco, USA, Paper: 7610.

Gumerov N.A., and Duraiswami R. (**2001**). "Modeling the effect of a nearby boundary on the HRTF," in Proceedings of IEEE 2001 International Conference on Acoustics, Speech, and Signal Processing, **Vol.(5)**, 3337-3340.

Gumerov N.A., and Duraiswami R. (**2002a**). "Computation of scattering from N spheres using multipole reexpansion," J. Acoust. Soc. Am. **112(6)**, 2688-2701.

Gumerov N.A., Duraiswami R., and Tang Z. (**2002b**). "Numerical study of the influence of the torso on the HRTF," in Proceeding of 2002 IEEE International Conference on Acoustics, Speech and Signal Processing, Orlando, FL, USA, **Vol.(2)**, 1965-1968.

Gumerov N.A., and Duraiswami R. (**2004**). *Fast multipole methods for the Helmholtz equation in three dimensions* (Elsevier, Amsterdam, The Netherlands).

Gumerov N.A., and Duraiswami R. (**2005**). "Computation of scattering from clusters of spheres using the fast multipole method," J. Acoust. Soc. Am. **117(4)**, 1744-1761.

Gumerov N.A., and Duraiswami R. (**2009**). "A broadband fast multipole accelerated boundary element method for the three dimensional Helmholtz equation," J. Acoust. Soc. Am. **125(1)**, 191-205.

Gumerov N.A., O'Donovan A.E., and Duraiswami R., et al. (**2010**). "Computation of the head-related transfer function via the fast multipole accelerated boundary element method and its spherical harmonic representation," J. Acoust. Soc. Am. **127(1)**, 370-386.

Gupta N., Barreto A., and Ordonez C. (**2002**). "Spectral modification of head-related transfer functions for improved virtual sound spatialization," in Proceeding of 2002 IEEE International Conference on Acoustics, Speech and Signal Processing, Orlando, FL, USA, **Vol.(2)**, 1953-1956.

Hacihabiboglu H., Gunel B., and Murtagh F. (**2002**). "Wavelet-based spectral smoothing for head-related transfer function filter design," in AES 22nd International Conference, Espoo, Finland.

Hacihabiboglu H., Gunel B., and Kondoz A.M. (**2005**). "Head-related transfer function filter interpolation by root displacement," in Proceeding of 2005 IEEE Workshop on Applications of Signal Processing to Audio and Acoustics, New Paltz, NY, USA, 134-137.

Halkossari T., Vaalgamaa M., and Karjalainen M. (**2005**). "Directivity of artifical and human speech," J. Audio. Eng. Soc. **53(7/8)**, 620-631.

Hamada H., Ikeshoji N., and Ogura Y., et al., (**1985**). "Relation between physical characteristics of ortho-stereophonic system and horizontal plane localization," J. Acoust. Soc. Japan. (in English) **6(3)**, 143-154.

Hammershøi D., and Møller H. (**1996**). "Sound transmission to and within the human ear canal," J. Acoust. Soc. Am. **100(1)**, 408-427.

Hammershøi D., and Møller H. (**2008**). "Determination of noise immission from sound sources close to the ears," Acta Acoustica united with Acoustica **94(1)**, 114-129.

Han H.L. (**1994**). "Measuring a dummy head in search of pinna cues," J. Audio Eng. Soc. **42(1/2)**, 15–37.

Haneda Y., Makino S., and Kaneda Y., et al. (**1994**). "Common acoustical pole and zero modeling of room transfer functions," IEEE Trans. Speech and Audio Processing **2(2)**, 320-328.

Haneda Y., Makino S., and Kaneda Y., et al. (**1999**). "Common-acoustical-pole and zero modeling of head-related transfer functions: IEEE Trans. on Speech and Audio Processing **7(2)**, 188-196.

Härmä A., Karjalainen M., and Savioja L., et al. (**2000**). "Frequency-warped signal processing for audio applications," J. Audio Eng. Soc. **48(11)**, 1011-1031.

Härmä A., Jakka J., and Tikander M., et al. (**2004**). "Augmented reality audio for mobile and wearable appliances," J. Audio. Eng. Soc. **52 (6)**, 618-639.

Hartmann W.M., and Wittenberg A. (**1996**). "On the externalization of sound images," J. Acoust. Soc. Am. **99(6)**, 3678–3688.

Hartung K., and Raab A. (**1996**). "Efficient modeling of head-related transfer function," Acta Acustica **82(suppl 1)**, 88.

Hartung K., Braasch J., and Sterbing S.J. (**1999a**). "Comparison of different methods for the interpolation of head-related transfer functions," in AES 16th International Conference, Rovaniemi, Finland.

Hartung K., Sterbing S.J., and Keller C.H., et al. (**1999b**). "Applications of virtual auditory space in psychoacoustics and neurophysiology (abstract)," J. Acoust. Soc. Am. **105(2)**, 1164.

Hayakawa Y., Nishino T., and Takeda K.(**2007**). "Development of small sound equipment with microdynamic-type loudspeakers for HRTF measurement," in The 19th International Congress on Acoustics, Madrid, Spain.

Hayes M.H. (**1996**). *Statistical signal processing and modeling* (John Wiley & Sons Inc., New York, USA).

Hawksford M.J. (**2002**). "Scalable multichannel coding with HRTF enhancement for DVD and virtual sound systems," J. Audio. Eng. Soc. **50(11)**, 894-913.

He P., Xie B.S., and Rao D. (**2006**). "Subjective and objective analyses of timbre equalized algorithms for virtual sound reproduction by loudspeakers," Applied acoustics (in Chinese) **25(1)**, 4-12.

He P., Xie B.S., and Zhong X.L. (**2007**). "Virtual sound signal processing using HRTF without pinnae," Applied acoustics (in Chinese) **26(2)**, 100-106.

He Y.J., Xie B.S., and Liang S.J. (**1993**). "Extension of localization equation for stereophonic sound image," Audio Engineering (in Chinese) **1993(10)**, 2-4.

Hebrank J., and Wright D. (**1974**). "Spectral cues used in the localization of sound sources on the median plane," J. Acoust. Soc. Am. **56(6)**, 1829-1834.

Henning G.B. (**1974**). "Detectability of interaural delay in high-frequency complex waveforms," J. Acoust. Soc. Am. **55(1)**, 84-90.

Henry Dreyfuss Assoc. (**1993**). *The measure of man and woman* (Whitney library of design, New York, USA).

Hess W. (**2012**). "Head-tracking techniques for virtual acoustics applications," in AES 133rd Convention, San Francisco, CA, USA, Paper: 8782.

Hetherington C., and Tew A.I. (**2003**). "Parameterizing human pinna shape for the estimation of head-related transfer functions," in AES 114th Convention, Amsterdam, The Netherlands, Preprint: 5753.

Hiekkanen T., Mäkivirta A., and Karjalainen M. (**2009**). "Virtualized listening tests for loudspeakers," J. Audio. Eng. Soc. **57(4)**, 237-251.

Hiipakka M., Kinnari T., and Pulkki V. (**2012**). "Estimating head-related transfer functions of human subjects from pressure–velocity measurements," J. Acoust. Soc. Am. **131(5)**, 4051-4061.

Hill P.A., Nelson P.A., and Kirkeby O., et al. (**2000**). "Resolution of front–back confusion in virtual acoustic imaging systems," J. Acoust. Soc. Am. **108(6)**, 2901-2910.

Hirahara T., Sagara H., and Toshima I., et al. (**2010**). "Head movement during head-related transfer function measurement," Acoust. Sci. & Tech. **31(2)**, 165-171.

Hoffmann P.F., and Møller H. (**2005a**). "Audibility of time switching in dynamic binaural synthesis," in AES 118th Convention, Barcelona, Spain, Preprint: 6326.

Hoffmann P.F., and Møller H. (**2005b**). "Audibility of spectral switching in head-related transfer functions," in AES 119th Convention, New York, USA, Preprint: 6537.

Hoffmann P.F., and Møller H. (**2006a**). "Audibility of spectral differences in head-related transfer functions," in AES 120th Convention, Paris, France, Preprint: 6652.

Hoffmann P.F., and Møller H. (**2006b**). "Audibility of time differences in adjacent head-related transfer functions," in AES 121st Convention, San Francisco, USA, Preprint: 6914.

Hoffmann P.F., and Møller H. (**2008a**). "Some observations on sensitivity to HRTF magnitude," J. Audio. Eng. Soc. **56(11)**, 972-982.

Hoffmann P.F., and Møller H. (**2008b**). "Audibility of differences in adjacent head-related transfer functions," Acta Acustica united with Acustica **94(6)**, 945-954.

Hoffmann P.F., and Møller H. (**2008c**). "Audibility of direct switching between head-related transfer functions," Acta Acustica united with Acustica **94(6)**, 955-964.

Hofman P.M., Van Riswick J.G., and Van Opstal A.J. (**1998**). "Relearning sound localization with new ears," Nature Neuroscience **1(5)**, 417–421.

Hom R.C.M., Algazi V.R., and Duda R.O. (**2006**). "High-frequency interpolation for motion-tracked binaural sound," in AES 121st Convention, San Francisco, USA, Preprint: 6963.

Honda A., Shibata H., and Gyoba J., et al. (**2007**). "Transfer effects on sound localization performances from playing a virtual three-dimensional auditory game," Applied Acoustics **68(8)**, 885-896.

Horbach U. (**1997**). "New techniques for the production of multichannel sound," in AES 103th Convention, New York, USA, Preprint: 4624.

Horbach U., Karamustafaoglu A., and Pellegrini R., et al. (**1999**). "Design and applications of a data-based auralization system for surround sound," in AES 106th Convention, Munich, Germany, Preprint: 4976.

Hosoe S., Nishino T., and Itou K., et al. (**2005**). "Measurement of Head-related transfer functions in the proximal region," in Proceeding of Forum Acusticum 2005, Budapest, Hungary, 2539-2542.

Hosoe S., Nishino T., and Itou K., et al. (**2006**). "Development of micro-dodecahedral loudspeaker for measuring head-related transfer functions in the proximal region," in Proceeding of IEEE 2006 International Conference on Acoustics, Speech and Signal Processing, Toulouse, France, **Vol. (5)**, 329-332.

Humanski R.A., and Butler R.A. (**1988**). "The contribution of the near and far ear toward localization of sound in the sagittal plane," J. Acoust. Soc. Am. **83(6)**, 2300-2310.

Huopaniemi J., and Karjalainen M. (**1997a**). "Review of digital filter design and implementation methods for 3D sound," in AES 102nd Convention, Munich, Germany Preprint: 4461.

Huopaniemi J., Savioja L., and Karjalainen M. (**1997b**). "Modeling of reflections and air absorption in acoustical spaces—a digital filter design approach," in Proceeding of IEEE 1997 Workshop on Applications of Signal Processing to Audio and Acoustics, New Paltz, NY, USA.

Huopaniemi J., and Smith J.O. (**1999a**). "Spectral and time-domain preprocessing and the choice of modeling error criteria for binaural digital filters," in AES 16th International Conference, Rovaniemi, Finland.

Huopaniemi J., Zacharov N., and Karjalainen M. (**1999b**). "Objective and subjective evaluation of head-related transfer function filter design," J. Audio. Eng. Soc. **47**(**4**), 218-239.

Hwang S., Park Y. (**2007**). "HRIR customization in the median plane via principal components analysis," in AES 31st International Conference, London, UK.

Hwang S., and Park Y. (**2008a**). "Interpretations on principal components analysis of head-related impulse responses in the median plane," J. Acoust. Soc. Am. **123**(**4**), EL65-EL71.

Hwang S., Park Y., and Park Y.S. (**2008**). "Modeling and customization of head-related impulse responses based on general basis functions in time domain," Acta Acustica united with Acustica **94**(**6**), 965-980.

IEC (**1990**). *IEC 60959, Provisional head and torso simulator for acoustic measurement on air conduction hearing aids* (International Electrotechnical Commission, Geneva, Switzerland).

Inoue N., Kimura T., and Nishino T., et al. (**2005**). "Evaluation of HRTFs estimated using physical features," Acoust. Sci. & Tech. **26**(**5**), 453-455.

IRCAM Lab (**2003**). *Listen HRTF database*, http://recherche.ircam.fr/equipes/salles/listen/

Ise S., and Otani M. (**2002**). "Real time calculation of the head related transfer function based on the boundary element method," in Processing of The 2002 International Conference on Auditory Display, Kyoto, Japan, 1-6.

ISO (**2002**). *ISO 11904-1, Acoustics—determination of sound immission from sound sources placed close to the ear-Part 1: technique using a microphone in a real ear (MIRE technique)*, (International Standardization Organization, Geneva, Switzerland).

ISO (**2003**). *ISO 226, Acoustics—normal equal-loudness-level contours* (International Organization for Standardization, Geneva, Switzerland).

ISO (**2004**). *ISO 11904-2, Acoustics—determination of sound immission from sound sources placed close to the ear-Part 1: Technique using a manikin* (International Standardization Organization, Geneva, Switzerland).

ITU (**1994**). *ITU-R BS 775-1: Multichannel stereophonic sound system with and without accompanying picture*, Doc 10/63 (International Telecommunication Union, Geneva, Switzerland).

ITU (**1997**). *ITU-R BS 1116-1: Methods for the subjective assessment of small impairments in audio systems including multichannel sound system* (International Telecommunication Union, Geneva, Switzerland).

Iwahara M., and Mori T. (**1978**). "Stereophonic sound reproduction system," United States Patent 4,118,599.

Iwaya Y., and Suzuki Y. (**2007**). "Rendering moving sound with the doppler effect in sound space," Applied Acoustics **68**(**8**), 916-922.

Jeffress L.A. (**1948**). "A place theory of sound localization," J. Comp. Physiol. Psych. **41**(**1**), 35-39.

Jenison R.L., Neelon M.F., and Reale R.A., et al. (**1998**). "Synthesis of virtual motion in 3D auditory space," in Proceeding of the 20th Annual International Conference of the IEEE Engineering in Medicine and Biology Society, Hong Kong, China, **Vol**.(**3**), 1096-1100.

Jeong J.W., Lee J., and Park Y.C., et al. (**2005**). "Design and implementation of IIR crosstalk cancellation filters approximating frequency warping," in AES 118th Convention, Barcelona, Spain, Preprint: 6490.

Jin C., Leong P., and Leung J., et al. (**2000**). "Enabling individualized virtual auditory space using morphological measurements," in Proceedings of the First IEEE Pacific-Rim Conference on Multimedia, Sydney, Australia, 235-238.

Jin C., Corderoy A., and Carlile S., et al. (**2004**). "Contrasting monaural and interaural spectral cues for human sound localization," J. Acoust. Soc. Am. **115**(**6**), 3124-3141.

Jin C., Tan T., and Kan A., et al. (**2005**). "Real-time, head-tracked 3D audio with unlimited simultaneous sounds," in the Eleventh Meeting of the International Conference on Auditory Display (ICAD 05), Limerick, Ireland, 305-311.

Jin L.X., Ding S.H., and Hou H.Q. (**2000**). "Cephalo-facial anthropometry of the students of Han nationality in Shandong," Acta Aacademiae Medicinae Qingdao Universitatis (in Chinese) **36**(**1**), 43-45.

Jo H., Park Y., and Park Y.S. (**2008**). "Approximation on head related transfer function using prolate spheroidal head model," in Proceeding of 15th International Congress on Sound and Vibration, Daejeon, Korea, 2963-2970.

Jones D.L., Stanney K.M., and Foaud H. (**2005**). "An optimized spatial audio system for virtual training simulations: design and evaluation," in Proceedings of Eleventh Meeting of the International Conference on Auditory Display (ICAD 05), Limerick, Ireland, 223-227.

Jot J.M., and Chaigne A. (**1991**). "Digital delay networks for designing artificial reverberators," in AES 90th Convention, Paris, France, Preprint: 3030.

Jot J.M., Larcher V., and Warusfel O. (**1995**). "Digital signal processing issues in the context of binaural and transaural stereophony," in AES 98th Convention, Paris, France, Preprint: 3980.

Jot J.M., Wardle S., and Larcher V. (**1998**). "Approaches to Binaural Synthesis," in AES 105th Convention, San Francisco, California, USA, Preprint: 4861.

Jot J.M., Larcher V., and Pernaux J.M. (**1999**). "A comparative study of 3D audio encoding and rendering techniques," in AES 16th International Conference, Rovaniemi, Finland.

Jot J.M., Walsh M., and Philp A. (**2006a**). "Binaural simulation of complex acoustic scenes for interactive audio," in AES 121st Convention, San Francisco, USA., Preprint: 6950.

Jot J.M., and Trivi J.M. (**2006b**). "Scene description model and rendering engine for interactive virtual acoustics," in AES 120th Convention, Paris, France, Preprint: 6660.

Kahana Y., Nelson P.A., and Petyt M., et al. (**1998**). "Boundary element simulation of HRTFs and sound fields produced by virtual sound imaging system," in AES 105 Convention, San Francisco, CA, USA, Preprint: 4817.

Kahana Y., Nelson P.A., and Petyt M., et al. (**1999**). "Numerical modeling of the transfer functions of a dummy-head and of the external ear," in AES 16th International Conference, Rovaniemi, Finland.

Kahana Y., and Nelson P.A. (**2000**). "Spatial acoustic mode shapes of the human pinna," in AES 109th Convention, Los Angeles, CA, USA, Preprint: 5218.

Kahana Y., and Nelson P.A. (**2006**). "Numerical modelling of the spatial acoustic response of the human pinna," J. Sound and Vibration **292(1/2)**, 148-178.

Kahana Y., and Nelson P.A. (**2007**). "Boundary element simulations of the transfer function of human heads and baffled pinnae using accurate geometric models," J. Sound and Vibration **300(3/5)**, 552-579.

Kan A., Pope G., and Jin C., et al. (**2004**). "Mobile spatial audio communication system," in the Tenth Meeting of the International Conference on Auditory Display (ICAD 04), Sydney, Australia.

Kan A., Jin C., and Tan T., et al. (**2005**). "3-D Ape: A real-time 3D audio playback engine," in AES 118th Convention, Barcelona, Spain, Preprint: 6343.

Kan A., Jin C., and Schaik A.V. (**2006a**). "Distance variation function for simulation of near-field virtual auditory space," in Proceedings of IEEE 2006 International Conference on Acoustics, Speech and Signal Processing, Toulouse, France, **Vol.(5)**, 325-328.

Kan A., Jin C., and Schaik A.V. (**2006b**). "Psychoacoustic evaluation of a new method for simulating near-field virtual auditory space," in AES 120th Convention, Paris, France, Preprint: 6801.

Kan A., Jin C., and Schaik A.V. (**2009**). "A psychophysical evaluation of near-field head-related transfer functions synthesized using a distance variation function," J. Acoust. Soc. Am. **125(4)**, 2233-2242.

Kang S.H., and Kim S.H. (**1996**). "Realistic audio teleconferencing using binaural and auralization techniques," ETRI Journal **18(1)**, 41-51.

Karamustafaoglu A., and Spikofski G. (**2001**). "Binaural room scanning and binaural room modelling," in AES 19th International Conference, Schloss Elmau, Germany.

Katz B.F.G. (**2000**). "Acoustic absorption measurement of human hair and skin within the audible frequency range," J. Acoust. Soc. Am. **108(5)**, 2238–2242.

Katz B.F.G. (**2001a**). "Boundary element method calculation of individual head-related transfer function. I. Rigid model calculation," J. Acoust. Soc. Am. **110(5)**, 2440-2448.

Katz B.F.G. (**2001b**). "Boundary element method calculation of individual head-related transfer function.II. Impedance effects and comparisons to real measurements," J. Acoust. Soc. Am. **110(5)**, 2449-2455.

Katz B.F.G., and Begault D.R. (**2007**). "Round robin comparison of HRTF measurement systems: preliminary results," in The 19th International Congress on Acoustics, Madrid, Spain.

Katz B.F.G., and Parseihian G. (**2012**). "Perceptually based head-related transfer function database optimization," J. Acoust. Soc. Am. **131(2)**, EL99-105.

Kawano S., Taira M., and Matsudaira M., et al. (**1998**). "Development of the virtual sound algorithm," IEEE Trans. Consumer Electronics **44(3)**, 1189-1194.

Kendall G.S., and Rodgers C.A.P. (**1982**). "The simulation of three-dimensional localization cues for headphone listening," in Proceeding of the International Computer Music Conference, Denton, Texas, USA., 225-243.

Kendall G.S., and Martens W.L. (**1984**). "Simulating the cues of spatial hearing in natural environment," in Proceeding of International Computer Music Conference, Denton, Texas, USA, 115-125.

Kendall G.S. (**1995**). "The decorrelation of audio signals and its impact on spatial imagery," Computer Music Journal **19(4)**, 71-87.

Keyrouz F., and Diepold K. (**2008**). "A new HRTF interpolation approach for fast synthesis of dynamic environmental interaction," J. Audio. Eng. Soc. **56(1/2)**, 28-35.

Kim S., Lee J., and Jang S., et al. (**2004**). "Virtual sound algorithm for wide stereo sound stage," in AES 117th Convention, San Francisco, USA, Preprint: 6290.

Kim S., Kong D., and Jang S. (**2008**). "Adaptive virtual surround sound rendering system for an arbitrary listening position," J. Audio. Eng. Soc. **56(4)**, 243-254.

Kim S.M., and Choi W. (**2005**). "On the externalization of virtual sound images in headphone reproduction: A Wiener filter approach," J. Acoust. Soc. Am. **117(6)**, 3657-3665.

Kim Y., Kim S., and Kim J., et al. (**2005**). "New HRTFs (head related transfer functions) for 3D audio applications," in AES 118th Convention, Barcelona, Spain, Preprint: 6495.

Kim Y., Deille O., and Nelson P.A. (**2006**). "Crosstalk cancellation in virtual acoustic imaging systems for multiple listeners," J. Sound and Vibration **297(1/2)**, 251-266.

Kirkeby O., Nelson P.A., and Hamada H. (**1998a**). "The 'Stereo Dipole'—a virtual source imaging system using two closely spaced loudspeakers," J. Audio Eng. Soc. **46(5)**, 387-395.

Kirkeby O., Nelson P.A., and Hamada H. (**1998b**). "Local sound field reproduction using two closely spaced loudspeakers," J. Acoust. Soc. Am. **104(4)**, 1973-1981.

Kirkeby O., Rubak P., and Johansen L.G., et al. (**1999a**). "Implementation of cross-talk cancellation networks using warped FIR filters," in AES 16th International Conference, Rovaniemi, Finland.

Kirkeby O., Rubak P., and Nelson P.A., et al. (**1999b**). "Design of cross-talk cancellation networks by using fast deconvolution," in AES 106th Convention, Munich, Germany, Preprint: 4916.

Kirkeby O., and Nelson P.A. (**1999c**). "Digital filter design for inversion problems in sound reproduction," J. Audio. Eng. Soc. **47(7/8)**, 583–595.

Kirkeby O. (**2002**). "A balanced stereo widening network for headphones," in AES 22nd International Conference, Espoo, Finland.

Kirkeby O., Seppälä E.T., and Kärkkäinen A.,et al. (**2007**). "Some effects of the torso on head-related transfer functions," in AES 122nd Convention, Vienna, Austria, Preprint: 7030.

Kistler D.J., and Wightman F.L. (**1992**). "A model of head-related transfer functions based on principal components analysis and minimum-phase reconstruction," J. Acoust. Soc. Am. **91(3)**, 1637-1647.

Kleiner M., Dalenbäck B.I., and Svensson P. (**1993**). "Auralization—an overview," J. Audio. Eng. Soc. **41(11)**, 861-875.

Kopčo N., and Shinn-Cunningham B.G. (**2003**). "Spatial unmasking of nearby pure-tone targets in a simulated anechoic environment," J. Acoust. Soc. Am. **114(5)**, 2856-2870.

Köring J., and Schmitz A. (**1993**). "Simplifying Cancellation of cross-talk for playback of head-related recordings in a two-speaker system," Acustica **79(3)**, 221-232.

Krebber W., Gierlich H.W., and Genuit K. (**2000**). "Auditory virtual environments: basics and applications for interactive simulations," Signal Processing **80(11)**, 2307-2322.

Kreuzer W., Majdak P., and Chen Z. (**2009**). "Fast multipole boundary element method to calculate head-related transfer functions for a wide frequency range," J. Acoust. Soc. Am. **126(3)**, 1280-1290.

Kuhn G.F. (**1977**). "Model for the interaural time difference in the azimuthal plane," J. Acoust. Soc. Am. **62(1)**, 157-167.

Kulkarni A., and Colburn H.S. (**1995**). "Efficient finite-impulse-response filter models of the head-related transfer function," J. Acoust. Soc. Am. **97(5)**, 3278.

Kulkarni A. (**1997**). "Sound localization in real and virtual acoustical environments," Doctor dissertation of Boston University, Boston, USA.

Kulkarni A., and Colburn H.S. (**1998**). "Role of spectral detail in sound-source localization," Nature **396**, 747-749.

Kulkarni A., Isabelle S.K., and Colburn H.S. (**1999**). "Sensitivity of human subjects to head-related transfer-function phase spectra," J. Acoust. Soc. Am. **105(5)**, 2821-2840.

Kulkarni A., and Colburn H.S. (**2000**). "Variability in the characterization of the headphone transfer-function," J. Acoust. Soc. Am. **107(2)**, 1071-1074.

Kulkarni A., and Colburn H.S. (**2004**). "Infinite-impulse-response models of the head-related transfer function," J. Acoust. Soc. Am. **115(4)**, 1714-1728.

Kuttruff H. (**1991**). "On the audibility of phase distortions in rooms and its significance for sound reproduction and digital simulation in room acoustics," Acustica **74(1)**, 3–5.

Kuttruff H. (**1993**). "Auralization of impulse responses modeled on the basis of ray-tracing results," J. Audio. Eng. Soc. **41(11)**, 876-880.

Kuttruff H. (**2000**). *Room Acoustics* (Fourth edition, Spon Press, London, UK).

Kyriakakis C. (**1998**). "Fundamental and technological limitations of immersive audio systems," Proceedings of the IEEE **86(5)**, 941–951.

Laakso T.I., Valimaki V., and Karjalainen M., et al. (**1996**). "Splitting the unit delay—tools for fractional delay filter design," IEEE Signal Process. Mag. **13(1)**, 30-60.

Langendijk E.H.A., and Bronkhorst A.W. (**2000**). "Fidelity of three-dimensional-sound reproduction using a virtual auditory display," J. Acoust. Soc. Am. **107(1)**, 528-537.

Langendijk E.H.A., and Bronkhorst A.W. (**2002**). "Contribution of spectral cues to human sound localization," J. Acoust. Soc. Am. **112(4)**, 1583-1596.

Larcher V., Vandernoot G., and Jot J.M. (**1998**). "Equalization methods in binaural technology," in AES 105th Convention, San Francisco CA, USA, Preprint: 4858.

Larcher V., Jot J.M., and Guyard J., et al. (**2000**). "Study and comparison of efficient methods for 3D audio spatialization based on linear decomposition of HRTF data," in AES 108th Convention, Paris, France, Preprint: 5097.

Leakey D.M. (**1959**). "Some measurements on the effects of interchannel intensity and time difference in two channel sound systems," J. Acoust. Soc. Am. **31(7)**, 977-986.

Lee K., Son C., and Kim D. (**2009**). "Low complexity binaural rendering for multi-channel sound," in AES 126th Convention, Munich, Germany, Preprint: 7687.

Lee S.L., Kim L.H., and Sung K.M. (**2003**). "Head related transfer function refinement using directional weighting function," in AES 115th Convention, New York, USA, Preprint: 5918.

Lee T., Jang D., and Kang K., et al. (**2004**). "3D Audio acquisition and reproduction system using multiple microphones on a rigid sphere," in AES 116th Convention, Berlin, Germany, Preprint: 6135.

Lee T., Park G.Y., and Jang I., et al. (**2005**). "An object-based 3D audio broadcasting system for interactive services," AES 118th Convention, Barcelona, Spain, Preprint: 6384.

Lehnert H., and Blauert J. (**1992**). "Principles of binaural room simulation," Applied Acoustics **36(3/4)**, 259-291.

Leitner S., Sontacchi A., and Höldrich R. (**2000**). "Multichannel sound reproduction system for binaural signals—The Ambisonic approach," in Proceedings of the COST G-6 Conference on Digital Audio Effects (DAFX-00), Verona, Italy.

Lemaire V., Clerot F., and Busson S., et al. (**2005**). "Individualized HRTFs from few measurements: a statistical learning approach," in Proceedings of International Joint Conference on Neural Networks, Montreal, Canada, 2041-2046.

Lentz T., and Schmitz O. (**2002**). "Realisation of an adaptive cross-talk cancellation system for a moving listener," in AES 21st International Conference, St. Petersburg, Russia.

Lentz T., Assenmacher I., and Sokoll J. (**2005**). "Performance of spatial audio using dynamic cross-talk cancellation," in AES 119th Convention, New York, USA, Preprint: 6541.

Lentz T., Assenmacher I., and Vorländer M., et al. (**2006a**). "Precise near-to-head acoustics with binaural synthesis," Journal of Virtual Reality and Broadcasting **3(2)**, urn:nbn:de:0009-6-5890.

Lentz T. (**2006b**). "Dynamic crosstalk cancellation for binaural synthesis in virtual reality environments," J. Audio. Eng. Soc. **54(4)**, 283-294.

Leong P., and Carlile S. (**1998**). "Methods for spherical data analysis and visualization." J Neurosci Met **80(2)**, 191-200.

Li A.Q., Jin C., and Schaik A.V. (**2006**). "Listening through different ears in the Sydney Opera House," in Proceeding of IEEE 2006 International Conference on Acoustics, Speech and Signal Processing, Toulouse, France, **Vol.(5)**, 333-336.

Li M. (**2003**). "Implementation of a model of head-related transfer functions based on principal components analysis and minimum-phase reconstruction," Dissertation of Degree of Master of Science at the University of Kaiserslautern, Germany.

Li Z., and Duraiswami R. (**2006**). "Headphone-based reproduction of 3D auditory scenes captured by spherical/hemispherical microphone arrays," in Proceeding of IEEE 2006 International Conference on Acoustics, Speech, and Signal Processing, Toulouse, France, **Vol.(5)**, 337-340.

Liang Z.Q., and Xie B.S. (**2012**). "A head-related transfer function model for fast synthesizing multiple virtual sound sources," Acta Acoustica (in Chinese) **37(3)**, 270-278.

Liu C.J., and Hsieh S.F. (**2001**). "Common-acoustic-poles/zeros approximation of head-related transfer functions," in Proceedings of IEEE International Conference on Acoustics, Speech, and Signal Processing, **Vol.(5)**, 3341-3344.

Liu Z.Y., Zheng Q, and Chao D.N. (**1999**). "Correlation study of measuring items between head facial part and bodily part in human," Journal of Tianjin Normal University **19(1)**, 40-44.

Lokki T., Hiipakka J., and Savioja L. (**2001a**). "A framework for evaluating virtual acoustic environments," in AES 110th Convention, Amsterdam, The Netherlands, Preprint: 5317.

Lokki T., and Järveläinen H. (**2001b**). "Subjective evaluation of auralization of physics-based room acoustics modeling," in The 2001 International Conference on Auditory Display, Espoo, Finland.

Lokki T., Nironen H., and Vesa S., et al. (**2004**). "Application scenarios of wearable and mobile augmented reality audio," in AES 116th Convention, Berlin, Germany, Preprint: 6026.

Lokki T., and Gröhn M. (**2005**). "Navigation with auditory cues in a virtual environment," IEEE Multimedia **12(2)**, 80-86.

Loomis J.M., Hebert C., and Cicinelli J.G. (**1990**). "Active localization of virtual sounds," J. Acoust. Soc. Am. **88(4)**, 1757-1764 .

Loomis J.M., Golledge R.G., and Klatzky R.L., et al. (**1998**). "Navigation system for the blind: Auditory display modes and guidance," Presence **7(2)**, 193-203.

Lopez J.J., and González A. (**1999**). "3-D audio with dynamic tracking for multimedia environments," in The 2nd COST-G6 Workshop on Digital Audio Effects (DAFX99), Trondheim, Norway.

Lopez J.J., and Gonzalez A. (**2001**). "Experimental evaluation of cross-talk cancellation regarding loudspeakers' angle of listening," IEEE Signal Processing Letters **8(1)**, 13-15.

Lopez-Poveda E. A., and Meddis R. (**1996**). "A physical model of sound diffraction and reflections in the human concha," J. Acoust. Soc. Am. **100(5)**, 3248–3259.

Lorho G., Isherwood D., and Zacharov N., et al. (**2002**). "Round robin subjective evaluation of stereo enhancement system for headphones," in AES 22nd International Conference, Espoo, Finland.

Lorho G., and Zacharov N. (**2004**). "Subjective evaluation of virtual home theatre sound systems for loudspeakers and headphones," in AES 116th Convention, Berlin, Germany, Preprint: 6141.

Lorho G. (**2005**). "Evaluation of spatial enhancement systems for stereo headphone reproduction by preference and attribute rating," in AES 118th Convention, Barcelona, Spain, Preprint: 6514.

Maa D. Y., and Shen H. (**2004**). The handbook of Acoustics (Revised Edition, in Chinese, Science Press, Beijing, China).

Mackensen P., Fruhmann M., and Thanner M., et al. (**2000**). "Head-tracker based auralization systems: Additional consideration of vertical head movements," AES 108th Convention, Paris, France, Preprint: 5135.

Mackenzie J., Huopaniemi J., and Valimaki V., et al. (**1997**). "Low-order modeling of head-related transfer functions using balanced model truncation," IEEE Signal Processing Letters **4(2)**, 39-41.

Macpherson E.A. (**1991**). "A computer model of binaural localization for stereo image measurement," J. Audio. Eng. Soc. **39(9)**, 604-622.

Macpherson E.A., and Sabin A.T. (**2007**). "Binaural weighting of monaural spectral cues for sound localization," J. Acoust. Soc. Am. **121(6)**, 3677-3688.

Maher R.C. (**1997**). "Single-ended spatial enhancement using a cross-coupled lattice equalizer," in 1997 IEEE Workshop on Application of Signal Processing to Audio and Acoustics, New Paltz, NY, USA.

Majdak P., Balazs P., and Laback B. (**2007**). "Multiple exponential sweep method for fast measurement of head related transfer functions," in AES 122nd Convention, Vienna, Austria, Preprint: 7019.

Majdak P., Goupell M.J., and Laback B. (**2010**). "3-D localization of virtual sound sources: Effects of visual environment, pointing method, and training," Attention, Perception, & Psychophysics **72(2)**, 454-469.

Mannerheim P., Nelson P.A., and Kim Y. (**2006**). "Filter update techniques for adaptive virtual acoustic imaging," in AES 120th Convention, Paris, France, Preprint: 6715.

Mannerheim P., and Nelson P.A. (**2008**). "Virtual sound image using visually adaptive loudspeakers," Acta Acustica united with Acustica **94(6)**, 1024-1039.

Marques de Sá J.P. (**2007**). *Applied statistics using SPSS, STATISTICA, MATLAB and R* (Springer-Verlag, Berlin, Heidelberg, New York).

Martens W.L. (**1987**). "Principal component analysis and resynthesis of spectral cues to perceived direction," in Proceeding of the International Computer Music Conference, San Francisco, CA, USA, 274-281.

Martin R.L., McAnally K.I., and Senova M.A. (**2001**). "Free-field equivalent localization of virtual audio," J. Audio. Eng. Soc. **49(1/2)**, 14-22.

Matsumoto M., and Tohyama M. (**2003**). "Algorithms for moving sound images," in AES 114th Convention, Amsterdam, The Netherlands, Preprint: 5770.

Matsumoto M.,Yamanaka S., and Tohyama M. (**2004**). "Effect of arrival time correction on the accuracy of binaural impulse response interpolation, interpolation methods of binaural response," J. Audio. Eng. Soc. **52(1/2)**, 56-61.

MacCabe C.J., and Furlong D.J. (**1991**). "Spectral stereo surround sound pan-pot," in AES 90th Convention, Paris, France, Preprint: 3067.

McAnally K.I., and Martin R.L. (**2002**). "Variability in the headphone-to-ear-canal transfer function," J. Audio. Eng. Soc. **50(4)**, 263-266.

Mehrgardt S., and Mellert V. (**1977**). "Transformation characteristics of the external human ear," J. Acoust. Soc. Am. **61(6)**, 1567–1576.

Melchior F., Thiergart O., and Galdo G.D., et al. (**2009**). "Dual radius spherical cardioid microphone arrays for binaural auralization," in AES 127th Convention, New York, USA, Paper 7855.

Mendonça C., Campos G., and Dias P., et al. (**2012**). "On the improvement of localization accuracy with non-individualized HRTF-based sounds," J. Audio. Eng. Soc. **60(10)**, 821-830.

Melick J.B., Algazi V.R., and Duda R.O., et al. (**2004**). "Customization for personalized rendering of motion-tracked binaural sound," in AES 117th Convention, San Francisco, CA, USA, Preprint: 6225.

Menzies D. (**2002**). "W-Panning and O-format, tools for object spatialization," in AES 22nd International Conference, Espoo, Finland.

Menzies D., and Marwan A.A. (**2007**). "Nearfield binaural synthesis and ambisonics," J. Acoust. Soc. Am. **121(3)**, 1559-1563.

Merimaa J. (**2009**). "Modification of HRTF filters to reduce timbral effects in binaural synthesis," in AES 127th Convention, New York, NY, USA, Preprint: 7912.

Merimaa J. (**2010**). "Modification of HRTF filters to reduce timbral effects in binaural synthesis, part 2: individual HRTFs," in AES 129th Convention, San Francisco, CA, USA, Paper: 8265.

Mertens H. (**1965**). "Directional hearing in stereophony theory and experimental verification," EBU Rev., Part A, **92(Aug.)**, 146-158.

Mickiewicz W., and Sawicki J. (**2004**). "Headphone processor based on individualized head related transfer functions measured in listening room," in AES 116th Convention, Berlin, Germany, Preprint: 6067.

Middlebrooks J.C., Makous J.C., and Green D.M. (**1989**). "Directional sensitivity of sound-pressure levels in the human ear canal," J. Acoust. Soc. Am. **86(1)**, 89-108.

Middlebrooks J.C. (**1992a**). "Narrow-band sound localization related to external ear acoustics," J. Acoust. Soc. Am. **92(5)**, 2607-2624.

Middlebrooks J.C., and Green D.M. (**1992b**). "Observations on a principal components analysis of head-related transfer functions," J. Acoust. Soc. Am. **92(1)**, 597-599.

Middlebrooks J.C. (**1999a**). "Individual differences in external-ear transfer functions reduced by scaling in frequency," J. Acoust. Soc. Am. **106(3)**, 1480-1492.

Middlebrooks J.C. (**1999b**). "Virtual localization improved by scaling nonindividualized external-ear transfer functions in frequency," J. Acoust. Soc. Am. **106(3)**, 1493-1510.

Middlebrooks J.C., Macpherson E.A., and Onsan Z.A. (**2000**). "Psychophysical customization of directional transfer functions for virtual sound localization," J. Acoust. Soc. Am. **108(6)**, 3088-3091.

Miller J.D., and Wenzel E.M. (**2002**). "Recent developments in SLAB: A software-based system for interactive spatial sound synthesis," in the 2002 International Conference on Auditory Display, Kyoto, Japan.

Mills A.W. (**1958**). "On the minimum audible angle," J. Acoust. Soc. Am. **30(4)**, 237-246.

Minnaar P., Christensen F., and Møller H., et al. (**1999**). "Audibility of all-pass components in binaural synthesis," in AES 106th Convention , Munich, Germany, preprint: 4911.

Minnaar P., Plogsties J., and Olesen S.K., et al. (**2000**). "The interaural time difference in binaural synthesis," in AES 108th Convention, Paris, France, Preprint: 5133.

Minnaar P., Olesen S.K., and Christensen F., et al. (**2001**). "Localization with binaural recordings from artificial and human heads," J. Audio Eng. Soc. **49(5)**, 323-336.

Minnaar P., Plogsties J., and Christensen F. (**2005**). "Directional resolution of head-related transfer functions required in binaural synthesis," J. Audio. Eng. Soc. **53(10)**, 919-929.

Moldrzyk C., Ahnert W., and Feistel S.,et al. (**2004**). "Head-tracked auralization of acoustical simulation," in AES 117th Convention, San Francisco, CA, USA, Preprint: 6275.

Møller H. (**1989**). "Reproduction of artificial-head recordings through loudspeakers," J. Audio. Eng. Soc. **37(1/2)**, 30-33.

Møller H. (**1992**). "Fundamentals of binaural technology," Applied Acoustics **36(3/4)**, 171-218.

Møller H., Hammershøi D., and Jensen C.B., et al. (**1995a**). "Transfer characteristics of headphones measured on human ears," J. Audio. Eng. Soc. **43(4)**, 203-217.

Møller H., Sørensen M.F., and Hammershøi D.,et al. (**1995b**). "Head-related transfer functions of human subjects," J. Audio. Eng. Soc. **43(5)**, 300-321.

Møller H., Jensen C.B., and Hammershøi D., et al. (**1996a**). "Using a typical human subject for binaural recording," in AES 100th Convention, Copenhagen, Denmark, Preprint: 4157.

Møller H., Sørensen M.F., and Jensen C.B., et al. (**1996b**). "Binaural technique: Do we need individual recordings? "J. Audio. Eng. Soc. **44(6)**, 451-469.

Møller H., Hammershøi D., and Jensen C.B., et al. (**1999**). "Evaluation of artifical heads in listening tests," J. Audio. Eng. Soc. **47(3)**, 83-100.

Mokhtari P., Takemoto H., and Nishimura R., et al. (**2007**). "Comparison of simulated and measured HRTFs: FDTD simulation using MRI head data," in AES 123rd Convention, New York, NY, USA, Preprint: 7240.

Moore B.C.J., Oldfield S.R., and Dooley G.J. (**1989**). "Detection and discrimination of spectral peaks and notches at 1 and 8 kHz," J. Acoust. Soc. Am. **85(2)**, 820–836.

Moore B.C.J., Glasberg B.R., and Bear T. (**1997**). "A model for the prediction of thresholds, loudness,and partial loudness," J. Audio. Eng. Soc. **45(4)**, 224-240.

Moore B.C.J. (**2003**). *An introduction to the psychology of hearing* (Fifth edition, Academic Press, San Diego, USA).

Moore B.C.J., and Glasberg B.R. (**2008**). "Modeling binaural loudness," J. Acoust. Soc. Am. **121(3)**, 1604-1612.

Moore A.H., Tew A.I., and Nicol R. (**2010**). "An initial validation of individualized crosstalk cancellation filters for binaural perceptual experiments," J. Audio. Eng. Soc. **58(1/2)**, 36-45.

Morimoto M., and Ando Y. (**1980**). "On the simulation of sound localization," J. Acoust. Soc. Jpn.(E) **1(3)**, 167–174.

Morimoto M. (**2001**). "The contribution of two ears to the perception of vertical angle in sagittal planes," J. Acoust. Soc. Am. **109(4)**, 1596-1603.

Morse P.M., and Ingrad K.U. (**1968**). *Theoretical Acoustics* (McGraw-Hill, New York, USA).

Mouchtaris A., Reveliotis P., and Kyriakakis C. (**2000**). "Inverse filter design for immersive audio rendering over loudspeakers," IEEE Trans. on Multimedia **2(2)**, 77-87.

Musicant A.D., Chan J.C.K., and Hind J. E. (**1990**). "Direction-dependent spectra properties of cat external ear: New data and cross-species comparisons," J. Acoust. Soc. Am. **87(2)**, 757-781.

Nam J., Kolar M.A., and Abel J.S. (**2008**). "On the minimum-phase nature of head-related transfer functions," in AES 125th Convention, San Francisco, CA, USA, Preprint: 7546.

Naylor G.M. (**1993**). "ODEON–Another hybrid room acoustical model," Applied Acoustics **38(2-4)**, 131-143.

Nelson P.A., Hamada H., and Elliott S.J. (**1992**). "Adaptive inverse filters for stereophonic sound reproduction," IEEE Trans. Signal Processing **40(7)**, 1621-1632.

Nelson P.A., Bustamante O.F., and Engler D., et al. (**1996**). "Experiments on a system for the synthesis of virtual acoustic sources," J. Audio. Eng. Soc. **44(11)**, 990-1007.

Nelson P.A., and Rose J.F.W. (**2005**). "Errors in two-point sound reproduction," J. Acoust. Soc. Am. **118(1)**, 193-204.

Nicol R., Lemaire V., and Bondu A. (**2006**). "Looking for a relevant similarity criterion for HRTF clustering: a comparative study," in AES 120th Convention, Paris, France, Preprint: 6653.

Nielsen S. H. (**1993**). "Auditory distance perception in different rooms," J. Audio Eng. Soc. **41(10)**, 755-770 .

Nishino T., Mase S., and Kajita S., et al. (**1996**). "Interpolating HRTF for auditory virtual reality," J. Acous. Soc. Am. **100(4)**, 2602.

Nishino T., Kajita S., and Takeda K., et al. (**1999**). "Interpolating head related transfer functions in the median plane," in Proceeding of 1999 IEEE Workshop on Applications of Signal Processing to Audio and Acoustics, New Paltz, New York, USA, 167-170.

Nishino T., Nakai Y., and Takeda K., et al. (**2001**). "Estimating head related transfer function using multiple regression analysis," IEICE Trans A. **84(A)**, 260-268.

Nishino T., Hosoe S., and Takeda K., et al. (**2004**). "Measurement of the head related transfer function using the spark noise," in The 18th International Congress on Acoustics, Kyoto, Japan.

Nishino T., Inoue N., and Takeda K., et al. (**2007**). "Estimation of HRTFs on the horizontal plane using physical features," Applied Acoustics **68(8)**, 897-908.

Noisternig M., Sontacchi A., and Musil T., et al. (**2003**). "A 3D Ambisonic based binaural sound reproduction system," in AES 24th International Conference, Banff, Canada.

Noisternig M., Katz B.F.G., Siltanen S., et al. (**2008**). "Framework for real-time auralization in architectural acoustics," Acta Acustica united with Acustica **94(6)**, 1000-1015.

Olive S. (**2001**). "Evaluation of five commercial stereo enhancement 3D audio software plug-ins," in AES 110th Convention, Amsterdam, The Netherlands, Preprint: 5386.

Olive S.E., and Nind T. (**2007**). "Important methodological issues in the subjective evaluation of automotive audio system sound quality," in Proceeding of 2007 International Electroacoustics Technology Symposia, Nanjing, China, 145-152.

Oppenheim A.V., Schafer R.W., and Buck J.R. (**1999**). *Discrete-time signal processing* (Second Edition, Prentice-Hall, Englewood Cliffs, NJ, USA).

Orduna F., Lopez J.J., and Gonzdlez A. (**2000**). "Robustness of acoustic crosstalk cancellation as a function of frequency and loudspeaker separation," in AES 109th Convention, Los Angeles, CA, USA, Preprint: 5219.

Otani M., and Ise S. (**2003**). "A fast calculation method of the head-related transfer functions for multiple source points based on the boundary element method," Acoust. Sci. & Tech. **24(5)**, 259-266.

Otani M., and Ise S. (**2006**). "Fast calculation system specialized for head-related transfer function based on boundary element method," J. Acoust. Soc. Am. **119(5)**, 2589-2598.

Otani M., Hirahara T., and Ise S. (**2009**). "Numerical study on source-distance dependency of head-related transfer functions," J. Acoust. Soc. Am. **125(5)**, 3253-3261.

Ou K.Y., and Bai M.R. (**2007**). "A perceptual approach for interaural transfer function calculations based on Wiener filtering," J. Audio. Eng. Soc. **55(9)**, 752-761.

Paavola M., Karlsson E., and Page J. (**2005**). "3D audio for mobile devices via Java," in AES 118th Convention, Barcelona, Spain, Preprint: 6472.

Papadopoulos T., and Nelson P.A. (**2010**). "Choice of inverse filter design parameters in virtual acoustic imaging systems," J. Audio. Eng. Soc. **58(1/2)**, 22-35.

Park M.H.,Choi S.I., and Kim S.H., et al. (**2005**). "Improvement of front-back sound localization characteristics in headphone-based 3D sound generation," in Proceeding of The 7th IEEE International Conference on Advanced Communication Technology, **Vol.(1)**, 273-276.

Park M., Nelson P.A., Kim Y. (**2006**). "An auditory process model for the evaluation of virtual acoustic imaging systems," in AES 120th Convention, Paris, France, Preprint: 6854.

Parodi Y.L., and Rubak P. (**2010**). "Objective evaluation of the sweet spot size in spatial sound reproduction using elevated loudspeakers," J. Acoust. Soc. Am. **128(3)**, 1045-1055.

Parodi Y.L., and Rubak P. (**2011**). "A subjective evaluation of the minimum channel separation for reproducing binaural signals over loudspeakers," J. Audio. Eng. Soc. **59(7/8)**, 487-497.

Parseihian G., and Katz B.F.G. (**2012**). "Rapid head-related transfer function adaptation using a virtual auditory environment," J. Acoust. Soc. Am. **131(4)**, 2948-2957.

Paul S. (**2009**). "Binaural recording technology: a historical review and possible future developments," Acta Acustica united with Acustica **95(5)**, 767-788.

Pedersen J.A., and Minnaar P. (**2006**). "Evaluation of a 3D-audio system with head tracking," in AES 120th Convention, Paris, France, Preprint: 6654.

Pellegrini R.S. (**1999**). "Comparison of data- and model-based simulation algorithms for auditory virtual environments," in AES 106th Convention, Munich, Germany, Preprint: 4953.

Pernaux J.M., Emerit M., and Daniel J., et al. (**2002**). "Perceptual evaluation of static binaural sound synthesis," in AES 22nd International Conference, Espoo, Finland.

Pernaux J.M., Emerit M., and Nicol R. (**2003**). "Perceptual evaluation of binaural sound synthesis: the problem of reporting localization judgments," in AES 114th Convention, Amsterdam, The Netherlands, Preprint: 5789.

Perrett S., and Noble W. (**1997**). "The effect of head rotations on vertical plane sound localization," J. Acoust. Soc. Am. **102(4)**, 2325-2332.

Plenge G. (**1974**). "On the differences between localization and lateralization," J. Acoust. Soc. Am. **56(3)**, 944-951.

Plogsties J., Minnaar P., and Olesen S., et al. (**2000a**). "Audibility of all-pass components in head-related transfer functions," in AES 108th Convention, Paris, France, Preprint: 5132.

Plogsties J., Minnaar P., and Christensen F., et al. (**2000b**). "The directional resolution needed when measuring head-related transfer functions," in DAGA 2000, Oldenburg, Germany.

Poletti M.A. (**2000**). "A unified theory of horizontal holographic sound systems," J. Audio. Eng. Soc. **48(12)**, 1155-1182.

Poletti M.A. (**2005**). "Effect of noise and transducer variability on the performance of circular microphone arrays," J. Audio. Eng. Soc. **53(5)**, 371-384.

Poletti M.A., and Svensson U.P. (**2008**). "Beamforming synthesis of binaural responses from computer simulations of acoustic spaces," J. Acoust. Soc. Am. **124(1)**, 301-315.

Pollow M., Masiero B., and Dietrich P., et al. (**2012**). "Fast measurement system for spatially continuous individual HRTFs," in AES 25th UK Conference, York, UK.

Pollow M., Nguyen K.V., and Warusfel O., et al. (**2012**). "Calculation of head-related transfer functions for arbitrary field points using spherical harmonics decomposition," Acta Acustica united with Acustica **98(1)**, 72-82.

Porschmann C. (**2007**). "3-D audio in mobile communication devices: methods for mobile head-tracking," J. Virtual Reality and Broadcasting **4(13)**, urn:nbn:de:0009-6-11833.

Posselt C., Shroter J., and Opitz M., et al., (**1986**). "Generation of binaural signals for research and home entertainment," in The 12nd International Congress on Acoustics, Toronto, Canada.

Pralong D., and Carlile S. (**1996**). "The role of individualized headphone calibration for the generation of high fidelity virtual auditory space," J. Acoust. Soc. Am. **100(6)**, 3785-3793.

Prodi N., and Velecka S. (**2003**). "The evaluation of binaural playback systems for virtual sound fields," Applied Acoustics **64(2)**, 147-161.

Pulkki V. (**1997**). "Virtual sound source positioning using vector base amplitude panning," J. Audio. Eng. Soc. **45(6)**, 456-466.

Pulkki V., Karjalainen M., and Huopaniemi J. (**1999**). "Analyzing virtual sound source attributes using a binaural auditory model," J. Audio. Eng. Soc. **47(4)**, 203-217.

Qian J., and Eddins D.A. (**2008**). "The role of spectral modulation cues in virtual sound localization," J. Acoust. Soc. Am. **123(1)**, 302-314.

Qiu X., Masiero B., and Vorländer M. (**2009**). "Channel separation of crosstalk cancellation systems with mismatched and misaligned sound sources," J. Acoust. Soc. Am. **126(4)**, 1796-1806.

Rafaely B. (**2005**). "Analysis and design of spherical microphone arrays," IEEE Trans. Speech and Audio Processing **13(1)**, 135-143.

Rafaely B., and Avni A. (**2010**). "Interaural cross correlation in a sound field represented by spherical harmonics," J. Acoust. Soc. Am. **127(2)**, 823-828.

Rao D., and Xie B.S. (**2005**). "Head rotation and sound image localization in the median plane," Chinese Science Bulletin **50(5)**, 412-416.

Rao D., and Xie B.S. (**2006**). "Repeatability analysis on headphone transfer function measurement," Technical Acoustics (in Chinese) **25(supplement)**, 441-442.

Rao D., and Xie B.S. (**2007**). "Influence of sound source directivity on binaural auralization quality," Technical Acoustics (in Chinese) **26(5)**, 899-903.

Rao D. (**2008a**). "Comparison of head-related transfer functions for male and female subjects," Technical Acoustics (in Chinese) **27(5, Pt.2)**, 392-393.

Rao D., and Wu S.X. (**2008b**). "Auralization difference between individualized and non-individualized binaural room impulse response," J. South China University of Technology **36(8)**, 123-127.

Rao D., and Xie B.S. (**2010**). "The equivalence between principal components analysis of head-related transfer functions in the time domain and frequency domain," in Proceeding of 2010 3rd International Congress on Image and Signal Processing (IEEE) **8**, 3895-3898, Yantai, China.

Raykar V.C., Duraiswami R., and Yegnanarayana B. (**2005**). "Extracting the frequencies of the pinna spectral notches in measured head related impulse responses," J. Acoust. Soc. Am. **118(1)**, 364-374.

Richard L., Burden J., and Douglas Faries (**2005**). *Numerical Analysis* (eighth edition, Thomson Learning Inc., Stamford, USA).

Riederer K.A.J. (**1998a**). "Head-related transfer function measurement," Master thesis of Helsinki University of Technology, Finland.

Riederer K.A.J. (**1998b**). "Repeatability analysis of head-related transfer function measurements," in AES 105th Convention, San Francisco, USA, Preprint: 4846.

Riederer K.A.J. (**2000**). "Computational quality assessment of HRTFs," in proceeding of 2000 European Signal Processing Conference, Tampere, Finland, **Vol. (4)**, 2241-2244.

Riederer K.A.J., and Niska R. (**2002**). "Sophisticated tube headphones for spatial sound reproduction," in AES 21st International Conference, St. Petersburg, Russia.

Rife D.D., and Vanderkooy J. (**1989**). "Transfer function measurements with maximum-length sequence," J. Audio Eng. Soc. **37(6)**, 419–444.

Rodríguez S.G., and Ramírez M.A. (**2005**). "HRTF individualization by solving the least squares problem," in AES 118th Convention, Barcelona, Spain, Preprint: 6438.

Rose J., Nelson P., and Rafaely B., et al. (**2002**). "Sweet spot size of virtual acoustic imaging systems at asymmetric listener locations," J. Acoust. Soc. Am. **112(5)**, 1992-2002.

Rumsey F. (**2001**). *Spatial Audio* (Focal Press, Oxford, England).

Runkle P.R., Blommer M.A., and Wakefield G.H. (**1995**). "A comparison of head related transfer function interpolation methods," in Proceeding of IEEE 1995 Workshop on Application Signal Processing to Audio and Acoustics, New Paltz, NY, USA, 88-91.

Sakamoto N., Gotoh T., and Kimura Y. (**1976**). "On 'Out-of-head localization' in headphone listening," J. Audio Eng. Soc. **24(9)**, 710-716.

Sakamoto N., Gotoh T., and Kogure T., et al. (**1981**). "Controlling sound-image localization in stereophonic sound reproduction, part 1," J. Audio Eng. Soc. **29(11)**, 794-799.

Sakamoto N., Gotoh, T., and Kogure T., et al. (**1982**). "Controlling sound-image localization in stereophonic sound reproduction, part 2," J. Audio Eng. Soc. **30(10)**, 719-722.

Sandvad J, and Hammershøi D. (**1994**). "Binaural auralization: comparison of FIR and IIR filter representation of HRIRs," in AES 96th Convention, Amsterdam, The Netherlands, Preprint: 3862.

Sandvad J. (**1996**). "Dynamic aspects of Auditory virtual environments," in AES 100th Convention, Copenhagen, Denmark, Preprint 4226.

Satarzadeh P., Algazi V.R., and Duda R.O. (**2007**). "Physical and filter pinna models based on anthropometry," in AES 122nd Convention, Vienna, Austria, Preprint: 7098.

Saviojia L., Huopaniemi J., and Lokki T., et al. (**1999**). "Creating interactive virtual acoustic environments," J. Audio. Eng. Soc. **47(9)**, 675-705.

Saviojia L, Lokki T., and Huopaniemi J. (**2002**). "Auralization applying the parametric room acoustic modeling technique—The DIVA Auralization system," in The 2002 International Conference on Auditory Display, Kyoto, Japan.

Scheirer E.D., Väänänen R., and Huopaniemi J. (**1999**). "AudioBIFS: Describing audio scenes with the MPEG-4 multimedia standard," IEEE Trans. on Multimedia **1(3)**, 237-250.

Schiff L.I. (**1968**). *Quantum mechanics* (third edition, Mcgraw-Hill Book Company, New York, USA).

Schönstein D., and Katz B.F.G. (**2012**). "Variability in perceptual evaluation of HRTFs," J. Audio. Eng. Soc. **60(10)**, 783-793.

Schroeder M.R. (**1962**). "Natural sounding artificial reverberation," J. Audio. Eng. Soc. **10(3)**, 219-223.

Schroeder M. R., and Atal B.S. (**1963**). "Computer simulation of sound transmission in rooms," Proceeding of IEEE **51(3)**, 536-537.

Schroeder M.R. (**1965**). "New method of measuring reverberation time," J. Acoust. Soc. Am. **37(3)**, 409-412.

Schroeder M.R. (**1979**). "Integrated-impulse method measuring sound decay without using impulse," J. Acoust. Soc. Am. **66(2)**, 497-500.

Searle C.L., Braida L.D., and Cuddy D.R., et al. (**1975**). "Binaural pinna disparity: another localization cue," J. Acoust. Soc. Am. **57(2)**, 448-455.

Seeber B.U., and Fastl H. (**2003**). "Subjective selection of non-individual head-related transfer functions," in Proceedings of the 2003 International Conference on Auditory Display, Boston, MA, USA, 259-262.

Seo J., Jang D.Y., and Park G.Y., et al. (**2003**). "Implementation of interactive 3D audio using MPEG-4 multimedia standards," in AES 115th Convention, New York, USA, Preprint: 5980.

Sevona M.A., McAnally K.I., and Martin R.L. (**2002**). "Localization of virtual sound as a function of head-related impulse response duration," J. Audio. Eng. Soc. **50(1/2)**, 57-66.

Shaw E. A. G. (**1966**). "Ear canal pressure generated by a free sound field," J. Acoust. Soc. Am. **39 (3)**, 465-470.

Shaw E.A.G, and Teranishi R. (**1968**). "Sound pressure generated in an external-ear replica and real human ears by nearby point source," J. Acoust. Soc. Am. **44(1)**, 240-249.

Shaw E. A. G. (**1974**). "Transformation of sound pressure level from the free field to the eardrum in the horizontal plane," J. Acoust. Soc. Am. **56(6)**, 1848-1861.

Shaw E.A.G., and Vaillancourt M.M. (**1985**). "Transformation of sound-pressure level from the free field to the eardrum presented in numerical form," J. Acoust. Soc. Am. **78(3)**, 1120-1123.

Shi L.Z. (**2007a**). "Research on four-loudspeaker reproduction system for virtual surround sound," MS thesis, South China University of Technology, Guangzhou, China.

Shi L.Z., and Xie B.S. (**2007b**). "An improved signal processing algorithm for virtual sound using four loud-speaker," Audio Engineering (in Chinese) **31(4)**, 41-44.

Shimada S., Hayashi N., and Hayashi S. (**1994**). "A clustering method for sound localization transfer functions," J. Audio. Eng. Soc. **42(7/8)**, 577-584.

Shinn-Cunningham B.G., Durlach N.I., and Held R.M. (**1998a**). "Adapting to supernormal auditory localization cues. I. Bias and resolution," J. Acoust. Soc. Am. **103(6)**, 3656-3666.

Shinn-Cunningham B.G., Durlach N.I., and Held R.M. (**1998b**). "Adapting to supernormal auditory localization cues. II. Constraints on adaptation of mean response," J. Acoust. Soc. Am. **103(6)**, 3667-3676.

Shinn-Cunningham B.G. (1998c). "Applications of virtual auditory displays," in Proceedings of the 20th International Conference of the IEEE Engineering in Biology and Medicine Society, Hong Kong, China, **Vol.(3)**, 1105-1108.

Shinn-Cunningham B.G., Schickler J., and Kopčo N., et al. (**2001**). "Spatial unmasking of nearby speech sources in a simulated anechoic environment," J. Acoust. Soc. Am. **110(2)**, 1118–1129.

Shiu Y.M., Chang T.M., and Chang P.C. (**2012**). "Realization of surround audio by a quadraphonic headset," in Proceeding of 2012 IEEE International Conference on Consumer Electronics, Las Vegas, NV, USA, 13-14.

Silzle A., Novo P., and Strauss H. (**2004**). "IKA-SIM: A system to generate auditory virtual environments," in AES 116th Convention, Berlin, Germany, Preprint: 6016.

Sivonen V.P., and Ellermeier W. (**2006**). "Directional loudness in an anechoic sound field, head-related transfer functions, and binaural summation," J. Acoust. Soc. Am. **119(5)**, 2965-2980.

Sivonen V.P., and Ellermeier W. (**2008**). "Binaural loudness for artificial-head measurements in directional sound fields," J. Audio. Eng. Soc. **56(6)**, 452-461.

So R.H.Y., Ngan B., and Horner A., et al. (**2010**). "Toward orthogonal non-individualised head-related transfer functions for forward and backward directional sound: cluster analysis and an experimental study," Ergonomics **53(6)**, 767–781.

Sodnik J., Susnik R., and Tomazic S. (**2006**). "Principal components of non-individualized head related transfer functions significant for azimuth perception," Acta Acustica united with Acustica **92(2)**, 312-319.

Song W., Ellermeier W., and Hald J. (**2008**). "Using beamforming and binaural synthesis for the psychoacoustical evaluation of target sources in noise," J. Acoust. Soc. Am. **123(2)**, 910-924.

Sontacchi A., Noisternig M., and Majdak P., et al. (**2002**). "An objective model of localization in binaural sound reproduction systems," in AES 21st International Conference, St. Petersburg, Russia.

Spezio M.L., Keller C.H., and Marrocco R.T., et al. (**2000**). "Head-related transfer functions of the Rhesus monkey," Hearing Research **144 (1/2)**, 73-88.

Spikofski G., and Fruhmann M. (**2001**). "Optimization of binaural room scanning (BRS): considering inter-individual HRTF-characteristics," in AES 19th International Conference, Schloss Elmau, Garmany.

Spors S., Wierstorf H., and Ahrens J. (**2011**). "Interpolation and range extrapolation of head-related transfer functions using virtual local wave field synthesis," in AES 130th Convention, London, UK, Paper: 8392.

Stan G.B., Embrechts J.J., and Archambeau D. (**2002**). "Comparison of different impulse response measurement techniques," J. Audio. Eng. Soc. **50(4)**, 249-262.

Strauss H. (**1998**). "Implementing Doppler shifts for virtual auditory environments," in AES 104th Convention, Amsterdam, The Netheland, Preprint: 4687.

Sugiyama K., Sakaguchi T., and Aoki S., et al. (**1995**). "Calculation of acoustic coefficients between two ears using spheroids," J. Acoust. Soc. Jpn. **51(2)**, 117–122.

Susnik R., Sodnik J., and Tomazic S. (**2008**). "An elevation coding method for auditory displays," Applied Acoustics **69(3)**, 333-341.

Suzuki Y., Kim H.Y., and Takane S., et al. (**1998**). "A modeling of distance perception based on auditory parallax model (abstract)," J. Acoust. Soc. Am. **103(5, Pt2)**, 3083.

Svensson U.P., and Kristiansen U.R. (**2002**). "Computational modeling and simulation of acoustic spaces," in AES 22nd International Conference, Espoo, Finland.

Takala, T., Hanninen R., and Valimaki V., et al., (**1996**). "An integrated system for virtual audio reality," in AES 100th Convention, Copenhagen, Denmark, Preprint 4229.

Takane S., Arai D., and Miyajima T., et al. (**2002**). "A database of head-related transfer functions in whole directions on upper hemisphere," Acoust. Sci. & Tech. **23(3)**, 160-162.

Takane S., Suzuki Y., and Miyajima T., et al. (**2003a**). "A new theory for high definition virtual acoustic display named ADVISE," Acoust. Sci. & Tech. **24(5)**, 276-283.

Takane S., Takahashi S., and Suzuki Y., et al. (**2003b**). "Elementary real-time implementation of a virtual acoustic display based on ADVISE," Acoust. Sci. & Tech. **24(5)**, 304-310.

Takemoto H., Mokhtari P., and Kato H., et al. (**2012**). "Mechanism for generating peaks and notches of head-related transfer functions in the median plane," J. Acoust. Soc. Am. **132(6)**, 3832-3841.

Takeuchi T., Nelson P.A., and Kirkeby O., et al. (**1998**). "Influence of individual head related transfer function on the performance of virtual acoustic imaging systems," in AES 104th Convention, Amsterdam, The Netherlands, Preprint: 4700.

Takeuchi T., Nelson P.A., and Hamada H. (**2001a**). "Robustness to head misalignment of virtual sound image system," J. Acoust. Soc. Am. **109(3)**, 958-971.

Takeuchi T. (**2001b**). "Systems for virtual acoustic imaging using the binaural principle," Ph.D. Thesis of Doctor Degree of Philosophy, University of Southampton, Southampton, UK.

Takeuchi T., Nelson P.A., and Teschl M. (**2002a**). "Elevated control transducers for virtual acoustic imaging," in AES 112th Convention, Munich, Germany, Preprint: 5596.

Takeuchi T., and Nelson P.A. (**2002b**). "Optimal source distribution for binaural synthesis over loudspeakers," J. Acoust. Soc. Am. **112(6)**, 2786-2797.

Takeuchi T., and Nelson P.A. (**2008**). "Extension of the optimal source distribution for binaural sound reproduction," Acta Acustica united with Acustica **94(6)**, 981-987.

Tan C.J., and Gan W.S. (**1998**). "User-defined spectral manipulation of HRTF for improved localization in 3D sound systems," Electronics letters **34(25)**, 2387-2389.

Tan C.J., and Gan W.S. (**1999**). "Wavelet packet decomposition for spatial sound condition," Electronics Letters **35(21)**, 1821-1823.

Tan S.E., Yang J., and Liew Y.H., et al. (**2000**). "Elevated speakers image correction using 3-D audio processing," in AES 109th Convention, Los Angeles, CA, USA, Preprint: 5204.

Tao Y., Tew A.I., and Porter S.J. (**2002**). "Interaural time difference estimation using the differential pressure synthesis method," in AES 22nd International Conference, Espoo, Finland.

Tao Y., Tew A.I., and Porter S.J. (**2003a**). "The differential pressure synthesis method for efficient acoustic pressure estimation," J. Audio. Eng. Soc. 51(7/8), 647-656.

Tao Y., Tew A.I., and Porter S.J. (**2003b**). "A study on head-shape simplification using spherical harmonics for HRTF computation at low frequencies," J. Audio. Eng. Soc. **51(9)**, 799-805.

Theile G., and Plenge G. (**1977**). "Localization of lateral phantom sources," J. Audio. Eng. Soc. 25 (4), 196-200.

Thomas M.V. (**1977**). "Improving the stereo headphone sound image," J. Audio. Eng. Soc. **25(7/8)**, 474-478.

Xiao T., and Liu Q.H. (**2003**). "Finite difference computation of head-related transfer function for human hearing," J. Acoust. Soc. Am. **113(5)**, 2434-2441.

Tikander M., Härmä A., and Karjalainen M. (**2004**). "Acoustic positioning and head tracking based on binaural signals," in AES 116th Convention, Berlin, Germany, Preprint: 6124.

Toh C.W., and Gan W.S. (**1999**). "A real-time virtual surround sound system with bass enhancement," in AES 107th Convention, New York, USA, Preprint: 5052.

Tong J., and Xie B.S. (**2005**). "An improvement three loudspeakers virtual reproduction system for 5.1 channel surround sound," Applied Acoustics (in Chinese) **24(6)**, 381-388.

Toole F.E. (**1991**). "Binaural record/reproduction systems and their use in psychoacoustic investigation," in AES 91st Convention, New York, USA, Preprint: 3179.

Torres J.C.B., Petraglia M.R., and Tenenbaum R.A. (**2004**). "An efficient wavelet-based HRTF model for auralization," Acta Acustica united with Acustica **90(1)**, 108-120.

Torres J.C.B., and Petraglia M.R. (**2009**). "HRTF interpolation in the wavelet transform domain," Proceeding of 2009 IEEE Workshop on Applications of Signal Processing to Audio and Acoustics, New Paltz, NY, USA., 293-296.

Travis C. (**1996**). "Virtual reality perspective on headphone audio," in AES 101st Convention, Los Angeles, USA, Preprint: 4354.

Treeby B.E., Paurobally R.M., and Pan J. (**2007a**). "The effect of impedance on interaural azimuth cues derived from a spherical head model," J. Acoust. Soc. Am. **121(4)**, 2217-2226.

Treeby B.E., Pan J., and Paurobally R.M. (**2007b**). "The effect of hair on auditory localization cues," J. Acoust. Soc. Am. **122(6)**, 3586-3597.

Trivedi U., Dieckman E., and Xiang N. (**2009**). "Reciprocal maximum-length and related sequences for artificial spatial enveloping reverberation (abstract)" J. Acoust. Soc. Am. **125(4, Pt2)**, 2735.

Trivedi U., and Xiang N. (**2009**). "Utilizing reciprocal maximum length sequences within a multichannel context to generate a natural, spatial sounding reverberation (abstract)," J. Acoust. Soc. Am. **126(4, Pt2)**, 2155.

Tsujino K., Kobayashi W., and Onoye T., et al. (**2006**). "Automated design of digital filters for 3-D sound localization in embedded applications," in Proceedings of 2006 IEEE International Conference on Acoustics, Speech, and Signal Processing, Toulouse, France, **Vol.(5)**, 349-352.

Turku J., Vilermo M., and Seppala E., et al. (**2008**). "Perceptual evaluation of numerically simulated head-related transfer function," in AES 124th Convention, Amsterdam, The Netherlands, Preprint: 7489.

Vää nänen R. (**2000**). "Synthetic audio tools in MPEG-4 standard," in AES 108th Convention, Paris, France, Preprint: 5080.

Väänänen R., and Huopaniemi J. (**2004**). "Advanced audioBIFS: virtual acoustics modeling in MPEG-4 scene description," IEEE Trans. Multimedia **6(5)**, 661-675.

Vanderkooy J. (**1994**). "Aspects of MLS measuring systems," J. Audio Eng. Soc. **42(4)**, 219–231.

Ven E.V.D., Aarts R.M., and Sommen P.C.W. (**2007**). "Blind Cancellation of stereo-base widening," J. Audio Eng. Soc. **55(4)**, 227-235.

Virette D., Philippe P., and Pallone G., et al. (**2007**). "Efficient binaural filtering in QMF domain for BRIR," in AES 122nd Convention, Vienna, Austria, Preprint: 7095.

Vorländer M. (**1995**). "International round robin on room acoustical computer simulations," in Proceedings of 15th International Congress on Acoustics, Trondheim, Norway, 577–580.

Vorländer M. (**2004**). "Past, present and future of dummy heads," in 2004 conference of Federation of the iberoamerican acousical societies, Portugal.

Vorländer M. (**2008**). *Auralization, fundamentals of acoustics, modelling, simulation, algorithms and acoustic virtual reality* (Springer-Verlag Berlin Heidelberg).

Walker R. (**2000**). "Approximation functions for virtual acoustic modeling," in AES 108th Convention, Paris, France, Preprint: 5138.

Wallach H. (**1940**). "The role of head movement and vestibular and visual cue in sound localization," J.Exp. Psychol. **27(4)**, 339-368.

Walsh M., and Jot J.M. (**2006**). "Loudspeaker-based 3-D audio system design using the M-S shuffler matrix," in AES 121st Convention, San Francisco, USA, Preprint: 6949.

Walsh T., Demkowicz L., and Charles R. (**2004**). "Boundary element modeling of the external human auditory system," J. Acoust. Soc. Am. **115(3)**, 1033-1043.

Ward D.B., and Elko G.W. (**1998**). "Optimum loudspeaker spacing for robust crosstalk cancellation," in Proceeding of the 1998 IEEE International Conference on Acoustics, Speech and Signal Proceesing, Seattle, WA, USA, **Vol.(6)**, 3541-3544.

Ward D.B., and Elko G.W. (**1999**). "Effect of loudspeaker position on the robustness of acoustic crosstalk cancellation," IEEE Signal Processing Letters **6(5)**, 106-108.

Watanabe K., Takane S., and Suzuki Y. (**2003**). "Interpolation of head-related transfer functions based on the common-acoustical-pole and residue model," Acoust. Sci. & Tech. **24(5)**, 335-337.

Watanabe K., Takane S., and Suzuki Y. (**2005**). "A new interpolation method of HRTF based on the common pole-zero model," Acta Acoustica United with Acoustica **91(6)**, 958-966.

Watanabe K., Ozawa K., and Iwaya Y., et al. (**2007**). "Estimation of interaural level difference based on anthropometry and its effect on sound localization," J. Acoust. Soc. Am. **122(5)**, 2832-2841.

Watkins A.J. (**1978**). "Psychoacoustical aspects of synthesized vertical locale cues," J. Acoust. Soc. Am. **63(4)**, 1152–1165.

Wenzel E.M., Wightman F.L., and Kistler D.J. (**1988**). "Acoustic origins of individual differences in sound localization behavior," J. Acoust. Soc. Am. **84(Suppl. 1)**, S79.

Wenzel E.M. (**1991**). "Localization in Virtual Acoustic Displays," Presence **1 (1)**, 80–107.

Wenzel E.M., Arruda M., and Kistler D.J., et al. (**1993a**). "Localization using nonindividualized head-related transfer functions," J. Acoust. Soc. Am. **94(1)**, 111-123.

Wenzel E.M., and Foster S.H. (**1993b**). "Perceptual consequences of interpolating head-related transfer functions during spatial synthesis," in Proceeding of IEEE 1993 Workshop on Applications of Signal Processing to Audio and Acoustics, New Paltz, NY, USA, 102-105.

Wenzel E.M. (**1995**). "The relative contribution of interaural time and magnitude cues to dynamic sound localization," in Proceedings of the IEEE 1995 Workshop on Applications of Signal Processing to Audio and Acoustics, New Paltz, NY, USA, 80-83.

Wenzel E.M. (**1996**). "What perception implies about implementation of interactive virtual acoustic environments," in AES 101st Convention, Los Angeles, CA, U.S.A., Preprint: 4353.

Wenzel E.M. (**1997**). "Analysis of the role of update rate and system latency in interactive virtual acoustic environments," in AES 103rd Convention, New York, U.S.A., Preprint: 4633.

Wenzel E.M. (**1999**). "Effect of increasing system latency on localization of virtual sounds," in AES 16th International Conference: Spatial Sound Reproduction, Rovaniemi, Finland.

Wenzel E.M., Miller D.J., and Abel J.S. (**2000**). "Sound Lab: a real-time, software-based system for the Study of Spatial hearing," in AES 108th Convention, Paris, France, Preprint: 5140.

Wersenyi G., and Illenyi A. (**2003**). "Test signal generation and accuracy of turntable control in a dummy-head measurement system," J. Audio. Eng. Soc. **51(3)**, 150-155.

Wersenyi G. (**2009**). "Effect of emulated head-tracking for reducing localization errors in virtual audio simulation," IEEE Trans.Audio and Speech Processing **17(2)**, 247-252.

Wersenyi G. (**2012**). "Virtual localization by blind persons," J. Audio. Eng. Soc. **60(7/8)**, 568-579.

Wiener F.M., and Ross D.A. (**1946**). "The pressure distribution in the auditory canal in a progressive sound field," J. Acoust. Soc. Am. **18(2)**, 401-408 .

Wiener F.M., (**1947**). "On the Diffraction of a Progessive Sound Wave by the Human Head," J. Acoust. Soc. Am. **19(1)**, 143-146.

Wightman F.L., and Kistler D.J. (**1989a**). "Headphone simulation of free-field listening, I: stimulus synthesis," J. Acoust. Soc. Am. **85(2)**, 858-867.

Wightman F.L., and Kistler D.J. (**1989b**). "Headphone simulation of free-field listening, II: psycho-physical validation," J. Acoust. Soc. Am. **85(2)**, 868-878.

Wightman F.L., and Kistler D.J. (**1992a**). "The dominant role of low-frequency interaural time difference in sound localization," J. Acoust. Soc. Am. **91(3)**, 1648-1661.

Wightman F.L., Kistler D.J., and Arruda M. (**1992b**). "Perceptual consequences of engineering compromises in synthesis of virtual auditory objects (abstract)," J. Acoust. Soc. Am. **92(4)**, 2332.

Wightman F.L., and Kistler D.J. (**1993**). "Multidimensional scaling analysis of head-related transfer functions," in Proceeding of IEEE 1993 Workshop on Applications of Signal Processing to Audio and Acoustics, New Paltz, NY, USA, 98–101.

Wightman F.L., and Kistler D.J. (**1997**). "Monaural sound localization revisited," J. Acoust. Soc. Am. **101(2)**, 1050-1063.

Wightman F.L., and Kistler D.J. (**1999**). "Resolution of front-back ambiguity in spatial hearing by listener and source movement," J. Acoust. Soc. Am. **105(5)**, 2841-2853.

Wightman F.L., and Kistler D.J. (**2005**). "Measurement and validation of human HRTFs for use in hearing research," Acta Acoustica united with Acoustica **91(3)**, 429–439.

Woodworth R.S., and Schlosberg H. (**1954**). *Experimental Physchology* (Second edition, Holt, Rinehart & Winston Inc., New York, USA).

Wu S.X., and Zhao Y.Z. (**2003**). *Room and environmental acoustics* (in Chinese, Guangdong science and Technology Press, Guangzhou, China).

Wu Z.Y., Chan F.H.Y., and Lam F.K., et al. (**1997**). "A time domain binaural model based on spatial feature extraction for the head-related transfer function," J. Acoust. Soc. Am. **102(4)**, 2211-2218.

Xiang N., and Blauert J. (**1991**). "A miniature dummy head for binaural evaluation of tenth-scale acoustic models," Applied Acoustics **33(2)**, 123-140.

Xiang N., and Blauert J. (**1993**). "Binaural scale modeling for auralisation and prediction of acoustics in auditoria," Applied Acoustics **38(2-4)**, 267-290.

Xiang N., and Schroeder M.R. (**2003**). "Reciprocal maximum-length sequence pairs for acoustical dual source measurements," J. Acoust. Soc. Am. **113(5)**, 2754-2761.

Xiang N. (**2009**). *Digital Sequences* (in *Handbook of Signal Processing in Acoustics*, edit by Havelock D., et al., Springer New York, USA).

Xie B.S., and Xie X.F. (**1996**). "Analyse and sound image localization experiment on multi-channel plannar surround sound system," Acta Acustica (in Chinese) **21(4, Supplement)**, 648-660.

Xie B.S., and Zhang C.Y. (**1999**). "A simple method for stereophonic image stage extension," Technical Acoustics (in Chinese) **18(supplement)**, 187-188.

Xie B.S. (**2000**). "Principle and misunderstanding on virtual surround sound," Audio Engineering (in Chinese) **2000(2)**, 8-13.

Xie B.S. (**2001a**). "Signal mixing for a 5.1 channel surround sound system—Analysis and experiment," J. Audio. Eng. Soc. **49(4)**, 263-274.

Xie B.S., Wang J., and Guan S.Q. (**2001b**). "A simplified way to simulate 3D virtual sound image," Audio Engineering (in Chinese) **2001(7)**, 10-14.

Xie B.S., and Guan S.Q. (**2002a**). "Development of Multi-channel surround sound and its psychoacoustic principle," Audio Engineering (in Chinese) **2002(2)**, 11-18.

Xie B.S. (**2002b**). "Interchannel time difference and stereophonic sound image localization," Acta Acustica (in Chinese) **27(4)**, 332-338.

Xie B.S. (**2002c**). "Effect of head size on virtual sound image localization," Applied acoustics (in Chinese) **21(5)**, 1-7.

Xie B.S. (**2003**). "Reproducing of stereophonic sound by using nonstandard loudspeaker arrangement," Audio Engineering (in Chinese) **2003(2)**, 68-70.

Xie B.S., and Guan S.Q. (**2004**). "Virtual sound and its application," Applied acoustics (in Chinese) **23(4)**, 43-47.

Xie B.S. (**2005a**). "Correction on the character of HRTF at low frequency," Technical Acoustics (in Chinese) **24(Supplement)**, 510-512.

Xie B.S. (**2005b**). "Rotation of head and stability of virtual sound image," Audio Engineering (in Chinese) **2005(6)**, 56-59.

Xie B.S., Shi Y., and Xie Z.W., et al. (**2005c**). "Virtual reproducing system for 5.1 channel surround sound," Chin J. Acoust. **24(1)**, 76-88.

Xie B.S., Wang J., and Guan S.Q. et al. (**2005d**). "Virtual reproduction of 5.1 channel surround sound by headphone," Chin J. Acoust. **24(1)**, 63-75.

Xie B.S., Wang J., and Guan S.Q., et al. (**2005e**). "Headphone virtual 5.1 channel surround sound signal processing method," China patent No.ZL02134415.9.

Xie B.S., Shi Y., and Xie Z.W., et al. (**2005f**). "Two-loudspeaker virtual 5.1 channel surround sound signal processing method," China patent No.ZL02134416.7.

Xie B.S. (**2006a**). "Analysis on the symmetry of interaural time difference," Technical Acoustics (in Chinese) **25(supplement)**, 411-412.

Xie B.S., Zhang L.S., and Guan S.Q. et al. (**2006b**). "Simplification and subjective evaluation of filters for virtual sound using loudspeakers," Technical Acoustics **25(6)**, 547-554.

Xie B.S. (**2006c**). "Spatial interpolation of HRTFs and signal mixing for multichannel surround sound," Chin J. Acoust. **25(4)**, 330-341.

Xie B.S., Zhong X.L., and Rao D., et al. (**2007a**). "Head-related transfer function database and analyses," Science in China Series G, Physics, Mechanics & Astronomy **50(3)**, 267-280.

Xie B.S. (**2008a**). "Principal components analysis on HRTF and the effect of pinna," Technical Acoustics (in Chinese) **27(5, Pt.2)**, 374-375.

Xie B.S. (**2009**). "On the low frequency characteristics of head-related transfer function," Chin J. Acoust. **28(2)**, 116-128.

Xie B.S., and Zhang T.T. (**2010**). "The audibility of spectral detail of head-related transfer functions at high frequency," Acta Acustica united with Acustica **96(2)**, 328-339.

Xie B.S., and Zhang C.Y. (**2012a**). "An algorithm for efficiently synthesizing multiple near-field virtual sources in dynamic virtual auditory display," in AES 132nd Convention, Budapest, Hungary, Paper: 8646.

Xie B.S. (**2012b**). "Recovery of individual head-related transfer functions from a small set of measurements," J. Acoust. Soc. Am. **132(1)**, 282-294.

Xie B.S., Shi B., and Xiang N. (**2012c**). "Audio signal decorrelation based on reciprocal-maximal length sequence filters and its applications to spatial sound," in AES 133rd Convention, San Francisco, USA, Paper: 8805.

Xie B.S., and Zhong X.L. (**2012d**). "Similarity and cluster analysis on magnitudes of individual head-related transfer functions (abstract)," J. Acoust. Soc. Am., **131(4, Pt.2)**, 3305.

Xie X.F. (**1981**). *The principle of stereophonic sound* (Science Press, Beijing), pp. 1-433.

Xie X.F. (**1982**). "The 4-3-N matrix multi-channel sound system," Chin J. Acoust. **1(2)**, 210-218.

Xie X.F. (**1988**). "A mathematical analysis of three dimensional surrounding sound field," Acta Acustica (in Chinese) **13(5)**, 321-328.

Xie Z.W., Yin J.X., and Rao D. (**2006**). "The experiment study of spatial mask effect," Chin J. Acoust. **25(1)**, 75-86.

Xie Z.W., and Jin J. (**2008a**). "The influence of time delay on forward masking when spatially separated sound image reproduced with headphone," Acta Acustica (in Chinese) **33(3)**, 283-287.

Xie Z.W., and Jin J. (**2008b**). "A preliminary study on spatial unmasking of virtual separated sources," Science in China Series G, Physics, Mechanics & Astronomy **51(10)**, 1565-1572.

Yairi S., Iwaya Y., and Suzuki Y. (**2008**). "Influence of large system latency of virtual auditory display on behavior of head movement in sound localization task," Acta Acustica united with Acoutica **94(6)**, 1016-1023.

Yang J., and Gan W.S. (**2000**). "Speaker Placement for robust virtual audio display system," Electronics Letters **36(7)**, 683-685.

Yang J., Tan S.E., and Gan W.S. (**2002**). "Observation on the robust performance of elevated sound reproduction system," in Proceeding of IEEE 6th International Conference on Signal Processing, Beijing, China, **Vol.(2)**, 1007-1010.

Yasuda Y., Ohya T., and McGrath D., et al. (**2003**). "3-D audio communications services for future mobile networks," in AES 23rd International Conference, Copenhagen, Denmark.

Young E.D., Rice J.J., and Tong S.C. (**1996**). "Effects of pinna position on head-related transfer functions in the cat," J. Acoust. Soc. Am. **99(5)**, 3064-3076.

Yu R., Robinson C.Q., Cheng C. (2007). Low-complexity binaural decoding using time/Frequency domain HRTF equalization," Lecture notes in computer science, Springer-Verlag, 545-556.

Yu G.Z., Xie B.S., and Rao D. (**2008**). "Effect of sound source scattering on measurement of near-field head-related transfer functions," Chinese Physics Letter **25(8)**, 2926-2929.

Yu G.Z., Xie B.S., and Rao D. (**2009**). "Directivity analysis on spherical regular polyhedron sound source used for near-field HRTF measurements," in The 10th Western Pacific Acoustics Conference, Beijing, China.

Yu G.Z., Xie B.S., and Rao D. (**2010a**). "Characteristics of Near-field head-related transfer function for KEMAR," in AES 40th International Conference, Tokyo, Japan.

Yu G.Z., Xie B.S., and Rao D. (**2010b**). "Analysis of effect of torso scattering on near-field head-related transfer functions by using perturbation method," J. South China University of Technology (in Chinese) **38(3)**, 143-147.

Yu G.Z., Xie B.S., and Chen Z.W., et al. (**2012a**). "Analysis on error caused by multi-scattering of multiple sound sources in HRTF measurement," in AES 132nd Convention, Budapest, Hungary, Paper: 8643.

Yu G.Z., Liu Y., and Xie B.S., et al. (**2012b**). "Fast measurement system and super high directional resolution head-related transfer function database (Abstract)," J. Acoust. Soc. Am. **131(4, Pt.2)**, 3304.

Yu G.Z., Xie B.S., and Chen X.X. (**2012c**). "Analysis on minimum-phase characteristics of measured head-related transfer functions affected by sound source responses," Computers and Electrical Engineering **38(1)**, 45-51.

Yu G.Z., Xie B.S., and Rao D. (**2012d**). "Near-field head-related transfer functions of an artificial head and its characteristics," Acta Acustica (in Chinese) **37(4)**, 378-385.

Yu G.Z., and Xie B.S. (**2012e**). "Analysis on multiple scattering between the rigid-spherical microphone array and nearby surface in sound field recording," in AES 133rd Convention, San Francisco, CA, USA, Paper: 8710.

Zacharov N., Tuomi O., and Lorho G. (**2001**). "Auditory periphery, HRTF's and directional loudness perception," in AES 110th Convention, Amsterdam, The Netherlands, Preprint: 5315.

Zahorik P., Wightman F., and Kistler D. (**1995**). "On the discriminability of virtual and real sound sources," in Proceeding of 1995 IEEE Workshop on Applications of Signal Processing to Audio and Acoustics, New York, USA, 76-79.

Zahorik P., Wightman F.L., and Kistler D.J. (**1996**). "The fidelity of virtual auditory display (abstract)," J. Acoust. Soc. Am. **99(4, Pt2)**, 2596.

Zahorik P. (**2000**). "Limitations in using Golay codes for head-related transfer function measurement," J. Acoust. Soc. Am. **107(3)**, 1793-1796.

Zahorik P., Tam C., and Wang K., et al. (**2001**). "Effects of visual-feedback training in 3D sound displays (abstract)," J. Acoust. Soc. Am. **109(5, Pt2)**, 2487.

Zahorik P. (**2002a**). "Assessing auditory distance perception using virtual acoustics," J. Acoust. Soc. Am. **111(4)**, 1832-1846.

Zahorik P. (**2002b**). "Auditory display of sound source distance," in Proceedings of the 2002 International Conference on Auditory Display, Kyoto, Japan, 326-332.

Zahorik P., Brungart D.S., and Bronkhorst A.W. (**2005**). "Auditory distance perception in humans: a summary of past and present research," Acta Acustica United with Acustica **91(3)**, 409–420.

Zhang C.Y., Xie B.S., and Xie Z.W. (**2000**). "Elimination of effect of inside-the-head localization in sound reproduction by stereophonic earphone," Audio Engineering (in Chinese) **2000(8)**, 4-6.

Zhang C.Y., and Xie B.S. (**2012**). "Platform for Virtual Auditory Environment Real Time Rendering System (abstract)," J. Acoust. Soc. Am. **131(4, Pt.2)**, 3269.

Zhang M., Tan K.C., and Er M.H. (**1998**) "Three-dimensional sound synthesis based on head-related transfer functions," J. Audio. Eng. Soc. **46(10)**, 836-844.

Zhang T.T., and Xie B.S. (**2008a**). "Principal Component analysis and spatial Fourier Reconstruction of head-related transfer functions," Audio Engineering (in Chinese) **32(8)**, 48-56.

Zhang T.T., and Xie B.S. (**2008b**). "Experiment on HRTF reduction and continuous reconstruction," Technical Acoustics (in Chinese) **27(5, Pt.2)**, 388-389.

Zhang W., Abhayapala T.D., and Kennedy R.A., et al. (**2010**). "Insights into head-related transfer function: Spatial dimensionality and continuous representation," J. Acoust. Soc. Am. **127(4)**, 2347-2357.

Zhang W., Zhang M., and Kennedy R.A., et al. (**2012**). "On high-resolution head-related transfer function measurements: an efficient sampling scheme," IEEE Trans. on Audio, Speech and Language Processing **20(2)**, 575-584.

Zhong X.L., and Xie B.S. (**2004**). "Progress in the research of head-related transfer function," Audio Engineering (in Chinese) **2004(12)**, 44-46.

Zhong X.L., Xie B.S. (**2005a**). "Extrapolating the head-related transfer function in the median plane using neural network," Technical Acoustics (in Chinese) **24(Supplement)**, 513-515.

Zhong X.L., and Xie B.S. (**2005b**). "Spatial characteristics of head related transfer function," Chinese Physics Letter **22(5)**, 1166-1169.

Zhong X.L., and Xie B.S. (**2006a**). "Overall influence of clothing and pinnae on shoulder reflection and HRTF," Technical Acoustics **25(2)**, 113-118.

Zhong X.L. (**2006b**). "Research on head-related transfer function," Dissertation of doctor degree, South China University of Technology, Guangzhou, China.

Zhong X.L., and Xie B.S. (**2007a**). "A novel model of interaural time difference based on spatial Fourier analysis," Chinese Physics Letter **24(5)**, 1313-1316.

Zhong X.L. (**2007b**). "Criterion selection in the leading-edge method for evaluating interaural time difference," Audio Engineering (in Chinese) **31(9)**, 47-52.

Zhong X.L., and Xie B.S. (**2007c**). "Spatial symmetry of head-related transfer function," Chin J. Acoust. **26(1)**, 73-84.

Zhong X.L., and Xie B.S. (**2008**). "Reconstructing azimuthal continuous head-related transfer functions under the minimum-phase approximation," in Proceeding of Internoise 2008, Shanghai, China.

Zhong X.L., and Xie B.S. (**2009a**). "Maximal azimuthal resolution needed in measurements of head-related transfer functions," J. Acoust. Soc. Am. **125(4)**, 2209-2220.

Zhong X.L., and Xie B.S. (**2009b**). "A Continuous model of interaural time difference based on surface spherical Harmonic," in The 10th Western Pacific Acoustics Conference, Beijing, China.

Zhong X.L, Liu Y., and Xiang N., et al. (**2010**). "Errors in the measurements of individual headphone-to-ear-canal transfer function," in DAGA 2010, Berlin, Germany.

Zhou B., Green D.M., and Middlebrooks J.C. (**1992**). "Characterization of external ear impulse responses using Golay codes," J. Acoust. Soc. Am. **92(2)**, 1169–1171.

Zotkin D.N., Duraiswami R., and Davis L.S. (**2002**). "Creation of virtual auditory spaces," in Proceedings of the IEEE 2002 International Conference on Acoustics, Speech and Signal Processing, Orlando, FL, USA, **Vol.(2)**, 2113-2116.

Zotkin D.N., Hwang J., and Duraiswami R.,et al. (**2003**). "HRTF personalization using anthropometric measurements," in Proceedings of the 2003 IEEE Workshop on Applications of Signal Processing to Audio and Acoustics, New Paltz, NY, 157-160.

Zotkin D.N., Duraiswami R., and Davis L.S. (**2004**). "Rendering localized spatial audio in a virtual auditory space," IEEE Trans. on Multimedia **6(4)**, 553-564.

Zotkin D.N., Duraiswami R., and Grassi E., et al. (**2006**). "Fast head related transfer function measurement via reciprocity," J. Acoust. Soc. Am. **120(4)**, 2202-2215.

Zotkin D.N., Duraiswami R., and Gumerov N.A. (**2007**). "Efficient conversion of XY surround sound content to binaural head-tracked form for HRTF-enabled playback," in Proceeding of IEEE 2007 International Conference on Acoustics, Speech and Signal Processing, Honolulu, HI, USA, **Vol.(1)**, 21-24.

Zotkin D.N., Duraiswami R., and Gumerov N.A. (**2009**). "Regularized HRTF fitting using spherical harmonics," in Proceeding of 2009 IEEE Workshop on Applications of Signal Processing to Audio and Acoustics, New Paltz, NY, USA, 257-260.

Zurek P.M. (**1987**). *The precedence effect* (in *Directional Hearing*, edited by Yost W.A. and Gourevitch G., Springer-Verlag, New York, USA).

Zwicker E., and Fastl H. (**1999**). *Psychoacoustics: facts and models* (The second Edition, Springer-verlag, Berlin, Germany).

Index